# Lecture Notes in Computer Science    12791

More information about this subseries at http://www.springer.com/series/7409

Heidi Krömker (Ed.)

# HCI in Mobility, Transport, and Automotive Systems

Third International Conference, MobiTAS 2021
Held as Part of the 23rd HCI International Conference, HCII 2021
Virtual Event, July 24–29, 2021
Proceedings

 Springer

*Editor*
Heidi Krömker
Technische Universität Ilmenau
Ilmenau, Germany

ISSN 0302-9743            ISSN 1611-3349  (electronic)
Lecture Notes in Computer Science
ISBN 978-3-030-78357-0       ISBN 978-3-030-78358-7  (eBook)
https://doi.org/10.1007/978-3-030-78358-7

LNCS Sublibrary: SL3 – Information Systems and Applications, incl. Internet/Web, and HCI

This Springer imprint is published by the registered company Springer Nature Switzerland AG
The registered company address is: Gewerbestrasse 11, 6330 Cham, Switzerland

# Foreword

Human-Computer Interaction (HCI) is acquiring an ever-increasing scientific and industrial importance, and having more impact on people's everyday life, as an ever-growing number of human activities are progressively moving from the physical to the digital world. This process, which has been ongoing for some time now, has been dramatically accelerated by the COVID-19 pandemic. The HCI International (HCII) conference series, held yearly, aims to respond to the compelling need to advance the exchange of knowledge and research and development efforts on the human aspects of design and use of computing systems.

The 23rd International Conference on Human-Computer Interaction, HCI International 2021 (HCII 2021), was planned to be held at the Washington Hilton Hotel, Washington DC, USA, during July 24–29, 2021. Due to the COVID-19 pandemic and with everyone's health and safety in mind, HCII 2021 was organized and run as a virtual conference. It incorporated the 21 thematic areas and affiliated conferences listed on the following page.

A total of 5222 individuals from academia, research institutes, industry, and governmental agencies from 81 countries submitted contributions, and 1276 papers and 241 posters were included in the proceedings to appear just before the start of the conference. The contributions thoroughly cover the entire field of HCI, addressing major advances in knowledge and effective use of computers in a variety of application areas. These papers provide academics, researchers, engineers, scientists, practitioners, and students with state-of-the-art information on the most recent advances in HCI. The volumes constituting the set of proceedings to appear before the start of the conference are listed in the following pages.

The HCI International (HCII) conference also offers the option of 'Late Breaking Work' which applies both for papers and posters, and the corresponding volume(s) of the proceedings will appear after the conference. Full papers will be included in the 'HCII 2021 - Late Breaking Papers' volumes of the proceedings to be published in the Springer LNCS series, while 'Poster Extended Abstracts' will be included as short research papers in the 'HCII 2021 - Late Breaking Posters' volumes to be published in the Springer CCIS series.

The present volume contains papers submitted and presented in the context of the 3rd International Conference on HCI in Mobility, Transport and Automotive Systems (MobiTAS 2021), an affiliated conference to HCII 2021. I would like to thank the Chair, Heidi Krömker, for her invaluable contribution to its organization and the preparation of the proceedings, as well as the members of the Program Board for their contributions and support. This year, the MobiTAS affiliated conference has focused on topics related to urban mobility, cooperative and automated mobility, UX in intelligent transportation systems, and mobility for diverse target user groups.

I would also like to thank the Program Board Chairs and the members of the Program Boards of all thematic areas and affiliated conferences for their contribution towards the highest scientific quality and overall success of the HCI International 2021 conference.

This conference would not have been possible without the continuous and unwavering support and advice of Gavriel Salvendy, founder, General Chair Emeritus, and Scientific Advisor. For his outstanding efforts, I would like to express my appreciation to Abbas Moallem, Communications Chair and Editor of HCI International News.

July 2021                                                 Constantine Stephanidis

# HCI International 2021 Thematic Areas and Affiliated Conferences

**Thematic Areas**

- HCI: Human-Computer Interaction
- HIMI: Human Interface and the Management of Information

**Affiliated Conferences**

- EPCE: 18th International Conference on Engineering Psychology and Cognitive Ergonomics
- UAHCI: 15th International Conference on Universal Access in Human-Computer Interaction
- VAMR: 13th International Conference on Virtual, Augmented and Mixed Reality
- CCD: 13th International Conference on Cross-Cultural Design
- SCSM: 13th International Conference on Social Computing and Social Media
- AC: 15th International Conference on Augmented Cognition
- DHM: 12th International Conference on Digital Human Modeling and Applications in Health, Safety, Ergonomics and Risk Management
- DUXU: 10th International Conference on Design, User Experience, and Usability
- DAPI: 9th International Conference on Distributed, Ambient and Pervasive Interactions
- HCIBGO: 8th International Conference on HCI in Business, Government and Organizations
- LCT: 8th International Conference on Learning and Collaboration Technologies
- ITAP: 7th International Conference on Human Aspects of IT for the Aged Population
- HCI-CPT: 3rd International Conference on HCI for Cybersecurity, Privacy and Trust
- HCI-Games: 3rd International Conference on HCI in Games
- MobiTAS: 3rd International Conference on HCI in Mobility, Transport and Automotive Systems
- AIS: 3rd International Conference on Adaptive Instructional Systems
- C&C: 9th International Conference on Culture and Computing
- MOBILE: 2nd International Conference on Design, Operation and Evaluation of Mobile Communications
- AI-HCI: 2nd International Conference on Artificial Intelligence in HCI

# List of Conference Proceedings Volumes Appearing Before the Conference

1. LNCS 12762, Human-Computer Interaction: Theory, Methods and Tools (Part I), edited by Masaaki Kurosu
2. LNCS 12763, Human-Computer Interaction: Interaction Techniques and Novel Applications (Part II), edited by Masaaki Kurosu
3. LNCS 12764, Human-Computer Interaction: Design and User Experience Case Studies (Part III), edited by Masaaki Kurosu
4. LNCS 12765, Human Interface and the Management of Information: Information Presentation and Visualization (Part I), edited by Sakae Yamamoto and Hirohiko Mori
5. LNCS 12766, Human Interface and the Management of Information: Information-rich and Intelligent Environments (Part II), edited by Sakae Yamamoto and Hirohiko Mori
6. LNAI 12767, Engineering Psychology and Cognitive Ergonomics, edited by Don Harris and Wen-Chin Li
7. LNCS 12768, Universal Access in Human-Computer Interaction: Design Methods and User Experience (Part I), edited by Margherita Antona and Constantine Stephanidis
8. LNCS 12769, Universal Access in Human-Computer Interaction: Access to Media, Learning and Assistive Environments (Part II), edited by Margherita Antona and Constantine Stephanidis
9. LNCS 12770, Virtual, Augmented and Mixed Reality, edited by Jessie Y. C. Chen and Gino Fragomeni
10. LNCS 12771, Cross-Cultural Design: Experience and Product Design Across Cultures (Part I), edited by P. L. Patrick Rau
11. LNCS 12772, Cross-Cultural Design: Applications in Arts, Learning, Well-being, and Social Development (Part II), edited by P. L. Patrick Rau
12. LNCS 12773, Cross-Cultural Design: Applications in Cultural Heritage, Tourism, Autonomous Vehicles, and Intelligent Agents (Part III), edited by P. L. Patrick Rau
13. LNCS 12774, Social Computing and Social Media: Experience Design and Social Network Analysis (Part I), edited by Gabriele Meiselwitz
14. LNCS 12775, Social Computing and Social Media: Applications in Marketing, Learning, and Health (Part II), edited by Gabriele Meiselwitz
15. LNAI 12776, Augmented Cognition, edited by Dylan D. Schmorrow and Cali M. Fidopiastis
16. LNCS 12777, Digital Human Modeling and Applications in Health, Safety, Ergonomics and Risk Management: Human Body, Motion and Behavior (Part I), edited by Vincent G. Duffy
17. LNCS 12778, Digital Human Modeling and Applications in Health, Safety, Ergonomics and Risk Management: AI, Product and Service (Part II), edited by Vincent G. Duffy

**http://2021.hci.international/proceedings**

# 3rd International Conference on HCI in Mobility, Transport and Automotive Systems (MobiTAS 2021)

Program Board Chair: **Heidi Krömker,** *TU Ilmenau, Germany*

- Angelika C. Bullinger, Germany
- Bertrand David, France
- Marco Diana, Italy
- Christophe Kolski, France
- Lutz Krauss, Germany
- Josef F. Krems, Germany
- Lena Levin, Sweden

- Matthias Rötting, Germany
- Lionel P. Robert Jr., USA
- Philipp Rode, Germany
- Thomas Schlegel, Germany
- Ulrike Stopka, Germany
- Xiaowei Yuan, China

The full list with the Program Board Chairs and the members of the Program Boards of all thematic areas and affiliated conferences is available online at:

**http://www.hci.international/board-members-2021.php**

# HCI International 2022

The 24th International Conference on Human-Computer Interaction, HCI International 2022, will be held jointly with the affiliated conferences at the Gothia Towers Hotel and Swedish Exhibition & Congress Centre, Gothenburg, Sweden, June 26 – July 1, 2022. It will cover a broad spectrum of themes related to Human-Computer Interaction, including theoretical issues, methods, tools, processes, and case studies in HCI design, as well as novel interaction techniques, interfaces, and applications. The proceedings will be published by Springer. More information will be available on the conference website: http://2022.hci.international/:

General Chair
Prof. Constantine Stephanidis
University of Crete and ICS-FORTH
Heraklion, Crete, Greece
Email: general_chair@hcii2022.org

**http://2022.hci.international/**

# Contents

## Cooperative and Automated Mobility

**Studies on Intelligent Transportation Systems**

# Urban Mobility

# Requirement Analysis for Personal Autonomous Driving Robotic Systems in Urban Mobility

Kathrin Bärnklau[1,2]($\boxtimes$) (iD), Matthias Rötting[1] (iD), Eileen Roesler[1] (iD),
and Felix Wilhelm Siebert[3] (iD)

[1] Department of Psychology and Ergonomics, Technische Universität Berlin, Berlin, Germany
k.baernklau@campus.tu-berlin.de, roetting@mms.tu-berlin.de,
eileen.roesler@tu-berlin.de
[2] Institute of Innovation and Entrepreneurship, Tongji University, Shanghai, China
[3] Department of Psychology, Friedrich-Schiller University Jena, Jena, Germany
felix.siebert@uni-jena.de

**Abstract.** Urban mobility is changing due to the emergence of new technologies like autonomously navigating robots. In the future, various transport operators and micro mobility services will be integrated in an increasingly complex mobility system, potentially realizing benefits such as a reduction of congestion, travel costs, and emissions. The field of personal robotic transport agents is projected to increasingly play a role in urban mobility, hence in this study, prospective target groups and corresponding user needs concerning human-following robots for smart urban mobility applications are investigated. Building on an extensive literature review, three focus groups with a total of 19 participants are conducted, utilizing scenario-based design and personas. Results show clearly definable user needs and potential technological requirements for mobile robots deployed in urban road environments. The two most mentioned potential applications were found in the fields of leisure applications and in healthcare for elderly people. Based on these focus group results, two personal automated driving robots which differ in function, operation and interaction were designed. The focus group-based results and derived requirements shed light on the importance of context-sensitivity of robot design.

**Keywords:** Autonomous driving · Human-machine-interaction · Personal human-following robots · Urban mobility

## 1 Introduction

In a rapidly urbanizing world, autonomous driving has gained a lot of attention and importance in recent years and has created new opportunities for city dwellers. Self-driving cars are considered as a well-known and widely noted forthcoming driving technology. However, outside of the broad public view, other forms of smaller automated vehicles are becoming more attainable and reachable for personal use [1]. These automated driving robotic vehicles can potentially support users by carrying goods and thus minimize physical load of humans [2]. While robots are already increasingly integrated in daily and

© Springer Nature Switzerland AG 2021
H. Krömker (Ed.): HCII 2021, LNCS 12791, pp. 3–18, 2021.
https://doi.org/10.1007/978-3-030-78358-7_1

social life [3], there is a need to further investigate beneficial product characteristics for smooth and effective human-robot-interaction (HRI) for personal autonomous driving robotic vehicles. However, the assessment of demands and the actual users' standards for personal self-driving robots are often speculative due to varying operational challenges [1, 4–6]. Although a number of studies have evaluated technical-functional and HRI requirements of personal robots, a consensus of distinct features and characteristics has not been reached [1, 4, 6, 7]. Thus, it is questionable how the market for personal mobile human-following robots will manifest and how these autonomous systems would be developed. The current research body [1, 3] focuses on technical issues dealing with how to create safe and functional person-following behavior. Only a few studies [2, 5, 8] included user surveys that indicate the number of potential users or record the personal needs from the user's perspective. Honig et al. [3] identified this gap in current research activities in terms of social interaction and user needs. Considering these limitations in the knowledge on personal human-following robots, the goal of this study was to define optimal conditions for a potential personal robot vehicle launch from a user's point of view by conducting focus group surveys. After the requirements of potential users are identified and considered, a human centered product design is developed incorporating the identified user needs.

## 2    Current State of Research

### 2.1    Autonomous Mobility

Small automated driving vehicles such as human-following robots were developed by two major research disciplines: Autonomous driving and Robotics. The research object represents an (electric) human-following robot operating in urban mobility applications. Thus, this personal vehicle includes characteristics of the field of research of *Autonomous driving systems* and *Robotics*. Both terminologies are consolidated and the actual state of current research on various human-following robots are examined.

Humans are already increasingly confronted with self-driving systems in the road environment. Some vehicles are already equipped with technologies that enable supervised autonomous driving. For these types of vehicles, there are a number of applicable definitions in the field of autonomous driving and robotics. According to the norm SAE J3016 (Society of Automotive Engineers), automated systems are classified in six different levels of automation depending on the area of application. These six levels encompass the following: No automation (level 0), driver assistance (level 1), partial automation (level 2), conditional automation (level 3), high automation (level 4) and finally full automation (level 5) [9].

For each level, the autonomy of the driving system increases while the need for surveillance through the user decreases. Autonomous systems consist of components of sensor technology, artificial intelligence, and actuators. Sensors such as cameras and radar systems, capture the environment surrounding the vehicle. Artificial intelligence is used to evaluate the recorded data and subsequently to change the state of the environment in order to achieve the objectives via actuators [10]. However, while first automated vehicles are tested on public roads, aspects concerning the liability for damages and the role of data privacy protection have not been sufficiently clarified.

In addition to existing taxonomies on autonomous cars, the field of robotics has also conducted research on smart mobility [6, 11]. In general, robots are artificial technical objects designed and built by human beings [12]. In literature, the distinction between industry and service robots is commonly used [4, 6, 12]. A first definition of industrial robots was given according to the VDI 2860 (The Association of German Engineers) [13]. According to the research of Hertzberg et al. [12], mobile service robots differ from industrial robots in the fact that all their actions depend on the actual environment. Since the surroundings are constantly changing, the programming of the robot has to be adaptive to a wide array of environments. Thus, mobile robots have to record their environment by using sensors, evaluate their data and finally choose their action accordingly. A major challenge in designing robot systems is achieving a seamless interaction between humans and machines as social and technical constraints are to be taken into account.

In robotics research, mobile robots are often referred to as personal service robots [3]. The ability of tracking and following users is projected as a main task for personal service robots and the design of human robot interaction in for these robotic systems has been evaluated (e.g. [14, 15]). In general, tracking is defined as following or detecting of a target, e.g. a human that can occur in different application areas such as in traffic systems. A specific application comprises the assistance of people and contribution to their daily life system by ground service robots. These robots are able to transport goods (e.g. groceries) by following their users and hence participate in urban mobility scenarios. Most of the existing human-following robots that are identified in literature represent simple first prototypes that were built and tested mainly for research purposes [2, 7, 16]. In order to develop user-centric robots, the specific market and user requirements of potential user groups have to be further analyzed.

## 2.2 Legal Background

However, the question arises which legal market peculiarities have to be respected for the implementation of mobile robots as traffic attendees in urban mobility. In principle, vehicles are only permitted for use on public roads if they meet the requirements of the Road Traffic Licensing Regulations (StVZO). This regulation regulates the conditions for registration of all vehicles on the road in Germany according to §16 Sect. 2 StVZO, excluding certain vehicles such as wheelchairs, baby strollers and similar transport vehicles equipped with an auxiliary drive system that do not exceed a maximum speed of six km/h nor are motor operated [18]. In order to obtain registration approval for an automated driving robot, it must be clarified whether the robot qualifies as a vehicle in terms of road traffic law and, if so, what specific registration requirements would apply to it. As per §1 Sect. 2 of the Road Traffic Act (StVG), ground vehicles which are moved by engine-power without being attached to railway tracks are considered as vehicles. Thereby, the type of motion and purpose of use are irrelevant. Moreover, pedestrians must use sidewalks according to §25 of the Road Traffic Regulations (StVO). Only in cases where pedestrians are accompanied by a vehicle that obstructs others, the roadway must be used (§25 Sect. 2StVO). The electrically operating parcel delivery robot of a project in Hamburg represents an example for registration. The robotic system was classified as a vehicle in the sense of §1 Sect. 2StVG and thus was licensed for use in traffic. The delivery robot moved on the sidewalk with a maximum speed of six km/h [19].

Furthermore, if mobile robots as examined in this work are classified as small electric vehicles (similar to e-scooters), they have to meet the legal requirements and regulations of the Ordinance on Small Electric Vehicles (eKFV). Vehicles that move at no less than 6 km/h and not more than 20 km/h plus are electrically powered could be considered as micro-electric vehicles according to § 1 Sect. 1 eKFV. However, micro-electric vehicles are explicitly not allowed to be operated on the sidewalk, as stated by the Federal Ministry of Transport and Digital Infrastructure [20]. Moreover, micro-electric vehicles must comply with the corresponding speed limits or, for instance, must install a handlebar. In addition, the legal basis for autonomous driving must be taken into account. In line with the framework for autonomous and automated driving [21], assisted and semi-automated driving, representing level 1 and 2 are currently in conformity with German traffic law. Due to legal changes at a national and international stage, the implementation of high-level and fully automated driving functions at levels 3 and 4 are now enabled [21]. Autonomous driving according to level 5 is not covered by the German Road Traffic Regulation. Depending on the concrete requirements identified by the users, appropriate legal prerequisites must be met.

## 2.3 Existing Personal Robot

To better understand existing mobile companions that could influence the transportation infrastructure in urban cities, best practice use cases are given. Not only automotive manufacturers promote advanced digital solutions for self-driving technologies, but also other technology and mobility service providers like Piaggio Fast Forward [27], Microsoft [22] or Toshiba [23] have developed mobile robotic vehicles that follow humans and assist with various services. Only a few are available for sale and the majority represent prototypes for the purpose of research.

With regard to treatment applications, robot ROREAS [24], for instance, assists stroke patient constantly by providing walking exercises and guiding them. For senior care, a robot that follows, watches and monitors the safety of a patient by detecting falls was developed [16]. In addition, oxygen therapy patients that must carry their tank continuously can be remedied by a robot assisting as portable oxygen supplier [25]. Accompanying elderly people by providing walking assistance such as the cane robot [26] helps to prevent falls. Another application field comprises transportation, respectively cargo carrier uses. The most novel application is Gita [27], a robot that follows humans and supports them by carrying heavy items (e.g. groceries). Furthermore, robotic suitcases offered by various companies [28] assist people in carrying and thus enable hands-free operation. And finally, the company Smart be Intelligent Stroller [29] developed a stroller that moves independently. Moreover, several prototypes of personal service robots in the area of sports were developed to assist runners during training or while participating in a marathon [30] by carrying the runner's water, food or clothes. This idea is as well applied via a robotic golf caddy [31] which was designed to assist golfers by carrying their golf bags. Besides transportation of personal belongings, another application field is the communication sector. Microsoft enhanced the communication field by developing a telepresence robot that provides video conferencing. Remote users can participate as a presence-robot in any conversation anywhere [22].

The variety of different areas of application and related tasks emphasizes the need for context and task sensitive design. As the particular field of application sets general conditions of human users (e.g. homogeneity of healthcare workers in contrast to the heterogeneity of personal transport robot users) [6], it is crucial to learn more about the specific needs and requirements of potential users, multiple focus groups were consecutively conducted.

## 3 Methodology

The core part of this study comprises the implementation of three focus groups for which a total of $n = 19$ participants were recruited. All attendees were purposely selected according to specific socio-demographic data and assigned to either a heterogenous or homogenous focus groups respectively application field. A questionnaire prefilled from all discussion participants enabled the evaluation of various socio-demographic data. Due to the COVID-19 pandemic, the focus group format was changed from in-person meetings to an online format. The first focus group $(n = 6)$ explored possible applications and respectively user groups for automated driving small vehicles in general. The subsequent second $(cargo carrier group, n = 7)$ and third $(healthcare group, n = 6)$ focus group specified the most frequently discussed use cases, resulting from the first discussion. In each focus group, scripted questions were asked by the moderator to initiate the discussion. The guideline of all focus groups started with an introductory block where attendees introduced themselves. In addition, further specified possible application fields were investigated. Afterwards, this collection of ideas was evaluated and compared with existing application fields from literature by revealing a virtual flipchart via Google Docs, in which previous use case scenarios had been demonstrated. The question of the concrete need and added value of automated-driving robots was reviewed again by asking about the specific benefits of the technology. Furthermore, another question block explicitly referred to the research question regarding requirements in terms of function, interaction and operation. In addition, the second and third focus group subsequently illustrated individual scenarios in order to visualize the use-case of an automated robot more precisely based on results from the first focus group. An overview of the used methodologies is shown in the Fig. 1.

## 4 Result Evaluation

Possible use case scenarios and user requirements were generally and openly discussed in the first focus group. In order to accurately evaluate findings of the focus groups, the program MAXQDA was applied. Thus, data was systematically classified, and participants' statements were grouped for similar meaning in categories via codings and qualitative frequency evaluation. Moreover, the second and third focus group analyzed user desired requirements of automated-driving robots. After discussing openly various technical-functional, psychological and interactional needs, the users ranked the most relevant five requirements. With the help of ratings, a quantitative measurement was executed to analyze the gathered data.

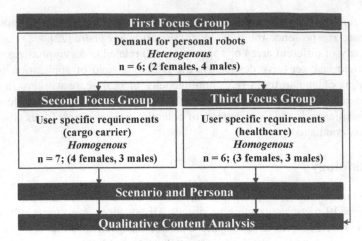

**Fig. 1.** Research methodologies overview

## 4.1 Need Assessment (1st Focus Group)

The main purpose of the first focus group was to prioritize prospective use case scenarios and to ascertain whether a demand for such mobile companions exists. The evaluated use case scenarios expected to show the most value are listed in Fig. 2, where most frequent mentions are listed in descending order. It can be observed that the respondents perceived the greatest need in the application of mobile cargo carrier in daily life or during leisure as well as in healthcare. Each of the six participants ranked the top three use cases, thus 18 total assertions resulted. Participants ascribed the greatest value in the application of mobile freight carriers such as *cargo carriers* for daily life of leisure ($n = 8$). Furthermore, the application field of healthcare ($n = 7$) was considered to be reasonably practical by providing e.g. medical support while carrying medical appliances or personal items of the person as well as the application sport ($n = 3$). Not considered at all during the evaluation were application fields with regard to children nor the telepresence robot despite the fact that use case scenarios of children were broadly discussed during the integrative session of the focus group.

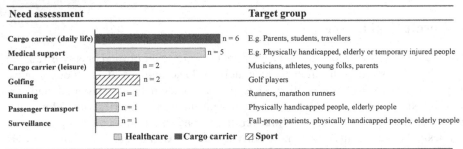

Calculation based on evaluation of use cases of the first focus group ($n = 6$). Each participant ranked top three use cases, thus 18 total assertion exist, results rounded

**Fig. 2.** Need assessment for automated driving human-following robotic systems

Therefore, the following target groups resulted for the two prioritized application fields:

- **For mobile cargo carrier applications:** Parents, students, travelers, elderly people, physically handicapped people, musicians, athletes and the young individuals
- **For healthcare applications:** Physically handicapped people, elderly people, temporary injured people, people after surgery, fall-prone patients

Particularly in the application field of a self-driving cargo carrier, the user group could represent people without having a car or a driver's license, especially in urban areas. This could lead to a replacement of the car in some situations. In this context, the topic of environmental pollution was mentioned by several participants. Many of the attendees perceived this technology as an alternative to transport cargo emission-free and considered mobile robots as an opportunity to not cause environmental pollution. Regardless of the application, the function ranked most valuable of the robot was to carry and transport items. Healthcare could therefore be considered as a separate subordination of cargo carriage. The subsequently conducted two focus groups explored specific scenarios and user requirements in the respective application area.

## 4.2  Requirement Engineering (2$^{nd}$ and 3$^{rd}$ Focus Group)

When designing urban self-driving cargo carrier, it is crucial to firstly understand the needs, desires and requirements of the users. Limited awareness in existing literature occurs as actual developments do not comprehensively provide sufficient user studies [1]. However, one major study worth mentioning conducts in-depth research about developing socially acceptable person-following robots [3]. The model is based on the premise that psychological needs are sorted and classified hierarchically, derived from Maslow's pyramid of needs which may serve as model for optimizing human satisfaction. Accordingly, a system should be developed in a manner that is primarily safe, functional and usable (ergonomic features) before it becomes pleasurable or even individualistic (hedonistic features) [32].

Thus, requirements from the focus groups were categorized into technical-functional, psychological and human-machine-interaction (HMI) needs. A total of 21 basis requirements and 11 application specific requirements (cargo carrier: 5, healthcare: 6) were elaborated within the conducted focus groups. An overview is provided in the following Fig. 3 and described in greater detail thereafter.

While a comprehensive list of potentially relevant requirements can be derived from the literature, it is unclear which priority is given to each individual need by potential users. To create a valid foundation, the results were quantified by a ranking. The evaluation was carried out by two subsequent additional focus groups. Thus, the seven cargo carrier participants of the second focus group and the six healthcare participants of the third focus group defined the five most significant needs. The conclusion of the evaluation results is given by the following ranking as shown in Table 1. On the one hand, the figure shows the requirements considered most important by both the cargo carrier and the healthcare group. On the other hand, large differences in preferences between the two user groups appear.

| | | Requirement | Explanation |
|---|---|---|---|
| Requirement catalogue | Technical-functional | Filling volume | Filling capacity |
| | | Robot weight | Total vehicle weight |
| | | Load capacity | Maximum load |
| | | Battery life | Duration or exchangeable battery |
| | | Charging conditions | Charging time, station, usability, battery weight |
| | | Safety and reactivity | Collision avoidance, independent traffic integration |
| | | Elevation gain | Climbing altitude differences (sidewalk, stairs) |
| | | Lock function | Robot closing and locking |
| | | Weather resistance | Rain protection, rain cape |
| | Psychological considerations | Reliability and accuracy | No sporadic sheering off |
| | | Social acceptance | Dispose of negative social stigma (shame), trust |
| | | Adjustable following direction | Following from behind, side-by-side, front |
| | | Height adjustability | Modular volume expansion, ergonomics |
| | | Price | Moderate retail price |
| | | Loss of autonomy | Free power of decision, appreciable enrichment |
| | | Legal clarification | Traffic permit, liability |
| | | Proximity setting | Situational geographical distance |
| | | Theft protection | Unambiguous authentication, cyber security |
| | | Data privacy and protection | Video recordings, surveillance |
| | | Optics | Discreet, slim, practical, modern (vehicle-like) |
| | HMI | Handling | Intuitive, easy, self-explanatory operation, |
| | | Connectivity | Smooth interaction with humans, robots, smartphones |
| | | Mobile control | Charge status display, actual fill weight |
| Application field specific | Cargo carrier | Children seat | Folding seat |
| | | Refrigeration | Cooling function (groceries, drinks) |
| | | Installation of light | Installation of lighting / flashlight |
| | | Modular coupling | Linking several robots |
| | | Situational interface | Operator tailored user interface |
| | Healthcare | Seat | Rollator seat |
| | | Emergency call system | Contact to a doctor/ambulance service |
| | | Surveillance function | Intervention mechanisms, reminder, warning |
| | | Medical data measurement | E.g. blood pressure, temperature by means of a bracelet or surface |
| | | Entertainment | Communication (video, telephone), radio, news |
| | | Haptic user surface | Large display, few buttons, activation by token or classic key |

**Fig. 3.** Categorized requirements as resulted from the focus group discussions

Overall, a long-time battery life represented the most important requirement ranked by in total nine of all focus group participants. The battery must certainly last for a prolonged time as this ensures the operation of the vehicle. In addition, an intuitive handling $(n = 8)$ was expected to be met by all potential users of the healthcare robot. In contrast, this requirement represented a lower important need when examining the cargo carrier vehicle $(n = 2)$. Regarding the development of mobile cargo carries, six of those surveyed people considered a large filling volume to be important, for example, for stowing children, work materials, sports gear or groceries after shopping. Thus, the capacity volume must be sufficiently sized considering the specific use case that all personal belongings fit into the vehicle. For elderly or physically disabled people, the filling volume represented a lower value $(n = 1)$. The ability of the robot to climb stairs was likewise interpreted differently. Respondents from the healthcare field regarded

**Table 1.** Prioritization of requirements between the cargo carrier and healthcare target group

**Table 1.** Prioritization of requirements between the cargo carrier and healthcare target group

| Requirements | Cargo Carrier (n = 7) Absolute | Health-care (n = 6) Absolute | Total (n = 13) |
|---|---|---|---|
| Battery life | 4 | 5 | 4 5 — n = 9 |
| Handling | 2 | 6 | 2 6 — n = 8 |
| Filling volume | 6 | 1 | 6 1 — n = 7 |
| Safety and Reactivity | 2 | 2 | 2 2 — n = 4 |
| Elevation gain | 4 | 0 | 4 — n = 4 |
| Emergency call system | 0 | 4 | 4 — n = 4 |
| Weather resistance | 2 | 1 | 2 1 — n = 3 |
| Social acceptance | 2 | 1 | 2 1 — n = 3 |
| Price | 0 | 3 | 3 — n = 3 |
| Robot weight | 2 | 0 | 2 — n = 2 |
| Reliability and accuracy | 1 | 1 | 1 1 — n = 2 |
| Adjustable following direction | 2 | 0 | 2 — n = 2 |
| Height adjustability | 1 | 1 | 1 1 — n = 2 |
| Seat | 2 | 0 | 2 — n = 2 |
| Entertainment | 0 | 2 | 2 — n = 2 |
| Haptic user surface | 0 | 2 | 2 — n = 2 |
| Load capacity | 0 | 1 | 1 — n = 1 |
| Lock function | 1 | 0 | 1 — n = 1 |
| Loss of autonomy | 1 | 0 | 1 — n = 1 |
| Legal clarification | 1 | 0 | 1 — n = 1 |
| Connectivity | 1 | 0 | 1 — n = 1 |
| Refrigeration | 1 | 0 | 1 — n = 1 |
| **Sum** | Σ 35 | Σ 30 | ■ Cargo Carrier ☐ Healthcare |

potential applications in e.g. nursing homes or clinics. This environment is barrier-free accessible, thus none of the users rated the robot's ability to climb stairs as of primary relevance. In comparison, four of the mobile cargo carrier users considered it as vital to overcome altitude differences, especially in urban areas. In addition, specific requirements could be distinguished, being only viewed as essential by the respective target group. In case of cargo carrying, these included, for instance, the robot's weight $(n = 2)$, the individually adjustable following direction $(n = 2)$ or the installation of a refrigeration module $(n = 1)$. For the healthcare robot, this comprised the implementation of an emergency call system $(n = 4)$, an element of entertainment (e.g. telephone, radio) $(n = 2)$ or the consideration of a haptic surface with few and big buttons $(n = 2)$.

## 5 Discussion

In this study, three focus groups were conducted to analyze potential demands of personal robots under consideration of user group heterogeneity [6]. After participants in a first focus group identified broad areas where robots could be useful, two subsequent focus groups identified potential requirements of a cargo carrier robot, and a healthcare focused personal robot. Overall, a majority of the $2^{nd}$ and $3^{rd}$ focus group attendees evaluated a high battery capacity as the most essential need $(n = 9)$. However, the duration of the battery life affects the battery's weight. The greater the battery life, the heavier the battery becomes which ultimately affects the total weight of the robot. The provision of an additional portable battery could possibly offer a remedy. Another widely argued request depicts the robot's ability to overcome elevation. All applications identified in literature were not capable of climbing stairs, except one military robot developed by Boston Dynamics [33]. With regard to cargo carrying, urban areas were considered as especially conceivable terrain for mobile robots instead of rural areas. Thereby, according to the potential target group, the robot must either be able to climb stairs $(n = 4)$ or be handy and light enough $(n = 2)$ to be carried. Another restriction arises in respect to legal requirements and the robot's ability to follow humans. Regulations for registration approval must be addressed. By using focus groups, it became obvious that the robotic vehicle is intended to follow a human on the sidewalk according to the user's speed.

Consequently, the robot cannot be categorized according to the Ordinance on Small Electric Vehicles. Accordingly, micro-electric vehicles are only allowed to drive on roadways. It can be concluded that automated driving vehicles are permitted for operation if fulfilling the correspondent registration approval criteria that include e.g. a speed limit of more than six km/h. The development of this technology, however, would lead to a comprehensive registration procedure with no immediate guarantee for success. This could be remedied by a separate authorization regulation, similar to the current requirements for motor vehicles. In addition, no concrete road safety obligations exist regarding healthcare robots and legal clarification is still unclarified. Consequently, due to an uncertain legal situation, the approval of the to be developed robotic vehicle is classified as to be further investigated.

With regard to HCI it can be concluded that the user-robot interaction is perceived as an interdisciplinary field that investigates social behavior, communication and intelligence in natural and artificial systems. In other words, interaction requires interchange and communication (verbal or non-verbal) between the robot and humans. Thereby, interaction refer to direct and indirect human-robot communication, stated as explicit and implicit mode of interaction [1]. Furthermore, the appearance of a robot also has an effect on the interaction. Derived from this, it can be concluded that the mobile cargo carrier represented a functional vehicle, thus is only capable of displaying emotions via e.g. flashing lights or sounds. Potential users of the focus group described this robot as rather functional, practical and handy. The automated driving robot was perceived by the users rather as a means of automated transportation vehicle and less as robot providing social functions. On the other hand, the designed healthcare robot provided richer movement and gesture via voice control, speech or body language. In contrast, the healthcare robot aimed to simulate human cognition and interaction by basic physical gestures and facial expressions, thus representing a socially mobile companion. Likewise, if the robot gives warnings during the use. Natural human-robot communication was intended even if the majority of studies does not provide communication during the human-following interaction [3]. Moreover, the healthcare robot needs to be connected to a doctor, the emergency service or the retirement home in case the SOS button is pressed. Both robots must trigger an alarm in case of theft. A loud signal should appear when an unauthorized attempt is made to steal the robot. The cargo carrier should be capable of being linked to other robots. A smart, smooth connectivity must therefore be provided. The healthcare robot must follow, observe and monitor the safety of the patient as similarly being previously researched in a study by Tomoya et al. [17]. If the user falls and does not move anymore, the robot triggers a signal and contacts the pre-defined emergency contact. In general, simple voice control between robot and user has to be implemented to interact socially with the user or other road users. In conclusion, the healthcare robot not only takes on the role of a carrier but takes over the role of a supervisor by providing further direct human-robot communication. In both application fields, there exist interfaces between the robot and the user, as well as between other pedestrians, traffic participants, obstacles and additional appliances.

## 5.1  Conceptual Modelling

In order to get an impression how the robots could look like, two different mobile companions were conceptually developed, and modelling drafts were made. The aim of the subsequent modelling was to design two adequate robots that meet the main requirements and constraints as compiled in the focus groups. This can help future researchers to gain an understanding of interacting with the designed systems. Both design drafts were created as a result of the intensive dialogue with the potential users. The pencil sketches particularly incorporated requirements that were ranked as highly relevant in the focus groups. Thus, the second focus group desired a vehicle to transport goods and children for long excursions. In comparison, the third focus group desired a healthcare robot that assists and helps elderly people during their daily life. The following Fig. 4 illustrates the drawn pencil sketches of the designed cargo carrier.

A key characteristic of the robotic vehicle was its large filling volume as requested by six of the respective users. Thus, a seat for children including safety belts was also installed as demanded by the users. The robot can be locked via a flap. In addition, the attachable rain cape protects the transported goods as well as the children from rain or strong sunlight. The vehicle is mounted on several robust wheels which have the required grip to overcome curbs. The height of the robot can be extended via an axle and bar. At the bottom there is a kind of a drawer which represents a refrigeration system to cool for example food or beverages. The robot is equipped all around with cameras and sensors to capture the environment holistically. Inside the casing, a replaceable battery is located. The operation of the robot is executed by means of a smartphone. It can be concluded that the mobile cargo carrier is primarily functionally designed and resembles a mobile transport vehicle with heavily focus on family application fields.

In contrast, the healthcare robot had a minimalistic creature-like appearance and is illustrated in the following Fig. 5. Thus, the healthcare robot differs significantly from the mobile cargo carrier. Special attention was paid to ensure the surveillance functions as the robot was intended to accompany the user by giving warnings and by providing guidance. The healthcare robot should appear friendly and create humanity. Thus, the robot is able to use cues to interact and communicate with the user. At the back of the robot, a resting area was installed. Thus, the robot can drive the user back home while

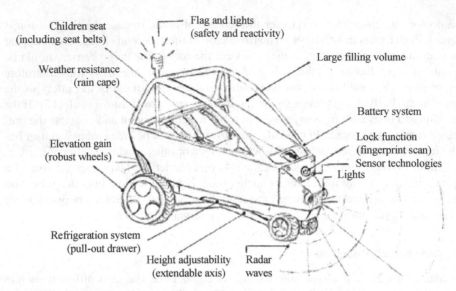

Children seat
(including seat belts)

Flag and lights
(safety and reactivity)

Large filling volume

Weather resistance
(rain cape)

Battery system

Elevation gain
(robust wheels)

Lock function
(fingerprint scan)

Sensor technologies

Lights

Refrigeration system
(pull-out drawer)

Height adjustability
(extendable axis)

Radar
waves

**Fig. 4.** Pencil sketch of the mobile cargo carrier robot

sitting. The operation is carried out with the help of an intuitive and easy to use operating panel, mentioned by all of the participants in the healthcare focus group $(n = 6)$ as the most important requirement. The robot is further equipped with sensors and actuators in order to accurately perceive the environment and to react appropriately. A compartment for recording the user's health data (e.g. blood pressure) is also integrated in the armrest.

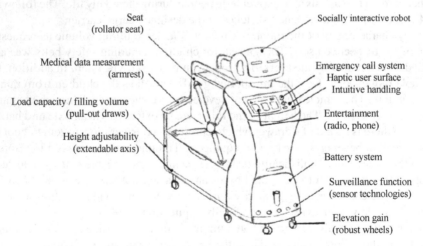

Seat
(rollator seat)

Socially interactive robot

Medical data measurement
(armrest)

Emergency call system
Haptic user surface
Intuitive handling

Load capacity / filling volume
(pull-out draws)

Entertainment
(radio, phone)

Height adjustability
(extendable axis)

Battery system

Surveillance function
(sensor technologies)

Elevation gain
(robust wheels)

**Fig. 5.** Pencil sketch of the healthcare robot

Ultimately, it can be concluded that the automated-driving cargo carrier robot's primary function is to follow and carry the load (e.g. grocery shopping) of humans. In comparison, the healthcare robot additionally adopts further functions, as the robot not only carries the user's personal belongings, but also monitors, guides, supports and, if necessary, transports the user home. Thus, the healthcare robot no longer just follows the human, in fact it could autonomously drive the passenger home. The healthcare robot extends the capabilities of the initially examined research object (human-following robot) to rather an autonomously driving micro-vehicle, which can also carry and transport passengers. With regard to operation, it became clear that the two robots were dissimilar concerning the operating device and individual setting options. The automated driving cargo carrier was aimed to be operated via an app. Focus was set on a high usability and adjustability of the selectable parameters. In contrast, the interface of the healthcare robot seemed more tactical, conventional and was characterized by a large-scale display.

## 5.2 Mobility As-A-Service

In addition, it became clear that the function has to exceed the capabilities of the human beings as stated by the focus group participants. Urban areas are particularly suitable for the technology as stated by many interviewed users and as indicated in literature [34]. Therefore, electrically driven mobile robots could relieve the roads due to less car traffic and further could represent a new form of mobility for people that have no car. Extensive testing will be necessary for the implementation to ensure the required safety. In this context, economic efficiency calculations must be carried out in order to compare anticipated earnings with presumed expenses. Participants of the focus groups discussed from the very beginning that it would probably be less economically viable to purchase and own such an expensive technology. In addition, it must be ascertained whether the healthcare robot can be subsidized, e.g. by health insurance companies. However, the participants saw great potential in renting mobile robots, leasing them for a short time or using pay-per-use / pay-per-time models for only a limited period of time. In general, shared mobility such as bike sharing, e-scooters or carsharing already has a great impact on urban mobility. Thus, the range of different mobility options available needs to be aligned to the passenger's interest.

In addition, it is recommended to not only focus on automated driving systems but also to investigate autonomous driving scenarios. In each focus group, autonomous use cases made a subject of discussions. Furthermore, the scenario of linking a mobile cargo carrier with supermarkets so that they transport groceries home after shopping appeared to be full of potential. The use case of shopping was particularly further examined in a research study conducted by Dautzenberg et al. [34].

However, legal issues arising with regard to autonomous driving robots as previously mentioned have to be considered, as to date, the implementation of fully autonomous driving systems are in Germany so far not enabled [21]. The necessity of a manual vehicle control system persists, hindering the full roll-out of autonomous systems.

## 6  Conclusion

The present work was conducted to give researchers and providers in the field of user-centric personal robotic vehicles insights into potential user needs in two main areas of application, transport and healthcare. For this reason, the knowledge gained specifically in this work serves as a basis for further research activities. An evaluation and concretization of requirements have to be carried out by designers, technical specialists and further traffic stakeholders. To sum up, the vast majority of respondents recognize a benefit for small vehicles operating automatically. The way road users will use and access transportation will possibly change in the future. As consequence, self-driving private or commercial mobile robots could represent a completely new form of mobility for people living and participating in urban traffic.

# References

1. Islam, M.J., Hong, J., Sattar, J.: Person-following by autonomous robots: a categorical overview. Int. J. Robot. Res. **38**(14), 1581–1618 (2019). https://doi.org/10.1177/027836491 9881683
2. Siebert, F.W., Pickl, J., Klein, J., Rötting, M., Roesler, E.: Let's not get too personal – distance regulation for follow me robots. In: Stephanidis, C., Antona, M., Ntoa, S. (eds.) HCII 2020. CCIS, vol. 1293, pp. 459–467. Springer, Cham (2020). https://doi.org/10.1007/978-3-030-60700-5_58
3. Honig, S.S., Oron-Gilad, T., Zaichyk, H., Sarne-Fleischmann, V., Olatunji, S., Edan, Y.: Toward socially aware person-following robots. IEEE Trans. Cogn. Dev. Syst. **10**(4), 936–954 (2018). https://doi.org/10.1109/TCDS.2018.2825641
4. Decker, M., Fischer, M., Ott, I.: Service robotics and human labor: a first technology assessment of substitution and cooperation. Robot. Auton. Syst. **87**, 348–354 (2017). https://doi.org/10.1016/j.robot.2016.09.017
5. Frank, B., Schvaneveldt, S. J.: Quality attributes of robotic vehicles and their market potential. In: 2017 IEEE International Conference on Industrial Engineering and Engineering Management (IEEM), pp. 750–754. IEEE (2017). https://doi.org/10.1109/IEEM.2017.828 9991
6. Onnasch, L., Roesler, E.: A taxonomy to structure and analyze human–robot interaction. Int. J. Social Robot. (2020). https://doi.org/10.1007/s12369-020-00666-5
7. Siebert, F.W., Klein, J., Rötting, M., Roesler, E.: The influence of distance and lateral offset of follow me robots on user perception. Front. Robot. AI **7**(74) (2020). https://doi.org/10.3389/frobt.2020.00074
8. Olatunji, S., et al.: User Preferences for socially acceptable person-following robots. In: 2018: Assistance and Service Robotics in a Human Environment: From Personal Mobility Aids to Rehabilitation-Oriented Robotics. A workshop in conjunction with the 2018 IEEE/RSJ International Conference on Intelligent Robots and Systems (2018)
9. SAE J3016: Taxonomy and definitions for terms related to on-road motor vehicle automated driving systems (2018). https://doi.org/10.4271/J3016_201401
10. Wahlster, W.: Künstliche Intelligenz als Grundlage autonomer Systeme. Informatik-Spektrum **40**(5), 409–418 (2017). https://doi.org/10.1007/s00287-017-1049-y
11. Olaverri Monreal, C.: Autonomous vehicles and smart mobility related technologies. Infocommunications J. **8**(2), 17–24 (2016)
12. Hertzberg, J., Lingemann, K., Nüchter, A.: Mobile Roboter. Springer, Heidelberg (2012). https://doi.org/10.1007/978-3-642-01726-1
13. VDI 2860: Assembly and handling; handling functions, handling units; terminology, definitions and symbols. Technical norm. Beuth Verlag GmbH (1990). https://www.vdi.de/richtl inien/details/vdi-2860-montage-undhandhabungstechnik-handhabungsfunktionen-handha bungseinrichtungen-begriffe-definitionen-symbole, Accessed 19 Feb 2020

14. Kim, M., e al.: An architecture for person-following using active target search. Robotics, 1–7 (2018)
15. Pradeep, B.V., Rahul, E.S., Bhavani, R. .: Follow me robot using bluetooth-based position estimation. In: 2017 International Conference on Advances in Computing, Communications and Informatics (ICACCI), pp. 584–589. IEEE (2017). https://doi.org/10.1109/ICACCI.2017.8125903
16. Tomoya, A., Nakayama, S., Hoshina, A., Sugaya, M.: A mobile robot for following, watching and detecting falls for elderly care. Procedia Comput. Sci. **112**, 1994–2003 (2017). https://doi.org/10.1016/j.procs.2017.08.125
17. Hanson, S., Kaul, A.: Executive summary consumer robotics. (2019). https://tractica.omdia.com/research/consumer-robotics/, Accessed 12 March 2020
18. Bundesamt für Justiz: Straßenverkehrs-Zulassungs-Ordnung (StVZO) (2012). https://www.gesetze-im-internet.de/stvzo_2012/BJNR067910012.html, Accessed 03 March 03 2020
19. Brandt, C., Böker, B., Bullinger, A., Conrads, M., Duisberg, A., Stahl-Rolf, S.: Fallstudie: Delivery Robot Hamburg für KEP Zustellung (2019). https://www.bmwi.de/Redaktion/DE/Downloads/C-D/delivery-robot-hamburg.pdf?__blob=publicationFile&v=4, Accessed 21 May 2020
20. Bundesministerium für Verkehr und digitale Infrastruktur: Elektrokleinstfahrzeuge – Fragen und Antworten (2020). https://www.bmvi.de/SharedDocs/DE/Artikel/StV/Strassenverkehr/elektrokleinstfahrzeuge-verordnung-faq.html, Accessed 21 May 2020
21. Deutscher Bundestag: Autonomes und automatisiertes Fahren auf der Straße – rechtlicher Rahmen (2018). https://www.bundestag.de/resource/blob/562790/c12af1873384bcd1f8604334f97ee4b9/wd-7-111-18-pdf-data.pdf, Accessed 16 March 2020
22. Cosgun, A., Florencio, D.A., Christensen, H.I.: Autonomous person following for telepresence robots. In: 2013 IEEE International Conference on Robotics and Automation (ICRA), pp. 4335–4342. IEEE (2013). https://doi.org/10.1109/ICRA.2013.6631191
23. Sonoura, T., Yoshimi, T., Nishiyama, M., Nakamoto, H., Tokura, S., Matsuhir, N.: Person following robot with vision-based and sensor fusion tracking algorithm. In Zhihui, X. (ed.) Computer Vision, pp. 519–538. InTech (2008). https://doi.org/10.5772/6161
24. Gross, H.-M., et al.: Mobile robotic rehabilitation assistant for walking and orientation training of stroke patients: a report on work in progress. In: 2014 IEEE International Conference on Systems, Man, and Cybernetics (SMC), pp. 1880–1887. IEEE (2014). https://doi.org/10.1109/SMC.2014.6974195
25. Tani, A., Endo, G., Fukushima, E.F., Hirose, S., Iribe, M., Takubo, T.: Study on a practical robotic follower to support home oxygen therapy patients-development and control of a mobile platform. In: 2011 IEEE/RSJ International Conference on Intelligent Robots and Systems, pp. 2423–2429. IEEE (2011). https://doi.org/10.1109/IROS.2011.6094633
26. Itadera, S., Watanabe, T., Hasegawa, Y., Fukiida, T., Tanimoto, M., Kondo, I.: Coordinated movement algorithm for accompanying cane robot. In: 2016 International Symposium on Micro-Nano-Mechatronics and Human Science (MHS), pp. 1–3. IEEE (2016). https://doi.org/10.1109/MHS.2016.7824241
27. Piaggio Fast Forward: Meet gita (2020). https://mygita.com/, Accessed 10 Feb 2020
28. Ferreira, B.Q., Karipidou, K., Rosa, F., Petisca, S., Alves-Oliveira, P., Paiva, A.: A study on trust in a robotic suitcase. In: Agah, A., Cabibihan, J.-J., Howard, A.M., Salichs, M.A., He, H. (eds.) ICSR 2016. LNCS (LNAI), vol. 9979, pp. 179–189. Springer, Cham (2016). https://doi.org/10.1007/978-3-319-47437-3_18
29. smartbe Intelligent Stroller. New revolutionary concept. https://www.weinvent.global/smartbe, Accessed 2 Mar 2020

30. Jung, E.-J., Lee, J.H., Yi, B.-J., Park, J., Yuta, S., Noh, S.-T.: Development of a laser-range-finder-based human tracking and control algorithm for a marathoner service robot. IEEE/ASME Trans. Mechatron. **19**(6), 1963–1976 (2014). https://doi.org/10.1109/TMECH. 2013.2294180
31. Caddy Trek: The most compact and lightweight remote control golf card in the world (2020). https://www.caddytrek.com/, Accessed 3 Mar 2020
32. Hancock, P.A., Pepe, A.A., Murphy, L.L.: Hedonomics: the power of positive and pleasurable ergonomics. Ergon. Des. Q. Hum. Factors Appl. **13**(1), 8–14 (2005). https://doi.org/10.1177/ 106480460501300104
33. Raibert, M., Blankespoor, K., Nelson, G., Playter, R.: BigDog, the rough-terrain quadruped robot. IFAC Proc. **41**(2), 10822–10825 (2008). https://doi.org/10.3182/20080706-5-KR-1001.01833
34. Dautzenberg, P., Köhler, A.L., Reske, M., Depner, N., Ladwig, S.: Less urban private transport through intelligent, micro-mobile transport solutions? a user-centered investigation of relevant use cases and requirements. In: Conference: 13th ITS European Congress (2019)

# Visualization of Zero Energy Bus Implementation Through Effective Computer Interaction

Jeremy Bowes[1], Sara Diamond[1(✉)], Greice C. Mariano[1,2],
Mona Ghafouri-Azar[1], Sara Mozafari-Lorestani[1], Olufunbi Disu-Sule[1],
Jacob Cram[1], Zijing Liu[1], and Zuriel Tonatiuh Ceja De La Cruz[1]

[1] Visual Analytics Lab, OCAD University, Toronto, Canada
sdiamond@ocadu.ca
[2] AI R&D Lab, Samsung R&D Institute Brazil, Campinas, SP, Brazil

**Abstract.** This paper outlines our process of researching, developing, and implementing a data analysis and visualization framework for Electric Bus (eBus) implementation – battery and hydrogen – in Canada, described as Zero Energy Buses (ZEB) using Human Computer Interaction strategies. Our research team works in collaboration with the Canadian Urban Transit & Innovation Research Consortium (CUTRIC), the coordinating agency for research and eBus trials across Canada. Our paper reviews factors considered to provide a meaningful systems analysis of eBus implementation, relevant data resources, and user-centric design approaches that will mitigate risks to create a successful eBus system analytics tool able to support diverse users. We are developing innovative techniques able to manage and represent the large volume of multisource spatial and temporal data in this new field of transit electrification including 2D/3D space-time and 4D visualization; real-time and predictive visualizations resulting from generative algorithms. Visual analytics will allow the careful tracking of multiple data sources, to measure the process of adoption, providing tools for monitoring of services, consumer attitudes, user experience, and prediction of impacts. Visual analytics provides a comprehensive overview of implementation nationally, enabling opportunities to share data, assess progress, identify challenges, and to explore challenges in this new field. Based on user consultations we are developing a Sustainable Transit Planning Index which considers multiple environmental, economic and equity/access factors in supporting ZEB implementation in Canada and applicable to other jurisdictions. It will extract and link local ZEB implementation data into a national framework.

**Keywords:** Data visualization · Visual analytics · Transportation planning · Transit planning · Electric bus · Zero energy bus · Urban systems

© Springer Nature Switzerland AG 2021
H. Krömker (Ed.): HCII 2021, LNCS 12791, pp. 19–38, 2021.
https://doi.org/10.1007/978-3-030-78358-7_2

# 1  Introduction

Zero Emissions Bus (ZEB) transit infrastructure – battery and hydrogen – is identified by all levels of government in Canada and regional transit authorities as a critical step in meeting clean growth goals and Zero emissions targets. The importance of transport for climate action is underscored in the Paris Agreement because almost 25% of energy that is related to global greenhouse gas emissions is emitted by transport systems [1]. Canada has set the goals of 5000 transit and school eBuses on the roads by 2025. However, according to a study developed by Mohamed et al. [2] with Canadian transit service providers, finding the right electric bus for implementation in cities across Canada is not simple. There are some factors that hinder the adoption of electric buses in Canada, at technological, operational, decision-making, and business levels. Examples include bus operations, electrification logistics, environmental and cost impacts [3–5]. Thus, meeting the target requires a national strategy to mobilize resources, share data, and enable effective knowledge sharing and collaboration between municipalities, who must draw on provincial resources.

Our research team works in collaboration with the Canadian Urban Transit & Innovation Research Consortium (CUTRIC) who are the coordinating agency to support industry-academic collaborations in the development of next-generation low-carbon smart mobility technologies to make Canada a global leader in zero-emissions [6,7]. CUTRIC membership includes government policy leaders municipal transit authorities transit system designers and operators, electric bus manufacturers such as New flyer, Nova Bus, BYD and Proterra; charging station and software providers such as Siemens, VeriCiti and Giro; utilities, such as hydro, government agencies; infrastructure providers and groups representing transit passenger users. CUTRIC and its members are contributing and supporting the implementation of several ZEB demonstration projects across Canada, which include battery-electric buses (BEBs) and hydrogen-fuel-cell electric trials. Some examples can be found in Vancouver, Edmonton, Toronto, Brampton, and York Region. CUTRIC has helped the Toronto Transit Corporation (TTC), Canada's largest transit system, to evaluate and plan to implement a zero-emissions fleet by 2040. Currently, the TTC is running and evaluating the performance of 59 electric buses.

Research in the Canadian context bears relevance to other federated jurisdictions implementing ZEB systems where collaboration between levels of governments and industry is required. We are developing innovative techniques able to manage and represent the large volume of multisource spatial and temporal data in this new field of transit electrification including 2D/3D space-time and 4D visualization; real-time and predictive visualizations [8] resulting from generative algorithms. Visual analytics will allow the careful tracking of multiple data sources, to measure the process of adoption, provide tools for monitoring of services, changing consumer attitudes, user experience, and prediction of impacts. As Canada moves forward with ZEB implementation, visual analytics will provide a comprehensive overview of implementation nationally, enabling opportunities to share data, assess progress, and to identify and explore challenges in this new field. Based on user consultations we are developing a Sustainable Development Transit Index which considers multiple environmental, economic and equity/access factors in support of ZEB implementation in Canada.

## 2   Related Work

### 2.1   Literature Review

We reviewed and summarized related work on visualization in transportation categories; Road and Rail Transportation Systems; Predictive and Prescriptive Data Insights; Spatial Analytics, GIS, and Cartography; IOT sensors; Energy Efficiency and Optimization; and Data Analysis and Visual Analytics and Design. The literature review identified many areas of visualization support that would benefit future decision making, system and equipment planning, and investment.

This work is framed by our analysis of HCI approaches to urban analytics [10]. Analysis expanded our previous taxonomies of urban transit analytics tools [11, 12]. For instance, the inclusion of GPS sensors in different devices (e.g., cell phones) makes it possible to acquire data reflecting movements of passengers and their behaviours under different circumstances such as the choice of routes during rush hours [13]. Zhang et al. [2018] introduce the importance of space-time visualization analysis stating, "the analysis of bus passenger distribution studied from Beijing IC bus data can help to understand the rules of human behavior and provide reliable data guidance for reasonable decision-making around Beijing passenger traffic planning" [13, p. 813]. The analysis of data of bus passenger behaviors collected by smart cards survey correlated with demographic data support the analysis of potential customers, allows customer engagement in eBus planning, and analysis of satisfaction as implementation rolls out [14].

Some studies address the challenge of creating spatial and temporal data visualizations and analytics methods related to transportation mobility [15–18] as users must manage a huge volume of real-time traffic data obtained from different sources. According to Ma and Chen [2019], "Massive data are generated during the public transportation operation process. These data are highly continuous, have wide coverage, contain comprehensive information, and update dynamically and quickly (for example real time data); thus, they have high application value to public transportation operation and planning" [15, p. 176]. Bouillet et al. [19] propose dashboards as the most effective way to enter the data and direct multiple user types. An important feature of Bouillet's Public Transport Awareness Application Architecture is "its ability to provide an operator with a detailed representation of the public transit status, including the results of analytics which is less than 10 s old on average" [19, p. 798]. We also reviewed existing visualization systems such as VeriCiti [20] and Zight[21] which municipalities are using to plan routes, charging and manage visualization of the resulting data.

The literature review identified principles for eBus route visualization, and guidelines to begin to build low-fidelity prototypes. Visualization design must take a user-centric approach by working with the eventual users of the tools (transportation planners, scheduling managers, and agency administrators) to identify use cases, and groupings of related tasks, and the associated data required.

We highlighted decision-making needs of users as represented by CUTRIC's membership and considered their need to integrate diverse types of data and then correlated these users with our literature review categories, identifying a list of task objectives and data management needs. For example, route and scheduling managers require vehicle routing, travel times, real time GPS vehicle tracking, load information, carrying capacity, and passenger counts. All contexts require tools to analyze costs of implementation in relation to immediate and long-term impacts, including cost-savings in direct and indirect environmental benefits which may take time to play out [22].

The exploration of cartography and visualization approaches should support exploratory analysis of highly scalable and dynamic data. There is a need to integrate 2D and 3D visualization outcomes into a meaningful visualization system design and prototype of eBus Routes. For example, spatial data display could be coupled with 2D cartographic base maps, and layered display performance characteristics charts, tables, or graphs [23]. Mashups of geographical representations create new applications tailored to specific tasks by integrating current applications and data sources [24,25]. Wood proposes an alternative approach to the exploration of data, "inspired by the geographical mashup, which takes a set of open and friendly resources, and combines them using *de facto* standards often based on XML. The use of general-purpose scripting offers flexibility in interactive visual encoding specification and the filtering and processing of data according to spatial temporal and attribute criteria" [25, p. 1176]. Systems for transportation planning require real-time data that can be compared to historical patterns, to identify changing usage patterns. A predictive 2D/3D visual framework of eBus route and prescriptive analytics would provide historical, real-time, predictive, visual, video and image information; and allow data normalization, brokering, and storage for visualization, and event mapping as described by Neilson et al. [16] and Bouillet et al. [19]. Systems for transit operators should provide visual exploration of eBus Routes in a real-time and dynamic scalable dashboard environment. These tools can be used to optimize aspects of the system services, and to change the user experience. Dashboards can couple scheduling and trip data, with incident and real-time reporting. These could be converted to passenger-friendly visualizations that provide route choices for passengers creating an enhanced informational experience for them as outlined by Bouillet et al. [19] in the Dublin case study.

With these factors in mind, we identified groupings of related tasks, and the associated data required, building on the initial user and stakeholder group categories within CUTRIC. We developed a set of principles with which to apply visual analytics in order to optimize user experience:

- Recognize patterns and other meaningful information by transforming and mapping data onto visual forms.

- Provide multiple, coordinated views to allow users to manage the complexity of data.
- Optimize aspects of the system services and provide a more responsive enhanced informational experience for passengers.
- Deploy the 'mash-up' approach to data integration to exploratory geo-visualization of large spatiotemporal datasets including open-source data sources; and
- Analyze data reflecting movements of passengers and their behaviours under different circumstances such as pricing, weather, transit type availability, to allow analysis of potential customers, customer engagement in eBus planning, and analysis of satisfaction as implementation rolls out.

The sustainable impacts of eBus technological innovations need to be visualized, and the related performance metrics identified and developed through consultation with experts. eBus adoption has a direct impact on $CO_2$ emissions and pollution [26,27, p. 2829]. Furthermore, technological innovations adopted by urban bus companies can improve cities' sustainability through the effective use of information and communication technologies that will reduce the need for travel and traffic volume [30], and enable appropriate route planning, reducing traffic congestion, traffic collisions, and travel distance and time [31]. In many instances eBuses will be integrated alongside diesel or hydrogen buses hence system wide analysis is needed. The following are principles for visual analytics that can support sustainability:

- Integrate means to monitor, analyze, and respond to the events in the smart grid relevant to ZEB implementation and overall transit system operation [32].
- Provide analytics to support optimal energy management to improve fuel economy.
- Determine whether the transition to a "clean fleet" is financially viable and worthwhile based on expected emissions reductions; and
- Investigate the impact of uncertainty in predicting the energy consumption of electric buses.

This list of principles provided us with the following data types needed, related to task objectives which support a breadth of visualization tasks; spatio-temporal data, GTFS feed data, spatial data, telecommunication data, financial data, demographics data, and visualization to support decision making.

## 2.2  Case Studies

**Best Practices.** Our goal was to identify the best practices through an analysis of case studies. We sought to outline the strategies that integrate a variety of information and users who would require support for visualization and analysis. Two key questions that we considered were: What are the appropriate visualizations for each stage of implementation, and what aspects need to be integrated for a low-fidelity prototype?

The case studies were divided into Canadian examples such as Manitoba [33], Toronto, Brampton, Vancouver, or Edmonton [7], and international examples [19,28,34–37] with the intention of identifying these best practices, the scope and type of data queries that stakeholders needed, and the types of tools and visualizations required for decision support. The international case studies helped us to compare the visualizations and data types needed for routes data, fares data, cost data [9], and electric performance data, as many of the systems in place were at further stages of implementation. These had created open platforms and dashboards outlining the factors that were desirable to compare. In our case study review we identified stages of implementation, to understand the data that was needed before, during and after implementation and whether different kinds of service analysis reports were required according to implementation stages. For example, planners need to ensure at least the same level of service in replacing diesel buses with eBuses, understand the number of buses required, drivers, hence charging and labour costs, and balance this against gains in energy efficiency. Once eBuses are implemented agencies need to monitor the actual data against planning projections.

The first stage is the planning stage where planning occurs, then pilots are run and equipment is sourced, and the second stage is the on-going deployment of a pilot implementation of at least twelve and as many as twenty eBuses. The final stage is full scale deployment that requires the measurement of system metrics during implementation and afterwards. One goal of our national case study analysis was to identify a case study municipality at each stage (before, during and after implementation) as points of comparison. Accessing historical data from the transit agency users who have implemented eBus systems in cities like Winnipeg, Edmonton, Toronto, and Montreal will provide viable case study information. A deeper dive into the data being generated by the national transit deployment of eBus case studies will aid any comparative visualization of the related factors outlined above and provide more direction for the creation of informational dashboards to aid in decision - making and operational planning. We have now begun work with three Greater Toronto and Hamilton transit authorities to access data and develop our mid-fidelity prototypes.

**Dashboard Strategies.** We completed a review of applicable dashboard types and features, which allowed us to plan and select key eBus performance metrics. Our prototypes build on prior related work where we designed a bus bridging visualization system [38] for the iCity Complete Streets project; and a dashboard visualization tool for urban planners to assess the completeness of streets and measure interventions for change over time [39–41].

The Bus Bridging Comparative Visualization tool was a collaborative work between OCAD University and University of Toronto teams. When a disruption occurs in any point of a subway line, buses from different routes have to be reallocated to attend to shuttle subway line passengers from the places that are effected. In this context, our challenge was to build a user-centred dashboard to help subway operators to make quick decisions in relation to how many buses they need to select and from which routes, so that the overall transportation system is minimally impacted. Passengers should not have to wait long or encounter a full bus. Figure 1 presents our first mock-ups of the proposed dashboard. Together with University of Toronto team, we validated the information to be shown and the main interactions. Based on their feedback, we developed a new dashboard using the ArcGIS tool from EsriCanada, where the user can simulate two scenarios at the same time. Figure 2 presents our implemented dashboard and addresses a disruption at the Keele Subway Station (green line) which is part of the TTC System in Toronto. In our proposed dashboard, there are two map views and two data visualization views where users can interact and compare the scenarios. For example, in Fig. 2a both maps represent two different scenarios, indicating the source routes where buses can be obtained (indicated by colored lines in the maps), and in the bottom, as a chart, the number of buses they need to pull to support each route. Similarly, Fig. 2b shows other variables that can be visually explored on this dashboard. The colored lines indicate the out of service duration and the charts show the bus riders delay (passengers per minute). Subway operators will make a decision about the best scenario, with the lowest system impact, based on the lowest total bus riders delay. Other analysis available on this tool are deadhead time; comparison between out of service delay and bus riders delay; and a sum of total delay.

The Complete Streets tool included transportation types from active to public transit, street widths, landscape and built transportation space characteristics. The interface used 3D and 2D maps and graphs, bringing together multi variant data sets which were placed against an index that measured the completeness of streets. This project bears relevance to the development of a possible index and measurement tool that would assess the implementation of ZEB transit systems. Users can compare the qualities of a street and change the data informing factors, with each change indicating its impact on other factors. Such an approach could indicate energy consumption, pollutants, costs, and access, or any number of different factors (Fig. 3).

## 3   Methodology

We compiled a visualization needs analysis identified by types of users, tasks, and stages of eBus implementation from the literature review and case studies. In collaboration with CUTRIC we gathered knowledge on the state of implementation and tools in use. We created a consulting user group comprised of municipalities of different scales (205,960 to 5,982,000 residents) and a group of commercial technology supporters who provide specific visualization tools. From this we were able to develop a comprehensive and detailed overview of the types of data typically collected in electric bus planning, implementation, and monitoring, and the means by which these data are collected (sensors, cameras etc.) and the variable scale required of the data, from the individual vehicle to systems analysis (emissions for example). We have also begun to correlate data types with how the data is captured with who is responsible for data collection

26     J. Bowes et al.

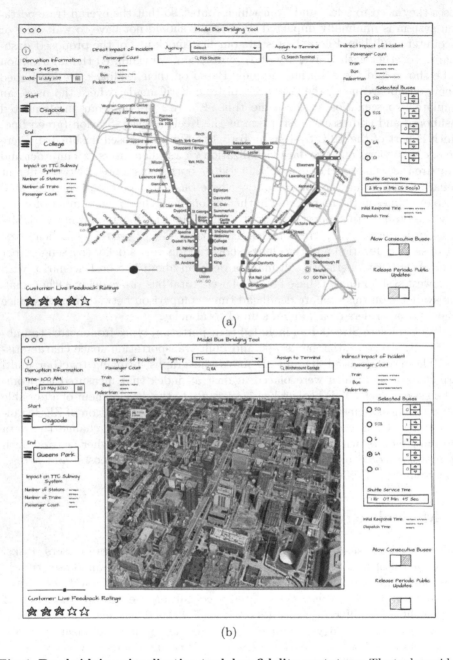

(a)

(b)

**Fig. 1. Bus bridging visualization tool, low fidelity prototype**. The tool provides information on subway lines that may be affected and require bus bridging and a 3D image of a route in relation to density of the built space. VAL team, CUTRIC project: Olufunbi Disu-Sule. [38]

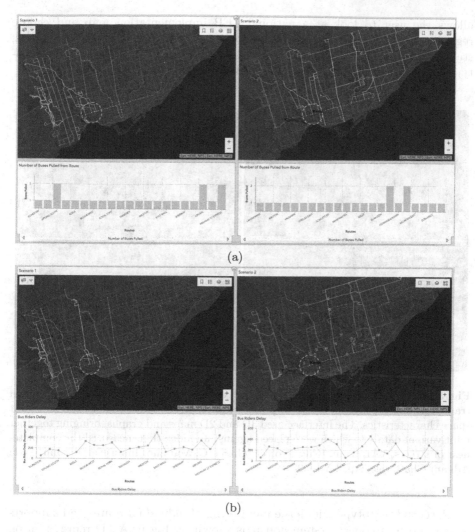

(a)

(b)

**Fig. 2. Comparative visualizations for bus bridging scenarios.** Provides alternate sources of buses for subway route planners who hope to have the least negative impact on the subway system. VAL team, CUTRIC project: Olufunbi Disu-Sule [38] (Color figure online)

and administration. Utilizing CUTRIC as our expert agency, we highlighted priority concerns, and identified use cases and are beginning to build prototypes with users from individual agencies.

## 4   Prototypes

As a first stage we developed a series of prototypes that support municipal, regional, and national analysis of implementation progress. The first prototype allows the user to look at a national ZEB implementation map and choose regions, and then drill down to municipalities to understand GHG emissions, state of deployment, and macro-scale trade-offs. Figure 4 presents those prototypes, which provide a 2D map perspective and associated graphs (Fig. 5).

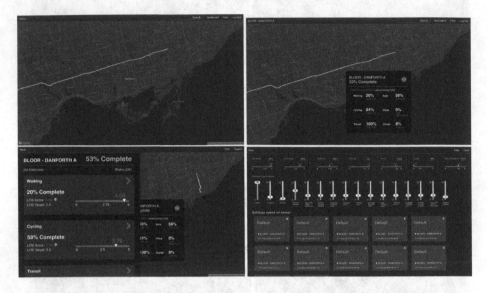

**Fig. 3. ICITY dashboards/complete streets**. Example of criteria that included transportation types from active to public transit, street widths, landscape and built space characteristics. The interface used 3D and 2D maps and graphs, bringing together multivariant data sets which were placed against an index that measured the completeness of streets, VAL team, iCity Project: Greice C. Mariano, Iman Kewalarami, Veda Adnani Chatterjee, April Yu [39]

A second prototype allows the monitoring of individual routes and supports a scale up to a transit system and it is shown by Fig. 6. A 2D representation allows the view of eBus implementation in the context of the built environment.

Our most recent dashboard prototype, which was developed using the ArcGIS platform from EsriCanada's toolkit, integrates and correlates multivariate variables, reflecting our previous proposed prototypes with some improvements, see Fig. 6. Users can interact with a map and visualize the ZEB implementation throughout Canada, or they can select a specific city and access detailed information about the implementation in that city, such as, energy consumption, eBus routes, and battery capacity. Figure 6a shows the first screen with a macro map view for all of Canada's provinces with all the cities that currently have an eBus demonstration running. In Fig. 6b, the map provides details of the cities within a selected province that are implementing eBuses. Users can interact with the map and visualize (right top view) information about the electrical routes and in the right bottom view, users can see some charts that display information related

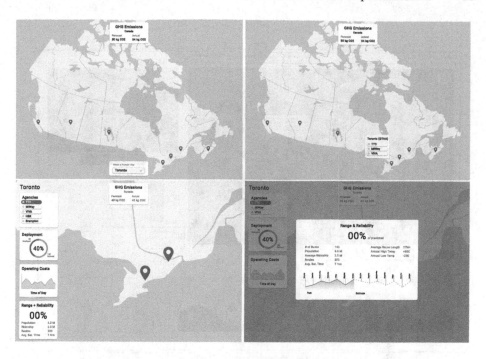

**Fig. 4.** This example shows the ways that researchers and implementation teams can accurately track the progress and impacts of eBus implementation. Design and prototyping of 2D visualizations integrated with maps. Development of user-centered dashboard to support implementation of electrical buses in Canada, Maps & Charts, VAL team, CUTRIC project: Jacob Cram

to energy consumption. These display comparative results between cities; time taken to complete a one-way trip; and summary related to sequential charging points in the routes. Users can analyze the total number of garages available for buses in the city, and the average nominal charge rates by city in current map view. Users can select a city to visualize the implementation of ZEB. All the information represented in this dashboard refers to a provincial or municipal maps, displaying the types of data that the user is interested in.

**User Consultations.** Our expert group of local GTHA transit agencies and technology providers responded to the prototypes we had built. Response was positive and consultations identified several significant analysis gaps that municipalities are facing, two at the municipal implementation level and the third shared by municipalities and tool providers at the systems level. First, municipal transit authorities need practical tools to identify and visualize changes in work process, labour skills and transition requirements; to consider the number and costs of operators for ZEB implementation given that more buses and shorter routes may be required, and to model all labour costs into an economic analysis of ZEB implementation. Second, they need to model ZEB implementation

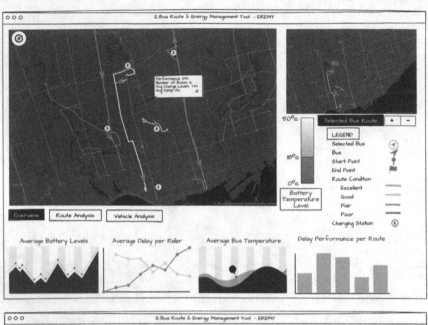

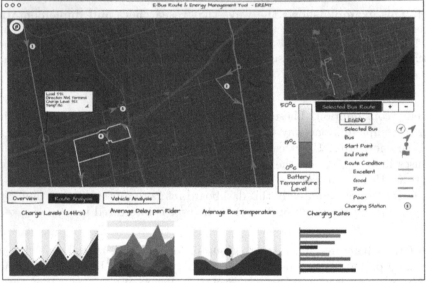

**Fig. 5. Design and prototyping of 2D and 3D visualizations integrated with maps.** First iteration of the tool provided with addition of a color-coded route system for the readability of the tools map. Inclusion of more comparative charts and graphs help visualize the data more appropriately. VAL team, CUTRIC project: Olufunbi Disu-Sule

(a)

(b)

**Fig. 6. Design and prototyping** of local, regional and national maps for all provinces and cities in Canada. that currently have an electrical bus demonstration running. VAL team, CUTRIC project: Olufunbi Disu-Sule and Jacob Cram

in relation to traffic and congestion management, and to visualize approaches to increase, measure, and implement rider attraction through improved and environmentally sustainable transit, which would have a positive impact on congestion.

Third, municipalities and other stakeholders, such as technology companies, need contextual data analysis and visualization beyond the immediate environmental and cost analysis provided by current commercial systems. Building on our local/regional and national map, transit agencies, municipalities and other levels of government are seeking an interactive system that can identify and quantify the impacts of electrification across environmental, economic, and social factors.

## 5   Conclusions and Next Steps

### 5.1   Conclusions

In response to user feedback, we are developing a common Sustainable Transit Planning Index (STPI) and interactive visualization tool. The index and tool will allow municipalities, transit authorities, provinces, and national governments to set goals according to their priorities and measure implementation. Tool providers also expressed value in understanding implementation across interoperable systems as ZEB buses come into effect across Canada (Fig. 7).

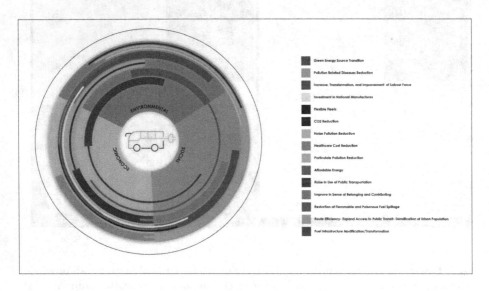

**Fig. 7. Diagram of the STPI factors and impacts.** This mock-up illustrates the overlap of impacts in Environmental, Economic, and Social sustainability with respect to GHG reductions and strategies. VAL team, CUTRIC project: Sara Mozafari-Lorestani, Mona Ghafouri-Azar

The most common definition of a sustainable development system is stated by the Bruntland Commission, "Sustainable development fulfils the needs of the present without compromising the ability of future generations to meet their own needs" [42, p. 11]. Zito and Salvo have extended this definition to propose an urban transport sustainability framework, "Satisfying current transportation and mobility needs without compromising the ability of future generations to meet their own needs" [43, p. 180]. This implies that a sustainable transportation system must address impacts of transportation on the environment [44, 45], support stable economic growth and employment [46], and improve societal inequities [36, 47–53]. These three dimensions have equal relevance for measuring progress towards sustainable transportation [37, 54]. Related qualities are safety, affordability, accessibility, equitable access, efficiency, and resilience. Sustainable transit must provide a choice of transportation modes; minimize consumption of non-renewable resources, reuse and recycle its components and minimize the emissions of carbon and other pollutants [55], including noise, to the atmosphere and environment [29, 56, 57] hence impacted residents' immediate future health. The diagram below includes the detailed areas within the STPI that require an integrated analysis such as environmental improvement, labour, transit access, equity, and improved health benefits.

## 5.2   Next Steps

We will also undertake focused visualization of data for municipal transit authorities of labour costs and congestion impacts and remedies and consider means to integrate these factors into broader analysis, drawing on holistic cases studies provided by Chandler et al. [58] in relation to San Francisco, California and Barriga [59] regarding Belo Horizonte, Brazil. As we develop dashboard prototypes, test the levels of interactivity, information filtering, and correlation of data with transit and industry users, we will be able to prioritize the human – computer interaction crucial for inter operational decision support. We will engage municipal transit authorities, a newly formed ZEB implementation working group of multiparty stake holders and CUTRIC national staff who will serve as our test group in assessing and testing the index and the prototypes.

We will select key factors to measure progress toward a sustainable development transit system (STPI) which considers multiple environmental, economic, and social factors in supporting ZEB implementation in Canada, within the context of global relevance. Current indexes for environment, social and economic aspects will be selected, and an index matrix will be constructed. Assume there are $n$ years, and $m$ indexes. The original matrix is show by Eq. (1), where where, $X_{ij}$ defines the index $j$ in year $i$. This will be a significant algorithmic task, for example six key indices over five years would amount to 30 indices requiring integration.

$$X = \{X_{ij}\}_{(n*m)}(1 \leq i \leq n, 1 \leq j \leq m) \tag{1}$$

There is some inconsistency in units among different indices. It is necessary to transform indexes to dimensionless numbers in order to have reasonable comparison between them. This process is called Standardization [60]. Standardization of indexes means that each value for each index will be converted to the same unit so that different indexes can be compared [61]. The procedure for standardization of each index is explained in Eqs. (2) for positive indexes and (3) for negative indexes. $Z_{ij}$ are the standardized indexes and lies between 0 and 1.

$$Z_{ij} = \frac{X_{ij} - X_{min}}{X_{max} - X_{min}} \tag{2}$$

$$Z_{ij} = \frac{X_{max} - X_{ij}}{X_{max} - X_{min}} \tag{3}$$

Then, standardized indexes with different dimensions will be normalized to have dimensionless indexes using Eq. (4):

$$P_{ij} = \frac{Z_{ij}}{\sum_{i=1}^{n} Z_{ij}} \tag{4}$$

We will then weight and combine indexes into a composite index for environmental, social, and economic impacts. We will develop a weighting system that allows effective balance of factors through computation and integration [62,63].

We will prototype interactive visualization dashboards using a dashboard approach similar to Complete Streets [39–41] and benefiting from our case study and literature review. It interprets the combined index and meet the needs of different stakeholders in evaluating sustainable transit data at various points of planning, implementation, and evaluation. This methodology will support decision makers to make realistic data-driven decisions and to design future policies and practices.

# References

1. High-level advisory group on sustainable transport. Mobilizing sustainable transport for development, United Nations (2016). https://www.un.org/development/desa/undesavoice/more-from-undesa/2016/11/29738.html. Accessed 8 Feb 2021
2. Mohamed, M., Ferguson, M., Kanaroglou, P.: What hinders adoption of the electric bus in Canadian transit? Perspectives of transit providers. Transport. Res. Part D Transp. Environ. **64**, 1–18 (2018)
3. Mahmoud, M., Garnett, R., Ferguson, M., Kanaroglou, P.: Electric buses: a review of alternative powertrains. Renew. Sustain. Energy Rev. **62**, 673–684 (2016)
4. Noel, L., McCormack, R.: A cost benefit analysis of a V2G-capable electric school bus compared to a traditional diesel school bus. Appl. Energy **126**, 246–255 (2014)
5. Tzeng, G.-H., Lin, C.-W., Opricovic, S.: Multi-criteria analysis of alternative-fuel buses for public transportation. Energy Policy **33**(11), 1373–1383 (2005)
6. Petrunic, J., Abotalebi, E., Raj, A.: Best Practices and Key Considerations for Transit Electrification and Charging Infrastructure Deployment to Deliver Predictable, Reliable, and Cost-Effective Fleet Systems. Canadian Urban Transit Research and Innovation Consortium (CUTRIC), Toronto (2020)
7. Petrunic, J.: Supporting Zero-Emissions Technology Innovation, Research, Development, Demonstration & Integration (RDD&I). Canadian Urban Transit Research and Innovation Consortium (CUTRIC), Toronto (2016)

8. Sobral, T., Galvão, T., Borges, J.: Visualization of urban mobility data from intelligent transportation systems. Sensors **19**(2), 332 (2019)
9. Perera, T., Gamage, C.N., Prakesh, A., Srikanthan, T.: A simulation framework for a real-time demand responsive public transit system. In: 2018 21st International Conference on Intelligent Transportation Systems (ITSC), Maui, Hawaii, USA , pp. 608–613. (2018)
10. Skelton, C., et al.: Citizen informatics: integrating urban data and design for future stakeholders. Int. J. Electron. Gov. **11**(1), 23–43 (2019)
11. Bowes, J., Diamond, S., Juneja, M., Gordon, M., Skelton, C., Gunatilleke, M., Carnevale, M., Zheng, M.D.: User-centered taxonomy for urban transportation applications. In: Nah, F.F.-H., Xiao, B.S. (eds.) HCIBGO 2018. LNCS, vol. 10923, pp. 577–593. Springer, Cham (2018). https://doi.org/10.1007/978-3-319-91716-0_46
12. Gordon, M., Diamond, S., Zheng, M., Carnevale, M.: Compara. Encounters Theor. Hist. Educ. **19**, 163–185 (2018)
13. Zhang, J., Chen, Z., Liu, Y., Du, M., Yang, W., Guo, L.: Space-time visualization analysis of bus passenger big data in Beijing. In: Cluster Computing, pp. 1–13 (2018)
14. Kwon, Y., Kim, S., Kim, H., Byun, J.: What attributes do passengers value in electrified buses? Energies **13**(10), 1–14 (2020)
15. Ma, X., Chen, X.: Chapter 7 - Public transportation big data mining and analysis. In: Wang, Y., Zeng, Z. (eds.) Data-Driven Solutions to Transportation Problems, pp. 175–200. Elsevier (2019). https://doi.org/10.1016/B978-0-12-817026-7.00007-2
16. Neilson, A., Indratmo, B.D., Tjandra, S.: Systematic review of the literature on big data in the transportation domain: concepts and applications. Big Data Res. **17**, 35–44 (2019)
17. Andrienko, G., Andrienko, N., Chen, W., Maciejewski, R., Zhao, Y.: Visual analytics of mobility and transportation: state of the art and further research directions. IEEE Trans. Intell. Transp. Syst. **18**(8), 2232–2249 (2017)
18. Nusrat, S., Kobourov, S.: Task taxonomy for cartograms. In: Bertini, E., Kennedy, J., Puppo, E. (eds.) Eurographics Conference on Visualization 2015, EuroVis-Short2015. The Eurographics Association, Cagliari (2015)
19. Bouillet, E., Gasparini, L., Versheure, O.: Towards a real time, public transport awareness system: case study in Dublin. In: Proceedings of the 19th ACM International Conference on Multimedia, MM 2011, pp. 797–798. Association for Computing Machinery, New York (2011). https://doi.org/10.1145/2072298.2072463
20. VeriCiti Fleet Management Homepage. https://viriciti.com/. Accessed 8 Feb 2021
21. Zight transit Homepage. https://zight.nl/clients/. Accessed 8 Feb 2021
22. Freedman, D.: Market-driven considerations affecting the prospects of alternative road fuels. Philos. Trans. R. Soc. A Math. Phys. Eng. Sci. **372**, 20120326 (2014)
23. Sorger, J., Ortner, T., Piringer, H., Hesina, G., Gröller, E.: A taxonomy of integration techniques for spatial and non-spatial visualizations. In: Bommes, D., Ritshel, T., Schultz, T. (eds.) Vision, Modeling, and Visualization, VMV 2015. The Eurographics Association, Aachen(2015). https://doi.org/10.2312/vmv.20151258
24. Gu, Y., Kraak, M.-J., Engelhardt, Y.: Revisiting flow maps: a classification and a 3D alternative to visual clutter. In: Proceedings of the International Cartographic Association (ICA), Washington, USA, vol. 1, p. 51 (2017)
25. Wood, J., Dykes, J., Slingsby, A., Clarke, K.: Interactive visual exploration of a large spatio-temporal dataset: reflections on geovisualization mashup. IEEE Trans. Visual Comput. Graphics **13**(6), 1176–1183 (2007)

26. Varga, B.O., Mariasiu, F., Miclea, C.D., Szabo, I., Sirca, A.A.: Direct and indirect environmental aspects of an electric bus fleet under service. Energies **13**(2), 1–12 (2020)
27. Holland, S.P., Mansur, E.T., Muller, N.Z., Yates, A.J.: The environmental benefits from transportation electrification: urban buses. Working Paper Series, Paper No. 27285. National Bureau of Economic Research (NBER), Cambridge, MA, USA (2020)
28. Fortini, P.M., Clodoveu, A.: Analysis, integration and visualization of urban data from multiple heterogeneous sources. In: Proceedings of the 1st ACM SIGSPATIAL Workshop on Advances on Resilient and Intelligent Cities (ARIC 2018), pp. 17–26. Association for Computing Machinery, Seattle (2018)
29. C40 Cities Climate Leadership Group Homepage. https://www.c40knowledgehub.org/s/article/Electric-Buses-in-Cities-Driving-Toward-Cleaner-Air-and-Lower-CO2. Accessed 8 Feb 2021
30. Clairand, J.-M., Guerra-Terán, P., Serrano-Guerrero, X., González-Rodríguez, M., Escrivá-Escrivá, G.: Electric vehicles for public transportation in power systems: a review of methodologies. Energies **12**(16), 1–22 (2019)
31. Lopez, C., Ruíz-Benítez, R., Vargas-Machuca, C.: On the environmental and social sustainability of technological innovations in urban bus transport: The EU case. Sustainability. **11**(5), 1–22 (2018)
32. Sanchez-Hidalgo, M.-A., Cano, M.-D.: A survey on visual data representation for smart grids control and monitoring. Sustain. Energ. Grids Netw. **16**, 351–369 (2018)
33. Manitoba Government and City of Winnipeg Joint Task Force on Transit Electrification. The Future is ahead of you: battery electric-bus zero emissions. Manitoba Government, City of Winnipeg, province of Manitoba (2016)
34. Pihlatie, M.: Planning of electric bus systems. VTT Technical Research Centre of Finland (2017)
35. Sisson, P.: How a Chinese city turned all of its 16,000 buses electric. Curbed Journal transportation (2018). https://www.curbed.com/2018/5/4/17320838/china-bus-shenzhen-electric-bus-transportation Accessed 11 Feb 2021
36. Teunissen, T., Sarmiento, O., Zuidgeest, M., Brussel, M.: Mapping equality in access: the case of Bogotá's sustainable transportation initiatives. Int. J. Sustain. Transp. **9**(7), 457–467 (2015)
37. Todoruţ, A., Cordo, N., Iclodean, C.: Replacing diesel buses with electric buses for sustainable public transportation and reduction of CO2 emissions. Polish J. Environ. Stud. **29**(5), 3339–3351 (2020)
38. Disu-Sule, O., et al.: Designing a bus bridging tool for team iCity. In: Fourth Bi-Annual GIS in Education and Research Conference, Toronto, Canada (2020)
39. Mariano, G.C., et al.: Designing a dashboard visualization tool for urban planners to assess the completeness of streets. In: Yamamoto, S., Mori, H. (eds.) HCII 2020. LNCS, vol. 12184, pp. 85–103. Springer, Cham (2020). https://doi.org/10.1007/978-3-030-50020-7_6
40. Kasraian, D., et al.: Evaluating complete streets with a 3D stated preference survey. In: Proceedings 12th International Conference on Travel Survey Methods, Lisbon, Portugal, May 31–June 5, 2020. (Conference delayed due to COVID-19) (2020)
41. Kasraian, D., et al.: Designing a complete streets dashboard for team iCity. In: Fourth Bi-annual GIS in Education and Research Conference, Toronto, Canada (2020)
42. Bruntland Commission. Report of the World Commission on Environment and Development. General Assembly Resolution 42/187 (1987)

43. Zito, P., Salvo, G.: Toward an urban transport sustainability index: an European comparison. Eur. Transp. Res. Rev. **3**, 179–195 (2011)
44. Faulin, J., Grasman, S.E., Juan, A.A., Hirsch, P.: Chapter 1 - Sustainable transportation: concepts and current practices. Sustain. Transp. Smart Logistics, pp. 3–23. Elsevier (2019)
45. Grijalva, E.R., López Martínez, J.M.: Analysis of the reduction of $CO_2$ emissions in urban environments by replacing conventional city buses by electric bus fleets: Spain case study. Energies **12**(3), 1–31 (2019)
46. Thomas, C.E.S.: Sustainable Transportation Options for the 21st Century and Beyond. A Comprehensive Comparison of Alternatives to the Internal Combustion Engine. Springer, Cham (2015). https://doi.org/10.1007/978-3-319-16832-6
47. Welch, D.: Electrified transportation for all: how electrification can benefit low income communities. Centre for Climate and Energy Solutions, Virginia (2017)
48. Mahady, J.A., Octaviano, C., Bolaños, O.S.A., López, E.R., Kammen, D.M., Castellanos, S.: Mapping opportunities for transportation electrification to address social marginalization and air pollution challenges in greater Mexico city. Environ. Sci. Technol. **54**, 2103–2111 (2020)
49. Farber, S., Bartholomew, K., Li, X., Páez, A., Habib, K.M.N.: Assessing social equity in distance-based transit fares using a model of travel behavior. Transp. Res. Part A Policy Pract. **67**, 291–303 (2014)
50. Ollier, M.: At the crossroads of sustainable transportation and social inclusion: the potential of public transit to create inclusive and equitable cities A research case on transport-related social exclusion in Montreal. Social Connectedness Fellow 2018, Samuel Centre for Social Connectedness (2018)
51. Mercado, R.G., Páez, A., Farber, S., Roorda, M.J., Morency, C.: Explaining transport mode use of low-income persons for journey to work in urban areas: a case study of Ontario and Quebec. Transportmetrica **8**(3), 157–179 (2012)
52. Kaplan, S., Popoks, D., Giacomo, C.P., Ceder, A.: Using connectivity for measuring equity in transit provision. J. Transp. Geogr. **37**, 82–92 (2014)
53. Litman, T.: Sustainable transportation indicators. A recommended research program for developing sustainable transportation indicators and data. Paper presented at the 2009 Transportation Research Board Annual Conference, Washington, DC, 11–15 January 2008
54. Mameli, F., Marletto, G.: Can national survey data be used to select a core set of sustainability indicators for monitoring urban mobility policies? Int. J. Sustain. Transp. **8**(5), 336–359 (2014)
55. Cooper, E., Kenney, E., Velásquez, J.M., Li, X., Tun, T.H.: Costs and emissions appraisal tool for transit buses. World Resources Institute, Washington (2019). Technical Note
56. Borén, S.: Electric buses' sustainability effects, noise, energy use, and costs. Int. J. Sustain. Transp. **14**(12), 956–971 (2020)
57. Adheesh, S.R., Vasisht, M.S., Ramasesha, S.K.: Air-Pollution and economics: diesel bus versus electric bus. Curr. Sci. **110**(5), 858–862 (2016)
58. Chandler, S., Espino, J., O'Dea, J.: Electrification of Trucks and Buses: Delivering Opportunity Report - How Electric Buses and Trucks Can Create Jobs and Improve Public Health in California. Union of Concerned Scientists, California (2016)
59. Barriga, M.R.M.: Electric Mobility in Belo Horizonte. Wuppertal Institute, Berlin (2018)
60. Nardo, M., Saisana, M., Saltelli, A., Tarantola, S.: Tools for Composite Indicators Building. Institute for the Protection and Security of Citizen, Italy (2005)

61. Gjolberg, M.: Measuring the immeasurable? Constructing an index of CSR practices and CSR performance in 20 countries. Scand. J. Manag. **25**(1), 10–22 (2009)
62. Jeon, C.M., Amekudzi-Kennedy, A.A.: Addressing sustainability in transportation systems: definitions, indicators, and metrics. J. Infrastruct. Syst. **11**(1), 31–50 (2005)
63. Lee, Y.-J., Huang, C.-M.: Sustainability index for Taipei. Environ. Impact Assess. Rev. **27**(6), 505–521 (2007)

# Modeling of Onboard Activities: Public Transport and Shared Autonomous Vehicle

Jamil Hamadneh(✉) and Domokos Esztergár-Kiss

Department of Transport Technology and Economics, Faculty of Transportation Engineering and Vehicle Engineering, Budapest University of Technology and Economics (BME), 1111 Budapest, Hungary
jhamadenh@edu.bme.hu, esztergar@mail.bme.hu

**Abstract.** The continuous development in technology allows to have fully autonomous vehicles (AVs) on the market. Travelers are interested in maximizing their utility onboard by involving themselves in multitasking. Does choosing a particular type of multitasking determine what type of transport mode to be used? Several studies are conducted on conventional transport modes (CTMs); however, scarcely can be found on AVs. A stated preference (SP) survey including sociodemographic and trip characteristics as well as discrete choice experiments (DCEs) is used. The impact of multitasking on travel behavior is assessed, where multitasking is divided into six types of activities based on the characteristics of an activity (i.e., active, or passive activity). The random utility theory approach including the discrete choice modeling is applied, where transport choice models for the shared autonomous vehicle (SAV) and public transport (PT) are developed considering time, cost, and multitasking availability. The results demonstrate that each onboard activity has a different impact on the transport mode choice, and the social media activity has the largest positive, while the only writing activity shows a negative impact on the transport mode choice. Moreover, the impact of travel time on multitasking is higher than that of the travel cost. Additionally, the changes in the travel time and the travel cost do not show strong and high differences between onboard activities. SAV is more affected by a change in the onboard activities, the travel time, and the travel cost than PT. In conclusion, the results show that inside urban areas, PT is more likely to be used than SAV concerning the multitasking possibility, the travel time, and the travel cost.

**Keywords:** Travel time · Multitasking · Shared autonomous vehicle · Onboard activities · Discrete choice modeling

## 1 Introduction

Autonomous vehicles (AVs) are soon to arrive on the market due to the rapid advancement in technology. The impact of AVs on people's behavior deserves to be studied. Generally, depending on how important travel time is for them, travelers assign an indirect cost to travel time [1]. Waiting time, crowding, and other unpleasant parts of the travel are wished to be compensated by paying money to reduce the travel time or to use a more beneficial

© Springer Nature Switzerland AG 2021
H. Krömker (Ed.): HCII 2021, LNCS 12791, pp. 39–55, 2021.
https://doi.org/10.1007/978-3-030-78358-7_3

transport mode [2]. The amount of money that a traveler pays represents the value of travel time (VOT) in a given journey [3]. The VOT is not a fixed value for all travelers, but it depends on several factors, such as the multitasking availability onboard, the availability of information and communication technology (ICT), the transport mode type, the trip purpose, the demographics, the time of the journey [2, 4–6]. The travel time contains the walking time, the waiting time, and the in-vehicle time [7]. People evaluate the in-vehicle time lowest and the waiting the largest [7, 8]. The valuation of the upcoming new technology is required in various studies that examine the VOT concerning the unique characteristics of AVs. Some distinct characteristics of AVs include that no driver is required, it is a door-to-door service, it provides higher safety, and no driving license is necessary [9]. This paper studies the in-vehicle time of shared autonomous vehicles (SAVs) and public transport (PT), where a traveler selects the appropriate alternative (i.e., SAV or PT) based on the environment and the possibility of multitasking. The method where people select a transport mode based on the possibility of conducting certain onboard multitasking is called travel-based multitasking (multitasking means conducting at least one activity during the travel) [10].

Varghese and Jane (2018) as well as Ettema and Verschuren (2007) conduct studies which examine multitasking on conventional transport modes (CTMs), where the changes in the VOT due to multitasking are evaluated [4, 11]. Keseru et al. (2020) demonstrate that the availability of ICT influences the trip purpose, and the trip purpose impacts the use of ICT [12]. Malokin et al. (2019) use a revealed preference (RP) survey and examine the modal shift based on multitasking, where ICT earns high importance [13]. Rhee et al. (2013) and Mokhtarian et al. (2015) study the effects of onboard activities on the attitudes of travelers during the travel [14, 15]. Etzioni et al. (2020) study people's acceptance of AVs across six European countries [16]. The results show conservation on the use of AVs (i.e. 70% choose CTMs, and only 30% choose AVs).

The preferences of travelers determine the selection of a transport mode for a certain journey. Travelers seek alternatives that maximize their profits (i.e., utility), such as saving time or reducing the negative impact of travel like crowding in case of PT [2]. People are likely to change transport mode to have a better environment where they can work, study, or relax thus to have productive travel time [8]. Across generations, the advancement of ICT affects the behavior of travelers differently, for instance, people who were born after 1980 have a higher demand to use ICT than older generations [17]. The availability of ICT motivates travelers to multitask onboard, which impacts the VOT positively (i.e., lower VOT) [18]. Additionally, the VOT decreases once the comfort level increases [19, 20]. The characteristics of a transport mode affect the type of onboard activity and vice versa, in addition to other factors that impact the travel behavior (i.e., multitasking) such as socio-economic variables [21].

Multitasking on the travel behavior is a hot topic. However, scarcely can be found studies on multitasking on the board of AVs. This research acts as an addition to the literature concerning the impact of multitasking on travel behavior studying AVs. Previous studies focus primarily on the analysis of CTMs, while this paper examines the impact of the various types of multitasking on the travel behavior of people in case of SAVs and PT. The contributions of this research are (1) to set transport mode choice models where travel time, travel cost, and multitasking options are included and (2) to examine the impact of each onboard activity on the transport mode choice.

## 2 Literature Review

The fast advancement of technology leads to the possibility of having AVs on the market [22]. The characteristics of AVs might change the behavior of travelers gradually based on people's acceptability of a transport mode driven by a machine and the possible gained benefits regarding such factors as travel cost and travel time [21, 22]. One of the motives of travelers for changing a transport mode is the degree of importance attached to certain factors, such as multitasking, travel time, and travel cost [23]. It is known that each traveler is more willing to pay money to have a pleasant journey, or in other words, to reduce the time spent on traveling [1]. The time allocation theory states that traveling is like paying money rather than gaining money as in the case of activity time [20]. The amount of time obtained from reducing the travel time is a good opportunity to allocate it to other activities (i.e., positive utility) [24–26].

Some scholars study the travel behavior, and they evaluate the VOT based on quantitative and qualitative attributes that have a significant impact on the transport choice models. Litman (2008) suggests that the improvements of quality are less costly compared to infrastructure solutions, for instance, a reduction of the VOT by constructing a new road. The scholar shows that travelers are more likely to change their transport modes if certain developments are considered, such as enhancing convenience and comfort. These qualitative measures provide travelers with a better environment, which makes them use the travel time productively or at least have a pleasant journey [8]. Furthermore, the author demonstrates that the VOT is affected not only by a reduction in the travel time but by improvements gained by qualitative measures, such as the development of the service, the comfort, the convenience, and safety. For example, the advanced technology is an improvement in qualitative and quantitative measures on the board of a transport mode, such as the promising technology of AVs. Belenky (2011) shows that multitasking is a good tool that makes travelers utilize their travel and have a small perceived travel time [2]. Perk et al. (2012) demonstrate a change in travel behavior from one person to another and across time for the same person [27]. Cirillo and Axhausen (2006) say that travelers are more likely to extend their travel time when they find benefits onboard, such as a pleasant journey [28]. Moreover, the researchers find that trip purpose, and time constraints (i.e., departure and arrival time) impact the travel behavior not in a fixed pattern.

People try to maximize their utility by minimizing the travel time through involving themselves in multitasking, to transform some parts of the travel time into a productive time. For example, people with high income are more likely to be influenced by the activities onboard compared to low-income people as the economic theory says [29]. Considering the in-vehicle time, the waiting time, and the transfer time; the trip purpose, like work and leisure, is studied by Xumei et al. (2011) [30]. The findings show that the VOT of leisure trips are less than the VOT of work trips. Furthermore, the scholars find that the value of the waiting time is the highest. Using SAV might provide travelers with a pleasant time when the accompanying stress of the driving and the transfer time is removed, and the possibility of multitasking during traveling is increased on the board of SAV [25, 31, 32].

The estimation of the subjective VOT requires the application of a discrete choice model based on the factors that affect the travel time from the travelers' perspectives [20]. Some studies which use the discrete choice modeling for AVs by applying a stated preference (SP) survey are presented. Simoni et al. (2018) show that the VOT decreases when travelers use AVs [33]. Gurumurthy & Kockelman (2020) demonstrate that people prefer using AVs in long-distance travel compared to short-distance [34]. Lavieri and Bhat (2019) present that the companion's type impacts the use of SAVs, while the accompanying delay of dropping off the passengers along the way has the largest impact on the use of SAVs [35]. The impact of AVs on certain groups of users are presented in the literature. As an example, Bozorg and Ali (2016) state that high-income groups have 35% less VOT when AVs are used [36]. Steck et al. (2018) show that for commuters, AVs and SAVs reduce the value of travel saving (VOTS). The authors declare that the AVs (i.e., not shared use) are more attractive to commuters than the SAVs [5]. Kolarova et al. (2019) demonstrate that the value of travel time saving (VOTTs) of AVs are 41% less than that of the conventional cars; on the contrary, it is found that the VOTS of SAV is higher than conventional cars' [37]. Other studies reveal a reduction in the VOT once travelers are able to multitask onboard, and they get rid of the driving task as well as its accompanying stress and mental tensions [21, 38]. Other factors affecting multitasking are the vehicle design and the in-vehicle features that facilitate the use of traveling [21]. In long-distance travel, the reduction of the VOT is rather significant; thus, people need a better opportunity to multitask [21]. It is worth mentioning that multitasking is affected by the trip purpose (i.e., the direction of the trip) [38]. However, empirical studies are necessary to evaluate the impact of AVs and SAVs on the VOT.

In this study, the travelers' behavior onboard of SAV and PT as well as the variables that impact the choice of a transport mode are discussed. In the literature, scarcely, can be found any studies about the choice of a transport mode based on the availability of multitasking in case of SAVs. This study provides an addition to the literature by evaluating different onboard activities on the transport choice.

## 3   Methodology

The aim of this section is to present the methodological approach that is applied in this study. Travelers conduct onboard activities, and these activities affect the behavior of travelers, and these activities affect the transport mode choice and the perceived travel time. Therefore, understanding the impact of various onboard activities on the behavior of people when they travel to their main destinations are examined in this paper. An SP survey including sociodemographic, economic, and journey variables are constructed. Moreover, the SP survey includes questions about the preferences of people toward two transport modes (i.e., AV and PT) considering time, cost, and multitasking availability. Those questions are built to ensure consistency with the design choice experiment (DCE) method. An SP survey method is used to extract responses that detect the priorities, the preferences, and the relative importance of travelers' features associated with the transport modes' characteristics [39]. The discrete choice modeling approach is applied to develop mode choice models for SAV and PT. Figure 1 presents the methodological app-roach. The DCE contains two transport modes (a.k.a. alternatives) and three attributes such as travel time, travel cost, and multitasking with different levels. Descriptive statis-tics contain the sociodemographic, economic, and trip variables including gender, age, income, education, car ownership, job, transport mode, and trip purpose. Finally, based on the developed model, the VOT is estimated to measure how travelers evaluate their travel time.

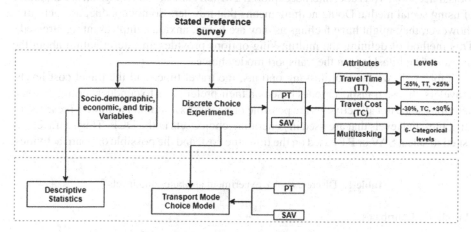

**Fig. 1.** The methodological approach

Discrete choice modeling is a method used to capture preferences of people in various aspects of life, for example social sciences and transportation, while it is consistent with the probability framework (i.e., the random utility theory) [39]. DCEs reflect the existing and the proposed conditions (i.e., hypothetical market) of a problem, which enables the decision-makers to have rich data sources [40]. The DCE gives an estimation for multiple alternatives rather than for one alternative, and it provides information about the willingness of the travelers to pay for a characteristic of an alternative [40].

The alternatives in the DCE include SAV and PT. The two alternatives are mutually exclusive and exhaustive [40]. The attributes of each alternative are travel time, travel cost, and multitasking with certain levels for each, as shown in Table 1. The levels are selected for each attribute to ensure that a traveler can make a trade-off among the alternatives. Lancsar and Louviere (2008) say that the quantitative attributes and the attribute levels are assumed to be linear in the utility, while qualitative attributes are presented as dummy variables or effect-coded variables [40]. In this research, those multitasking options which provide categorical/qualitative data are represented as effect-coded variables. The target trip in this study is the main trip in the urban area. An average examined trip is 4 km and 30 min in the case of Budapest, where a study is conducted. This average trip represents the average trip distance/time of the main trip of travelers in Budapest based on the simulations of the daily activity plans of household survey data by using MATSim open-source software [25]. In the survey, travelers are asked to choose one alternative based on their preferences.

Furthermore, multitasking options are formulated to indicate the possibility of conducting one activity onboard. Six onboard activities are taken that summarize what people can possibly conduct onboard, these are the followings: (1) reading, (2) writing, (3) talking, (4) using social media (including gaming), (5) eating/drinking, and (6) doing nothing onboard. In the case of reading and writing, one can use either paper-based or technological tools (e.g., mobile, laptop or tablet) during the travel. Talking is conducted with strangers, family, acquaintances, or friends either face to face or via technological tools. Facebook, YouTube, internet explorer, Twitter, games, and Instagram are examples of using social media. Doing nothing means that travelers do not conduct any activities (however, they might have feelings as they are bored, anxious, unpleasant, or stressed). This method of defining the multitasking options provides an interpretation about the impact of multitasking on the transport mode choice.

Table 1 shows the multitasking options, the travel time, and the travel cost levels. Travelers choose transport modes based on their preferences to conduct certain onboard activities since not all activities are possible onboard. As a summary, Table 1 presents the attributes and levels that are used in generating choice sets in the DCE. Thus, a traveler is asked to select SAV or PT based on the time, the cost, and the possible onboard activities.

**Table 1.** Discrete choice experiment attributes and levels [a]

| Levels | Attributes | | |
|--------|------------|--------------------|--------------------|
| | Multitasking options | Total trip cost (HUF)* | Total trip time (minute) |
| L 1 | Reading | 1.30*(TC) | 1.25*(TT) |
| L 2 | Writing | TC | TT |
| L 3 | Talking | 0.70*(TC) | 0.75*(TT) |
| L 4 | Using social media | | |
| L 5 | Eating and drinking | | |
| L 6 | Doing nothing | | |

a TC: travel cost, and TT: travel time. * 1Euro = 340HUF, and HUF: Hungarian forint.

- Random utility theory

In the theory of random utility, every individual is a rational decision-maker, and travelers seek a choice to maximize their utilities [41]. The certainty in the choice prediction is not usable, but the probability is applicable, where a traveler can choose one alternative out of few alternatives. This is shown in Eq. 1:

$$Pr(c_j/C) = \Pr(U_{c_j} > U_k) \forall k \neq c_j, K \in C \tag{1}$$

Where Pr $(c_j/C)$ is the probability of a traveler to select alternative $(c_i)$ from a choice set (C), $U_{c_j}$ is the utility of selecting the alternative $(c_j)$ from the choice set C, where the obtained utility is bigger than the utility of all other alternatives (K) in the choice set

C [41]. The utility function contains two parts a deterministic and a stochastic part as given in Eq. 2:

$$U_j = V_j + \varepsilon_j \tag{2}$$

Where V is the deterministic part and ε is the stochastic part. The deterministic part is the mean perceived utility of travelers when they select an alternative [41]. The random part is the unknown deviation of a traveler's utility from the mean value, and it captures the uncertainty in the choice modeling [41]. The characteristics of the stochastic part determine the type of model that fits the data and makes the produced model more accurate. Hereafter, one of the models of random utility is presented.

- Mixed Logit (ML) model

The Mixed Logit (ML) model overcomes the taste variation issues (i.e., the marginal utility is random) that exists in the Multinomial Logit model [42]. The ML model is selected to be used for modeling traveler utilities since the ML combines the properties of the Conditional and the Multinomial Logit models [42]. The ML model is the integral of the standard logit probabilities over a density of parameters; Eq. 3 shows the general form of the utility function of the ML model.

$$U_{ijc} = V(\beta_i / X_{ijc}) + \varepsilon_{ijc} \tag{3}$$

Where $X_{ijc}$ is the vector of the independent variables including the attributes of the alternatives and the socio-demographic characteristics of the travelers, which is an important part of the choice selection. $\varepsilon_{ijc}$ is a random error, an independent and identically distributed (IID) extreme value type 1 (a.k.a. Gumbel distribution) across all alternatives, individuals, and choice situations. Both $\beta_i$ and $\varepsilon_{ijc}$ are treated as stochastic parameters, which influence the accuracy of the model because they are not observed directly [42]. The IID is restrictive since it does not let the alternatives to be correlated across each other because of the information that is unseen to the analyst that might cause correlation. Therefore, the stochastic component is divided into two parts. One part is the IID and the other part is correlated over the alternatives, and it is heteroskedastic as shown in Eq. 4.

$$U_{ijc} = V(\beta_i / X_{ijc}) + \eta_{ijc} + \varepsilon_{ijc} \tag{4}$$

Where $\eta_{ijc}$ is a random term with zero mean property, whose distribution over individuals and alternatives depends in general on the underlying parameters and the observed data related to alternatives (i) and individuals (q); and $\varepsilon_{iq}$ is a random term with zero mean, which is the IID over the alternatives and does not depend on the underlying parameters or data. The ML assumes a general distribution for η and an IID extreme value type 1 distribution for ε. That is, η can take on several distributional forms including normal, uniform, lognormal, and triangular [43].

The panel data-mixed logit model, which is known as the heteroskedastic error component type model, is used in this research to take the unobserved factors and to capture the correlations between the choices selected by the same participant (as it is shown in the alternative constant) [44]. The modeling of each alternative's selection in each scenario is done by using panel data rather than by choosing a single probability of choice. A discrete choice experiment is prepared to develop models without high errors and to avoid biases.

• Model specifications and design

The utility function of selecting one transport mode over others is presented considering three factors such as travel time, travel cost, and multitasking. The choice model specification can be formulated based on Eq. 3 and Eq. 4.

$$V_m = \beta_0 + \beta_{Travel\,cost} * Travel\,cost + \beta_{Travel\,time} * Travel\,time + \beta_{Multitasking\,(i)} * Multitasking\,(i) + \quad (5)$$
$$(\varepsilon + \eta),\ where\ i \leq 6$$

Where m stands for the transport mode, $\beta_0$, $\beta_{Travel\,cost}$, $\beta_{Travel\,time}$, and $\beta_{Multitasking\,(i)}$ stand for the parameters to be evaluated from the collected data, and $\varepsilon + \eta$ means an indeterministic error. It is worth mentioning that the VOT is calculated as the marginal utility of travel time is divided by the marginal utility of travel cost ($\beta_{Travel\,time}/\beta_{Travel\,cost}$) multiplied by the cost of the time unit.

In this study, the estimates measure the relative value of one alternative to the others. It is worth mentioning that the description of the results is referred to as a selected reference category (i.e., base alternative) [39]. The discrete choice models are influenced by some attributes, such as the socio-economic characteristics of the travelers, the selected attributes, and the environmental variables of the chosen alternative [39].

## 4  Results

LimeSurvey tool is used to distribute an SP survey in Budapest, Hungary [45]. The collected data contains 346 respondents, who register their sociodemographic and travel characteristics, as shown in Fig. 2. The data includes 12.14% high school diploma holders, 51.45% undergraduate degree holders, 32.08% graduate degree holders, and 4.34% with other educational degrees, as shown in Fig. 2(a). People with different monthly income are represented in the survey, where 35.55% of the respondents are low-income people (i.e., less than 500 Euro per month), 15.61% are with middle income (i.e., more than 500 Euro but less than 950 Euro), 24.86% are high-income people (i.e., more than 950 Euro), while 23.99% indicate others (i.e., do not declare about their income levels), as shown in Fig. 2(b). The job categories affect the preferences of respondents, as well. The jobs are grouped into seven categories, as shown in Fig. 2(c). There are full-time workers represented by 46.53%, 6.36% of the participants are part-time workers, students are represented by 36.71%, while 6.65% are unemployed, 2.02% are self-employed, 1.16% are retired, and finally, the category of other job types is indicated by 0.58%. The age categories show that 38.15% of the participants are 15–24 years old, and 56.07% are 25–54 years old. 4.34% of the respondents are 55–64 years old, and 1.45% are more than 65 years old, as shown in Fig. 2(d). Regarding the age category, the majority of the participants in the collection represent young people who are older than 15 years but younger than 55 years. The car ownership percentages demonstrate that 36.99% own personal cars, while 63.01% do not own personal cars, as shown in Fig. 2(e). There is a rather balanced representation of the gender in the survey as 45.66% are females, while 54.34% are males, as shown in Fig. 2(f).

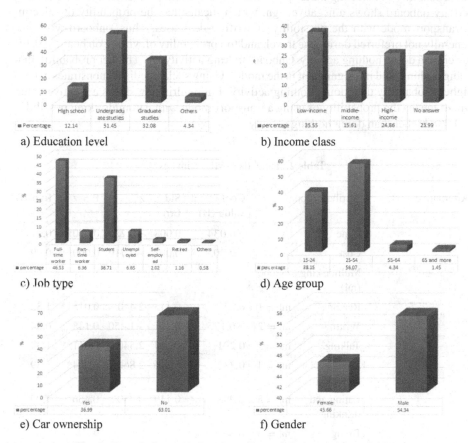

a) Education level

b) Income class

c) Job type

d) Age group

e) Car ownership

f) Gender

**Fig. 2.** Descriptive statistics about the sociodemographic variables

The DCE part in the SP survey shows that 49.5% select SAV, while 50.5% choose PT. The Panel data-mixed logit model simulates the travelers' behavior when they select a transport mode based on travel time, travel cost, and the availability of one of the six onboard activities, as shown in Table 2. The model with the lowest Bayesian information criterion (BIC) is chosen. The statistics of the model are written at the bottom of the table. The model is statistically significant (i.e., probability > Chi2 = 0.000) with 3892 observations, the log simulated-likelihood is −1256.146, and the BIC is 2777.293. The insignificant parameters are not removed from the model due to comparison reasons. It is presented in the results that the sign of the values of both the travel time and the travel cost is negative, which coincides with the economic theory which states that travelers lose money and time during the travel, while multitasking shows varied results. Based on the statistical results of the developed model, the VOT of the travelers is 660 HUF per hour (calculated as the marginal utility of the travel time is divided by the marginal utility of the travel cost). The reference attribute is the doing nothing option (i.e., it is a selected category). Onboard reading activity shows a positive value, which leads to that the probability of selecting a transport mode with reading possibility is 0.277. Writing activity onboard shows a negative sign, which means that the probability of selecting a transport mode with the possibility of writing decreases. Thus, onboard writing is generally not preferred during the travel, and the probability of writing on board is 0.175 lower than doing nothing activity. Onboard talking activity has a higher probability than doing nothing, and it is significant in the model. Using social media demonstrates a 0.332 higher probability than doing nothing activity. Eating/drinking shows a positive value, which means the probability of using a transport mode with eating/drinking possibility is 0.088 higher than doing nothing activity.

**Table 2.** Panel data-ML estimates

| Alternative | Attribute (code) | | Coefficient value (B) | Std. Error | z | P > z | Exp(B) |
|---|---|---|---|---|---|---|---|
| | Time | | −0.034 | 0.008 | −4.220 | 0.000* | 0.97 |
| | Cost | | −0.003 | 0.001 | −4.710 | 0.000* | 0.99 |
| | Multitasking (ml) | Symbol | | | | | |
| | Reading | ml = 1 | 0.277 | 0.113 | 2.440 | 0.015** | 1.32 |
| | Writing | ml = 2 | −0.175 | 0.121 | −1.450 | 0.148 | 0.84 |
| | Talking | ml = 3 | 0.291 | 0.114 | 2.540 | 0.011** | 1.34 |
| | Using social media | ml = 4 | 0.332 | 0.116 | 2.860 | 0.004* | 1.39 |
| | Eating and drinking | ml = 5 | 0.088 | 0.119 | 0.750 | 0.456 | 1.09 |
| | Doing nothing | ml = 6 | base | | | | |
| | ASC [a] | − | 0.385 | 0.398 | 0.970 | 0.333 | 1.47 |
| PT | | Base | | | | | |
| Number of observations = 3884 | | | | | | | |
| Chi2(2) = 153.74 | | | | | | | |
| Log simulated-likelihood = −1256.146 | | | | | | | |
| Prob > chi2 = 0.000 | | | | | | | |
| AIC = 2582.291 | | | | | | | |
| BIC = 2777.293 | | | | | | | |

* $p < 0.01$, ** $p < 0.05$, *** $p < 0.1$, [a] ASC: alternative specific constant

The marginal effect of changing the coefficient value on the probability of observing an alternative is done by using the estimation of the marginal effect, where the change in a response for a change in the attribute is examined, as shown in Fig. 3. The average marginal effect of multitasking on the probability of being in an alternative is either positive or negative. The marginal effect of the onboard activities on an alternative concerning SAV and PT is shown in graphs a and b, respectively, in Fig. 3. Graph (4.a) shows the conditional marginal effect of the multitasking coefficient of SAV and PT, when the outcome is SAV (i.e., the effects of multitasking on SAV when other alternatives are evaluated at the mean value). Graph (4.b) shows the conditional marginal effect of the multitasking coefficient of SAV and PT, when the outcome is PT (i.e., the effects of multitasking on PT when other alternatives are evaluated at the mean value). Compared to doing nothing, when eating/drinking activity is possible, the probability of an individual to choose SAV and PT increases by 0.02078 and 0.0207, respectively. Compared to doing nothing, the probability of an individual to choose SAV and PT increases by 0.0639 and 0.0637, respectively, when reading activity is possible. Compared to doing nothing, the probability of an individual to choose SAV and PT increases by 0.0764 and 0.0761, respectively, when social media activity is possible. Compared to doing nothing, when talking activity is possible, the probability of an individual to choose SAV and PT increases by 0.0668 and 0.0666, respectively. Compared to doing nothing, the probability of an individual to choose SAV and PT decreases by 0.03884 and 0.03881, respectively, when writing activity is possible. Based on the two graphs, the impacts of the six onboard activities on choosing the transport mode are presented and discussed. The result shows that the impact of multitasking on the SAV and PT are similar, with preferences to the SAV.

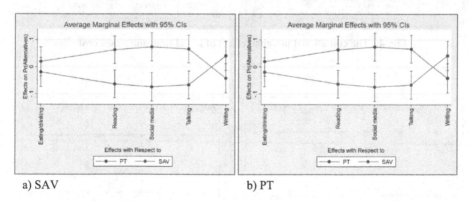

a) SAV                                    b) PT

**Fig. 3.** The marginal effects of multitasking on the predicted probability of SAV and PT

Figure 4, shows the predictive margins of each SAV and PT when travel time, and travel cost change. The value on the y-axis shows the margin of each alternative based on the mean value of the other alternative parameters, such as the margin of using PT is 54.1% at the 20 min travel time, while it is 46.5% when the travel time is 30 min. The margin of PT when the travel time of SAV is 20 min is 46.7%, and at 30 min of travel time, it is 54.4%. On the other hand, the margin of SAV is 53.3% when the travel time of SAV is 20 min, and it is 45.6% at 30 min of travel time. The margin of SAV when 20 min is the travel time of PT is 45.9%, while at 30 min travel time of PT, it is 53.5%. The margin of using PT is 54.9% at 175 HUF travel cost, while it is 46.0% when the travel cost is 325 HUF. The margin of PT when the travel cost of SAV is 175 HUF is 46.3%, and when 325 HUF is the travel cost of SAV, it is 55.3%. On the other hand, the margin of SAV is 53.7% at 175 HUF travel cost of SAV, and it is 44.7% when 325 HUF is the travel cost of SAV. The margin of SAV when 20 HUF is the travel cost of PT is 45.0%, while at 325 HUF travel cost of PT, it is 54.0%. From the results, it is shown that the impact of the travel time and the travel cost on the transport mode choice is similar, for example, changing the travel time from 20 min to 30 min decreases the margin of PT from 54.1% to 46.5%, and for SAV from 53.3% to 45.6%. Based on these results, it can be concluded that the impact of travel time on SAV is larger than on PT. Moreover, the impact of travel cost on the margin of PT is less than on the SAV's. Based on the result of the predictive margins, the impact of travel time and travel cost on SAV and PT is close.

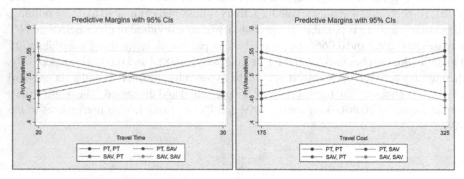

**Fig. 4.** The conditional marginal effect of travel time and travel cost

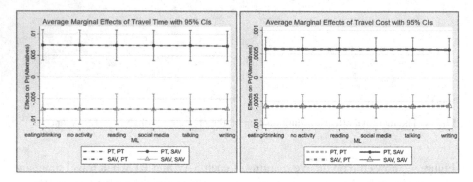

**Fig. 5.** The conditional marginal effect of travel time and travel cost on multitasking

The average marginal effects of travel cost and travel time on the six onboard activities are presented in Fig. 5. It is shown in the graphs that the impact of travel time on multitasking is higher than the impact of travel cost. It is shown that the impacts of travel time and travel cost on each onboard activity are almost equal. The results show that every activity from the six onboard activities of SAV and PT exposes to the same effects regarding the travel time and the travel cost.

# 5    Discussion

The effects of travel time, travel cost as well as some socio-economic and travel variables on the transport choice mode are analyzed. Stata software (version 16) is used in the analysis. The results present the transport choice models of SAV and PT. The models include travel time, travel cost, and multitasking options. In the analysis part, the superior model is selected from several models based on the BIC value. Different models are examined by, for example, using various forms of random error distribution (i.e., log normal, uniform, triangular, and Gaussian) and different variables in the model. The impact of the different variables on the transport mode choice is indicated in the graphs to see the associations among the variables (see Fig. 3, Fig. 4 and Fig. 5). The marginal utilities of travel time and travel cost are negative, which means that traveling is seen as a disutility. Based on the results, the VOT is around 660 HUF/hr as the specifications of the model indicate, and it might be larger without the possibility of multitasking. Each onboard activity shows a different impact on the choice. For example, reading, talking, using social media, and listening show positive value on the transport choice compared to doing nothing activity, while writing activity shows a negative value. The result demonstrates that writing onboard is not a preferred onboard task; however, people perceive the possibility of conducting other activities as an advantage on board.

Studying the impact of travel time, travel cost, and multitasking on the transport mode choice model is explained based on the mean values of the attributes of the target alternative and other accompanying alternatives. The results present the impact of each variable on choosing a transport mode, and the level of each variable's significance. Besides, the result demonstrates that there is not a considerable difference between the impact of multitasking on SAV and on PT, which means that travelers might conduct the same activities on the board of these transport modes. Furthermore, the results demonstrate that the impact of travel time on SAV is larger than on PT, the impact of travel cost on the margin of PT is less than SAV's. Multitasking is influenced much more by the travel time than by the travel cost because involvement in multitasking requires travel time rather than cost, especially in urban areas where the travel time is short and the travel cost is affordable. Besides, the results show that the conditional marginal effect of travel cost is lower than that of the travel time, which means that the travel cost has a lower impact on the transport mode choice once compared to the effects of travel time. The prediction of the alternatives is presented per travel time and travel cost, and the results show different values of predictions per time change and cost change. The results of this study are applied exclusively in the urban areas and for the main trips of the travelers. Therefore, it is recommended to expand this study and include AVs with individual ride, besides SAVs. Moreover, examining the developed models across various sociodemographic and trip characteristics is recommended. Besides, studying the preferences of people during traveling on the board of AVs and SAVs in long-distance trips might be beneficial. The methodology presented in this research can be applied variously by proposing a combination of two or more onboard activities. There is a potential to study the preferences of people across countries, which is a hot topic. In addition, examining the availability of ICT on the board of different transport modes is another topic to be studied.

## 6 Conclusion

The traveler preferences toward two transport modes (i.e., SAV and PT) concerning travel time, travel cost, and six types of onboard activities are studied. The traveler preferences are collected and analyzed by using the discrete choice modeling approach. An SP survey is distributed in Hungary to collect the participants' preferences on the appropriate transport mode once AVs appear on the market. Different variables related to the features of the people and the main trips such as trip purpose, transport mode, age, gender, car ownership, family size, income, and education are collected. A sample of 346 travelers with their preferences regarding their main trips is recorded. The data are analyzed by using Stata software (version 16), and the results demonstrate transport models for SAV and PT. The models include travel time, travel cost, and the six types of onboard activities. The six onboard activities include reading, writing, talking, using social media, eating/drinking, and doing nothing. Each of the six onboard activities shows various effects on the two transport modes (alternatives). Moreover, the impacts of the travel time and the travel cost changes on the travel behavior are analyzed. The results present that people are more likely to choose PT over SAV. The effect of the travel time on the onboard activities is higher than the travel cost, which indicates that having a longer travel time gives travelers more time to engage in multitasking. The variations on the impact of onboard activities are explained, such as only writing activity shows a negative impact on the transport choice compared to doing nothing.

**Acknowledgment.** The research reported in this paper and carried out at the Budapest University of Technology and Economics has been supported by the National Research Development and Innovation Fund (TKP2020 Institution Excellence Subprogram, Grant No. BME-IE-MISC) based on the charter of bolster issued by the National Research Development and Innovation Office under the auspices of the Ministry for Innovation and Technology.

## References

1. Becker, G.: A theory of the allocation of time. Econ. J. **75**(299), 493 (1965). https://doi.org/10.2307/2228949
2. Belenky, P.: The value of travel time savings: departmental guidance for conducting economic evaluations, revision 2. Department of Transportation (2011)
3. Mackie, P.J., Jara-Díaz, S., Fowkes, A.: The value of travel time savings in evaluation. Transp. Res. Part E Logist. Transp. Rev. **37**(2–3), 91–106 (2001). https://doi.org/10.1016/S1366-5545(00)00013-2
4. Varghese, V., Jana, A.: Impact of ICT on multitasking during travel and the value of travel time savings: Empirical evidences from Mumbai, India. Travel Behav. **12**, 11–22 (2018). https://doi.org/10.1016/j.jtrangeo.2012.02.007

5. Steck, F., et al.: How autonomous driving may affect the value of travel time savings for commuting. Transp. Res. Rec. J. Transp. Res. Board **2672**(46), 10 (2018). https://doi.org/10.1177/0361198118757980

6. Kouwenhoven, M., de Jong, G.: Value of travel time as a function of comfort. J. Choice Model. **28**, 97–107 (2018). https://doi.org/10.1016/j.jocm.2018.04.002

7. Cherlow, J.R.: Measuring values of travel time savings. J. Consum. Res. **7**(4), 360–371 (1981). https://doi.org/10.1086/208826

8. Litman, T.: Valuing transit service quality improvements. J. Public Transp. **11**(2), 3 (2008). https://doi.org/10.5038/2375-0901.11.2.3

9. Coppola, P., Esztergár-Kiss, D.: Autonomous Vehicles and Future Mobility. Elsevier, Amsterdam (2019)

10. Berliner, R.M., et al.: Travel-based multitasking: modeling the propensity to conduct activities while commuting. In: Transportation Research Board 94th Annual Meeting, Washington DC, United States (2015)

11. Ettema, D., Verschuren, L.: Multitasking and value of travel time savings. Transp. Res. Rec. J. Transp. Res. Board **2010**(1), 19–25 (2007). https://doi.org/10.3141/2010-03

12. Keseru, I., et al.: Multitasking on the go: an observation study on local public transport in Brussels. Travel Behav. Soc. **18**, 106–116 (2020). https://doi.org/10.1016/j.tbs.2019.10.003

13. Malokin, A., Circella, G., Mokhtarian, P.L.: How do activities conducted while commuting influence mode choice? Using revealed preference models to inform public transportation advantage and autonomous vehicle scenarios. Transp. Res. Part A Policy Pract. **124**, 82–114 (2019). https://doi.org/10.1016/j.tra.2018.12.015

14. Rhee, K.-A., et al.: Analysis of effects of activities while traveling on travelers' sentiment. Transp. Res. Rec. J. Transp. Res. Board **2383**(1), 27–34 (2013). https://doi.org/10.3141/2383-04

15. Mokhtarian, P.L., Papon, F., Goulard, M., Diana, M.: What makes travel pleasant and/or tiring? an investigation based on the French National Travel Survey. Transportation **42**(6), 1103–1128 (2014). https://doi.org/10.1007/s11116-014-9557-y

16. Etzioni, S., et al.: Modeling cross-national differences in automated vehicle acceptance. Sustainability **12**(22), 9765 (2020). https://doi.org/10.3390/su12229765

17. Malokin, A., Circella, G., Mokhtarian, P.L.: Do multitasking millennials value travel time differently? a revealed preference study of Northern California commuters. In: Transportation Research Board 96th Annual Meeting, Washington DC, United States (2017)

18. Banerjee, I., Kanafani, A.: The value of wireless internet connection on trains: implications for mode-choice models. In: UC Berkeley, U.o.C.T. Center, Editor, Berkeley (2008)

19. Evans, A.W.: On the theory of the valuation and allocation of time. Scot. J. Polit. Econ. **19**(1), 1–17 (1972). https://doi.org/10.1111/j.1467-9485.1972.tb00504.x

20. Jara-Díaz, S.R.: Allocation and valuation of travel time savings. Handb. Transp. **1**, 303–319 (2000). https://doi.org/10.1108/9780857245670-018

21. Singleton, P.A.: Discussing the "positive utilities" of autonomous vehicles: will travellers really use their time productively? Transp. Rev. **39**(1), 50–65 (2019). https://doi.org/10.1080/01441647.2018.1470584

22. Khaloei, M., Ranjbari, A., Mackenzie, D.: Analyzing the Shift in Travel Modes' Market Shares with the Deployment of Autonomous Vehicle Technology (2019)

23. Zhong, H., et al.: Will autonomous vehicles change auto commuters' value of travel time? Transp. Res. Part D: Transp. Environ. **83**, 102303 (2020). https://doi.org/10.1016/j.trd.2020.102303

24. Horni, A., Nagel, K., Axhausen, K.W.: The Multi-agent Transport Simulation MATSim, p. 620. Ubiquity Press, London (2016)

25. Hamadneh, J., Esztergár-Kiss, D.: Impacts of shared autonomous vehicles on the travelers' mobility. In: 2019 6th International Conference on Models and Technologies for Intelligent Transportation Systems (MT-ITS). IEEE, Poland (2019). https://doi.org/10.1109/MTITS. 2019.8883392
26. Hamadneh, J., Esztergár-Kiss, D.: Potential travel time reduction with autonomous vehicles for different types of travellers. Promet Traffic Transp. **33**(1), 61–76 (2021). https://doi.org/ 10.7307/ptt.v33i1.3585
27. Perk, V.A., et al.: Improving value of travel time savings estimation for more effective transportation project evaluation. Florida Department of Transportation, Center for Urban Transportation Research (2012)
28. Cirillo, C., Axhausen, K.W.: Evidence on the distribution of values of travel time savings from a six-week diary. Transp. Res. Part A Policy Pract. **40**(5), 444–457 (2006). https://doi. org/10.1016/j.tra.2005.06.007
29. DeSerpa, A.C.: A theory of the economics of time. Econ. J. **81**(324), 828–846 (1971). https:// doi.org/10.2307/2230320
30. Xumei, C., Qiaoxian, L., Guang, D.: Estimation of travel time values for urban public transport passengers based on SP survey. J. Transp. Syst. Eng. Inf. Technol. **11**(4), 77–84 (2011). https:// doi.org/10.1016/S1570-6672(10)60132-8
31. Janssen, C.P., Kenemans, J.L.: Multitasking in autonomous vehicles: ready to go? In: 3rd Workshop on User Experience of Autonomous Vehicles at AutoUI 2015. ACM Press, Nottingham (2015)
32. Ortega, J., et al.: Simulation of the daily activity plans of travelers using the park-and-ride system and autonomous vehicles: work and shopping trip purposes. Appl. Sci. **10**(8), 2912 (2020). https://doi.org/10.3390/app10082912
33. Simoni, M.D., et al.: Congestion pricing in a world of self-driving vehicles: an analysis of different strategies in alternative future scenarios. Transp. Res. Part C Emerg. Technol. **98**, 167–185 (2018). https://doi.org/10.1016/j.trc.2018.11.002
34. Gurumurthy, K.M., Kockelman, K.M.: Modeling Americans' autonomous vehicle preferences: A focus on dynamic ride-sharing, privacy & long-distance mode choices. Technol. Forecast. Social Change **150**, 119792 (2020). https://doi.org/10.1016/j.techfore.2019.119792
35. Lavieri, P.S., Bhat, C.R.: Modeling individuals' willingness to share trips with strangers in an autonomous vehicle future. Transp. Res. Part A Policy Pract. **124**, 242–261 (2019). https:// doi.org/10.1016/j.tra.2019.03.009
36. Bozorg, S.L., Ali, S.M.: Potential Implication of Automated Vehicle Technologies on Travel Behavior and System Modeling (2016)
37. Kolarova, V., Steck, F., Bahamonde-Birke, F.J.: Assessing the effect of autonomous driving on value of travel time savings: a comparison between current and future preferences. Transp. Res. Part A Policy Pract. **129**, 155–169 (2019). https://doi.org/10.1016/j.tra.2019.08.011
38. Wadud, Z., Huda, F.Y.: The potential use and usefulness of travel time in fully automated vehicles. In: Transportation Research Board 97th Annual Meeting, Washington DC, United States (2018)
39. Hauber, A.B., et al.: Statistical methods for the analysis of discrete choice experiments: a report of the ISPOR Conjoint analysis good research practices task force. Value Health **19**(4), 300–315 (2016). https://doi.org/10.1016/j.jval.2016.04.004
40. Lancsar, E., Louviere, J.: Conducting discrete choice experiments to inform healthcare decision making. Pharmacoeconomics **26**(8), 661–677 (2008). https://doi.org/10.2165/00019053-200826080-00004
41. Cascetta, E.: Random utility theory. In: Transportation Systems Analysis, pp. 89–167. Springer, Heidelberg (2009). https://doi.org/10.1007/978-0-387-75857-2_3
42. Hensher, D.A., Greene, W.H.: The mixed logit model: the state of practice. Transportation **30**(2), 133–176 (2003). https://doi.org/10.1023/A:1022558715350

43. Hoffman, S.D., Duncan, G.J.: Multinomial and conditional logit discrete-choice models in demography. Demography **25**(3), 415–427 (1988). https://doi.org/10.2307/2061541
44. Ben-Akiva, M.E., Lerman, S.R., Lerman, S.R.: Discrete Choice Analysis: Theory and Application to Travel Demand, vol. 9. MIT press, Cambridge (1985)
45. Schmitz, C.: LimeSurvey: an open source survey tool. LimeSurvey Project Hamburg, Germany (2012). http://www.limesurvey.org, http://www.limesurvey.org

# User Interface for Vehicle Theft Recovery System

Lawrence J. Henschen$^{(\boxtimes)}$ and Julia C. Lee

Northwestern University, Evanston, IL 60208, USA
henschen@eecs.northwestern.edu, j-lee@northwestern.edu

**Abstract.** We present a proposal for a generic vehicle anti-theft system. The system has two parts – a platform located in the protected object and a user interface on a user's phone or computer. A specific system is described in an XML fie, which is made by an anti-theft system designer using an interface described in this paper. High-level code to run on the protected object portion of the system is automatically generated from the XML description. After a system is deployed to a protected object, the user (owner, police, etc.) interacts with the protected device through a user interface described in this paper and can read status of the protected object as well as control devices on the object. The user interface is also generated automatically from the XML description. The system is generic and could be used to describe anti-theft systems for a variety of types of objects such as cars, trucks, boats, ATM machines, shipping containers, and others.

**Keywords:** Vehicle theft recovery · User interface · XML · Automatic code generation

## 1   Introduction

Motor vehicle theft continues to be a major worldwide problem. In 2018 there were over 700,000 motor vehicle thefts in the US with other countries experiencing even higher theft rates per 100,000 population [1]. Recovery of stolen vehicles is highly dependent on a variety of factors beyond the control of the owner – efforts and ability of local law enforcement, density of the area where stolen, intended use by thieves, recovery assistance devices in the vehicle, and others [1]. Recovery assistance devices, such as LoJack [2], are available but focus mainly on helping law enforcement track the vehicle. We propose a system with many more capabilities that would be much more valuable in recovering stolen vehicles and perhaps even preventing the theft in the first place. These improved capabilities include:

- Theft detection. For example, the vehicle can "call" the owner if the car is hotwired or a door opened without the key or without a security code being entered in a specified amount of time.

© Springer Nature Switzerland AG 2021
H. Krömker (Ed.): HCII 2021, LNCS 12791, pp. 56–72, 2021.
https://doi.org/10.1007/978-3-030-78358-7_4

- Vehicle control. For example, the owner or police could instruct the car to stop the engine or sound an alarm. The vehicle could be equipped with a still or video camera in the steering column, and the owner or police could activate it to take pictures for later identification of the thief.
- Communication. For example, the owner or police could instruct the car to turn the radio on and thereby allow voice communication with any occupants.
- Record and/or report. For example, the owner could initiate periodic transmission of the car's status and GPS location to the owner's cell phone.

Theft detection is extremely important because often the owner is not aware of the theft for some time; the car may be stolen while the owner is shopping or even overnight while the owner is sleeping. Successful recovery of vehicles is highly dependent on the time between the actual theft and when the owner becomes aware of the theft. Control, communication, and reporting are also extremely important because they can encourage the thief to abandon the theft.

Modern cars can easily implement the kinds of capabilities mentioned in the preceding paragraph. Remote starting has been available in cars for years. Remote locking and status check are options on some new cars (see, for example, [3]). The actual control of major automobile systems, like the engine, brakes, and steering, has been physically disconnected from the mechanical devices the driver uses (pedals, steering wheel, etc.) for many years and is accomplished instead by computers [4]. The kinds of control proposed in this paper can be easily accomplished. The technology for implementing such control is not of concern to the HCI community. Therefore, the remainder of this paper will be devoted to describing interfaces that would be used by car owners and police as well as by technicians defining the capabilities to be implemented in the car.

We note here and will illustrate in later sections that our proposal can be applied equally well to a large variety of types of objects, for example, ATMs, shipping containers, home safes, and many others. Each type of object has its unique set of capabilities. For example, ATMs have no engine or brakes, but they could be equipped in the future with an alarm, voice communication, a sensor to tell if it been removed from its installation, etc. With a system like we describe, an ATM machine could "call" the bank security if it were being tampered with. The bank security personnel could then take pictures of the thieves and even initiate communication with them to encourage them to abandon the crime. Similarly, shipping containers for valuable cargo could in the future be equipped with sensors to detect tampering and sirens, GPS locators and other hardware useful for recovery. An important feature of our proposed methodology is that it can be applied to any of these cases, and the generic interface used for defining the anti-theft capabilities as well as the interface for the users are the same for all cases.

We present a proposal for a generic anti-theft system. A system under our proposal would be described by an XML file. The use of XML is not critical. On the other hand, XML is human readable and can be used as the basis for the user interface, so it is a good choice. An implemented system has two parts – a platform located in the protected object and a user interface on a user's phone or computer. The description of such a system is made by an anti-theft system designer using an interface described later in this paper. High-level code to run on the protected object portion of the system is automatically generated from the XML description. After a system is deployed to a protected object, the user (owner, police, etc.) interacts with the protected device through a user interface described later in this paper. The user interface is also generated automatically from the XML description. The automatic generation of the code and user interface continues our work along these lines described in prior HCII conferences [5, 6].

The remainder of this paper is organized as follows. Section 2 gives an overview of the system. Section 3 shows two sample scenarios illustrating how our system could be used. Section 4 presents our proposal for an XML-based description language for defining the elements of an anti-theft system and the interface used by system designers to define a specific anti-theft system. Section 5 presents the user interface. Section 6 contains closing remarks and future work.

## 2   Overview of an Anti-theft System

A complete system has two parts – one in the protected object and one running on the user's PC, laptop, cell phone, or other connected device. "User" in this paper can refer to the owner of the protected item or to other agents such as police or other stakeholders.

The onboard system is the system in the protected object itself and would contain a processing element with a variety of inputs and outputs as well as communication circuitry for connection to the internet or cell phone network. For a car, the inputs could include status of various elements of the car (for example, was the engine running, was there was a key in the ignition, was the car was moving and how fast, etc.) and information from other devices in the car (for example, location from the GPS module, video from a special anti-theft camera installed in the steering column, etc.). The outputs might include a signal to disconnect power to the spark plugs, a signal to activate the camera, a signal to turn on the radio and play a warning message to the thief, etc. For an ATM, the inputs could include GPS location, if the ATM has been removed from its installation site, etc., and the outputs include a signal to turn on an alarm siren. The software on the processing element has two levels. The top level is the generic code automatically generated from the XML descriptions, as discussed in Sect. 4. This code monitors signals from the inputs, determines if an event has occurred, provides outputs in response to events, manages communication to the outside world, and handles messages received from the user. The low-level software consists of device handlers that interact directly with the input, output, and communication hardware. For example, the car system may include the capability to disable the engine. There would be some hardware module that accomplishes this inside the engine control module, and the software would contain a device driver that would handle all the details of interacting with that module. This separation of software into high-level operating system and low-level device drivers allows the automatic generation of the high-level code and follows a well-known principle of software engineering embodied, for example, in the UNIX [8] operating system.

The user system is the software running on the user's device, which can be any device (PC/laptop, cell phone, etc.) with browser capability. The following kinds of information and options would be displayed.

- Status appropriate for the kind of device. For a car, this could include if the engine was currently running, the car's speed and direction, etc. For an ATM, this could include things like whether the cash drawer had been opened, etc.

- Location and tracking information. Current location could be transmitted at regular intervals. The user's browser would save these transmissions over a period of time, for example twelve hours, and display the location history when requested. This could be useful to police, who could tell if the stolen vehicle had been at the site of a crime at some point. It might also be useful if the system were somehow disabled, for example the car had crashed and destroyed the communication circuit. The location history might give clues as to where the vehicle was heading at the time the crash occurred.
- Events. The onboard system would transmit information to the user when an event occurred. The information would include the name of the event (vehicle hotwired, vehicle started with key, etc.) and information related to the event.
- Commands. The user could ask for values from the inputs and activate relevant outputs.

The onboard system is a control system similar to building control systems. It has inputs that read various information from the environment and outputs that control various actuators in the system. For displaying these input/output parts of the onboard system on the user interface we adapt the display format we proposed in [6]. The main screen for a particular vehicle might appear as in Fig. 1. This screen contains the external name of the object and values from inputs that the system designers decided to always include when displaying an object, such as location and if the engine was running. Other sample screens are shown in Sect. 3 as we describe sample scenarios.

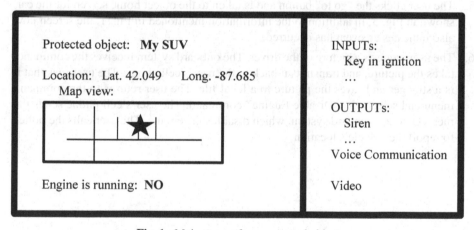

**Fig. 1.** Main screen for a protected object

## 3  Sample Scenarios

In this section we present two sample scenarios to illustrate how the system would behave.

### 3.1 A Thief Steals the Car

The owner of the car is shopping at a mall. A thief uses a coat hanger to reach in and undo the door lock and then hotwires the engine and drives away. The owner's cell phone gets an alert when the engine is started. The owner is busy paying for some purchased items and doesn't respond to the cell phone ringing until a few minutes later. When the owner does check the cell phone, the owner sees the event, brings up a list of commands and performs three actions – take a picture of the thief, disable the engine, and request location. When the user's device displays the current location of the car, the user calls the police. The following steps show the behavior of the system, the user, and the user's cell phone.

1. The onboard system detects the engine being started with no key in the ignition. The onboard system sends a message to the user's cell phone.
2. The user's cell phone receives an incoming call. The user does not answer, so the cell phone records the message.
3. The user eventually checks the cell phone. The following information is displayed through the normal cell phone answering service.
   Call received from "my car" at 10:45 am. Message "Car is being hotwired".
4. The user opens the anti-theft app and is immediately taken to the event screen for this event shown in Fig. 2.
5. The user clicks the "go to" button and is taken to the object home screen for the car, shown in Fig. 3. In addition to the information mentioned in Fig. 1, the screen now also indicates an event has occurred.
6. The user requests a picture of the driver. The onboard system receives the command, takes the picture, and transmits it back to the user's cell phone. The user sees that it is a stranger and saves the picture to a local file. The user returns to the command menu and selects the "Disable Engine" command. The user's cell phone sends the message to the onboard system, which disables the engine. The user calls the police to report the vehicle's location.

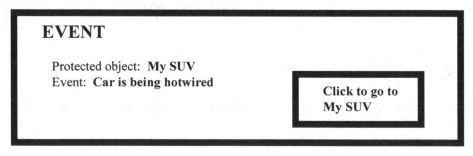

Fig. 2. Event screen for "Hotwire" event

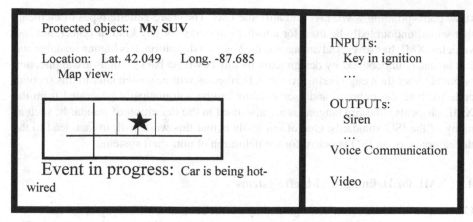

Protected object:  **My SUV**

Location:  Lat. 42.049    Long. -87.685
  Map view:

**Event in progress:** Car is being hot-wired

INPUTs:
  Key in ignition
  ...

OUTPUTs:
  Siren
  ...
  Voice Communication

  Video

**Fig. 3.** Protected object screen with event warning

## 3.2  The Owner's Child Takes the Car Without Permission

The owner sees the family car is missing from the driveway. The owner opens the app on the cell phone, selects the family car, and asks to see a picture of the driver. The owner sees it is a child who does not have permission to drive. The owner takes control of the car radio to tell the child to return home and monitors the progress until the child is home again. The initial steps are similar to the first scenario except that in this case the owner initializes the scenario instead of the onboard system sending an alert. After the owner requests voice communication from the command menu, the screen shown in Fig. 4 appears, and the owner starts a conversation with the driver to instruct the child to come home right away. After finishing the conversation, the owner requests location tracking from the command menu. The screen shown in Fig. 1 appears and is updated every 60 s.

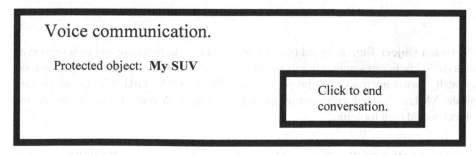

Voice communication.

  Protected object:  **My SUV**

  Click to end
  conversation.

**Fig. 4.**  Voice dialog screen

## 4  Defining an Anti-theft System

We propose an XML-based format for defining the elements of an anti-theft system. We describe some of the most common elements that would be included in such a system and show corresponding XML tags and attribute lists. These are generic types of elements that would undoubtedly be used for anti-theft systems for any kind of object, not just vehicles. XML has several advantages as the basis for definition. It is human readable and can be easily understood by design team members whose backgrounds are application oriented rather than engineering oriented. Further, as will be shown later in this section, code for both the onboard and user systems can be automatically generated from the XML elements. Markup languages are also used in the definition of standards, such as many of the ISO standards. One of our goals is that this work will, in fact, lead to the development of an ISO standard for the definition of anti-theft systems.

### 4.1   XML for Defining Anti-theft Systems

Our choice of XML elements is guided by our belief that anti-theft systems are event-driven systems. If nothing is happening to the car, then it is likely not being stolen. The system, then, should have inputs that can be used to detect events. Moreover, the events should cause actions to be taken, as illustrated in the scenarios presented earlier. The following list is a preliminary suggestion for a set of XML tags and corresponding attributes suitable for defining these inputs and outputs as well as additional functionalities.

Every XML element would have at least the following attributes.

- An internal name that can be used by other elements to refer to this one.
- An external name that can be used in the user's interface.

In addition, every input or output device should have an attribute that gives the name of the device handler, the low-level software function described in Sect. 2, that will be used to access the actual physical device. For example, a car might have an input sensor described as follows:

```
<INPUT internalname="engineOn" externalname="Engine is running"
        handler="checkEngineRunning()"
        ...
/>
```

**Protected Object Tag.** It would be common for a user to be interested in several protected objects. For example, a family might own two cars, and a trucking company would typically own many trucks. We propose the tag "PROTECTEDOBJECT" to encapsulate all the XML associated with a particular protected object. A complete definition for one object would then look like:

```
<PROTECTEDOBJECT internalname="car1" externalname="My SUV">
        ...
</PROTECTEDOBJECT>
```

The text between the opening and closing tags would define all the input, output, events, etc. contained in this protected object.

**Input Device Tag.** The protected object would have a variety of sensors and other sources of data that can provide input to the onboard processor. Each one of these input sources has a type, which can be digital (on/off, 1/0, etc.), analog (value in a range), complex data type (e.g., GPS reading, still picture), data stream, or one of many other types. For each of these data types there must be associated information, which is specified by an associated set of attributes. For example, for a digital input the XML element might include external display names to be used for the two possible values of the input. Finally, some inputs may be deemed important enough to always be displayed on the object's home screen whenever the object is selected by the user, while other inputs are less so and are displayed only on request from the user. Thus, the complete element for the engine running input mentioned above might look like.

```
<INPUT internalname="engineOn" externalname="Engine is running"
       handler="checkEngineRunning()"
       displayOnHomeScreen="NO"
       value1="YES" value0="NO"
/>
```

The attributes specify that the display name to be used when the device handler returns logic 1 is "YES" and when the device handler returns 0 is "NO". The value of this input would only be displayed if the user asked for it. For an analog input the element would include the range of values expected and the units of measure. For example, the input source that gives speed might be described by.

```
<INPUT internalname="speed" externalname="Vehicle speed"
       handler="getSpeed()"
       displayOnHomeScreen="NO"
       range="0-150" units="mph"
/>
```

Note, this implies that the device handler must translate whatever signal it receives from the vehicle hardware into miles-per-hour units. It is beyond the scope of this paper to give a complete list of attributes for all possible input types. Developing such a list is left for future work. Our goal here is to give the reader a sense of what can be done and of how easy it would be in a complete XML language for the anti-theft application.

**Output Device Tag.** The protected object would also have a variety of devices that can be controlled by outputs from the onboard processor. These outputs can be digital (e.g., turn a siren on or off), analog (e.g., decrease the speed to 20 mph), stream, and all the other types mentioned in the previous subsection. As for inputs, each data type would have an associated set of attributes. The output that disables the engine might be described with the following XML.

```
<OUTPUT internalname="engineDisable" externalname="Engine disable switch"
        handler="disableEngine(int)"
        val0="Disable engine" val1="Enable engine"
/>
```

The attribute identifying the handler for this output device indicates that the function takes a single argument of type integer. The attributes val0 and val1 associate names that appear on the user interface with values for the argument to the handler.

**Module Tag.** It is often the case in control systems that inputs and outputs are organized into groups. For example, in a car the sensor that reads current brake pressure and the actuator that applies pressure to the brake pads would likely be considered by auto engineers to be part of a single module that controls the brakes. While it would not be necessary to group these during the definition of the anti-theft system, we follow the lead of most control definition systems (e.g., BacNET [7]) and provide a tag that allows the design team to form such groups. A module element would then contain a list of the items contained in that module. For example,

```
<MODULE internalname="brakes" externalname="Brake system"/>
    <SUBMODULE  name="brakePressure" />
    <SUBMODULE  name="applyBrakePressure" />
    ...
</MODULE>
```

**Event Tag.** As noted, events are the driving concept for our proposed anti-theft systems. An event occurs when a set of conditions becomes satisfied. Basic conditions include an input having a specific current value, an input changing value, and receipt of a message from the user. Changes to analog inputs might involve the value moving into or out of a specified range. Changes to an input with complex data type, such as GPS location, would typically be more complex. A complete condition for an event is a Boolean combination of basic conditions. When the event is triggered there should be a set of actions to be performed. In some cases, the actions may be performed in any order; in other cases, the actions should be performed in a specific order. A complete list of possible actions is beyond the scope of this paper and is, in any case, still under development. Possible actions should include at least changing value of any output device, obtaining the value of any input device, transmitting a message to the user, and controlling the radio for voice communication with the user.

Our initial proposal for describing events is to use tags EVENT, CONDITION, ACTION, and tags for the Boolean operators like AND and OR. An EVENT element would include a single CONDITION element and a list of ACTION elements which are to be performed in the order listed. Inside the CONDITION element is a Boolean expression. Inside one ACTION element is a list of actions that can be performed in any order. Here is an example of a complete EVENT element.

```
<EVENT internalname="hotwire" externalname="Car is being hotwired." >
   <CONDITION>
      <AND>
         <CHANGE name="engineOn" oldvalue="NO" newvalue="YES" />
         <VALUE    name="keyInIgnition" val="NO" />
      </AND>
   </CONDITION>
   <ACTION>
      <SETOUTPUT name="siren" val="ON" />
      <TRANSMIT message="Car being stolen." />
      <SETOUTPUT name="radioControl" val="ON" />
   </ACTION
</EVENT>
```

This event is triggered if the engine starts (engineOn was NO the last time it was checked and is YES this time) with no key in the ignition (current value of keyInIgnition is NO). When the event does occur, the system turns the siren on, sends a message to the user, and turns the radio on for voice communication with the user, performing those actions in any order. If there were another ACTION element after the one shown, then the system would perform the first three actions in any order followed by whatever was specified in the second ACTION element.

**Other Tags.** The development of a full standard will undoubtedly produce more tags. For example, communication through a radio device, such as the car's radio, may be deemed special enough to warrant its own tag. To be general, an element with a communication tag would define the nature of the communication (voice, text, etc.), the nature of the device on the protected object (ordinary radio, special device, etc.), whether the communication is one-way or two-way, how the user opens the communication, etc. Similarly, a video tag could be included to specify the nature of a video device, which could be a video camera in the car's steering column or one recording the view seen out the front window. Attributes would include how recording is started and stopped, how the user can access recorded video, etc. It is beyond the scope of this paper to give a full list of all tags and attributes that would occur in a standard. Our goal is to show that it is feasible to use XML for generic anti-theft system definition and to show the designer's interface and the user's interface.

### 4.2  System Designer Interface

System designers need to develop a complete XML expression using the tags and attributes of the anti-theft standard. This allows for a simple designer interface, similar to other mark-up editor systems like Dreamweaver [9], consisting mainly of a work space window where the XML expression is shown, a sidebar with the allowed tags and relevant tools, and a top ribbon with standard editor function groups like FILE, EDIT, etc. Clicking a tag in the sidebar causes the editor to insert an empty element of the selected type into the work space at the current cursor position. The inserted element includes all the attributes for the selected tag with empty values.

The content of the sidebar changes depending on the content of the work space and the location of the cursor therein. For example, at the beginning the only allowable tag is PROTECTEDOBJECT, so that is the only tag shown in the sidebar when a new project is begun (see Fig. 5). Clicking that tag on the sidebar causes the system to insert an empty PROTECTEDOBJECT element with blank attribute values into the work space (see Fig. 6). When the designer moves the cursor inside the PROTECTEDOBJECT element (i.e., into a new context), the sidebar changes to indicate allowable tags for that context (see Fig. 7). As another example of context sensitivity, moving the cursor inside a CONDITION element causes the sidebar to list tags like AND, OR, NOT, CHANGE, VALUE, etc. as well as an icon that can be clicked if the designer wants help from an expression builder tool (see Fig. 8).

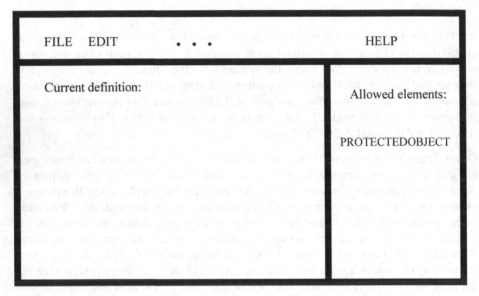

**Fig. 5.** Designer's screen at beginning of definition of a system

The sidebar should include tools for various ISO standards for data types, such as numbers, time, location, etc. These would be included whenever the cursor was in a context where the user needed to enter one of these data types. The editor would also automatically include an appropriate namespace reference if one of these standards was used.

```
FILE   EDIT            • • •                              HELP

Current definition:
                                                    Allowed elements:

   <PROTECTEDOBJECT
    internalname=""  externalname="" >




   </PROTECTEDOBJECT>
```

**Fig. 6.** Designer's screen after selecting PROTECTEDOBJECT

```
FILE   EDIT            • • •                              HELP

Current definition:                                 Allowed elements:

   <PROTECTEDOBJECT                                 INPUT
    internalname="car1" externalname="My SUV" >     OUTPUT
                                                    MODULE
                                                    EVENT
   —                                                • • •


   </PROTECTEDOBJECT>
```

**Fig. 7.** Designer's screen showing elements allowed inside PROTECTEDOBJECT

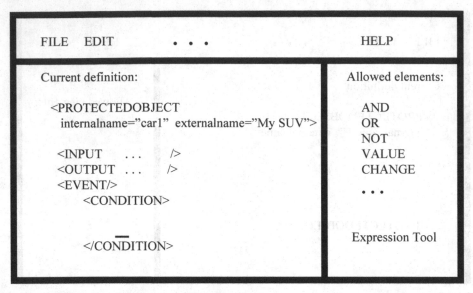

Current definition:                                  Allowed elements:

<PROTECTEDOBJECT                                        AND
  internalname="car1"  externalname="My SUV">          OR
                                                         NOT
<INPUT    . . .        />                               VALUE
<OUTPUT  . . .        />                                CHANGE
<EVENT/>                                                  . . .
   <CONDITION>


   ___
   </CONDITION>                                         Expression Tool

**Fig. 8.** Designer's screen with cursor inside CONDITION element

### 4.3  Automatic Generation of Code from XML

As noted in Sect. 2, the onboard software consists of two levels – high-level code derived from the XML and low-level device drivers that interact directly with the physical input and output devices. We believe the automatic generation of code will be useful in other application areas of interest to the HCI community, therefore we illustrate briefly how it can be done.

There are three main aspects to the automatic generation of the code – the generation of code to monitor for events, the generation of code to evaluate the event conditions, and the generation of code to access inputs and outputs.

Recall that the onboard system is event-driven. The system watches until an event occurs, at which time the system performs the actions specified for that event. One common way to manage this in software is by polling. The system constantly evaluates the condition expressions to see if any have become true. This is accomplished by a simple loop containing a set of if-statements. The form is:

```
while(1) {
    // Save old values for any inputs with a CHANGE test.
       Code for saving old values.  See next paragraph.
    for(j=1; j<=NUMBEROFEVENTS; j=j+1) {
        if(eventCondition1()) eventAction1();
        if(eventCondition2()) eventAction2();
        ...
    }
}
```

The function eventConditionj() evaluates the CONDITION expression for event j and returns true or false accordingly. The function eventActionj() executes the actions specified in the ACTION element of event j. The above code is automatically generated from the set of EVENT elements in the XML description. There is an if-statement in the loop for each EVENT element. There is an assignment statement (see below) in the code before the loop for each condition that tests for a change in an input value.

The Boolean expressions forming the conditions can be compiled using standard compiler techniques. References to values from input devices are compiled into calls to the device drivers named in the INPUT elements. Similarly, references to value names, such as ON or OFF, are translated into the corresponding integer values implied in the INPUT elements. For example, the INPUT element for engineOn associates 1 with "ON" (val1 ="ON") and 0 with "OFF (val0 ="OFF"). Finally, if a condition involves a CHANGE test, the code must keep track of the previously read value. For the condition of the event "hotwire" from Sect. 4.1, there would be code at the beginning of the while loop above to save the value for engineOn from the previous loop iteration.

```
oldEngineOn = engineOn;
```
The function to evaluate the condition for this event would look like the following.
```
int hotwireCondition() {
    engineOn     = checkEngineRunning();
    keyInIgnition = checkKey();
    if (
            ((oldEngineOn==0) && (engineOn==1))
        &&  (keyInIgnition==0)
    )    return 1;
        else return 0;
}
```

Actions specified in the ACTION portion of an event are similarly implemented by calls to the corresponding device handlers. For example, if a hotwire event occurs, the system is supposed to turn the siren on. This would be compiled into a call to the handler for the siren with argument value 1: sirenHandler(1);

## 5 User Interface

The main screens of the user interface are the home screen, the object home screen, the screens associated with various commands, and the event screen. Recall, "user" can be the owner of the object, police, security officers, or other stakeholders.

The home screen lists the protected objects owned by this user. Figure 9 shows a sample screen for a person with three vehicles. The user can click on any protected object to bring up the object home screen for that object.

The object home screen, as shown in Fig. 1, shows the name of the object and any other basic information the designers chose. For a car, the additional information would likely include location and possibly other information such as whether the doors were locked or the engine was running. This screen also contains a sidebar listing the commands that the user can execute. Commands include retrieving the current value

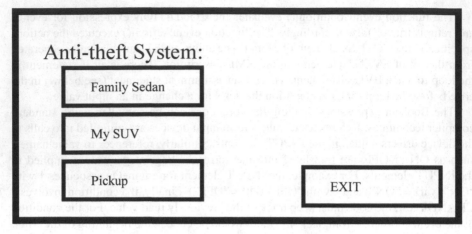

**Fig. 9.** Home screen for the user

of any INPUT device, setting the value of any OUTPUT device, and other commands relevant for that particular kind of protected object. For example, for a car, the additional commands might include open voice communication through the car's radio.

The screen for an individual command would include information and actions relevant for that command. For example, a command to change the value of an OUTPUT device, such as the door lock, would contain the description of the device (derived from the external name in the OUTPUT element), the current value of that device, and means for changing to a different value. For example, the screen for changing the value of a binary OUTPUT device might include two radio buttons with labels derived from the OUTPUT element, allowing the user to select one or the other of the two possible values. For the "Engine disable switch" OUTPUT device described in Sect. 4.1, the screen might look as shown in Fig. 10.

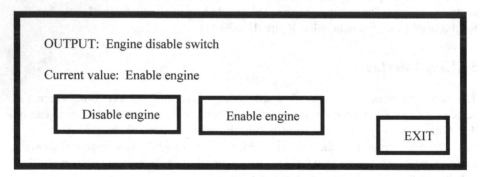

**Fig. 10.** User screen to change value of an OUTPUT

Finally, the event alarm screen is a pop-up that shows the user relevant information about the event, such as the name and other information deemed useful by the designers of the system. If the user's platform is a cell phone, the normal incoming-call method (vibration, ring tone, etc.) would indicate the arrival of an event alarm. On a PC or laptop, the normal pop-up mechanism would be used. From the event alarm screen the user can navigate to the screen for that protected object and perform operations. For example, for a family car the user might request a picture of the driver, as indicated in the scenarios in Sect. 2.

# 6 Closing Remarks and Future Work

## 6.1 Contributions

This work contributes to society in general by providing a generic framework for designing anti-theft systems for cars, boats, ATMs, large shipping containers, and other similar types of objects. The framework consists of a proposed XML-based standard for defining such systems and the organization of the onboard software into a high level generated automatically by the XML description and a low level consisting of the device handlers for interfacing directly to the hardware components of a protect device.

This work contributes to the HCI community in several ways. First, we have shown another example of how mark-up languages can conveniently and easily describe useful systems in different application areas. Moreover, for any given application area it is easy to define a generic set of tags and attributes that can be used to cover a wide variety of specific cases within the application area. In our case, the application area was protection against theft, and our proposed initial set of tags and attributes can be used for vehicles of all kinds, machines such as ATMs, shipping in general, and many other types of objects. Second, we have shown that when software is needed to implement that application, it can often be derived directly from the mark-up. Finally, once a standard has been established for an application area, generic presentation platforms can be easily derived to cover all specific cases within that application area, as this work and our prior work [5, 6] show.

## 6.2 Future Work

Continued study of anti-theft efforts in a variety of different areas is needed so that a full standard can eventually be developed. The standard would include a complete list of tags and their associated attributes, a complete list of capabilities that a protected device might include, and a more comprehensive study of what users should be able to do. This latter point should be conducted by experts in human-computer interaction, especially in the areas of user experience, and experts in the various application areas. As an example of the kinds of considerations that should be brought to bear in designing anti-theft systems for a particular area, consider the case of voice communications with a thief. In the case of a car, an owner talking to the thief might provoke the thief into driving more carelessly to escape and thereby increasing the risk of an accident that would damage the car or injure people. On the other hand, thieves who steal an ATM by ripping it from the wall to which it was attached have already damaged the protected object. Talking with them might frighten them into running away before they had a chance to break the ATM open. In this case, the device is already damaged anyway, but the money might be recovered. For each application area, experts in that area should work with HCI experts to determine which of the complete set of capabilities should be made available to the intended users.

# References

1. Motor_vehicle_theft. https://en.wikipedia.org/wiki/Motor_vehicle_theft, Accessed 18 Jan 2021
2. LoJack. https://www.lojack.com, Accessed 18 Jan 2021
3. Buick. https://www.buick.com/discover/connectivity/connected-services, Accessed 18 Jan 2021
4. New York Times. https://www.nytimes.com/2010/02/05/technology/05electronics.html, Accessed 18 Jan 2021
5. Henschen, L., Lee, J.: Human-computer interfaces for sensor/actuator networks. In: Kurosu, M. (ed.) HCI 2016. LNCS, vol. 9732, pp. 379–387. Springer, Cham (2016). https://doi.org/10.1007/978-3-319-39516-6_36
6. Henschen, L., Lee, J., Guthmann, R.: Automatic generation of human-computer interfaces from BacNET descriptions. In: Proceedings of the 20th HCII International Conference, vol. 21, pp. 71–84 (2018)
7. BacNET Home Page. http://www.bacnet.org, Accessed 04 Feb 2018
8. UNIX. https://en.wikipedia.org/wiki/Unix, Accessed 18 Jan 2021
9. Adobe Dreamweaver. https://en.wikipedia.org/wiki/Adobe_Dreamweaver, Accessed 18 Jan 2021

# Future of Urban Mobility – New Concepts Instead of New Technologies?

Katja Karrer-Gauß[✉] and Julia Seebode

VDI/VDE Innovation & Technik GmbH, Steinplatz 1, 10623 Berlin, Germany
{katja.karrer-gauss,julia.seebode}@vdivde-it.de

**Abstract.** Urban areas encounter a number of particular challenges due to increasing numbers of residents and thus strongly growing traffic with motorized private transport. Along with this, pollution degrades air quality, traffic noise affects residents and parking spaces take up a large part of urban public space. We need intelligent solutions that offer an attractive alternative to the use of private cars. In our view, technology has the potential to solve some of our current mobility challenges, but it is not a solution per se. We claim that as a first step we must define what kind of cities we do consider worth living in, what kind of mobility enhances our quality of life. Right from the start of the development process, we further advocate a continuous analysis of the psychological and sociological effects of the planned changes. We therefore, would like to give an overview of the topics that research and development projects are currently dealing with in the area of human-technology interaction in Germany and show some exemplary projects, which started with the analysis of user needs and work on socio-technological innovations with user-centered processes.

**Keywords:** Urban mobility · Intermodal mobility concepts · Shared mobility · Autonomous driving · Human-technology interaction

## 1 Introduction

### 1.1 A New Vision of Future Mobility

For a long time, the first association with a city of the future was an abundance of futuristic-looking vehicles that intersect on several levels, either through tubes, or flying on super-fast highways. This concept is based heavily on science fiction films and focuses on the technological advancement of individual (private) vehicles. The automation of these vehicles has long been considered the disruptive step into the future. From our view, future mobility refers not simply to a highly innovative technological development of transportation devices, but also to a rethinking of socially and environmentally compatible mobility. A transformation is necessary, a sustainable change in mobility behavior. It is not only a matter of increasingly automating means of transport and making them more context-sensitive, but also of concepts that encourage road users to rethink, to switch from using private cars to an intelligent combination of climate-neutral means of transport. We could conclude, many ideas of future mobility are already the past.

H. Krömker (Ed.): HCII 2021, LNCS 12791, pp. 73–88, 2021.
https://doi.org/10.1007/978-3-030-78358-7_5

Urban areas encounter a number of particular challenges. According to the study "Mobility in Germany 2008" [1], individual mobility in urban areas is strongly characterized by motorized private transport, which accounts for 71% of transport. The level of motorization in Germany has increased steadily in recent decades [2], and traffic jams on the way to and from work have become part of everyday life. Pollution degrades air quality, traffic noise affects residents and parking spaces take up a large part of urban public space. Further expansion of the transport infrastructure is hardly possible. We need intelligent solutions that offer an attractive alternative to the use of private cars.

Research has shown that various factors influence mobility behavior such as the transport system itself but also subjective socio-psychological as well as objective socio-demographic and socio-economic factors [3]. Objective and subjective factors complement each other and only the interaction of all influencing variables results in new, sustainable mobility behavior [4, 5]. In our view, the vision of future mobility consists of four important aspects (see Fig. 1): intermodal and connected solutions, shared and public means of transport, automated and flexible vehicles integrated, as well as a safe and secure interplay between all road users. These topics coincide with items on the European Strategic Transport Research and Innovation Agenda (STRIA)[1] and are prominent in current programs of national and international mobility conferences.

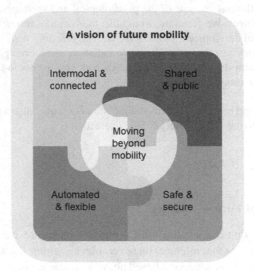

**Fig. 1.** Future mobility subjects

Some ideas to make local public transport more attractive refer to intermodal transport chains and to a combination of transport service providers. With mobile internet, the idea of Mobility as a Service (MaaS) appeared and is still an evolving concept [6]. This innovation allows users to plan trips and move with different mobility modes offered by different service providers, using a single interface. Hence, it becomes more comfortable to use different connected means of transportation, which is one factor for mobility

---

[1] https://ec.europa.eu/info/research-and-innovation/research-area/transport/stria_en.

behavior. The aim of moving in a more environmentally friendly way has also united a large number of people. This promoted the development of environmentally friendly micro vehicles "for the last mile" and sharing offers. These options need to provide a high level of comfort and meet mobility needs to replace the private use of individual vehicles [7]. Moreover, for the acceptance of all means of transport, safety and security issues are decisive. Even if the objective safety level not necessarily corresponds to how safe people feel within a traffic system or situation, feeling unsafe might affect the decision to participate in traffic in general as well as the decision for or against a certain means of transport and a certain mobility behavior [8].

Only all these aspects together, like pieces of a puzzle, give a complete picture. Another approach could be to move beyond mobility and enable telepresence or optimize home office work whilst thinking of how to reduce the needs for daily mobility. This cannot only help reduce $CO_2$ emissions, but likewise cut worker stress. Either way, it is crucial to get people to rethink. In order to achieve this, it is necessary to initiate a behavior change in transport and to support it via sophisticated and intelligent human-machine interfaces.

In the following, we would like to give an overview of the topics that research and development projects are currently dealing with in the area of human-technology interaction in Germany. We focus mostly on research projects funded by the German Federal Ministry of Education and Research (BMBF) and Federal Ministry of Transport and Digital Infrastructure (BMVI). One BMBF funding initiative "Individual and adaptive technologies for connected mobility" focuses on the connection of different mobility options and communication technologies in order to develop safe, individualized and flexible solutions for human mobility needs. The research projects have a strong emphasis on the human-technology interaction. The BMVI initiative "mFUND" is funding projects related to digital data-based applications for Mobility 4.0.

## 2 Intermodal Urban Mobility

### 2.1 Intermodal Transport Chains

Intermodal and connected mobility solutions are trending and seen as a key factor for moving people without using their own cars. With digital technologies come many opportunities to develop Mobility as a Service (MaaS) solutions. Schikofsky and colleagues studied the motivational mechanisms that encourage people to use those new services. They show that psychological needs like autonomy, competence and the feeling of being related to a social peer group play a crucial role in the acceptance of those services and can lead to behavioral changes [9].

Many researched projects focus on interlinking different means of transport, in particular individual transport, public transport and sharing services. They want to provide a seamless, comfortable and customized transport, for example by avoiding overcrowded buses and providing connection security. To reach this, digital services with well-designed user interfaces are necessary. Aim of the project "MaaS L.A.B.S"[2] is to

---

[2] https://www.maas4.de/maaslabs.

develop a MaaS platform, which offers integrated mobility planning for users and implements operational control and billing with cross-provider cooperation models. It will be possible to book different mobility options via the platform, which will include both public transport and sharing services. Payment will be possible by an integrated tariff model. Microbuses will operate in predefined service corridors in real time according to current passenger requests. The system will efficiently bundle the requests onto vehicles and at the same time secure connecting services. The central concern of the project is to promote an economically and ecologically sensible transformation of existing mobility systems with the help of attractive offers away from private ownership toward the shared use of mobility resources. In terms of broad transferability, the focus is on small and medium-sized cities. In an iterative user-centered process, the needs and requirements in two cities (with living labs in Potsdam and Cottbus) will be identified and implemented as demonstrations during the course of the project. The three living labs are used for user-centered research, they provide opportunities for workshops and tests where users can experience the future development and give feedback. In this way, the needs and requirements of potential users are comprehensively ascertained and flow directly into the iterative design and development of the app as well as mobility stations and microbuses.

Not only for end customers, the user interface is crucial for informed and thoughtful mobility decisions. Therefore, an important focus is also on the planners of the cities and transport systems. To help traffic managers and city planners, the project "allyMAP"[3], funded by BMVI, used data-based methods to enable more demand-oriented, economical, comfortable and environmentally friendly mobility solutions for people in urban and rural areas. The researchers implemented a solution as a mobility analysis platform for transport companies, private providers of mobility solutions, as well as city and transport planners. With this platform, it is possible to identify mobility needs, and accordingly plan a demand-based intermodal mobility service and simulate a planning by aggregating and intelligently processing a variety of different data (traffic, population, route network data, etc.). A challenge here was to design an interface that displays complex data and its interrelationships in an easy and understandable way and to develop models that can predict mobility behavior based on different data.

One of the projects that has already moved from research status to implementation of a beta phase is the "jelbi" mobility app for Berlin. A variety of shared mobility services in Berlin was connected with public transport in one single marketplace in 2019.

One important aspect of connected mobility services is that they inevitably collect location information from users. From the GPS location of a mobile phone, conclusions can be drawn about a person's activities, place of residence, inclinations, beliefs or social network. An important new research area is the so-called digital sovereignty of technology users. Most smartphones pass information via apps to companies by default, without explaining the risks to users in an understandable way or providing detailed control options. The "SIMPORT" project[4] aims to develop guidelines and software modules for the sovereign management of one's own location information on the mobile end device. A software will visualize possible conclusions from one's own location information and

---

[3] https://engineering.door2door.io/predicting-ridepooling-searches-using-pois-94e6b6cddc2f.
[4] https://simport.net.

explain the advantages and disadvantages of passing on location information. Users will be informed about the risks to pass on their data and can specify in detail which location data they want to share, when, with whom and in what detail.

In the future, the analysis must go beyond mere "user acceptance" in order to consolidate the adoption of more sustainable mobility options. Here, analysis of the factors that support a switch from private cars to more sustainable modes of transport will be important (see also [9]). Personal data can help create well-tailored and appropriately high-use services. However, safeguarding the privacy of users must also always be a priority.

## 2.2 Combined Transport of Goods and People

Carrying luggage, shopping bags or larger packages is a key barrier to the use of urban public transport [10]. A highly promising approach to solving this problem is the combination of passenger transport and delivery services for luggage or goods. Delivery services are common in food retailing. Customer-friendly delivery for on-site shopping is not yet established. Aim of the project "U-hoch-3"[5] is an intelligent mobility application that helps to organize the transport of goods while traveling by public transport. Users will have the option to leave purchases directly at the retailer and have them dropped off later at pick-up points near their home, for example, when they get off at their stop. In a survey, 30% of respondents (n = 357) completely or largely agreed that they would like to use such a delivery service [11]. In the project "U-hoch-3", there will be a first testing of unaccompanied autonomous delivery robots in two shopping centers that interact with customers to pick up purchases and take them to microdepots or city hubs for further delivery by electric micro mobiles or bicycles. To test the practicality and reliability of the developed services, there will be a one-year field test in 2023 in the city of Kassel.

The increasing number of parcels due to online retail poses logistical problems for courier, express, and parcel service (CEP) providers [12]. There is a lack of parking and storage facilities in the city center. An intelligent delivery concept in combination with electrified micro mobiles could lead to a reduction in noise and pollution as well as inner-city traffic density. The project "Ich ersetze ein Auto" (English: "I replace a car") [13] and its follow-up project "Ich entlaste Städte" ("I relieve cities") [14] have shown that by using electric cargo bikes, deliveries with a passenger car can be successfully substituted by more environmentally friendly vehicles. Courier service providers can reduce the use of combustion-engine vehicles and thus reduce greenhouse gas emissions. Half of the participants considered buying their own cargo bike at the end of the test phase or bought one immediately.

In Berlin, the project "KoMoDo"[6] combines a city hub with electric cargo bikes. Five of the largest national parcel service providers tested delivery of packages by using one microdepot cooperatively and cargo bikes.

Accordingly, one option is environmentally friendly delivery by bicycle delivery services. Another option is to automated delivery with other means. In the "FlowPro"

---

[5] https://u-hoch-3.de/.

[6] https://www.komodo.berlin/.

project[7], micro mobiles like multicopter drones and driverless transport systems are supposed to deliver goods on the base of a self-organizing logistics system that uses artificial intelligence (AI) algorithms. Similarly, the project "KEP-Town" analyzed the potential of automated micromobiles for courier, express and parcel delivery services in urban areas (2017–2018). Results suggest that automated delivery can be successful, from a technical, ethical, social or legal perspective. The project identified intention recognition of people on the sidewalks that may cross the trajectory of the autonomous delivery vehicle as essential. New interaction and display concepts are required to enable automated vehicles to communicate with pedestrians. Current robot concepts are not yet applicable in inner-city traffic.

While the transport options for goods described so far are spatially and temporally independent of the user, other projects are concerned with transport platforms that follow the user directly. The project "UrbANT"[8] develops a micromobile for transport of goods that automatically follows its user on sidewalks. The vehicle then drives back autonomously to a base station after completing the transport task. Based on the results of an extensive user study in the "UrbANT" project, a demonstrator of this transport platform will be developed and then tested in the field. The focus of the evaluation lies on the interaction between the micromobile and its users, but also passers-by on the sidewalk.

The challenge for the implementation of autonomous vehicles therefore clearly lies not only in the function of automated driving, but also in an appropriate interaction with the user and with all persons who come into contact with the new technology in their everyday life. Thus, research should continue to focus on the development of multimodal communication concepts of autonomous vehicles or robo cars with the environment. Again, user requirements, user acceptance and human factors, e.g. psychological aspects like trust or motivation should be the focus of research for a successful market launch.

### 2.3   Last Mile Mobility – (e-)Bikes, (e-)Scooters and Other Micro Mobiles

Personal Light Electric Vehicles (PLEV), e-scooters, pedelecs and e-bikes, but also traditional and cargo bikes are getting more and more popular. These small vehicles enable flexible individual mobility in urban environments, especially for short and mid distances. A suitable role of cycles and PLEVs in urban mobility may be to commute the first and last mile by complementing public transport. For successful integration, it needs good concepts of intermodal transport chains like examined above.

Even when electrically powered, these small vehicles small can contribute significantly to improving air quality in urban regions by replacing daily use of cars with combustion engine. Even compared to electric cars, they can be ecological due to the smaller moving mass and thus higher efficiency. Additionally, they can help reduce traffic jams by taking up less space. For the elderly or physically handicapped people, these vehicles can create new incentives and opportunities for individual "open-air mobility". Moreover, using those vehicles can support health and wellbeing for example by

---

[7] https://www.asti-insystems.de/en/forschung-und-entwicklung/flow-pro-mikrologistik-der-zuk unft/.

[8] https://www.ika.rwth-aachen.de/de/forschung/projekte/elektromobilität/2686-urbant.html.

reducing the risk of death from Cardiovascular Disease (CVD) and from cancer [15]. A study by Castro and colleagues found suggestions for e-bike use leading to significant increase in physical activity, especially for people switching from private motorized vehicle or public transport. Also people switching from bicycle to e-bike show more physical activity due to much longer trip distances [16].

Another important aspect is the acceptance of these new mobiles by other road users. Kampf and Constien therefore examined which aspects are necessary for a semi-autonomous PLEV in public to be accepted by other road users [17]. They found that people are in general open to use those new vehicles, and would like to use them e. g. for shopping, leisure trips or also to get to work. Challenges are both regulatory and technical in nature: PLEVs with sufficient engine power must be registered as motor vehicles and may then only be driven on the road. Many people find this too unsafe. It also leads to a loss of flexibility in case of traffic jams or construction sites. One proposed solution is the adaptive design of such vehicles that can adapt to the driving environment and limit speed accordingly. This is accompanied by the need to communicate the status of the vehicle to other road users. In the "Kamaeleon" project[9], the question on how the vehicle can show its current status to other road users is widely examined by field studies in different areas like parks, urban pedestrian walkways but also at exhibition centers or in industrial areas. An interesting research question here is the mechanism of "informal social control". If a rule violation such as speeding is displayed on the vehicle for all to see, compliant behavior could be achieved via this mechanism as the driver feels shame or guilty. In the project, the vehicle is developed tamper-proof to always drive automatically at adjusted speed according to the conditions and the surroundings.

In addition to these human factors issues, the discussion about where there is even room on the streets and sidewalks for PLEVs and all the people is in full swing and is crucial for the acceptance of new vehicles. This shows that not only the mere design of the vehicles themselves is a factor for acceptance. Urban mobility of the future must be thought of as a complex interplay of various factors to meet the challenges and enable a change to sustainable mobility behavior.

## 3 Shared Mobility

Part of intermodal transport chains as described earlier can be the shared use of vehicles like a car, bicycle, scooter or micro mobiles. Sharing economy services have become very popular in general, with shared mobility being a prime example. It has become common not only to share a ride in one's own car but also to share a taxi, or to borrow a vehicle on any street corner. A status report from Germany [18] describes the different concepts of goods sharing ranging from station-based or free floating car, roller, bike or kickscooter sharing or peer-to-peer carsharing, and of service sharing like carpooling, ridepooling or ridehaling. Most of the new sharing providers limit the offer to central urban areas, new business models and concepts are necessary, to expand those offers into suburban and rural areas.

Ridepooling is a still quite novel service of a collective transport, which combines different travel requests in one vehicle. Cities like Berlin, Hamburg or Hannover provide

---

[9] https://human-factors-consult.de/projekte/safety/kamaeleon/.

ridepooling services, which drive passengers in on-demand shuttles (Berlkönig, Clever-shuttle or MOIA). An empirical study could demonstrate that acceptance is a critical success factor and perceived compatibility has the strongest impact: The technology and service must be compatible to the individual lifestyle [19].

The project "EinfachTeilen"[10] (meaning "SimplySharing") developed a concept for intelligent car sharing in urban areas. While station-based or free-floating car sharing providers add their own vehicles to the existing amount of cars, peer-to-peer car sharing of private cars could help to reduce the number of vehicles in total. However, sharing private goods is strongly constrained by fears and trust issues. User studies within the project could show that some people are very open to new mobility concepts such as peer-to-peer car sharing. They show an increased willingness to change their own mobility behavior in an ecologically positive way. In order to put their ideas and intentions into practice, this group lacks sufficient information about the already existing concepts. In focus group interviews, it became clear that potential renters of private vehicles have strong concerns about disclosing their personal data. Shared car owners, on the other hand, have an interest in data that helps them better assess the driving behavior and thus the trustworthiness of potential renters. The operationalization of trustworthiness thus plays a central role in this concept. A solution could be rental to friends as in an approach by General Motors, or by ratings of other users [20].

Bikesharing is an eco-friendly alternative to the use of motorized vehicles that has become very popular in the past five years and is offered in mostly every larger city. Using the smartphone to locate a bike and then dropping the bike at any place after the ride is on one hand very user-friendly, but is accompanied by various problems like vandalism on the other hand. One of the biggest challenges in the operation of bike sharing systems is the redistribution or rebalancing problem, which results from the discontinuous demand and return of bicycles.

Recently, combining sharing concepts with autonomous components have been discussed as particularly advantageous. The use case is very simple: People can call an autonomous bike or cargo bike to their location, even in the suburbs, use it as a traditional bike and release it at their destination. However, this involves both technical hurdles and conceptual challenges. Even if the bike is fully functional and ready for use, people have to accept it to suit their mobility needs. Krause et al. therefore studied users' needs for an autonomous cargo bike as well as discussing aspects that are important for other road users who would encounter the bike in real life traffic [21].

The project "TRANSFORMERS" analyzed the acceptance of autonomous bicycles as a part of a bike sharing system. Here, the bike can also be called and returns to the station autonomously after use. In interviews, users showed an ambivalent attitude when evaluating video material; in actual contact with the autonomous e-bike, impressions were positive. Understandable signaling of bicycle behavior will be central to implementation, as bicycles will meet pedestrians. The perception of velocity and anticipation of reactions is an important issue for further research [22].

To sum up, sharing concepts provide a cost-effective and efficient utilization of vehicles, and reduce parking requirements. Trust issues and acceptance are the relevant human factor aspects to solve in the interaction with these new technologies. In addition,

---

[10] https://www.h-brs.de/de/einfach-teilen.

it will become necessary to develop standardized concepts for visual or acoustic signals with which (partially) autonomous vehicles interact with their environment.

## 4 Automated Vehicles for Public Transport

For some time now, autonomous vehicles are considered the disruptive step to profoundly change passenger and freight transport. Autonomous driving has the potential to make travel more comfortable, cheaper and safer, but it does not necessarily reduce traffic jam. Some even claim that fully autonomous vehicles allow more people to use a car, like people without a driving license or children [23]. Still, automated vehicles can also improve the public transport section, as with autonomous shuttle services to urban railway stations or as autonomous transport platforms for luggage or shopping bags. Some pilot projects in Germany have shown the advantages and challenges.

Since 2018, automated minibuses have been transporting passengers on the campuses of Berlin's Charité and Virchow-Klinikum hospitals (in cooperation with the Berlin transport association) along a defined route at a maximum speed of 12 km/h. Since 2017 the use of autonomously driving shuttles for the so-called "last mile" is tested in Bad Birnbach (by Deutsche Bahn) as a feeder to public transport. The design of human-technology interaction is in the focus of several research projects. The interaction with passengers, more precisely the situation of boarding and alighting when using autonomous shuttles and pods in the street space was analyzed by the "Hop-on_Hop-off" project [24]. User requirements were investigated on the EUREF campus in Berlin using the driverless shuttle "Emily". Passengers place high demands on safety and service quality. Willingness to use the service is particularly high if the shuttle picks up a person from home at his or her preferred time and the trip is included in a public transport subscription. The interviewed experts see the greatest obstacles in a lack of reliability and excessively low speeds due to regulatory and safety concerns. A recommendation for designing the interaction with passengers was, for example, that boarding the shuttle should be more personalized by indicating to the passenger that it is the correct and booked ride. The interaction with other road users was another core topic; for example, the shuttle should be clearly recognizable as a driverless vehicle.

The "KOLA" project [25] dealt with outward interaction of automated cars, but the findings are transferable to busses or other vehicles. Using light-based signals such as LEDs or laser headlights, the vehicle should communicate with the outside world in order to be able to act cooperatively in road traffic. This becomes particularly relevant if there is no more eye contact in the driverless vehicle. As a substitute, the vehicle can signal to pedestrians that they can cross the road, for example, by projecting a crosswalk. In the project, a number of those situations for cooperation in traffic was developed with the help of a diary study and an online survey. The designed light scenarios were than tested in two driving simulator studies. Results showed, that in these scenarios, appropriate communication through light could support the perception of safety and confidence in mobility, which can even contribute to a positive traffic atmosphere on the long term [26].

Some other projects focus on the changes brought by the introduction of automated driving in public transport, such as in the "RAMONA" project[11]. Trials in Braunschweig and Berlin were conducted with simulated self-driving buses, i.e. the driver was not visible to the outside or inside (so-called "Wizard of Oz" setting), and the reactions of pedestrians and passengers were observed. Similarly, the "UNICARagil" project[12] examined how passengers interact with an autonomous vehicle when entering and exiting the vehicle. Two areas were identified that would be significantly changed by the removal of a driver. A replacement is necessary for the many service tasks, such as giving passengers boarding information or helping with ticket selection. The "RAMONA" project uncovered that bus drivers drive extremely efficiently and react at bus stops under high concentration to pedestrians entering the safety areas. The distance between the bus and the waiting passengers is sometimes below the threshold of the safety protocol for autonomous driving. Consequently, an autonomous bus would have to enter the stop much more slowly for safety reasons. The demands on the sensor technology in the autonomous vehicle are thus very high.

For the implementation of autonomous vehicles in public transport, it is also necessary to consider central findings of automation psychology since the 1950s: when automated systems are implemented, the tasks for humans shift from executive to supervisory activities (monitoring) and this shift causes safety problems. Therefore, the interaction between driver and vehicle or operator and vehicle, has to be carefully designed on scientific knowledge. Likewise, interactions between the vehicle and passengers or other road users must be considered and redesigned. Not only with regard to automated vehicles is safety crucial for acceptance, adoption and impact.

## 5   Safe and Secure Solutions for Certainty and Acceptance

To promote the use of new mobility offers, it is necessary to create a secure transport infrastructure and surrounding. Agarwal et al. study the implications and possibilities of special bicycle highways in India [27] and found that physically segregated high-quality bicycle highways can attract previous non-cyclists and increase bicycle traffic in general. Also in many European cities the idea of special bike roads or highways is discussed, e.g. for Berlin with the first mobility law in Germany [28] and the first car-free bicycle highway in the Ruhr Region[13]. In Copenhagen and Amsterdam, a lot of effort was put into making the cities cyclist-friendly, with great success: Both are said to be some of the most bicycle-friendly cities in the world. Besides the interest in safe infrastructure, there are also security activities for bike users directly. The aim of research project "Safety4Bikes", funded by the BMBF was to develop modular assistance systems designed to detect imminent dangers based on the current traffic situation. In the event of acute hazards in the immediate vicinity or in potentially dangerous situations, the system warns the person riding the bike via acoustic, visual or haptic signals on the helmet or handlebars. The researchers examined which form of warning is most suitable for this

---

[11] https://www.dlr.de/vf/RAMONA.

[12] https://www.unicaragil.de/de/.

[13] https://www.rvr.ruhr/themen/mobilitaet/radschnellwege-ruhr/.

purpose. In addition, safety is to be improved via a communication interface to other people participating in traffic e.g. cars.

Another research project "ABALID" focused on supporting truck drivers in possible conflict situations with cyclists when turning right. Jürgensohn and project partners tested feedback principles of such a system for their effectiveness and user acceptance [29]. The technical recognition rate of the demonstrator was examined in real tests and determined to be about 80%. Increased safety for other road users through truck assistance systems is also the focus of the project "AugMir". Here, a system of robust environment sensors as well as the associated software is being developed to display processed data from the environment to drivers in augmented reality (AR). For this purpose, a network of wide-angle stereo cameras is used to continuously record the surroundings of the vehicle and any trailers that may be present in a redundant manner. This data is fused and displayed as a virtual mirror in the driver's cab using AR glasses. This not only makes it possible to point out obstacles and road users in the collision area, but also to provide a virtual view of the surroundings "through" the vehicle. Nevertheless, technical limitations and situational imponderables do not allow for one hundred percent system security, which makes acceptance by users difficult. Good system design is necessary to support drivers and bikers without leading to become overconfident.

In addition to assistance systems for individual road users such as vehicle drivers or cyclists, new types of systems are currently developed that can be integrated into the road infrastructure and thus reach all road users - regardless of the technological equipment. One example is the "MeBeSafe" project[14], which used various light elements in the road infrastructure to influence the individual driving speed of road users. In this way, it should be possible to avoid critical events in advance.

Even as semi-automated (micro) vehicles become part of our cities, there will long be a mix of diverse road users, most of whom are not automated. This is accompanied by unclear and dangerous situations in road traffic. Those can be avoided, if all parties involved communicate their intentions clearly and are thus able to correctly assess each other's behavior. For this kind of nonverbal communication we need new concepts and designs that are examined in different research projects (as the above mentioned project "KOLA"[15], or "atCity"[16]).

## 6 Moving Beyond Mobility

Another way to think of the future of urban life is to ask about the necessity of mobility itself. New applications with mixed reality (MR) technologies can enable collaborative intellectual but also physical work (e.g., new forms of remote maintenance) even over long distances. In addition to the advantages for commuters who have to travel longer distances every day for work reasons, virtual, augmented or mixed reality technologies could address social needs. Spatial distance and restricted time resources, social isolation or individual restrictions minimize the possibility of shared experiences. In times of globalization and urbanization, the social environment is often widely distributed. In

---

[14] https://www.mebesafe.eu/about/.

[15] http://www.experienceandinteraction.com/kola.

[16] https://www.atcity-online.de/partial_projects/sp4/_TP-4.html.

2018, around 10% of all couples in Germany lived apart [30]. Furthermore, health or financial limitations hinder some people from participating in cultural events. Other fields in which "proximity over distance" will be possible via AR and VR are education[17] as well as the participation of older people in social life. If people no longer had to travel, many hurdles in terms of work, social and cultural needs could be overcome.

Bridging spatial distances even without mobility has become a very strong focus, especially in 2020, due to the contact restrictions associated with the Corona pandemic. In healthcare, telemedicine approaches are booming, for example by implementing consultation, diagnostics and therapy in video consultations [31]. How to support telemedicine professionals is the objective of the "MEDIBILITY" project[18], which aims to hand over parts of patient communication to AI-supported bots. The medical staff will not be replaced but supported, so that the consultation process is divided hybrid between bot and doctor. Another approach is examined in the "EXGAVINE" project[19], which aim is to use VR exergames for the treatment of adults with dementia. One example game is the "Memory Journalist", were patients shall memorize locations of landmarks and photograph them using a camera-like controller by exploiting simple motor tasks. The results of a first user study were very promising as the exercises could be reproduced even by patients with medium dementia, who could use the camera controller, similar to what they were used to before [32].

Still there are several technical challenges as well as obstacles for a widespread use, primarily addressing the simultaneous work and movement of several people in both the virtual and the real environment as well as the safe implementation of manual operations from a distance. In addition to the technological development of new MR systems mentioned above, there are a number of human factors issues to be resolved. In order to actually experience participation or presence, mixed reality should be characterized by the highest possible level of immersion and offer opportunities for interaction. Research topics include place illusion (the feeling of actually being in a place in virtual space) and social presence illusion (the feeling of being in a place with another person). In addition, effects on social relationships as well as on group behavior by long-term usage have to be further examined. So far, it is unclear, how peoples' social behavior and cognitive abilities could change due to daily or extensive use of MR systems.

If these technical as well as social and psycho-physiological issues are solved and examined, the choice of place of residence is much less influenced by spatial proximity to the workplace. As a result, rural regions can become more attractive as business locations as well as places of residence and be strengthened overall. Additionally, cities could be designed for shorter ways with mixed areas of living, working and shopping.

---

[17] https://www.social-augmented-learning.de/.

[18] https://www.interaktive-technologien.de/perspektiven/forschung-in-zeiten-von-corona/corona-bot-uebernehmen-sie-medizinische-ersteinschaetzung-durch-einen-intelligenten-bot.

[19] https://www.exgavine.de/.

# 7  Conclusions for Future Research

Growing cities and traffic problems are a central social challenge. In the field of human factors in urban mobility, the research focus has thus shifted from traffic safety to new objectives such as quality of life in the city and climate protection.

Research on road safety will continue and has long driven technology development in the private passenger car. But now, new fields have emerged in which technology and digitalization are unfolding. These are intermodal and connected, shared and public mobility. In these areas, the goals are currently gains in convenience and efficiency for mobility users. Future research will certainly focus more on how to achieve behavioral changes through interaction with technology, i.e. how people can be convinced to switch to environmentally and socially friendly alternatives instead of using a private car. This is where the field of human factors can unfold its potential and, for example, take up concepts from motivational psychology. Serious games and gamification principles could be a key to encourage users to use public transportation or ride a bicycle instead of using a private car [33].

In the realization of connected mobility, the use of smartphones will continue to play a central role. Modern concepts such as car, bike or e-scooter sharing, carpooling or taxi shuttles would not be possible without the smartphone. However, we can move beyond concepts that solely rely on applications on our smartphones. For this purpose, we should discuss the potential of intelligent infrastructure. We might forgo smartphones, when bus stops or train stations as well as the vehicles are connected and interactive themselves.

In the latter areas, mobility will increasingly rely on the processing of data, including personal information. This is also accompanied by a new need for research. Private companies collected and analyzed data with little constraints until in Europe the General Data Protection Regulation (GDPR) came into force in May 2018. For interactive technologies, this is an opportunity to establish transparent and controllable data use in Europe that can compete internationally. User-friendly implementation of privacy can and should become a research focus also in the field of mobility services. Just like safety from accidents, data security plays a central role in the research field of trust in new, innovative means of transport, a key not only for autonomous mobility[20].

There is no doubt that automated vehicles could and will change our mobility dramatically. Large companies like Tesla, Ford or General Motors are working hard on their development, triggered by government funding, as well as Volkswagen, Daimler and BMW. The ethos of the maker, who works on visionary technologies in competition with others, is predominant here, as displayed in particular in the DARPA Challenges (see also [34]). However, early promises that autonomous cars would soon be on the roads have proved too optimistic. Autonomous driving on city streets is still in its infancy. It is noteworthy that decades of knowledge gained from human factors research on automation from the fields of aviation or production is now replicated with regard to mobility: It is not the mere replacement of humans that is the solution for more safety and comfort. It is about the successful support of humans by technology, which can only begin with a careful assessment of needs and a desirable vision of the future for society. Developers and engineers must not repeat the mistake of blindly realizing technology fantasies in a

---

[20] https://h2020-trustonomy.eu/.

highly competitive timeframe, putting the cart before the horse, so to speak. We should strive to avoid a strictly rational-technical worldview and be careful not to neglect "soft" factors such as sociological or psychological aspects of technological change, but – in contrast – to focus on them.

Technology has the potential to solve some of our current mobility challenges, but it is not a solution per se. We claim that as a first step we must define what kind of cities we do consider worth living in, what kind of mobility enhances our quality of life. Right from the start of the development process, we further advocate a continuous analysis of the psychological and sociological effects of the planned changes. Still often neglected, this should be the core of any research and development project.

From our point of view, the following research questions will remain important in the next years: It is urgently necessary to solve how individual mobility can be changed to be climate-friendly and environmentally compatible. We need to find out how people can be convinced to support this change. Everyone should be able to participate in mobility, regardless of physical impairments. We must find ways to ensure safety for all road users, independently of costly, individual assistance systems. Flagship projects are necessary to demonstrate how data protection can be guaranteed for strongly data-driven, connected mobility offers. And, as a final point, we need to find out if perhaps partially we can do without mobility because virtual and augmented reality technologies enable us to be close at a distance. These are exciting research questions that might help us to design liveable cities with sustainable infrastructure and mobility. Human factors and ergonomics should apply its relevant skills and knowledge to assist this process.

# References

1. Infas, D.: Mobilität in Deutschland 2008. [Mobility in Germany 2008]. Ergebnisbericht im Auftrag des Bundesministeriums für Verkehr, Bau und Stadtentwicklung. [Result report commissioned by the Federal Ministry of Transport, Building and Urban Development] (2010). http://mobilitaet-in-deutschland.de
2. Umweltbundesamt: Mobilität privater Haushalte (2020). https://www.umweltbundesamt.de/daten/private-haushalte-konsum/mobilitaet-privater-haushalte#verkehrsaufwand-im-personentransport
3. Gerike, R., Koszowski, C., Hubrich, S., Canzler, W., Epp, J.: Aktive Mobilität: Mehr Lebensqualität in Ballungsräumen. Texte I 226/2020, Umweltbundesamt (2020). https://www.umweltbundesamt.de/sites/default/files/medien/5750/publikationen/2020_12_03_t exte_226-2020_aktive_mobilitaet.pdf
4. Seebode, J., Wechsung, I., Greiner, S., Möller, S., West-ermann, T.: Sustainable mobility-How to overcome mobility behavior routines. In: Proceedings of the NordiCHI 2014: The 8th Nordic Conference on Hu-man-Computer Interaction: Fun, Fast, Foundational, pp. 1003–1006 (2014). https://doi.org/10.1145/2639189.2670261
5. Hunecke, M., Haustein, S., Grischkat, S., Böhler, S.: Psychological, sociodemographic, and infrastructural factors as determinants of ecological impact caused by mobility behavior. J. Environ. Psychol. **27**(4), 277–292 (2007)
6. Mola, L., Berger, Q., Haavisto, K., Soscia, I.: Mobility as a service: an exploratory study of consumer mobility behaviour. Sustainability. **12**, 8210 (2020). https://doi.org/10.3390/su12198210
7. Herberz, M., Hahnel, U.J., Brosch, T.: The importance of consumer motives for green mobility: a multi-modal perspective. Transp. Res. Part A: Policy Pract. **139**, 102–118 (2020)

8. Furian, G., Brandstätter, C., Kaiser, S., Witzik, A.: Subjective safety and risk perception. ESRA thematic report no. 5. ESRA project (European survey of road users' safety attitudes). Vienna, Austria, Austrian Road Safety Board KFV (2016). https://www.esranet.eu/storage/minisites/esra2018thematicreportno15subjectivesafetyandriskperception.pdf
9. Schikofsky, J., Dannewald, T., Kowald, M.: Exploring motivational mechanisms behind the intention to adopt mobility as a service (MaaS): Insights from Germany. Transp. Res. Part A: Policy Practice **131**, 296–312 (2019). https://doi.org/10.1016/j.tra.2019.09.022
10. Millonig, A., Haustein, S.: Human factors of digitalized mobility forms and services. Eur. Transp. Res. Rev. **12**(1), 1–4 (2020). https://doi.org/10.1186/s12544-020-00435-5
11. Klingauf, A., Hegenberg, J., Lambrecht, F., Bieland, D., Schmidt, L., Sommer, C.: Nutzeranforderungen an ein Assistenzsystem für den ÖPNV. In: Gesellschaft für Arbeitswissenschaft e. V. (Hrsg.): Arbeit interdisziplinär analysieren - bewerten - gestalten: 65. Kongress der Gesellschaft für Arbeitswissenschaft (Dresden 2019). Dortmund: GfA-Press, S. pp. 1–6 (2019) (B.2.6)
12. Rose, W.J., Bell, J.E., Autry, C.W., Cherry, C.R.: Urban logistics: establishing key concepts and building a conceptual framework for future research. Transp. J. **56**(4), 357–394 (2017)
13. Gruber, J.: Ich ersetze ein Auto (Schlussbericht). Elektro-Lastenräder für den klimafreundlichen Einsatz im Kuriermarkt, Vorhaben 03KSF029 der Nationalen Klimaschutzinitiative des BMUB, DLR Institut für Verkehrsforschung Berlin-Adlershof (2015)
14. Deutsches Zentrum für Luft – und Raumfahrt: Ich entlaste Städte (2018). https://www.lastenradtest.de/
15. Celis-Morales, C.A., Lyall, D.M., Welsh, P., Anderson, J., Steell, L., Guo, Y., et al.: Association between active commuting and incident cardiovascular disease, cancer, and mortality: prospective cohort study BMJ 2017 **357**, j1456 (2017)
16. Castro, A., et al.: Physical activity of electric bicycle users compared to conventional bicycle users and non-cyclists: Insights based on health and transport data from an online survey in seven European cities. Transp. Res. Interdis. Perspect. **1**, 100017 (2019)
17. Kampf, A., Constien, H.-P.: Kamäleon - Konstruktive adaptive Mobilität durch Leichtfahrzeuge ohne Grenzen: Abschlussbericht, Constin GmbH, Berlin (2019)
18. MOQO Shared Mobility Status Report 2020 (2020). https://moqo.de/shared-mobility-status-report-2020
19. Sonneberg, M.O., Werth, O., Leyerer, M., Wille, W., Jarlik, M., Breitner, M.H.: An Empirical Study of Customers' Behavioral Intention to Use Ridepooling Services–An Extension of the Technology Acceptance Model (2019)
20. Bossauer, P., Neifer, T., Stevens, G., Pakusch, C.: Trust versus privacy: using connected car data in peer-to-peer car sharing. In: Proceedings of the 2020 CHI Conference on Human Factors in Computing Systems, pp. 1–13 (2020)
21. Krause, K., Assmann, T., Schmidt, S., Matthies, E.: Autonomous driving cargo bikes – introducing an acceptability-focused approach towards a new mobility offer. Transp. Res. Interdisc. Perspect. **6**, 100135 (2020)
22. Zug, S., et al.: BikeSharing-System der 5. Generation 15. Smart Cities/Smart Regions–Technische, wirtschaftliche und gesellschaftliche Innovationen: Konferenzband zu den 10. BUIS-Tagen, p. 189 (2019)
23. Bultmann, M., Houben, A.J.H.: How autonomous vehicles could relieve or worsen traffic congestion. HERE Deutschland (2016)
24. Crössmann, L., Jonuschat, H., Filby, A., Power, M., Deibel, I., Steiner, J.: Hop-on Hop-off - Sichere Ein- und Ausstiegssituationen für autonome öffentliche Mikromobile : Schlussbericht des InnoZ: im Programm "Individuelle und adaptive Technologien für eine vernetzte Mobilität". [InnoZ GmbH]. (2018) https://doi.org/10.2314/KXP:166687485X

25. Powelleit, M., Winkler, S., Vollrath, M.: Cooperation through communication–using headlight technologies to improve traffic climate. In: Proceedings of the Human Factors and Ergonomics Society Europe, pp. 149–160 (2018)
26. Powelleit, M., Winkler, S., Vollrath, M.: Light-based communication to further cooperation in road traffic. In: Meixner, G. (eds.) Smart Automotive Mobility - Reliable Technology for the Mobile Human, pp. 305–346. Springer, Cham (2020) https://doi.org/10.1007/978-3-030-45131-8_6
27. Agarwal, A., Ziemke, D., Nagel, K.: Bicycle superhighway: an environmentally sustainable policy for urban transport. Transp. Res. Part A: Policy Pract. **137**, 519–540 (2020)
28. Senatsverwaltung für Umwelt, Verkehr und Klimaschutz (2018). https://www.berlin.de/sen/uvk/_assets/verkehr/verkehrspolitik/mobilitaetsgesetz/mobilitaetsgesetz_broschuere.pdf
29. Jürgensohn, T.: Projekt ABALID - Abbiegeassistent mit 3D-LIDAR-Sensorik : Schlussbericht. Human-Factors-Consult GmbH, Berlin (2016). https://doi.org/10.2314/GBV:883073943
30. Statista: Statistiken zum Thema Liebe (2018). https://de.statista.com/themen/142/liebe/
31. Ärztezeitung (2020). https://www.aerztezeitung.de/Wirtschaft/Fernbetreuung-von-Patienten-boomt-in-der-Corona-Pandemie-414414.html
32. Rings, S., Steinicke, F., Picker, T., Prasuhn, C.: Memory journalist: creating virtual reality Exergames for the treatment of older adults with dementia. In: 2020 IEEE Conference on Virtual Reality and 3D User Interfaces Abstracts and Workshops (VRW), pp. 687–688. IEEE (2020)
33. Gibert, A., Schäfer, P.K., Tregel, T., Göbel, S.: Förderung von umweltfreundlichen Verkehrsmitteln durch Gamification und Serious Games. In: Proff, H. (eds.) Neue Dimensionen der Mobilität, pp. 745–753. Springer, Wiesbaden (2020) https://doi.org/10.1007/978-3-658-29746-6_58
34. Davies, A.: Driven: The Race to Create the Autonomous Car. Simon & Shuster (2021)

# Assistive Systems for Mobility in Smart City: Humans and Goods

Yuhang Li[1] ⓘ, Chuantao Yin[1] ⓘ, Zhang Xiong[2] ⓘ, Bertrand David[3](✉) ⓘ,
René Chalon[3] ⓘ, and Hao Sheng[4,5] ⓘ

[1] Sino-French Engineer School, Beihang University, Beijing 100191, China
{yuhang.li,chuantao.yin}@buaa.edu.cn
[2] School of Computer Science and Engineering, Beihang University, Beijing 100191, China
xiongz@buaa.edu.cn
[3] Université de Lyon, CNRS, Ecole Centrale de Lyon, LIRIS, UMR5205, 69134 Lyon, France
{Bertrand.David,Rene.Chalon}@ec-lyon.fr
[4] State Key Laboratory of Software Development Environment, School of Computer Science
and Engineering, Beihang University, Beijing 100191, China
shenghao@buaa.edu.cn
[5] Beijing Advanced Innovation Center for Big Data and Brain Computing, Beihang University,
Beijing 100191, China

**Abstract.** Nowadays, the society is highly intelligent, and smart city has become the common expectation of people. Smart city is a large concept, which includes smart home, Smart Transportation, Smart public, Service and Social management, smart Urban Management and other aspects. This paper mainly discusses the problem of the mobility in cities. The mobility in cities can be divided into two categories in practice, namely, the movement of humans and the movement of goods. In this paper, we enumerate the typical cases in the mobility of humans and goods, and analyze the operation principle of these cases one by one. Some of these examples are implemented, some are not. In order to make urban mobility "smart", we propose a framework of assistive system, which has a four-tier structure and uses big data, cloud computing, AI, deep learning and other technologies. It is aimed at a variety of different scenes, to provide users with the most thoughtful advice and the most humanized service, to solve the problem of urban mobility.

**Keywords:** Smart city · Mobility of humans · Mobility of goods · Assistive systems

## 1 Introduction

### 1.1 Background

The rapid development of information technology has brought about a global trend of informatization. In the future, it is increasingly necessary to rely on information technology to promote the development of smart cities. Governments around the world have spontaneously proposed a plan to rely on the Internet and information technology to change the city's blueprint for the future.

© Springer Nature Switzerland AG 2021
H. Krömker (Ed.): HCII 2021, LNCS 12791, pp. 89–104, 2021.
https://doi.org/10.1007/978-3-030-78358-7_6

The United States took the lead in proposing the National Information Infrastructure (NII) and global Information Infrastructure (GII) programs. Then, the European Union focused on promoting the Information Society program and identified ten application fields of the European Information society as the main direction of the construction of the Information society of the European Union [1, 2].

As the whirlwind of smart cities intensifies, countries are constantly putting innovative ideas into action. In the process of industrial transformation and social development, many developed countries, such as the United States, the United Kingdom, Japan, the Netherlands, Sweden, South Korea, Singapore and Malaysia have recognized the importance of smart cities earlier and have achieved remarkable results [3, 4]. At present, countries all over the world have different degrees of realizing the idea of smart cities and different implementation efforts. However, there is no denying that in the 21st century, the age of smart has come.

## 1.2  What is Smart City?

The term 'smart city' appeared for the first time in the early 1990s, and researchers have emphasized technology, innovation and globalization in the process of urbanization [5]. Smart cities have attracted great attention since 2008, with the launch of IBM's Smarter Planet project [6]. Since then, the concept of smart cities has continued to grow and evolve.

Smart city originated in the field of media, which refers to the integration of urban systems and services by using various information technologies or innovative concepts, so as to improve the efficiency of resource utilization, optimize urban management, and improve the quality of life of citizens.

As a new mode and concept of modern city operation and governance, smart city is established on the basis of complete network infrastructure, massive data resources, multi-domain business process integration and other information and digital construction, which is an inevitable stage in the development process of modern city.

Smart city often intersects with regional development concepts such as digital city, perceptive city, wireless city, ecological city and low-carbon city, and even gets mixed up with industrial digitalization concepts such as e-government, intelligent transportation and smart power grid. People's interpretation of the concept of smart city often has different emphasis, some thought the key lies in the application of technology, some thought that the key lies in the network construction, some view is the key in the participation of people, some thought that the wisdom is the most important [7–11]. In a word, smart is more than intelligence. Smart city is not only another term for intelligent city, but also includes people's intelligent participation, people-oriented, sustainable development and other connotations.

## 1.3  Mobility in Smart City

Smart City is a very large concept, which contains many sub-fields. Su et al. [9] summarized eight main applications of Smart City in real life (shown in Fig. 1): wireless city, smart home, smart transportation, smart public service and social management, smart urban management, smart medical treatment, green city and smart tourism. Figure 1 shows the construction frame of application systems for smart city.

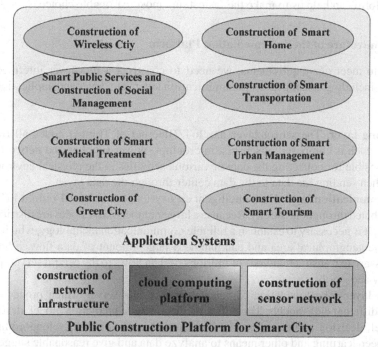

Fig. 1. Construction frame of application systems for smart city.

The architecture of smart cities is too large and it contains too many fields, so we cannot explain them all in detail in one article. Therefore, in this article, we mainly explore a sub-problem of the smart city, that is, the problem of urban mobility, which includes goods or humans with or without special needs. The description of other applications is available in [12].

## 2 Assistive System of Smart City

### 2.1 Intermediation Platform: A State-of-the-Art Assistive System

The mobility in cities is mainly devoted to two aspects: the mobility of people and the mobility of goods. Every second, thousands of people or goods in cities need to move. For people, there are many problems arising from the demand of mobility, such as the travel itinerary, route planning, how to park, and so on. Even things get more complicated when the people who travel are disabled. For the goods, how to choose the carrier is an obvious problem, in addition to this, the transfer of goods between different carriers and deliverymen cannot be ignored neither. It is very important to arrange and dispatch the traffic resources and human resources of city reasonably [13].

Based on these problems, we seem to need an intermediation platform to intelligently collect the data generated during the mobility. Through the analysis and simulation of these data, we can use scientific and information-based ways to meet our mobility needs. This platform can help us to make the fastest and most reasonable choice.

## 2.2 Architecture of the Intermediation Platform

In order to meet our requirements, we need to construct a four-layer intermediation platform, including sensing layer, communication layer, data layer and application layer (Fig. 2).

1. **Sensing layer.** The sensor layer has IoT (Internet of Things) nodes all over the city, which is composed of sensors, microchips, power supplies and networks. It is responsible for collecting the data of various activities in the physical environment, and then sending the data to the data center through the data layer.
2. **Communication layer.** Each intelligent city system has at least 1 billion IoT nodes distributed throughout the city. Because a large number of IoT nodes are distributed in a city, it is necessary to establish a reliable communication technology, which covers a wide geographical area and can handle a large amount of data flow, it becomes necessary to transmit hundreds of millions of data flows to the central server quickly, safely and stably.
3. **Data layer.** The foundation of data layer is a data server which processes data with different data models. These data models include prediction model, description model, decision model and so on. These models use big data, cloud computing, AI, deep learning and other means to analyze data and give reasonable suggestions. With the help of these models, the municipal government can make forward-looking decisions based on the data.
4. **Application layer.** As a visual platform that can directly interact with users, the application layer is the only level that is presented to the public. Typically, on the phone, the application layer is shown as apps. Users can enter their own mobility requirements, such as planning a destination, to get the desired path recommendation.

In general, when we talk about an intermediation platform, we only mean the application layer in a four-tier structure. The other three layers are invisible, but they support the application layer. To put it simply, for every citizen, our assistive system is a collection of a series of apps, that is, a multi-functional intermediation platform, which provides users with the most thoughtful suggestions and the most humanized services for various scenarios, so that people and goods can flow without barriers.

Without massive data and complex algorithms, our intermediation platform would be a beautiful shell. These invisible structures require huge investment in infrastructure and human resources, which each country needs to realize step by step according to the actual situation.

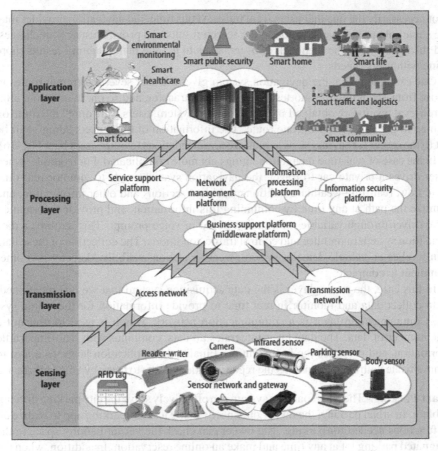

**Fig. 2.** Smart city architecture summarized by Liu and Peng [26].

## 3 Mobility of Humans and Goods

In the remainder of this article, we will focus on some of the classic examples of urban mobility in our lives, some of which have already been achieved, and some of which are not yet. We analyze the essential problems in each case and propose appropriate solutions through our mediation platform.

### 3.1 Mobility of Humans

**Intelligent Navigation.** Intelligent navigation, as a scenario achieved, the main solution is the path planning problem. When the user opens the app, the app will first locate the current location through GPS satellites. After entering the destination, the platform will plan an optimal route according to the user's starting position, destination location and travel mode.

Since road conditions in reality need to consider more factors such as road condition and environment, the "optimal" path obtained by the basic algorithm based on the graph is not necessarily the one that users want most. At this time, the navigation provider may give some suboptimal paths to add to the search results and provide users with choices. And then learn from the feedback of users, so as to get the path planning results more in line with users' needs.

One example is that the navigation system gives the user several different routes to choose from, including the shortest route, the least transfer, and so on. In addition, after the retrieval results are obtained in the navigation system, users can "drag" the route on the map to get a more appropriate route. The information of these users "drag" will be sent to the server for saving, so as to make recommendations to other users later [14–16].

In the case of real-time navigation, things get more complicated. For example, when a user drives a private car to travel, the navigation system needs to monitor real-time traffic flow, real-time weather conditions, accident sections and other information, and optimize the path at any time by combining this information, and provide information to the driver through variable information board or voice prompt. This requires a data acquisition system to monitor and control traffic conditions. The collection of these data mainly relies on vehicle detector system, weather station, visibility detector and other equipment to complete.

In this case, the sensor layer is the data acquisition system that we just mentioned. It can collect data and monitor in real time to provide information for the whole system. Communication layer is a communication network, which is used to transmit a large number of data. The data layer integrates all data, combines with real-time traffic information, and proposes the best solution. Finally, the application layer, as a tool to communicate with users, transmits the recommended results to users by voice broadcast.

**Smart Parking.** The smart parking system can effectively solve the city parking difficult problem. In urban parking, the owner only needs to use a smart parking app to realize all the functions needed for parking. The owner can check the vacant parking space in the designated parking lot at any time and make an online reservation. In addition, when the owner enters a destination location, the smart parking system will show him or her all available parking places near the destination and the number of available parking spaces in each location. For car owners who are not familiar with the parking routes, the APP also provides parking lot navigation services to help everyone find available parking spaces. The system also supports the online payment service; the car owner can leave the parking lot directly after the end of the parking, and pay online. This saves time in checkout lines and greatly improves the efficiency of parking lots. The following is the overall structure of the intelligent parking system:

- *Intelligent hardware layer*: The terminal intelligent device is the infrastructure for the operation of the city-level parking management system, providing parking services for users and business data for the platform
- *Network layer*: The network is the communication foundation of the whole Internet. For the on road parking service, one of the core business of urban parking, using IoT network has the characteristics of large connection, low power consumption, low cost and wide coverage, and its features of simple installation and high reliability are also more conducive to the construction and maintenance of on road parking projects.

- *Platform layer*: Platform layer is an open integration platform with multiple API interfaces, which is responsible for access management of network layer information and integration analysis of data.
- *Operation layer*: The operation layer is the core business layer of the urban parking platform, including subsystems such as in-road, off-road, induction, monitoring, member account, payment and settlement, marketing incentives, statistical analysis, operation management, information release, etc. The operation layer must fully realize the integrated business capability of parking business, constructed diversified payment channels, integrated member, payment, settlement, marketing and other links, and realized the closed-loop function of online and offline scenes.
- *Service layer*: Smart parking app, public account, background management system and open interface of resources provide more intelligent, more convenient and more efficient parking management services for car owners, organizations and businesses

**Smart Taxi-Hailing "Didi Chuxing".** An example of smart taxi-hailing is the Didi Chuxing App, which has changed the traditional way of taxi-hailing and established and cultivated a modern way of travelling led by users in the era of big mobile Internet. Compared with the traditional telephone call and roadside call, the birth of didi taxi has changed the traditional taxi market pattern, overturned the concept of roadside stop. Using the characteristics of mobile Internet, it integrates online and offline, from the initial stage of taxi to the use of online payment. It maximizes the passengers' taxi experience, changes the traditional taxi driver and other ways, and makes the driver more comfortable According to the passenger's destination, the master "receives the order" according to his will, which saves the communication cost between the driver and the passenger, reduces the empty driving rate, and saves the resources and time of the driver and the passenger as much as possible.

Users can use the Didi Chuxing app to book trips and use online payment at the end of the trip. In the process of taking a taxi, the user enters the starting point and destination, and the app uploads the user's request to the driver-pickup platform and recommends it to eligible drivers nearby. Those who are willing can grab the order online. After receiving the order successfully, the user can see the driver's car model, license plate number and driver's phone number in the app. The two parties can contact each other through the phone to further communicate about service requirements. At the same time, the user can also know the driver's real-time geographical location, as well as the estimated time of arrival to the starting point.

Didi Chuxing's drivers are not limited to full-time drivers. As long as the owners meet the requirements, they can register as a driver on the platform. The conditions are as follows:

- *Age:* 21–60 for males and 21–55 for females, with small variations depending on the city.
- *Driver's license:* Drivers should have relevant driving license and at least 3 years of driving experience.
- *Drivers need:* no criminal record, no traffic accident crime, no criminal record of dangerous driving, no drug record, no record of drunk driving.

On the one hand, Didi Chuxing enables car owners to make money by taking orders in their spare time; on the other hand, it enables users to reach their destinations quickly and efficiently at a lower price, which is convenient for both parties. This is a typical case of urban mobility in the smart city concept.

**Bike Sharing.** Another typical case is shared bikes. Bike sharing refers to the way in which enterprises provide bike-sharing services in campuses, subway stations, bus stations, residential areas, business districts and public service areas. It is a time-sharing rental model. This is also a new kind of environmental sharing economy.

At present, the existing transportation system cannot fully meet people's travel needs. Even in first-tier cities with well-developed transportation networks, there is still a shortage of short-haul transportation. Taking public transportation or driving private cars is not the most convenient way for people to travel between 1 and 3 km. Because of its small size, easy to park, convenient and quick to use, environmentally friendly characteristics, the bicycle can perfectly adapt to the travel needs. In response to this phenomenon, shared bikes have emerged.

The operation method of renting shared bikes is very simple: users need to pay a deposit of 99–299 yuan and get the right to use the bikes when they log into the APP client through their mobile phone number. Then, the vehicle location information is used to find the parked vehicle nearby, scan the code or enter the body number manually to open the lock, and then start riding. The app will start the timing after the lock is opened. After cycling, users need to park their bikes in a safe position on the roadside and re-lock them. When the user clicks "End of ride" in the app, or when the app recognizes that the vehicle is relocked, the ride ends and the billing ends. Fees are calculated according to the ride time, 0.1 to 1 yuan per 30 min, if less than 30 min will be charged according to the basic price.

Bike sharing offers users an excellent experience due to its convenience, environmental friendliness, economic benefits and high degree of freedom. However, shared bikes are currently facing too little profit. At present, the main source of revenue for bike-sharing companies is the rental of a single bike. If the cost of a bike is 3,000 yuan, renting it four to five times a day with each rental being 1 to 2 yuan, it will take about two years for a bike to recover its cost. The bicycle itself has a short life, which is easy to cause damage. The fluctuation of car demand caused by bad weather and other factors is also inevitable. Therefore, the current profit model of shared bikes cannot generate enough revenue, and enterprises must find a more suitable profit model to continue the development of this sharing economy model.

**Travel for the Disabled.** For the disabled, travel is much more difficult than that of normal people. Normal people only need to consider travel mode, travel route, destination and other information, while disabled people need to consider more information. For example, for blind people, our auxiliary system is their "eyes". The user can activate the navigation by voice. The system receives the voice message, plans the route and announces the voice. The GPS module can locate the starting position and destination position of the blind person, but such positioning is biased and not accurate enough to carry out accurate positioning. Therefore, the auxiliary system should be equipped with an ultrasonic module, which can measure obstacles on the path of the blind through ultrasonic waves and inform the blind of the specific location of obstacles by voice, so that the blind can effectively avoid obstacles.

However, for people with physical disabilities and mobility difficulties, the route planning during travel needs to take into additional consideration whether the site has barrier-free access. For people with limited mobility, going up and down stairs is extremely inconvenient, especially in the case of solo travel, so the need to use the elevator floor movement. Therefore, our system needs to establish a sound message network to obtain timely information about the availability of barrier-free channels, and then give a plan.

Considering that the users may be disabled, our assistive system needs to add a depth assistance mode. When logging into the app for the first time, the user can make a choice and switch the navigation mode at any time. When using depth assistance mode, the system will make more detailed path recommendation for specific types of users, including barrier free navigation, voice navigation and so on.

**Travel Planning.**  In order to enrich the demand of urban mobility, we can also consider the problem of travel planning. It belongs to the part of smart tourism in the structure of smart city, but also has an inseparable connection with urban mobility. By developing a smart travel app, we can provide users with personalized and personalized travel recommendations. When users log in the app for the first time, they can select travel tags they are interested in, such as "Beach" and "Aurora", etc. According to these tags, the platform can recommend destinations and routes to users through the recommendation algorithm. The travel plan obtained by the user includes the location of the play, the recommended time of play for each place, the recommended play items, the transportation to get from one scenic spot to the next, and so on. Users can also upload their own travel routes to the app at the end of a trip, as an alternative route recommended by the app.

Through cooperation with travel agencies, the App can include all kinds of popular travel routes as alternatives. If users determine travel routes through the App, they can enjoy preferential prices in cooperation with the platform. Similarly, the hotel users choose to cooperate on the platform can also enjoy platform discounts. Through this app, users will have a more affordable and convenient travel experience.

In order to make recommendations, the platform needs to use the recommendation algorithm in the field of deep learning. By analyzing the tags the user chooses to be interested in and his or her browsing history, the recommendation algorithm can calculate the similarity between users or between travel items to make recommendations. After a user has planned many trips through the app, the platform can extract the user's travel preferences to build the profile portrait of the user, and then make more accurate recommendations.

## 3.2  Mobility of Goods

Another aspect of urban mobility is the movement of goods. Unlike human movement, the movement of goods is usually passive and based on human will. In reality, the movement of goods needs to be supervised by a special person, that is, the movement needs to be carried out under supervision. Then the route of cargo transport, the way of movement and the planning of the custodians need to be carefully considered. Next, we will analyze several typical cases of the flow of goods, for example, takeaway delivery problems, supermarket order and delivery problems, long-distance mail problems and so on.

**Food Delivery.** With the impact of the epidemic worldwide, people are more and more inclined to enjoy food without leaving their homes. As a result, the food delivery industry has made great progress. The process of producing and finishing a complete takeout order is as follows.

First, the user browses the takeaway app, selects the food he or she wants, submits the order, confirms the address, pays online, and finally places the order successfully. After placing the order successfully, the merchant will receive the order notice. The merchant can choose whether to accept the order or not. If not, fill in the reason for rejection and inform the user that the order has been canceled. If the order is received, the user is notified of the receipt. When the merchant makes food, the system selects the rider who is free within a certain range and sends an order to him or her through the algorithm. If the rider refuses, the next rider in line will be notified. When there is a shortage of riders at the peak of meal, the system will plan the route for each rider to pick up and deliver food, find the optimal solution and inform the corresponding rider to pick up food. When multiple order recipients are close to each other or destination, the system will give priority to recommending the same rider to receive orders to save manpower. After receiving the order, the rider picks up the goods at the store and starts distribution. After the order is completed, the system notifies the user for delivery, and finally the user makes evaluation. During the delivery process, each step will be reported to the ordering user through the delivery app in real time. In the final docking stage between the rider and the ordering user, the rider will contact the user by phone and determine the specific place to pick up the meal.

In this case, we used a synchronous solution to the distribution problem. Due to the limited nature of the food itself, the rider and the ordering user usually meet face to face.

**Supermarket Delivery.** Different from the previous case, when the distribution of goods from food to daily necessities, the distribution mode can also be changed accordingly. In addition to the synchronous solution mentioned in the previous case, which is perfect for this case, we can also use the asynchronous approach to solve the distribution problem. The asynchronous means that in the process of distribution, the delivery personnel and the ordering personnel cannot appear at the same place, face to face, but choose to temporarily ship the goods to a designated location, the delivery personnel will deliver the goods to the designated location and notify the order, who will pick up the goods at their own time.

We are currently working on an auxiliary system based on the sharing economy to address pick-up needs by setting up a special supermarket aisle. The supermarket publishes the goods online for consumers to choose and buy. The consumers submit the order and fill in the expected pick-up time, and then go to the supermarket to pick up the ordered goods in their spare time. Supermarket aisles are usually located near the supermarket and organized to minimize loading time.

The supermarket aisles, which are open 24 h a day, include a supermarket lane and a supermarket sidewalk, which are designed for different needs. Among them, the supermarket lane is designed for consumers driving private cars. Consumers can get the ordered goods without getting off the bus, which greatly reduces the time cost. The supermarket sidewalk is designed for walking consumers, which helps users save a lot of shopping and queuing time [17].

Fresh products require special consideration because they will not last long without being refrigerated. If the consumer can pick up the goods quickly, then only need the merchant to carry out simple short-term preservation treatment, such as vacuum storage, add ice bags, etc. But if the consumer is unable to get the goods in a short time due to objective reasons, then they need to use the refrigerator. Refrigerated cabinets, provided by supermarkets, are specially designed to store fresh products and ensure that consumers can get the freshest goods at any time.

**Parcel Post.** Both of these cases are examples of short-haul freight. Whether it's takeout or supermarket orders, the goods are moving only within a city, or less. In this case, the delivery time of the goods and the pickup time of the order are predictable. But if our goods move to different cities, or even to different countries, the arrival time of goods becomes difficult to estimate in the case of long distance. This situation makes synchronous solutions no longer applicable, and we need to use asynchronous solutions to address these types of requirements.

When a user places an order online and buys an item from a store on the Internet, the merchant confirms the address to the consumer again, and arranges the delivery after confirming it is correct. Each cargo is marked with a unique QR code, which will be scanned and entered into the system when passing through each large distribution point and made available to consumers for inspection.

When the package is about to arrive at the final distribution center, the intermediary platform needs to obtain all the collection points registered in the system near the delivery address and further filter them to finally select the appropriate transporter.

The process of discovering a transporter who is a neighbor of the ordering client (receiver) can be based on different situations. The first is a totally open situation in which the goal is to find and create dynamically an association by checking the proximity of geographical locations between ordering persons and potential transporting persons. The presence of potential transporting persons can be discovered either by explicit declaration (signaling of presence in the shop or shopping mall) or by contextual location detection (by smartphone localization or other implicit identification). In this case, the intermediation algorithm must be able to take into account a huge amount of data.

In a more collaborative situation, parcel transportation can be an interesting functionality of a neighbor association collaborative system or an intergeneration collaborative system in which interpersonal help, information and cooperation are supported. Through this system, its members have access to a list of members and can either pre-establish potential relationships between the ordering person and the transporting person (one or more) or have access to less accurate information providing a list of potential transporting persons. These data can be supplied to the delivery service of the shop or mall. In this case, the intermediation algorithm works on a limited set of data [18].

When the package arrives, the system will send a message to the consumer, reminding him or her to pick it up as soon as possible.

## 4   Deep Learning Approach in Assistive System

In some situations, it seems that deep learning can be used to analyze existing data and lead to interesting solutions. We can take intelligent navigation and travel planning as examples. For intelligent navigation, the intermediation platform considers the starting position and the destination position, combines the time information, meteorological information and other interfering factors, and finally provide the most appropriate one or several recommended paths. As for travel planning, the focus is to intelligently analyze the user's intention of his or her behavior and finally make travel recommendations based on the tags he is interested in and his browsing history.

Deep learning is hugely popular today. The past few decades have witnessed its tremendous success in many applications. Academia and industry alike have competed to apply deep learning to a wider range of applications due to its capability to solve many complex tasks while providing state-of-the-art results [19]. Recently, deep learning has also revolutionized intermediation architectures, providing more opportunities to improve matching performance. Recent advances in deep learning-based intermediation platforms have gained significant attention by overcoming obstacles of conventional models and achieving high recommendation quality. Deep learning catches the intricate relationships within actual data, from abundant accessible data sources such as contextual, textual and visual information.

The classical recommendation system algorithm consists of two parts: one for candidate generation and one for sorting. As shown in the Fig. 3, taking video recommendation as an example and users' browsing history as input, candidate generation network can significantly reduce the number of recommended videos and select a group of most relevant videos from a large library. This generates candidate videos that are most relevant to the user, and then we predict the user ratings and recommend one or more of the highest scores to the user.

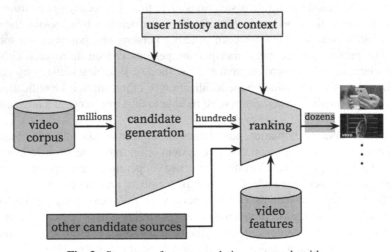

**Fig. 3.** Structure of recommendation system algorithm

First, some shallow distributed representation models are introduced. At present, the shallow distributed representation model has been widely used in the field of text, such as Word2VEc, GloVec, etc.. In contrast to the traditional model, the word-embedding model can map words or other information units (such as phrases, sentences, documents, etc.) into a low-dimensional implicit space. In this implicit space, the representation of each information unit is a dense eigenvector. The basic idea of the embedded word representation model is actually from the traditional "Distributional semantics", which can be summarized as the semantic meaning of the current word is closely related to its adjacent background words. Therefore, the modeling method of word embedding is to use embedded representation to construct the semantic association between the current word and the background word. Compared with the multi-layer neural network, the training process of word embedding model is very efficient, with good practical effect and good interpretability, so it has been widely used.

Corresponding to the neural network model, the most common models include multilayer perceptron [20], convolutional neural network [21], cyclic neural network [22], recursive neural network [23], etc. The multi-layer perceptron mainly uses the multi-layer neuron structure to construct the complex nonlinear feature transformation. The input can be various extracted features, and the output can be the label or value of the target task. In essence, a complex nonlinear transformation can be constructed. A convolutional neural network can be directly deployed on a multi-layer perceptron. The input features of the perceptron are likely to be indefinite or ordered. Through multiple convolutional layers and sub-sampling layers, a vector of fixed length is finally obtained. Cyclic neural network (NN) is a common model used for modeling sequential sequences, describing the correlation of implied states and capturing the data characteristics of the whole sequence. To solve the problem of long-term dependence ("gradient disappearance") on simple cyclic neural networks, which cannot make use of long interval historical information effectively, two improved models, are the long and short time memory neural network (LSTM) [24] and the cyclic unit (GRU) based on gate mechanism. Recursive neural network can be considered as a simplified recursive neural network, which recurses continuously to obtain a sequence representation according to a given external topology.

In practice, we can use RNN structure in path planning, including intelligent navigation, travel for the disabled, Parcel post and so on. And CNN model can be used for vehicle recognition to assist smart parking. For some scenarios with recommendation tasks, such as goods transportation and travel planning, we can use multi-layer perceptron, LSTM and other models [25].

Deep learning-based intermediation has proved very effective. However, the key problem for deep learning in different application scenarios is always data. Once data have been sufficiently collected and tagged, some deep learning-based solutions could be studied and applied.

## 5   Conclusion

In this paper, we mainly study the mobility in smart cities. First of all, we discuss the birth and development of smart cities in today's society. Smart city is a very big concept, which includes a series of fields such as Smart home, Smart Transportation, Smart public, Service and Social Management, smart Urban Management, etc. We cannot discuss each aspect in detail. Therefore, in this paper, we choose the issue of urban mobility for further in-depth discussion, which is also the focus of this paper.

Subsequently, in order to make the mobility of the city "smart", we proposed a framework of assistive system with a four-layer structure, which was concretely transformed into apps with different functions in the final application layer. To put it simply, for every citizen, our assistive system is a collection of a series of apps, that is, a multi-functional intermediation platform, which provides users with the most thoughtful suggestions and the most humanized services for various scenarios, so that people can travel barrier-free [26].

The mobility in cities can be divided into two categories in practice, namely, the mobility of humans and the mobility of goods. In this paper, we enumerate the typical cases in the flow of humans and goods, and analyze the operation principle of these cases one by one. Among them, there are 6 situations of human mobility, and the flow of goods is divided into 3 situations. Some of these examples have already been implemented, and some are the way forward.

Finally, according to the different scenarios that have been analyzed, the algorithm models that can be used for the assistive system are presented, from the shallow distributed representation model to the more complex neural network model.

From the case study, we can see that one of the most basic requirements is the popularity of online payment. Only when online payment is really popularized, can people's life gradually develop towards informatization and digitization. Secondly, in order to achieve various algorithms, data is really indispensable, and in order to obtain data, it needs a series of supporting infrastructure. Under the blueprint of sharing economy, everyone is not only the beneficiary of smart city, but also the provider of data.

**Acknowledgements.** This study is partially supported for Chinese part by the National Key R&D Program of China (No. 2019YFB2102200), the National Natural Science Foundation of China (No. 61977003, No. 61861166002, No. 61872025, No. 61635002), the Science and Technology Development Fund of Macau SAR (File no. 0001/2018/AFJ) Joint Scientific Research Project, the Macao Science and Technology Development Fund (No.138/2016/A3), the Fundamental Research Funds for the Central Universities and the Open Fund of the State Key Laboratory of Software Development Environment (No. SKLSDE2019ZX-04). For French part main supports were by the French Ministry of Ecology (No. 09MTCV37 and PREDIT-G02 CHORUS 2100527197).

# References

1. Bronstein, Z.: Industry and the smart city. Dissent Mag. **56**, 27–34 (2009)
2. Digital Agenda Scoreboard 2015: Most targets reached, time has come to lift digital borders. Website of Digital Agenda for Europe. http://ec.europa.eu/digital-agenda/en
3. Hosaka, T.A.: Japan creating 'smart city' of the future. San Francisco Chronicle. Associated Press, 11 October 2010
4. Ng, P.T.: Embracing emerging technologies: the case of the Singapore Intelligent Nation 2015 Vision. In: de Pablos, P.O., Lee, W.B., Zhao, J.Y. (eds.) Regional Innovation Systems and Sustainable Development: Emerging Technologies. IGI Global, Hershey, pp. 115–123 (2011)

5. Gibson, D.V., Kozmetsky, G., Smilor, R.W.: The Technopolis Phenomenon: Smart Cities, Fast Systems, Global Networks. Rowman & Littlefield Publishers, Boston (1992)
6. Palmisano, S.J.: A smarter planet: the next leadership agenda. IBM (2008)
7. Giffinger, R., Gudrun, H.: Smart cities ranking: an effective instrument for the positioning of the cities? Arch. City Environ. **4**(12), 7–25 (2010)
8. Washburn, D., Sindhu, U.: Helping CIOs understand "smart city" initiatives. Forrester Res. (2010)
9. Su, K., Li, J., Fu, H.: Smart city and the applications. In: Proceedings of IEEE International Conference on Electronics, Communications and Control (ICECC), Ningbo, pp. 1028–1031 (2011)
10. Mitton, N., Papavassiliou, S., Puliafito, A., et al.: Combining Cloud and sensors in a smart city environment. EURASIP J. Wireless Commun. Network. **2012**, 247 (2012)
11. Nam, T., Pardo, T.A.: Conceptualizing smart city with dimensions of technology, people, and institutions. In: Proceedings of 12th Annual International Digital Government Research Conference: Digital Government Innovation in Challenging Times, New York, pp. 282–291. ACM (2011)
12. Yin, C., Xiong, Z., Chen, H., Wang, J., Cooper, D., David, B.: A literature survey on smart cities. Sci. China Inf. Sci. **58**, 1–18 (2015). https://doi.org/10.1007/s11432-015-5397-4
13. Yin, C., David, B., Chalon, R., Sheng, H.: Assistive systems for special needs in mobility in the smart city. In: Krömker, H. (ed.) HCII 2020. LNCS, vol. 12213, pp. 376–396. Springer, Cham (2020). https://doi.org/10.1007/978-3-030-50537-0_27
14. Ziebart, B.D., Maas, A.L., Dey, A.K., et al.: Navigate like a cabbie: probabilistic reasoning from observed context-aware behavior. In: Proceedings of 10th International Conference on Ubiquitous Computing. New York, pp. 322–331. ACM (2008)
15. Li, B., Zhang, D., Sun, L., et al. : Hunting or waiting? Discovering passenger-finding strategies from a large-scale real-world taxi dataset. In: Proceedings of IEEE International Conference on Pervasive Computing and Communications Workshops, Seattle, pp. 63–68 (2011)
16. Sun, L., Zhang, D., Chen, C., et al.: Real time anomalous trajectory detection and analysis. Mob. Netw. Appl. **18**, 341–356 (2013)
17. David, B., Chalon, R.: Box/Lockers' contribution to Collaborative Economy in the Smart City. In; 2018 IEEE 22nd International Conference on Computer Supported Cooperative Work in Design (CSCWD 2018), 11 May 2018, Nanjing (China), pp. 802–807 (2018). https://doi.org/10.1109/CSCWD.2018.8465151
18. David, B., Chalon, R., Yin, C.: Collaborative systems & Shared Economy (Uberization): Principles & Case Study. In: International Conference on Collaboration Technologies & Systems (CTS), Orlando, IEEE (2016) https://doi.org/10.1109/CTS.2016.0029
19. Covington, P., Adams, J., Sargin, E.: Deep neural networks for YouTube recommendations. In: Recsys 2016: Proceedings of the 10th ACM Conference on Recommender Systems, pp. 191–198 (2016)
20. Dziugaite, G.K., Roy, D.M.: Neural network matrix factorization, preprint arXiv: 1511.06443 (2015)
21. Mikolov, T.: Using Neural Networks for Modeling and Representing Natural Languages. COLING (Tutorials), pp. 3–4 (2014)
22. Tang, J., Wang, K.: Personalized top-N sequential recommendation via convolutional sequence embedding. In: Proceedings of the Eleventh ACM International Conference on Web Search and Data Mining, pp. 565–573 (2018)
23. He, X., Liao, L., Zhang, H., Nie, L., Hu, X., Chua, T.-S.: Neural collaborative filtering. In: WWW 2017: Proceedings of the 26th International Conference on World Wide Web, pp. 173–182 (2017)
24. Yang, L., Zheng, Y., Cai, X., et al.: A LSTM based model for personalized context-aware citation recommendation. IEEE Access **6**, 59618–59627 (2018)

25. Huang, P.-S., He, X., Gao, J., Deng, L., Acero, A., Heck, L.: Learning deep structured semantic models for web search using click through data. In: CIKM 2013: Proceedings of the 22nd ACM international conference on Information & Knowledge Management, pp. 2333–2338 (2013)
26. Liu, P., Peng, Z.: China's smart city pilots: a progress report. Computer 10, 72–81 (2014)

# Usability Study of an Innovative Application in Public Transport by Using Hardware-Based Security Technology

Gertraud Schäfer$^{(\boxtimes)}$, Andreas Kreisel, and Ulrike Stopka

Technische Universität Dresden, Dresden, Germany
{gertraud.schaefer,andreas.kreisel,ulrike.stopka}@tu-dresden.de

**Abstract.** As part of the OPTIMOS 2.0 funding project of the German Federal Ministry for Economic Affairs and Energy (BMWi), a partner develops the TicketIssuance app for secure hardware-based storage of high-priced tickets. The app has implemented a previously unknown technology using the Secure Element and the NFC interface. It is therefore imperative to investigate the usability of the app for a successful market launch. For this purpose, user tests of a prototype of the app were conducted using the think-aloud method. This study analyses the results of five tasks. Test subjects rate the expected and perceived difficulty level for each task. That forms the basis for identifying improvement strategies. The test subjects' performance, the frequency of errors and problems encountered, and the need for moderator's support form the basis for prioritizing usability items within the tasks. The developed structure to determine the test tasks' prioritization and usability items, layout, navigation, handling, wording, system, and data economy offers improvements to increase usability.

Furthermore, the study investigates the determination of a suitable sample size for usability testing.

**Keywords:** Usability testing · Think-aloud method · Secure elements in mobile devices · Public transport

## 1 Motivation

The OPTIMOS 2.0 project aims to develop an open ecosystem providing technologies and infrastructure for security-critical services via mobile devices. It focuses on the possibility of secure, standardized hardware-based data storage on the available Secure elements (SE). They are either permanently installed in the mobile device, so-called embedded Secure Elements (eSE), or integrated at the Universal Integrated Circuit Card (UICC). Since a wide variety of mobile devices are offered on the market by different Original Equipment Manufacturers (OEM) and Mobile Network Operators (MNO), integrated hardware components are heterogeneous. OEMs and MNOs, as SE-owners, have to provide access to their SEs. They decide who gets access and provide the corresponding key material storing applets on a SE. Individual service providers need to purchase specific applets in order to gain access to the various SEs. For this purpose,

H. Krömker (Ed.): HCII 2021, LNCS 12791, pp. 105–125, 2021.
https://doi.org/10.1007/978-3-030-78358-7_7

the OPTIMOS 2.0 project develops the Trusted Service Manager (TSM). If OEMs and MNOs join the OPTIMOS2.0-ecosystem, the TSM can manage the access to the different SEs. Service providers get access to the various SEs via the TSM. It acts as the only contractual partner who manages the data or applets stored on all SEs in personal mobile devices.

One main developed application in the project provides the possibility to transfer the personal electronic identity (eID) from the ID card to a personal mobile device securely stored on its SEs. With this technology, it will be possible to transfr verified personal data from the derived eID directly in registration forms without using the physical ID card. Therefore, on the one hand, registration processes can be carried out much faster and more conveniently for customers in a single-step process without manual data entry. On the other hand, service providers using this technology receive verified data minimizing time-consuming verification processes.

Besides, the OPTIMOS 2.0 technology with its central TSM platform offers a wide range of other application scenarios for security-critical applications for mobile services, such as the storage of car keys for car-sharing services, room keys in hotels, or the storage of high-priced tickets in passenger transport.

This study dedicates to the usability of the TicketIssuance app prototype developed for Berlin's public transport operator BVG as part of the OPTIMOS 2.0 project. At the time, the app developers have not yet realized all final application scenarios. One of the primary objectives is to find out how potential users perceive the SE's configuration and the handling of the NFC interface for data transfer to smart cards.

Concerning Berlin's public transportation system, there is a general interest in using apps. In 2018 alone, People downloaded the Fahrinfo app of the Berliner Verkehrsbetriebe (BVG) around four million times [5]. Over a thousand subjects took place in a Germany-wide study surveying the use of public transport apps and electronic ticketing. It shows that about two-thirds of the Berlin study subjects use mobile apps for local public transport, but only 20% had already purchased electronic tickets via apps in 2018. However, 75% of all users see ticket sales as an essential part of public transport apps' functional scope [14]. Therefore, the question is how to design apps so that users increasingly purchase electronic tickets meeting the requirements for fast and secure purchasing and ticket control processes, including convenient payment processing and required data protection to reduce uncertainties and insecurities on the customer side.

New applications need to be developed with a high level of usability to ensure a good user experience. For successful market penetration, developers of innovative applications have to ensure user acceptance, not technical feasibility. Therefore, new apps must be examined about their usability during their development process.

## 2   Study Object - The BVG TicketIssuance App Based on OPTIMOS 2.0 Technology

The Berlin public transport company BVG wants to push the distribution of electronic tickets via mobile apps. The currently available BVG apps offer a selection of the most popular tickets in the lower price segment up to a maximum of 84 euros (monthly ticket Berlin tariff zone AB) by QR code. It is relatively easy to copy. Therefore the

BVG faces the risk of service fraud through duplicated tickets. To offer higher-priced tickets, such as annual or monthly once even in the broader tariff zone Berlin ABC, and to use NFC technology advantages for checking tickets, one OPTIMOS 2.0 project partner develops the TicketIssuance app. The app developers are in intense competition with other ones. Developers must be aware of how users can quickly and effortlessly substitute their app due to other ones. The YouGov report 2017 [28] surveyed that 89% of 2000 study subjects had already deleted apps at least once. The most common reason was an uninteresting or disappointing app performance. Therefore, before introducing the TicketIssuance app, it is crucial to ensure usability to achieve user acceptance through a positive user experience and reduce possible uncertainty when purchasing tickets.

As already described, the TicketIssuance app should offer the option to store tickets on the SE in the smartphone. This option requires that the integrated Smartphone SE supports the open OPTIMOS 2.0 interfaces and is addressable via the OPTIMOS 2.0 TSM. In summer 2020, only Samsung's Galaxy smartphones from model S9 and higher were ready to be part of the OPTIMOS 2.0 ecosystem. Other manufacturers do not give access to the SEs on their mobile devices up to now. The same applies to the MNO UICC.

The TicketIssuance app must be able to address the SE. Therefore, the first step is installing the OPTIMOS 2.0 TSM-API app for secure communication between the ticket provider and the SE on the personal smartphone. The one-time initialization process automatically downloads, installs, and personalizes the SE applet as a prerequisite for storing higher-priced tickets in the SE in a forgery- and copy-proof manner.

The checking of tickets stored in the SE occurs via NFC technology and works even on the smartphone's standby or low battery mode. In this way, public transport users can prove their ticket's validity even if the battery level is insufficient. The checking process is similar to one of the tickets stored on smartcards. This technology also allows storing tickets directly to appropriately configured smart cards via the TicketIssuance app. It is also possible to transfer not to copy tickets between SEs of different mobile devices or to a smart card via NFC. This function offers customers the possibility to buy tickets for other persons or to transfer already purchased tickets. One use case could be the purchase of electronic tickets for children or persons using smartcards. The study's prototype app provides low-priced tickets to store as QR code on the internal smartphone storage or a corresponding smart card and higher-priced tickets, such as monthly tickets or the so-called field test ticket, on the smartphone SE or a corresponding smart card.

In its final version, the TicketIssuance app will also provide users with a quick and easy way to personalize a user account by electronic transfer of personal data from the derived digital identity stored at the SE.

By integrating the OPTIMOS 2.0 technology, the final version of this app will offer the following functions:

1. Creation of user accounts with the integration of the eID stored in the smartphone
2. Purchase includes electronic payment and management of different ticket options, exceptionally high-priced season tickets such as monthly and annual tickets, to store securely on the SE.
3. Ticket transfer via the TicketIssuance app to an external smartcard or another smartphone.

4. Checking Ticket stored on the SE via the NFC interface - touching the smartphone to a checking device without opening the app-similarly the smartcards' ticket check; Check is also possible in switched-off mode or when the battery is low.

The usability test study object is the ticket purchase and storage as a central function of the TicketIssuance app prototype. However, subjects have to configure the SE and create and personalize a user account in the first step. The storage options will investigate whether the functions of managing tickets (purchase, show/check, delete) are easy to understand and perform. In particular, the study examines the different possibilities of ticket storage.

- Internal at the app of the smartphone,
- At the SE or
- At the smartcard.

## 3   Usability Testing Methods

Usability testing is "[…] activities that focus on observing users working with a product and performing tasks that are real and meaningful to them" [4]. Depending on the development of an application product, we can distinguish between formative and summative evaluations. The formative evaluation takes place as relatively small studies to identify fundamental usability problems and understand user requirements during the product development process. They provide developers customer-oriented input for the further development process. The focus is on questions such as:

- What works well for the user and what works poorly?
- What are the most critical problems use an application in terms of usability?

After the necessary adjustments to the product, a new evaluation usually occurs. Summative tests take place with a larger sample after completion of the product development. Here it is necessary to ensure the required statistical validity in order to be able to make reliable statements meeting the defined goals, such as a certain degree of efficiency, specific customer expectations, like error frequency, and the time required to fulfill the task or the comparison with competitor products [2, 4, 6]. Usability tests can occur in a laboratory, in a specific room, or under real conditions in the field, such as shopping malls, parks, vehicles, or the customer's home [4].

Various methods are possible to gather information about product characteristics. Card sorting - a participatory design method – is an often applied method in an early stage of product development, collecting user understanding and preference, for instance, developing a user-friendly structure and navigation of a software application [4].

The heuristic evaluation method takes a different approach. A group of usability experts independently tests a software product using defined simple and general criteria (heuristics) to uncover usability problems. The experts evaluated and prioritized their results jointly for processing [4]. One basis is the ten usability heuristics established by Nielsen in 1994 [17].

The think-aloud method can give a good insight into customer experiences in a more advanced application development stage. Test subjects, usually potential users, test the application and think aloud. Barnum summarizes it as follows: "... thinking out loud provides a rich source of information about the user's perceptions of the product's usability" [4]. Nielsen describes it as one of the essential methods for assessing usability [21]. It is also the applied method in this study.

## 3.1 Think-Aloud Method

Test subjects speak their thoughts aloud while performing the tasks. The think-aloud method follows synchronous verbalization of cognitive processes while performing a specific task performance. Researchers can observe and record mental processes affecting the test during actions or products' use through communication.

Compared to retrospective verbalization, the advantage is mainly the data's consistency and completeness [8]. The data collected are a snapshot regarding the subject- or customer-specific perception using a product or an application [7]. Deep insights into the subjects' problem-solving behavior become apparent. Thus, it becomes apparent which usability problems occur [9]. According to Henry et al. [13], thinking aloud does not affect performance levels. Alhadreti and Mayhew [3] also conclude that findings after analyzing three think-aloud studies.

According to Ericsson and Simon [8], there are different forms of verbalization.

Level 1 verbalizations are expressions as they occur in self-talk. Since they do not require any cognitive processes, there is no significantly higher effort and time requirement than situations without communication. In this way, the thoughts and information articulated are essential in detecting problems and errors in an application. They are expressed spontaneously and are not known to the subject in advance [8].

Level 2 verbalizations represent an extension of the articulated information from level 1, in that thoughts occurring intuitively in short-term memory are additionally explained and described. In this process, the subject converts thoughts into words. Due to the increased processing time involved, task performance may take longer during the test situation. It does not disturb the general course of the processing and influence the success of the task. It is to assume that the data collected for Level 2 verbalization are reliable [8].

Level 3 verbalizations reproduce thoughts themselves and explanations concerning the cognitive processes (e.g., behavioral descriptions or motives). Level 3 verbalizations link individual thoughts and memories, which change the process structure and lead to a change in the performance and correctness of the results and an increased time requirement [8].

Generally, the test design should formulate the think-aloud method's tasks on level 1 and level 2 verbalizations. Accordingly, test subjects should only think aloud and not explain. They should behave as if they were alone in the room. Before starting the test, the moderator should also point out that he will invoke an invitation to think aloud in the phase of prolonged silence. Any distraction should not occur to avoid level 3 verbalizations.

## 3.2  Determination of the Sample Size for a Usability Test

The question of the required sample size to determine a specific proportion of errors and problems in the context of usability testing is a much-discussed topic in practice and theory. Nielsen/Landauer developed the widely used and simplified model that "[...] is sufficiently accurate for practical purposes" [19]. The formula (1) calculates the number of errors detected by at least one of the subjects.

$$Found(i) = N(1 - (1 - \lambda)^i) \tag{1}$$

The result depends on the total number of errors N and the proportion of all usability problems a tester will detect ($\lambda$) [19]. Thus, researchers have to estimate the number of errors to determine a suitable sample size. Nielsen [20] assumes that one test subject reveals 31% of the errors, and 15 test subjects almost detect all errors. According to this calculation, three to four test subjects can perceive more than half of the errors.

In case the usability test is part of an iterative design process with regular test procedures for the adjustment and advancement of the product, a sample size with five test subjects gives the optimal cost-benefit ratio in practice. This sample size already considers that one or two people will not appear for the test even having agreed to do so [6, 20].

Virzi [27] supports these assumptions by the finding that four to five test subjects detect about 80% of the problems. It illustrates that as the number of test subjects increases, the probability of uncovering new information steadily decreases. Furthermore, it assumes that the first test subjects already detect problems that strongly influence the ease of use. The number and level of difficulty of newly uncovered information steadily decrease with increasing sample size.

Usability researchers question and criticize these theories from 1990 to 2000, especially when examining applications with increasing complexity. According to this theory, websites or applications have many functions and operating options, where users have different options at their disposal to achieve their personal goals. It can also influence the proportion of errors ($\lambda$) that a tester can uncover. For example, in their study of four different e-commerce websites, Spool/Schroeder [25] could not identify a $\lambda \geq 0.16$ in each case. According to Nielsen's model and this measure, a sample size of at least ten test subjects can detect 80% of the problems and errors. The results correspond with the study of Lewis [16], testing different office systems, according to which the average probability for the uncovering of an error lies with 0.16.

The different findings show there are not any binding statements for an optimal sample size. According to Cockton, this also depends on the following factors [6]:

- "[...] diversity of subjects,
- Test protocol design,
- Variety of task performance,
- Complexity of application,
- Design quality [...],
- Problem reporting procedures,
- Usability goals [...]".

# 4   Study Design

The study design follows the steps visualizes in Fig. 1.

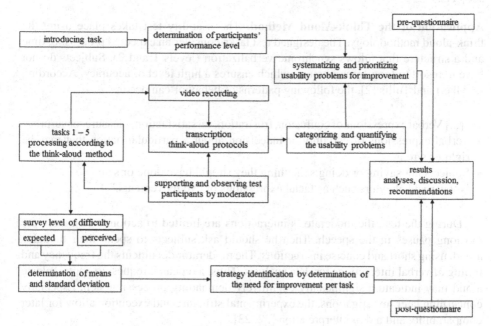

**Fig. 1.** Study design.

## 4.1   Workshop Design for Usability Testing of the TicketIssuancce App

**Determination of the Evaluation Procedure.** The TicketIssuance app for BVG to be tested is in the development phase. So far, up to now, no tests with potential users have taken place. This usability study follows the formative evaluation approach.

**Determining the Test Environment.** A standard method for usability tests following the formative evaluation approach is the laboratory test. For the present study, using a real test laboratory of the BVG is not possible. Nevertheless, the premises at BVG provide a reference to the company and offer a controlled environment for observing the test subjects' behavior during working on the structured tasks. Besides, supplementary data collection methods are used before, during, and after the experiment for extended analysis of user behavior and user requirements.

**Determining the Sample Size.** According to formula (1), an exact determination of the influencing variables (especially λ) for calculating the sample size is not feasible. No data is available so far since this is the first usability test. The usability test of the TicketIssuance app should reveal as many problems as possible and provide information for improvement.

The decision is to recruit 20 test subjects. It bases on Nielsens' assumption that a heterogeneous target group, such as public transport services users, will be covered by larger samples [20]. This number also takes into account the non-appearance of invited subjects.

**Application of the Think-Aloud Method.** The usability test takes place using the think-aloud methodology. The designed test tasks consider an expected processing time and a structure that allows synchronic verbalization (levels 1 and 2). Subjects do not have to resort to long-term memory, which ensures a high level of accuracy. According to Albert and Tullis [2], the following patterns of behavior can occur:

- "[...] Verbal expressions of confusion, frustration, dissatisfaction, pleasure, or surprise.
- Verbal expressions of certainty or indecision about a particular action that may be right or wrong
- Subjects not saying or doing something they should have done or said
- Nonverbal behaviors such as facial expressions or eye movements."

During the test, the moderator's interactions are limited to actions when there are too long pauses in the speech. Then he should ask subjects to speak their thoughts aloud, using short and concise instructions. The moderator documents the frequency and timing of verbal interactions for data analysis. Thus, a repeated request to communicate aloud may indicate a significantly cognitively demanding processing step. Especially concerning later investigations, the experimental structure and execution allow for later comparability and aid in interpretation [22, 23].

A test design with the think-aloud method shows apparent differences compared to everyday social communication. The test moderator receives instructions in a created moderator script for stringent execution. He explicitly points out that the object of investigation is the app itself and the associated recognition of problems using the app, not the subject's abilities.

**Design of Questionnaires.** A short standardized questionnaire before and after the usability-test of the app collects a few supplementary data.

Pre-questionnaire. Test subjects should get a comfortable and pleasant introduction to the unfamiliar test situation. Therefore, the pre-questionnaire serves to collect the sample's socio-demographic data and data on the previous use of BVG services. Furthermore, these data are the basis for describing the sample description and the result's analyses. Age cohorts base on empirical studies from the field of transportation for comparability [11].

Post-questionnaire. The post-questionnaire collects some supplementary information after the usability test's task processing has been completed. One criterion for evaluating usability is satisfaction. A questionnaire can best capture this [12]. Thus, one question focuses on the satisfaction of the tested app's usability using a five-point scale from one (very satisfied) to five (very dissatisfied). Other questions elicit the importance of using different media for electronic ticket storage and the subject's knowledge about the NFC interface of the personal smartphone.

**Test Tasks.** The concept of the tasks takes into account that no results from usability tests are available so far. There are no known indications where usability difficulties will exist. Therefore, the test tasks follow real future use cases.

The introducing task determines the test subjects' skills and abilities in dealing with mobile apps. For this task, subjects use their smartphones to determine their performance level. The procedure assumes that individuals with a comparatively low-performance have a less common approach to using apps and uncover a higher number of problems or errors. The basis for determining the performance level is the time required in each case.

In task 1, the test subjects get the instruction to install the app on the test smartphone. To do this, they must open, download and install the TicketIssuance app. Task 2 creates the precondition for ticket purchase and storage. Test subjects have to configure the app by initializing the SE in the first step. It is the prerequisite for the storage of high-priced tickets in the smartphone. In a further step, they create a user account and personalize it via manual data entry. With task 3, the test subjects are to purchase various tickets and store them on different storage media - in the internal database memory, on the SE, or the smartcard. Task 4 requires test subjects to retrieve the purchased and stored tickets, show them for a potential control operation and delete them afterward. In task 5, they shall remove the user account, uninstall the SE and delete the app.

### 4.2   Survey and Evaluation Methodology

**Identification of Strategies by Determination of the Need for Improvement per Task.**
One usability factor is satisfaction, which can vary widely based on different expectations and perceptions. In determining satisfaction, the confirmation/disconfirmation paradigm helps. It assumes that people compare the perceived performance level after using an application with their expectations before using it [1]. If the level of perception is above the expectation level (positive disconfirmation), this leads to a positive perception of quality and satisfaction. Dissatisfaction (negative disconfirmation) occurs if the perception level is lower than the expected one. If both levels are equal, there is neither explicit satisfaction nor dissatisfaction. Against this background, the test subjects indicate for each task how difficult they expect this task to be or how difficult they perceived the processing to be [1, 15].

There are four derived strategies, depending on the average values of expected and perceived level of difficulty. The strategy "No Modification" implies that users expect and perceive simple task handling. Thus, there is, for the time being, no need for modification. They will not result in an increased quality perception or customer satisfaction. The strategy "Improve Immediately" contains tasks for which the test subjects expected significantly easier processing than was possible in the end. There is a high potential for improvement in these tasks. It needs a priority effort to correct these identified errors and problems. If persons expect many difficulties and perceived much less, then these tasks can attract customers and lead to high satisfaction. These tasks belong to the strategy of "Advertise". They are starting points for communication with potential customers, particularly in the market penetration phase, highlighting potential benefits. However, expectations will change with frequent use of the app if users can draw on their experience. For tasks in the fourth strategy "Good Opportunity" the expected and perceived level of difficulty is generally higher than for the other tasks. Improving usability can lead to a more positive perception of quality and increased satisfaction. In this case, users will also adjust their expectations based on experience with the application and similar products. The individual strategies' delimitation uses the median of the values collected, assessing the expected or perceived level of difficulty based on a five-stage scale (1 = very simple to 5 = very difficult) [1].

Test subjects have very different abilities and skills in using apps, on the one hand, regarding smartphone use and, on the other hand, how often they use which apps for specific tasks, e.g., ticket purchase, connection search for public transport services. When interpreting the usability test results, the test subjects' performance levels can be significant. This determination of performance levels uses a standardized test based on completing the introducing task. The average of the individual times is the norm time. It is the basis to determine the test subjects' performance levels concerning their respective personal performance for classification and comparison of individual abilities [26].

Frühwald [10] calculates the degree of performance as follows:

$$performance\ level = \frac{norm\ time}{observed\ individual\ time} \times 100\ [\%] \qquad (2)$$

However, this norm-oriented test does not claim to define subjects' performance concerning their total smartphone use and handling ability.

**Systematizing and Prioritizing Usability Problems for Improvement.** The perceived problems during the test are the basis for recommendations improving the app's usability. One method is a discussion round in the test team, which determines the necessary adjustments' problem prioritization [4]. Even usability experts can classify the identified problems. Both methods have subjective influences depending on the diverse personal experiences, affecting the quality of the results [18].

The present study uses an own developed methodology based on Nielsen [19] to prioritize problems by importance. It considers two dimensions. The first one incorporates the average performance level scores of subjects who perceive the problem and the detected problems' frequency. A strong influence exists if the perceived problem occurs with above-average frequency during task processing or if the subjects' average performance level is low. The second dimension considers the possibility of independent task processing. A strong influence exists if the test subject needs the moderator's support for a task processing successfully.

Table 1 categorizes the collected problems and systematizes them by priority for improvement.

The basis for quantification and prioritization are the defined design areas (1) navigation, (2) layout, (3) handling, (4) wording, (5) system, and (6) data economy, to which the specific problems are assigned.

**Correlation Between Number of Errors and Number of Interactions.** In the context of usability testing by the think-aloud method, the test moderator should interact with the test subjects as little as possible. High usability does not need any intervention of the moderator. However, test users should detect as many usability problems as possible. The

**Table 1.** Systematization of the surveyed usability problems according to the importance for improvement.

| Dimension 2: Influence of problems or errors on the self-reliant task processing | Dimension 1: Influence of the test subjects' performance level plus the occurred error frequency | |
| --- | --- | --- |
| | **High** Low average performance level of test subjects or high frequency of errors | **Low** High average performance level of test subjects and high frequency of errors |
| **High** Successful task processing not possible without moderator's support | **Priority 1** Detection of errors very important Need action immediately | **Priority 2** Detection of errors important Need for remedial action before introduction into practice |
| **Low** Self-reliant successful task processing possible without moderator's support despite usability problem | **Priority 3** Detection of errors preferable | **Priority 4** Detection of errors after fixing the others |

tasks follow a workflow of real use. Therefore the test subjects have to finish each. This goal can sometimes require the intervention of the moderator. There is an expectation of a positive correlation between the number of problems and the number of necessary moderator interactions related to the individual tasks.

This analysis uses the empirical correlation coefficient according to Bravis and Pearson as a measure of linear correlation as well as the rank correlation coefficient of Spearman, which measures the monotonic correlation of two variables [24]. The correlation coefficients will help the result's analysis and interpretation.

# 5 Study Results

## 5.1 Sample Description

Of the 20 selected and invited test subjects, eleven appear. That corresponds to the minimum number of test subjects for uncovering more than 80% of the problems and errors. The group is sufficiently homogeneous. At least one subject from each of the four age cohorts participates. Most subjects belong to cohorts between 25 and 64 years. All test subjects are regular users of public transport services and use different types of tickets. Thus, they could draw on very different experiences and have different expectations regarding the purchase of tickets. The same applies to the experience of using apps for public transport. Almost all subjects know the four different BVG apps. However, only a few use them frequently or occasionally.

## 5.2 Analysis of the Test Task

The improvement of usability serves to achieve broad user acceptance and high user satisfaction. The expected and perceived levels of difficulty of the five test tasks form the basis for positioning them in the developed matrix. The subdivision of the two coordinates bases on the median by all 55 ratings from the eleven subjects. The respective average mean values determine the positioning of the tasks into the strategy fields (cf. Table 2, Fig. 2).

**Table 2.** Expected and perceived level of difficulty per task.

|  | Level of difficulty (1 = very simple ... 5 very difficult) | | | |
|  | Average mean | | Standard deviation | |
|  | Expected | Perceived | Expected | Perceived |
| Task 1 | 1,18 | 1,27 | 0,4045 | 0,6467 |
| Task 2 | 1,73 | 2,36 | 0,6467 | 0,9244 |
| Task 3 | 2,77 | 2,27 | 0,9840 | 0,7862 |
| Task 4 | 1,55 | 2,27 | 0,6876 | 0,6467 |
| Task 5 | 2,00 | 2,00 | 0,7746 | 0,8944 |

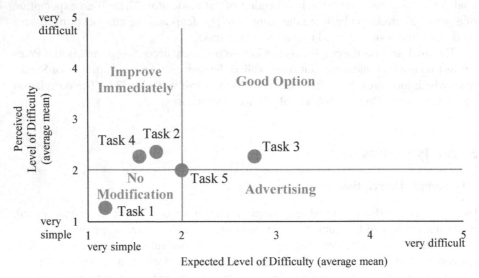

**Fig. 2.** Strategy identification by determination of the need for improvement per task.

**Task 1.** The test subjects do not have any particular difficulties with the installation of the app. Therefore, the task assignment falls to the strategy "No modification" (cf. Fig. 2). The mean values of the expected and perceived difficulty levels are the smallest compared to the other tasks. The likewise low standard deviations show a relatively homogeneous evaluation. In total, there are seven interactions between the test moderator and subjects. The test moderator intervenes once directly to ensure successful task completion. The tests identify one problem in the fields of action layout and wording (cf. Table 3). For test task 1, in general, there is no prioritized need for action.

**Table 3.** Task 1 - detected usability problems.

| Dimension 2: Influence of Problems or Errors on the Self-reliant Task Processing | Dimension 1: Influence of the test subjects' performance level plus the occurred error frequency | | | |
|---|---|---|---|---|
| | High | | Low | |
| | Field of action | Problems | Field of action | Problems |
| High | Priority 1 | | Priority 2 | |
| | | | Layout | 1 |
| Low | Priority 3 | | Priority 4 | |
| | | | Wording | 1 |

**Task 2.** Test task two involves processing steps to create the prerequisites for app use. That includes setting up and personalizing a user account and configuring the SE for storing high-priced tickets.

The observed usage behavior is quite different from task 1. The tests detect a total of 21 usability problems in all six fields of action. In total, there are 58 interactions between the test moderator and subjects. Direct intervention by the test moderator for successful task completion is necessary in 13 cases.

These results also reflect the ratings on the expected and perceived difficulty level. That leads to the positioning of the task in the matrix field "Improve immediately" (cf. Fig. 2). The test subjects expect the task processing to be significantly easier before the test than they perceive them during the test. The expected difficulty level's measurement scatters less around the mean with 0.65 than the perceived one's values with 0.92. The expectation is more homogeneous than the perception. The processing steps to be performed in task two involve the SE configuration, a process unknown to the test subjects. They have no previous experience or operating analogies to fall back on. However, for the app's full use, the users must configure the SE and set up a user account. The fixing of discovered usability problems should be prioritized according to Table 4 and corrected as quickly as possible to avoid acceptance barriers.

**Task 3.** Task three is dedicated to the purchase of tickets and their storage on different storage media. Storage on the different media via an app is not yet practice. Therefore, the test subjects expect some problems with these unfamiliar actions. However, as a result, subjects do not perceive the processing to be as challenging as expected. That leads to the positioning of the task in the matrix field "Good Option" (cf. Fig. 2). The expected

**Table 4.** Task 2 - detected usability problems.

| Dimension 2: Influence of problems or errors on the self-reliant task processing | Dimension 1: Influence of the test subjects' performance level plus the occurred error frequency | | | |
| | High | | Low | |
| | Field of action | Problems | Field of action | Problems |
| **High** | **Priority 1** | | **Priority 2** | |
| | Layout | 1 | Layout | 1 |
| | Navigation | 2 | | |
| | Handling | 1 | | |
| | Wording | 1 | | |
| | System | 1 | | |
| **Low** | **Priority 3** | | **Priority 4** | |
| | Layout | 3 | Layout | 3 |
| | Navigation | 1 | Navigation | 1 |
| | Handling | 2 | Handling | 1 |
| | Data economy | 1 | Wording | 3 |

difficulty level's measurement values scatter significantly more around the mean with 0.98 than the perceived one's values with 0.79. The more homogeneous evaluation of the perceived difficulty level underlines a positive user experience. That offers potential for external customer communication. The new ticket storage options can lead to a positive user experience.

In total, there are 43 interactions between the test moderator and subjects. In 15 cases, the test moderator had to intervene for successful task completion. The tests detect a total of twelve usability problems in the layout, wording, and handling fields of action. The fixing of the problems should be prioritized according to Table 5 and corrected to ensure a good user experience.

**Task 4.** Task four involves the processing steps of showing and deleting tickets. These are usage scenarios implement with different ticket types and storage media. On average, test subjects expect easier task processing than they perceive at the end, both on a lower value than in test task two. That leads to the positioning of the task in the matrix field "Improve immediately" (cf. Fig. 2). The measurement values of the expected difficulty level of 0.69 scatter around the mean value in a similar range as the values of the perceived difficulty of 0.64.

In total, there are 30 interactions between the test moderator and subjects. In seven cases, the test moderator had to intervene for successful task completion. The tests detect eleven usability problems in the layout, wording, and handling fields of action. The usability of these use cases, especially handling the smartcard and the layout and handling to delete tickets, must be improved before launching the app on the market. The fixing of the problems should be prioritized according to Table 6 and corrected to ensure a good user experience.

**Table 5.** Task 3 - detected usability problems.

| Dimension 2: Influence of problems or errors on the self-reliant task processing | Dimension 1: Influence of the test subjects' performance level plus the occurred error frequency | | | |
| | High | | Low | |
| | Field of action | Problems | Field of action | Problems |
| **High** | **Priority 1** | | **Priority 2** | |
| | Layout | 1 | Wording | 2 |
| | Handling | 1 | | |
| | Wording | 1 | | |
| **Low** | **Priority 3** | | **Priority 4** | |
| | Layout | 1 | Layout | 1 |
| | Wording | 1 | Wording | 2 |
| | | | System | 2 |

**Table 6.** Task 4 - detected usability problems.

| Dimension 2: Influence of problems or errors on the self-reliant task processing | Dimension 1: Influence of the test subjects' performance level plus the occurred error frequency | | | |
| | High | | Low | |
| | Field of action | Problems | Field of action | Problems |
| **High** | **Priority 1** | | **Priority 2** | |
| | Handling | 2 | Layout | 2 |
| **Low** | **Priority 3** | | **Priority 4** | |
| | Navigation | 2 | Layout | 2 |
| | | | Navigation | 1 |
| | | | Wording | 2 |

**Task 5.** The expected difficulty level of deleting or removing the user account from the device and uninstalling the app shows is the same value as the perceived one on average across all subjects. Therefore, task five lies precisely at the intersection of all four strategies (cf. Fig. 2). In total, there are 25 interactions between the test moderator and subjects. In five cases, the test moderator had to intervene for successful task completion.

The expected difficulty level values scatter around the mean of 0.77 and those perceived at 0.89 m, both at a relatively high level. These inconsistent ratings concerning the expected and perceived level of difficulty indicate the potential for improvement concerning the ten detected usability problems in fields of action: layout, navigation, wording, handling, and system. The fixing of the problems should be prioritized according to Table 7 and corrected to ensure a good user experience.

**Table 7.** Task 5 - detected usability problems.

| Dimension 2: Influence of problems or errors on the self-reliant task processing | Dimension 1: Influence of the test subjects' performance level plus the occurred error frequency | | | |
|---|---|---|---|---|
| | High | | Low | |
| | Field of action | Problems | Field of action | Problems |
| High | Priority 1 | | Priority 2 | |
| | Navigation | 2 | Layout | 1 |
| | System | 1 | | |
| Low | Priority 3 | | Priority 4 | |
| | Layout | 1 | Layout | 1 |
| | Navigation | 2 | Wording | 1 |
| | Handling | 1 | | |

**Correlation of the Number of Detected Usability Problems and the Moderator's Interactions.** A further step examines the correlation between the test moderator's number of interactions and the usability test's problems. Following the think-aloud method's methodological principles, the moderator should reduce his interactions to a minimum to enable independent task processing.

The analysis of the test results for the TicketIssuance app shows that most interactions (average 5.3 per test subject) between test moderator and test subjects took place in task two. The fewest interactions occurred in Task one, with an average of 0.6 per test subject. The number of interactions in tasks 4 and 5, with an average of 2.7 and 2.3, are similar. In task 3, there is a significantly higher number of interactions with an average of 4.3 per test subject.

The examination of the correlation concerning the five test tasks via the correlation coefficient of Bravis and Pearson results in a correlation coefficient of 0.9438. It shows a strong linear correlation between the average number of interactions per subject and the number of errors detected. The Spearman's rank correlation coefficient with the value of 1.0 shows a robust monotonic correlation, i.e., the more problems or errors occur, the more interactions are required.

**Results of Post-questionnaire.** The post-questionnaire responses show that four of the eleven test subjects are very satisfied (rating 1), and seven of them are satisfied (rating 2) with the use of the app.

The test subjects tend to attach greater relevance to the option of saving tickets on the smartphone than on the smart card. Nine out of eleven test subjects rate the option of saving tickets on the smartphone as very important. However, most test subjects also consider the additional option of using an external smart card to be an essential alternative. Four mentions, each of very important and important, clearly show this.

The prerequisite for saving tickets via the app on the SE and an extern smart card is an NCF-enabled smartphone. Seven test subjects state that they have smartphones equipped with NFC; two do not know, and two others answer this question negatively. The test subjects should indicate whether they regularly use mobile payments (e.g.,

Apple Pay, Samsung Pay), verifying NFC-equipped smartphone knowledge. Five test subjects answered this question in the affirmative, with one of these subjects stating that they do not own an NFC-enabled personal smartphone. It shows that customers often use mobile services as a matter of course without knowing the technical background systems and their devices' technical features.

# 6  Discussion

## 6.1  Implemented Measurement and Interpretation Methods

The subjects did not get information about the recording of the time taken to complete the introducing task. These times were the basis for determining each subject's performance level. Clarification about this could have potentially altered the results. The premise was that the test subjects get a practical and straightforward introduction to the test method. The recorded video material analysis from the individual tests formed the basis for determining the number of interactions between test subjects and the moderator and deriving the categories for problems and errors. On the one hand, the researchers' assessment may have led to measurement errors. On the other hand, there is room for interpretation in the problem analysis so that the results are not free of subjective influences.

## 6.2  Usability Tests Results

In total, eleven test subjects uncovered 56 usability problems and errors in the five test scenarios.

The developed evaluation methodology provides the framework for assigning the collected problems and errors to prioritize the required adjustments categorized in handling, wording, layout, navigation, system, and data economy.

In particular, there is a prior need to adjust the handling and navigation problems. Here, the test moderator often had to interact and intervene to complete the tasks successfully. It was challenging to find central functions such as initializing or configuring the Secure Element or selecting the various storage options. At present, this option is mainly the app's unique selling point and should therefore be intuitive and error-free to use. Potential users are not yet familiar with corresponding operating steps from other applications. It means that they cannot fall back on operating routines. Besides, there is a need for improvement in the layout and wording. The terminology used and the language used for explanations and notes have led to irritation.

## 6.3  Influence of Interactions by the Test Moderator

The calculated correlation coefficients show a strong correlation between the number of average interactions between the test subject and test moderator and the determined number of usability problems and errors per task. On the one hand, this may be since the test moderator influenced the problem and error detection due to unintentional suggestions. On the other hand, the results may also be due to the moderator's instructions. It required mandatory intervention in specific test scenarios to complete all the tasks successfully.

### 6.4 Sample Size

The usability test results make it possible to draw concrete conclusions about a suitable number of test subjects who uncover a certain proportion of the existing usability problems and errors. The following formula (3) calculates the proportion of usability problems uncovered by a tester ($\lambda$).

$$\gamma = 1 - \sqrt[i]{1 - \frac{Foud\ (i)}{N}} \tag{3}$$

Thereby, for the present study with 56 uncovered problems by eleven test subjects, an average $\lambda$ of 0.266 results. Accordingly, each test subject uncovered about 27% of all problems or errors during the usability test. This percentage is about five percentage points lower compared to Nielsen [20]. As Nielsen [20] and Virzi [27] specify, a sample size of five test subjects can elicit about 79% of the problems. The further correlations regarding the influence of the number of test subjects on the percentage of detected errors illustrate Fig. 3. Also shown is the influence at $\lambda = 0.176$, calculated based on two randomly selected test subjects, on the proportion of errors detected as a sample size function. It shows that together they uncover just under two-thirds, i.e., 18 out of 56 problems or errors. In the worst-case scenario, five subjects could find (62%), and eleven subjects could find 49 (88%) problems and errors.

The largest $\lambda$ (=0.385) calculation is possible with three randomly selected test subjects who uncover 43 of the total 56 items. In this best-case scenario, five subjects uncover 91%, and eleven subjects uncover 97% of the critical points.

For the TicketIssuance app usability study, this means that eleven subjects uncovered 88% to 97% of all problems or errors.

This calculation approach can determine the required number of test subjects for further usability tests to improve the TicketIssuance App. If further testing occurs as part of an iterative design process, it may be advantageous to recruit fewer subjects

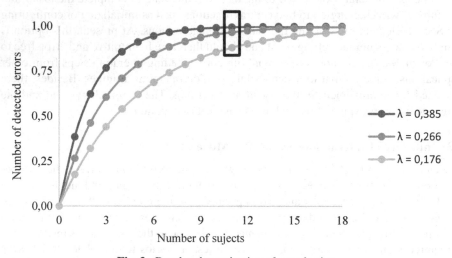

**Fig. 3.** Results: determination of sample size.

considering the cost-benefit ratio. According to the worst-case scenario, as few as five ones can uncover more than 60% of all problems and errors. However, if only one more usability test of the app will occur, twelve testers should uncover at least 90% of all problems. Ongoing technological developments usually require continuous adjustments to the app. New problems may arise and require usability testing to ensure user acceptance and a good user experience.

## 7 Conclusion

The OPTIMOS 2.0 project develops the TicketIssuance App for storing high-price tickets on a high-security level. For this purpose, the app offers new, not yet known functions. The study tested the usability of those new functionalities.

All study subjects completed test tasks. Sometimes the test moderator had to give support. Study subjects perceived the process of ticket purchase easier than expected before testing this function. The satisfaction with the new features for buying tickets and saving them on different external media like a smartcard was very high. Potential users very welcome the possibility of checking smartphone tickets in low battery mode. It was possible to elicit many indications and recommendations for prioritizing of redesigning, particularly regarding the layout, wording, navigation, and handling. The results of the usability tests give suitable bases for further adjustments. There is no need for changing the conception or design of the TicketIssuance app fundamentally, but further tests seem to be necessary. A field test with a larger user group should examine the functional capability of the implemented OPTIMOS 2.0 technology in connection with user-friendliness. Particular attention should lay to the handling of SE configuration and uninstalling processes. The goal must be to ensure a high acceptance level combined with a satisfying user experience for a successful market launch. It can ensure an intensive use of the TicketIssuance app in the future, particularly against the background that the demand for mobile electronic tickets will continue to increase.

## References

1. Albert, W., Dixon, E.: Is this what you expected? The use of expectation measures in usability testing. In: Proceedings of the Usability Professionals Association 2003 Conference, Scottsdale, AZ (2003)
2. Albert, W., Tullis, T.: Measuring the User Experience: Collecting, Analyzing, and Presenting Usability Metrics. Morgan Kaufmann, Burlington (2013)
3. Alhadreti, O., Mayhew, P.: Rethinking thinking aloud: a comparison of three think-aloud protocols. In: Proceedings of the 2018 CHI Conference on Human Factors in Computing Systems, pp. 1–12 (2018)
4. Barnum, C.M.: Usability Testing Essentials: Ready, Set … Test! Elsevier, Inc., Burlington (2011)
5. Berliner Verkehrsbetriebe: BVG Lagebericht & Jahresabschluss (2018). https://unternehmen. bvg.de/wp-content/uploads/2020/10/BVG-Lagebericht-2018.pdf. Accessed 23 Feb 2021
6. Bevan, N.: Classifying and selecting UX and usability measures. In: International Workshop on Meaningful Measures: Valid Useful User Experience Measurement, vol. 11, pp. 13–18 (2008)

7. Bilandzic, H., Trapp, B.: Die Methode des lauten Denkens: Grundlagen des Verfahrens und die Anwendung bei der Untersuchung selektiver Fernsehnutzung bei Jugendlichen. In: Paus-Haase, I., Schorb, B. (eds.) Qualitative Kinder-und Jugendmedienforschung, pp. 183–209, KoPäd - Kommunik. u. Päd., München (2000)
8. Ericsson, K.A., Simon, H.A.: Protocol Analysis: Verbal Reports as Data, 2nd edn. MIT Press, Cambridge (1993)
9. Fonteyn, M.A., Kuipers, B., Grobe, S.J.: A description of think aloud method and protocol analysis. Qual. Health Res. 3(4), 430–441 (1993)
10. Frühwald, A.: Das REFA-Gedankengut – Eine Darstellung für den Kaufmann. Gabler, Wiesbaden (1955)
11. Gerike, R., Hubrich, S., Ließke, F., Wittig, S., Wittwer, R.: Tabellenbericht zum Forschungsprojekt. In: Mobilität in Städten – SrV 2018, Berlin. Technische Universität Dresden (2020)
12. Harrison, R., Flood, D., Duce, D.: Usability of mobile applications: literature review and rationale for a new usability model. J. Interact. Sci. 1(1), 1–16 (2013)
13. Henry, S.B., Lebreck, D.B., Holzemer, W.L.: The effect of verbalization of cognitive processes on clinical decision making. Res. Nurs. Health 12(3), 187–193 (1989)
14. Jakobitz, D., Krüger, S.: Tickets auf dem Smartphone - das wünschen sich Fahrgäste im ÖPNV (2018). Unter: https://marktforschungsanbieter.de/files/profiles/249/123873544 37530.pdf. Accessed 23 Feb 2021
15. Kaiser, M.O.: Erfolgsfaktor Kundenzufriedenheit – Dimensionen und Messmöglichkeiten, 2nd edn. Erich Schmidt Verlag, Berlin (2005)
16. Lewis, J.R.: Sample sizes for usability studies: Additional considerations. Hum. Factors 36(2), 368–378 (1994)
17. Nielsen, J.: Heuristic evaluation. In: Nielsen, J., Mack, R.L. (eds.) Usability Inspection Methods, pp. 25–62. Wiley, New York (1994)
18. Nielsen, J.: Usability Engineering. Academic Press Inc., Cambridge (1993)
19. Nielsen, J., Landauer, T.K.: A mathematical model of the finding of usability problems. In: Proceedings of the INTERACT 1993 and CHI 1993 Conference on Human Factors in Computing Systems, pp. 206–213 (1993)
20. Nielsen Norman Group: Why you only need to test with 5 users. https://www.nngroup.com/articles/why-you-only-need-to-test-with-5-users. Accessed 23 Feb 2021
21. Nielsen Norman Group: Thinking Aloud: The #1 Usability Tool. https://www.nngroup.com/articles/thinking-aloud-the-1-usability-tool. Accessed 23 Feb 2021
22. Olmsted-Hawala, E.L., Murphy, E.D, Hawala, S., Ashenfelter, K.T.: Think-aloud protocols: a comparison of three think-aloud protocols for use in testing data-dissemination web sites for usability. In: Proceedings of the SIGCHI Conference on Human Factors in Computing Systems, pp. 2381–2390 (2010)
23. Rhenius, D., Deffner, G.: Evaluation of concurrent thinking aloud using eye-tracking data. Proc. Hum. Factors Soc. Annu. Meet. 34(17), 1265–1269 (1990)
24. Schlittgen, R.: Einführung in die Statistik – Analyse und Modellierung von Daten, 8th edn. Oldenbourg München Wien (1998)
25. Spool, J., Schroeder, W.: Testing web sites: five users is nowhere near enough. In: CHI 2001 Extended Abstracts on Human Factors in Computing Systems, pp. 285–186 (2001)
26. Tergan, S.O.: Grundlagen der Evaluation: ein Überblick. In: Schenkel, P., Tergan, S.O., Lottmann, A. (eds.): Qualitätsbeurteilung multimedialer Lern- und Informationssysteme – Evaluationsmethoden auf dem Prüfstand, pp. 22–51. BW Bildung und Wissen, Nürnberg (2000)

27. Virzi, R.A.: Streamlining the design process: Running fewer subjects. Proc. Hum. Factors Soc. Annu. Meet. **34**(4), 291–294 (1990)
28. YouGov: App in die Tonne – Wie Ihre App geladen, getestet und schließlich nicht wieder gelöscht wird. YouGov Reports, Köln (2017)

# Augmented Reality Passenger Information on Mobile Public Displays – an Iterative Evaluation Approach

Waldemar Titov[(✉)], Christine Keller, and Thomas Schlegel

Institute of Ubiquitous Mobility Systems, Karlsruhe University of Applied Sciences,
Moltkestrasse 30, 76133 Karlsruhe, Germany
{waldemar.titov,iums}@hs-karlsruhe.de

**Abstract.** The use of augmented reality is becoming more and more common-place. Whether via smartphone apps, on television or while driving a car, almost everyone has already encountered this technology more or less knowingly. In this paper, we investigate to what extent Augmented Reality (AR) can be used prof-itably in public transport, especially on semi-transparent display screens. These semi-transparent and interactive public displays are built-in as windows in public transport vehicles. In our research project SmartMMI these mobile public displays are called "SmartWindows". We implemented an iterative evaluation concept con-taining of four phases to assess different types of AR content to be displayed on such a SmartWindow as well as the special interest of individual types of users. Based on the results of two conducted online studies, specific user scenarios were created and evaluated in a lab-based user study. Finally, we designed an online survey to determine which AR concepts are most accepted by users. In this video-based survey, we evaluated the running speed of AR object inserts, the colors, the fade in's and the types of information displayed on a semi-transparent display. In this paper, we introduce all four stages of our iterative evaluation process, and focus further on study methodology, set-up and the results of the video-based online survey.

**Keywords:** Augmented Reality (AR) · Passenger information · Public displays

## 1 Introduction

Getting On, Traveling, Getting Off – Since the passenger is not busy driving the vehicle in public transport, there are numerous distraction possibilities during the journey: reading, eating, sleeping, etc. Especially by taking previous unknown trips or traveling as a tourist it is not uncommon that passengers lack information regarding objects passing by outside the train window. Questions like where we are right now, what is that tower and how much longer the journey will take arise. The technology of semi-transparent display screens as well as methods of Augmented Reality (AR) enable virtual content to crossfade objects in the field of vision. By using AR on the windowpane, additional information can be

© Springer Nature Switzerland AG 2021
H. Krömker (Ed.): HCII 2021, LNCS 12791, pp. 126–143, 2021.
https://doi.org/10.1007/978-3-030-78358-7_8

shown to passengers that can answer the questions mentioned before, thereby fulfilling the rising information need without the necessity of using private devices.

In our ongoing research project SmartMMI [1], we focus on improving passenger information along the travel chain. In this project, we research model and context-based mobility information on smart public displays and mobile devices in public transport. We want to improve information provision for passengers in any possible scenario. Depending on the situation, but also depending on the passenger, the need for information changes. Examples are disruptions, plan changes, the discovery of tourist destinations or of services available along the route. Our goal is to inform passengers appropriately according to their individual situation. To this purpose, we combine a variety of data sources to form a smart public transport data platform that integrates real-time public transport information, information about points of interests, but also information on bike, car sharing or other services along the route. The integrated data is then adapted to the user's context, and presented on semi-transparent mobile public interactive displays (PID) in public transport vehicles. These semi-transparent, multi-touch-enabled public displays are built-in as windows in public transport vehicles. They are called "SmartWindows". For the design of our SmartWindow prototype, we chose to follow the user centered design process. We therefore continually involved public transport passengers in our evaluation and in the further development process. Since public transport is widely used by passengers of very different backgrounds, we developed public transport personas to sort passengers in different user groups. We previously presented the utilized personas in [2]. In our evaluations, we considered these user groups, whenever possible, in order to consider their different requirements. The methods and results of the several conducted evaluations as well as the iteratively developed prototypes we introduced in [3].

This work addresses the application possibilities of information visualization on semitransparent panes using AR in public transport. We implemented an iterative evaluation process consisting of four iterations that are shown in [Fig. 1]. Each of the phases had different approach as well as a slightly different focus. Thereby focusing user's demand on information on the AR objects in public transport, the visualization and the interaction with it. In the next subsection, we will introduce our chosen evaluation approach, our applied evaluation methods and evaluation results of each phase.

## 2   Overview of the Evaluation Phases I, II, III and IV

Figure 1 shows an overview of the development and evaluation process we went through up until today. In phase I, we performed an online survey to determine the usefulness of different types of information that can be displayed on a SmartWindow. In the first online survey, we focused on the information need of different user groups. Therefore, we designed an online questionnaire to investigate the following questions. Which information is important to passengers during a train journey and which information should be displayed on the SmartWindow, specifically? The result of the evaluation demonstrated users wish for crossfading AR information on semitransparent displays with real points of interests. Based on the results of the evaluation in phase I we designed a second survey focusing further. We researched the informational content to be displayed in AR.

In phase II, we first designed augmented reality mockups, structuring and displaying most of the requested information from phase I. These mockups were evaluated in an online survey focusing the usability and intelligibility of the augmented reality visualizations. Based on the two previously conducted online studies, we created concrete user scenarios in phase III and evaluated them in a lab-based user study. In this evaluation, the test persons were asked to solve tasks for each user scenario without prior instruction, so that it is clear whether the visualizations are helpful and understandable. In phase IV, the results of both online surveys and the lab study were incorporated into video-based AR mockups. These mockups were evaluated using an online survey. In this study the focus was to find out an optimal running speed of the AR object inserts, evaluate the readability of the texts, the colors, the fade in's and the types of information displayed. In our future phase V, we will deploy the final prototype on a SmartWindow in a public transport vehicle of the local public transport provider and the SmartWindow will be available in general public transport outside our lab. In the consecutive evaluations of our AR designs for the SmartWindow, we went from low fidelity prototypes to high fidelity prototypes and we gradually increased public transport context of our prototype and study settings, in order to allow study participants to experience a situation as similar to traveling with public transport, as possible. We will elaborate on these steps, the evaluation settings and our results in the next subsections.

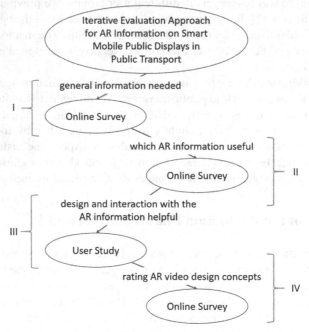

**Fig. 1.** Procedure of the iterative evaluation process for AR Information in Public Transport

## 3  Definition of Augmented Reality (AR)

The majority of the population has already encountered AR during sports broadcasts on television [4]. The insertion of the minimum distance of the wall from the free kick point in soccer or the world record time in swimming are classic examples. We summarize two very common used definitions for AR and use them to combine these to a working definition as a basis of our work. According to Klein G. 2006, augmented reality is a technology to enrich existing objects in reality with additional information. Mainly computer-generated graphics such as text, arrows or 3D objects are integrated into the user's field of vision to provide additional information about the environment [5]. According to Azuma et al. 2001, the technology of augmented reality is defined by the three following properties. First, it combines real and virtual objects in a real environment. Second, it runs interactively and in real-time. Third, it aligns real and virtual objects with each other [6]. In the context of this work, we will use a combination of both AR definitions that can be formulated as follows: "AR describes the visual integration of virtual graphics into a real environment. This integration takes place interactively and in real-time to provide additional information to the user."

## 4  Related Work

Today AR is used in almost all areas. Besides the use in television, as seen in Fig. 2, Augmented Reality is used in medicine and automotive engineering, as seen in Fig. 3. With AR, for the human eye invisible elements can be displayed, assisting medicine personal in difficult surgeries. This is made possible by superimposing image data from tomography or ultrasound equipment, simulating an "X-ray view" for the surgeon [7]. In the context of automotive engineering, AR is used in cars or aircraft. Here, for example, navigation instructions and information on speed limits are integrated into the user's field of vision by means of head-up displays.

**Fig. 2.**  AR technology in a TV studio [8] & AR technology during a TV broadcast [7]

**Fig. 3.** AR technology in medicine [9] & AR technology in the automobile industry [10]

### 4.1 Augmented Reality in Public Transport

We identified two projects that use AR technology in the context of public transport. The first project is called Connectram and applied AR in a public transport vehicle to provide additional information about future city development and connections to the user. Connectram was tested from October to December 2015 in Bordeaux, France and presented at the ITS World Congress 2015 [11]. Due to the lack of the semitransparent display technology, the video see-through AR method was applied. Thereby cameras installed on the roof of the tram captured live images of the surroundings and displays the recorded images on two monitors inside the tram. As shown in Fig. 4 the visualization allowed users to admire the future appearance of urban quarters. Additionally, real-time travel information on public transport lines and bike sharing stations were displayed overlaying the recorded video.

**Fig. 4.** Connectram representation of future buildings [11]

The second project applied AR outside the public transport vehicle, aiming to improve passenger information service at bus stops. The system is currently being tested in the Serbian city of Novi Sad. The user is provided with information about the expected arrival time of the buses, the best connections and the sights near the respective bus stop as shown in Fig. 5. The information is accessible via a smartphone app. Image and geo markers trigger the augmented information [12]. Due to the availability of the semi-transparent display technology in our research project, we will be able to inform public transport users about objects passing by using real AR technology.

**Fig. 5.** AR smartphone app in Novi Sad [12]

## 5 Iterative Approach Evaluating the AR-Passenger Information in Public Transport

### 5.1 Phase I: Online Survey Evaluating User's Information Demand

In the first online survey, we focused on the information need of different user groups. Therefore, we designed an online questionnaire to investigate the following questions. Which information is important to passengers during a train journey and what information should be displayed on the SmartWindow, especially? The online study was designed to give participants as much freedom of answering as possible. Therefore, the questionnaire included many free questions without given answering options. With this approach, we hoped that users bring up new ideas for information that could be displayed on the SmartWindow. We asked what information the participants would like to receive while riding the train, before changing and before alighting the vehicle. About 250 participants of different age groups, different experiences with public transport and affinity towards new technologies participated in the survey. The results give a good insight into the varying needs of the different passengers.

The results of the conducted online study pointed out a great amount of passenger information to be displayed on a SmartWindow. However, in this work we will focus on the information to be displayed crossfading existing sights and attractions only. The majority of the study participants (55%) would like to receive additional information about points of interest on the SmartWindow while riding the train. Asked about information they would like to see before alighting a vehicle, about 10% of respondents wished to receive information about points of interest at the alighting station, like historical or shopping information. Keller et al. 2019 describes the full result of this study in more detail. Based on the results of this online survey, the user's wish for augmented reality information on real points of interest was demonstrated. Therefore, we designed a second survey focusing further researching the informational content to be displayed in AR.

### 5.2 Phase II: Online Survey Evaluating User's Information Demand for Points of Interest – Presented in AR

The survey consists of 27 closed and 3 open questions. In 22 closed questions, we asked the subjects to rate AR information that potentially could be presented on the semi-transparent display serving as the train window. First, the age of the respondents is asked in order to classify the survey demographically. In the second block of the questionnaire the potential AR content to be displayed while riding were displayed. The AR content we picked for this study was based on results of the preliminary survey. In the third block, the subjects were asked to rate the information on an overview map based on their interests. The fourth questionnaire block is similar to questionnaire block two, with the difference that it deals with the time span of a few minutes before alighting the vehicle. Here, the subjects should evaluate different AR options according to how interesting they are. Additionally the subjects had the possibility to enter own ideas and further suggestions.

In this second online survey, more than 160 participants evaluated the information content of the points of interests and helped obtaining a more precise picture of the participant's interests. The results of the survey were clustered for further analysis. With 63% the majority of the respondents were in the age range between 20 and 39 years, due to the predominantly student distribution of the survey. The second strongest age group is the subsequent generation between 40 and 59 years with 25%. About 9% of the participants are from the group of persons under 20 years of age, 3% were between 60 and 79 years of age.

The study participants positively rated information about the surroundings like names of localities (83%), information about places of interest (72%), information about tunnels and bridges (46%) and future events and locations (35%).

Additionally about 87% of the participants expressed the wish for an AR-guidance concept while interchanging trains. Based on the results of the second survey, first augmented reality mockups, structuring and displaying most of the requested information were designed. These mockups were evaluated in a user study focusing the usability and intelligibility of the developed augmented reality visualizations. We created various scenarios for this study. The user study utilized the thinking aloud method, where participants were asked to perform these scenarios using our touch enabled SmartWindow prototype and to express their thoughts while doing so.

### 5.3 Phase III: User Study Evaluating Usability of Developed AR Mock-Ups

The designed mockups follow the three general requirements resulting from the conducted surveys in phase I and II. The three general requirements are:

- displaying traffic related real-time information during the journey,
- displaying information on surroundings during the journey and
- the SmartWindow being intuitive operable.

Thereby the real-time information should be positioned to be always visible and simultaneously not disturb the view through the SmartWindow. We positioned the traffic related informational containing: current delay, arrival time and connections at the bottom of the screen, as shown in Fig. 6, Above, the AR information on visible or nearby points of interest were arranged. Based on user's feedback, the AR insertions were kept short, consisting of a title and a thumbnail. In this way, the AR insertions are recognized in the landscape and visually link to real objects. Information regarding the train location is provided at the top edge of the screen. As shown in Fig. 7, interacting with the AR insertions additional information is displayed describing the selected object. More information about selected objects is available by following the QR-code.

Figure 8 displays the AR mockups created for scenarios in rural areas. Besides the real-time information, the focus here was on providing AR information on landscapes, rivers and infrastructure passing by outside the train.

Towards the end of the journey, the transfer information becomes more relevant for the users. As shown in Fig. 9, the visualization of the transfer information is intended to facilitate the transfer process by preparing the user for the situation. Passengers would like to be informed about how stations are structured, which way to go to connecting trains and whether there is time pressure. Nevertheless, the view through the transparent SmartWindow should be possible at any times despite the displayed information on it. The conducted user study evaluated the design of our AR mockups for usability and intelligibility.

Twelve participants evaluated the designed AR mockups. During the evaluation, the subjects were shown six different visualizations on an interactive monitor, one after another, for the information types: real-time information on connections, AR points of interest information and transfer information. Each visualization dealt with a specific scenario using personalized information. For each scenario, the subjects were given tasks to solve without previous explanations, to check their understanding. All subjects were able to solve the majority of the given tasks. However, it turned out that some terms were difficult to understand. After a short discussion with the subjects, these problems could be eliminated. To encounter this situation in real public transport a short introductory video at the beginning of the journey, explaining the main contents and functionality of the system, is recommended. Generally, the basic implementation of such an AR system is very positively received by the users, so that a further evaluation in phase IV of the AR concepts should be strived for.

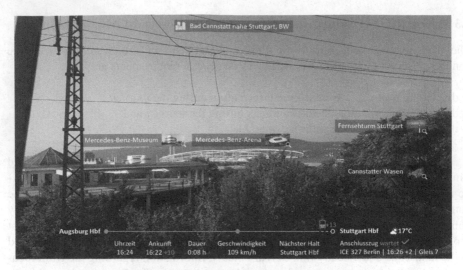

**Fig. 6.**  Visualization of real-time & AR surrounding information

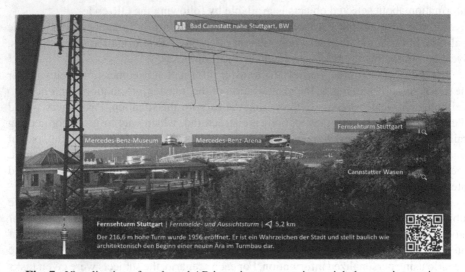

**Fig. 7.**  Visualization of a selected AR insertion representing a sight by user interaction

**Fig. 8.** Visualization of waters and mountains using AR technology

**Fig. 9.** Visualization of the transfer information using AR technology

### 5.4 Phase IV Video-Based Online Survey Evaluating Visualization Concepts of Developed AR Mock-Ups

In the fourth phase of the evaluation process, we have conducted a video-based online survey. The focus of the survey was to determine AR concepts that are most accepted by the users. Using videos to convey the concepts and an online questionnaire, we evaluated the following details, among other:

- the running speed of AR object inserts,

- the colors,
- the fade in's and
- the types of information.

**Study Methodology**

Online user studies evaluating the usability of public transport systems are often hard to set up. The usability of public transport systems is often depending on "real-life" context factors that are hard to reproduce in standard online surveys. Especially the feeling of a moving tram and changes of surroundings that lead to changing information needs and information provision is hard to simulate. We therefore have set up a video-based online study where the AR insertions are moving objects on top of a running video. The video was created by filming through the window during a tram ride on one specific and scenic line in our local public transport network. This video is shown in the background of the AR mockups and creates the illusion of sitting next to a window of a moving tram. The AR mockups developed for the online survey differ in:

- the used fade-in's,
- the moving concepts of the AR objects,
- the styles of backgrounds and
- the content of the information

In the next subchapter we will describe the design and set-up of our online survey, share the results of the conducted study as well as give an insight into the lessons we learned.

**Study Structure**

In the first part of the questionnaire, we asked the study participants for their age, their affinity for technology and their general interest regarding new technology and specifically the technology of augmented reality. We used this data to classify the collected survey outcome. Subsequently we need to know more about the participants public transport usage. We therefore asked the participants to provide information about the knowledge of their local public transport system as well as their usage frequency.

In the second step, we evaluated the video-based AR mockups that were developed based on the results of the phases I, II and III. We displayed three different visualization concepts shown in Fig. 10. The participants could choose from an AR insertion with a white, a black or a continuous gray background. After the participants have chosen the favorite visualization concept, the following questions were displayed only in the favorited visualization. In this paper, we use the gray background for demonstration. Among the favored visualization, we asked whether the passing AR representation of the sight is well readable and what other information about the sight should be displayed. The mockup shown in Fig. 10 shows the running video, thereby the menu and the (red and green icons) as well as the white rectangles demonstrates the users interaction are, where map and other functionality can be located.

Further investigating the readability of the AR objects and the receptivity of the information, we evaluated the running speeds of the AR insertions. Therefore, we selected a part of the background video where the tram passes by the city center. This way we

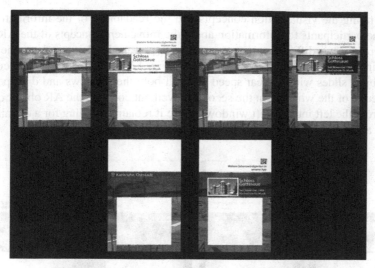

**Fig. 10.** The three different AR insertions visualization concepts

were able to have many sights in view of the camera. The resulting mockup containing three AR insertions within 30 s of video time is shown in Fig. 11.

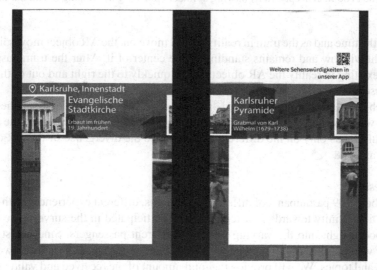

**Fig. 11.** AR insertions concept evaluating the readability of information

The decision for the favorite visualization concept is followed by two questions: How many points of interest should be displayed within the length of the video? Moreover, how many sights would you like to see at the same time in the whole window (left and right displays)?

After rating the visualization concepts and the readability of the information, we queried the participants for information about the movement concepts of the AR insertions. Therefore, two different concepts were developed and implemented into the video-based online survey. Both concepts are shown in Fig. 12. In the first movement concept, the AR object slides with a linear speed through both the windows and disappears on the right edge of the window. In the second movement concept, the AR object quickly fades in from the left into the left window, where it remains standing for a certain time. This time allow the user to read the information.

**Fig. 12.** The Fade-in concepts: (left) moving at linear speed (right) jumping in and out the view

After the time and as the tram in reality would move on, the AR object moves linearly to the right window and remains standing at the center of it. After the tram passed by the sight existing in reality the AR object moves quickly to the right and out of the view of the passengers.

Thereby the information provided in both concepts and the length of the video sequence were an equally long time. By replaying the videos multiple times, the participants could focus only on the differences and notice the divergence of the AR objects fading in and out.

## Study Results

Overall about 109 participants of different age groups, different experiences with public transport and affinity towards new technologies participated in the survey. The results give a good insight into the varying needs of different passengers. Since most of the questions were marked as not mandatory, the amount of participants varies between the questions and topics. We will provide the total amount of the received and valid answers to each topic. The results of the survey were clustered for further analysis. To get an idea about the participants we started by clustering the participant ages, technological affinity and the level of knowledge with the AR technology.

With 61% the majority of the respondents were in the age range between 18 and 24 years, due to the predominantly student distribution of the survey. The second strongest age group is the subsequent generation between 25 and 34 years with 15%. The age group 35 and 50 amounts to be 2%. About 5% of the participants are from the group of persons over 50 years of age. As shown in Fig. 13 about 17% of the total amount of 88 participants prefer not to specify their age.

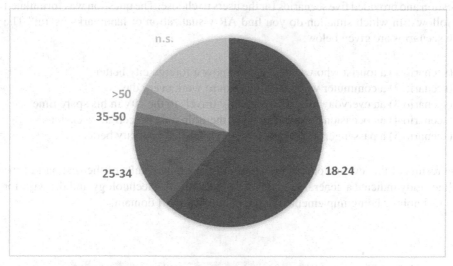

**Fig. 13.** Participants age distribution of online study of phase IV (n = 88)

The self-rated technological affinity as shown in Fig. 14 could be analyzed in 84 responses. Thereby 57% of the test persons chose a strong, 41% a medium and only 2% a weak technological affinity.

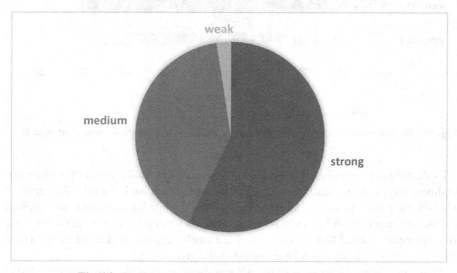

**Fig. 14.** Participant's self-rated technological affinity (n = 84)

In the next step, we analyzed the general use of AR technology in the public transport domain. Therefore, we asked the participants to rate to which user type of public transport AR information about sights would be most helpful. To simplify the task we formulated a question and provided five scenarios for the users to choose. The question was formulated as follows: In which situation do you find AR visualization of landmarks useful? The five scenarios are given below:

- (scenario 1) a tourist who wants to get to know a foreign city better
- (scenario 2) a commuter who takes the train to work every day
- (scenario 3) an everyday user who regularly travels to the city in his spare time
- (scenario 4) an occasional rider who takes the train only on certain occasions
- (scenario 5) a passenger who wants to get to know his own city better

The results of this evaluation can be seen in Fig. 15. The results of the first part of the online study indicate a general high interest toward the AR technology and the wish for this technology being implemented in the public transport domain.

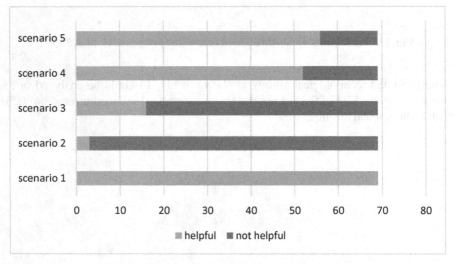

**Fig. 15.** Evaluation of user groups profiting most from AR tech. in public transport (n = 69)

Evaluating our video-based AR mockups we started by analyze users favorite visualization concept. The visualization concept with the black background color received with 14% the fewest votes. As indicated in Fig. 16 the white background color received 19% and the grey beam 67% out of 70 valid collected feedbacks. For the gray beam, the most participants stated that the AR inserts are easily readable and all of the provided information about the sight can be absorbed in the given time.

Analyzing the running speeds of the AR insertions, most of the participants stated to get displayed only two sights within the length of the video. About 5% wish to see only one sight during the video. On the other side, about 9% of the subjects would favorite

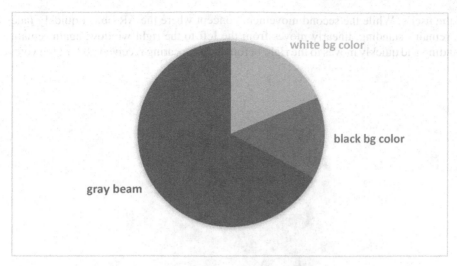

**Fig. 16.** Evaluation of visualization concepts (n = 70)

more than three sights. About 30% stated that the chosen speed in the mockup was fine. As shown in Fig. 17 over 56% of the study participants wished to get informed about two sights in the given time.

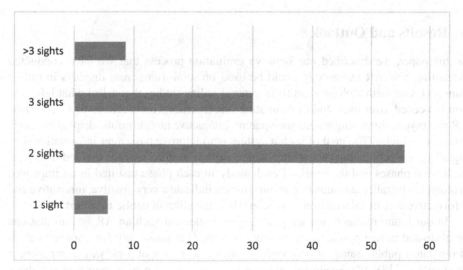

**Fig. 17.** Evaluation of running speeds and the readability of information (n = 70)

Finally, we analyzed the movement concept of the AR insertions. Especially the fading in and out where different in the both developed mockups. As shown in Fig. 18 the first movement concept where the AR object slides with a linear speed through both the windows and disappears on the right edge of the window is preferred by 45%

of the user's. While the second movement concept where the AR object quickly fades in, remains standing, linearly moves from the left to the right window, again remains standing and quickly moves to the right before it disappearing received 56% of the votes.

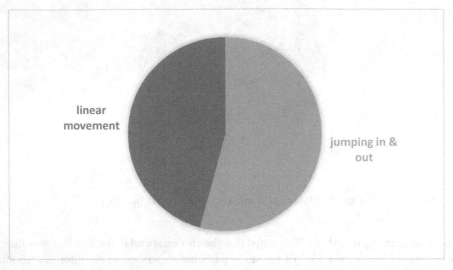

**Fig. 18.** Evaluation of the both AR movement concepts (n = 70)

## 6   Results and Outlook

In this paper, we described our iterative evaluation process that we have conducted evaluating how AR technology could be used on semi-transparent displays in public transport. Our methodology went from general online studies that asked what information is needed, over user studies evaluating AR insertions on mockups to video-based AR prototypes simulating a semi-transparent, interactive mobile public display in public transport vehicles. The goal of such a system is to improve passenger information during a trip in a public transport vehicle. Our development process was divided in several evaluation phases and the result of each study in each phase resulted in an improved prototype. Overall, the results of our four studies indicate a very positive, innovative and informative way of information provision while traveling in public transport.

In our future research, we are planning to implement such an AR system that can be evaluated in real public transport environment. One possibility for implementation is our latest public transport demonstrator that is a replica of a railway compartment as shown in [13]. We developed and constructed a mockup of a train that simulates a realistic public transport context for study participants while enabling us to control context factors and data. By applying the recorded train ride video in combination with the matching sound of the same train ride and the AR insertion displayed on interactive monitors, we will be able to relocate the participants into the context of using real public transport. Thus, we are hoping to achieve interesting result to be able to improve the passenger information on mobile public displays in public transport.

**Acknowledgements.** This work was conducted within the scope of the research project "Smart-MMI - model- and context-based mobility information on smart public displays and mobile devices in public transport" and was funded by the German Federal Ministry of Transport and Digital Infrastructure as part of the mFund initiative (Funding ID: 19F2042A). We would like to thank Julian Pohl and Nils Wenzel for the excellent contribution to this project.

# References

1. Research project Model- and Context-based Mobility Information on Smart Public Displays and Mobile Devices in Public Transport: https://smartmmi.de/project/
2. Keller, C., Struwe, S., Titov, W., Schlegel, T.: Understanding the usefulness and acceptance of adaptivity in smart public transport. In: Krömker, H. (ed.) HCI in Mobility, Transport, and Automotive Systems: First International Conference, MobiTAS 2019, Held as Part of the 21st HCI International Conference, HCII 2019, Orlando, FL, USA, July 26-31, 2019, Proceedings, pp. 307–326. Springer, Cham (2019). https://doi.org/10.1007/978-3-030-22666-4_23
3. Keller, C., Titov, W., Sawilla, S., Schlegel, T.: Evaluation of a smart public display in public transport. Mensch und Computer 2019-Workshopband (September 2019)
4. Tönnis, M.: Augmented Reality: Einblicke in die Erweiterte Realität. Springer, Berlin (2010)
5. Klein, G.: Visual tracking for augmented reality (Doctoral dissertation, University of Cambridge) (2006)
6. Azuma, R., Bailot, Y., Behringer, R., Feiner, S., Simon, J., MacIntyre, B.: Recent advances in augmented reality. IEEE Comput. Graph. Appl. **21**(6), 34–47 (2001)
7. Mehler-Bicher, A., Steiger, L.: Augmented Reality: Theorie und Praxis. Walter de Gruyter GmbH & Co KG (2014)
8. Swiss Radio and Television, Augmented Reality in ECO – Shows - SRF, 2013. http://www.srf.ch/sendungen/eco/erweiterte-realitaet-bei-eco. Accessed 09 Feb 2021
9. Pott, A.: Mbits imaging GmbH - Mobil in der Radiologie - Gesundheitsindustrie BW, 2016. https://www.gesundheitsindustriebw.de/de/fachbeitrag/aktuell/mbits-imaging-gmbh-mobil-in-der-radiologie/. Accessed 09 Feb 2021
10. Continental Automotive GmbH, Continental Head-up Display Augmented Reality HUD. http://continental-head-up-display.com/wpcontent/uploads/2014/08/slidernav3.jpg. Accessed 09 Feb 2021
11. Keolis Bordeaux Métropole, Connectram: the tram that takes you to the future – EN Tbm Interactive. http://www.tbminteractive.com/node/2890. Accessed 06 Feb 2021
12. Pokrić, B., Krco, S., Pokrić, M.: Augmented reality based smart city services using secure IoT infrastructure. In: 2014 28th International Conference on Advanced Information Networking and Applications workshops, pp. 803–808. IEEE (May 2014)
13. Keller, C., Titov, W., Trefzger, M., Kuspiel, J., Gerst, N., Schlegel, T.: An evaluation environment for user studies in the public transport domain. In: Krömker, H. (eds.) HCI in Mobility, Transport, and Automotive Systems. Driving Behavior, Urban and Smart Mobility. HCII 2020. Lecture Notes in Computer Science, vol. 12213, pp. 249–266. Springer, Cham (2020). https://doi.org/10.1007/978-3-030-50537-0_19

# Solve the Problem of Urban Parking Through Carpooling System and Blockchain Advertising

Sheng-Ming Wang[1]($\boxtimes$) and Wei-Min Cheng[2]($\boxtimes$)

[1] Department of Interaction Design, NTUT, Taipei, Taiwan
ryan5885@mail.ntut.edu.tw
[2] Doctoral Program in Design, College of Design, NTUT, Taipei, Taiwan
t108859003@ntut.edu.tw

**Abstract.** Currently, the problem of urban parking is getting serious. Limited by laws and regulation, cannot go through improving the parking fee to reduce the demands. This research tries to use the internet of things (IoT) and blockchain to provide a carpooling system to reducing the needs of parking and urban traffic congestion. The research design to build a platform which is through social internet of the vehicle to make a connection between urban of vehicle and need of customers. Making passengers who have the same destination can ride the car together at any time and reducing the needs of urban parking. The passenger needs to watching advertising and provide the basic information itself to exchange the service of riding. Through the simulation could find that the design mechanism could achieve the expecting, reducing the needs of parking problem during the urban peak hours. If the design mechanism and service pattern could attract most users and make a profit from user data, it is possible to let the system operation continued. Meanwhile, integrate the autopilot system could make the service system to become an important application of traffic service in the smart city.

**Keywords:** Car sharing system · Social internet of vehicle · Advertising of IoV · Carpooling system of IoV

## 1 Introduction

Following the development of the economy and ages, the vehicle is getting popular in life, and the number is growing every year. However, the space of urban is limited. Even cities with well-planned parking cannot cope with the growth of urban vehicles every year [1]. According to the normal planning of urban parking, the space for public parking is less than half the amount of cars. Currently, the problem of urban parking is getting serious. Even use all spaces completely, it still cannot be solved at popular spots during peak hours [2].

In order to protect fairness and justice in society, the government makes the law to limit the fee of parking, and one can have a right to parking no matter you rich or not in the city [3, 4]. Hence, Limited by laws and regulation, it is impossible to reduce the demands by upping the parking fee or planning more space for parking.

© Springer Nature Switzerland AG 2021
H. Krömker (Ed.): HCII 2021, LNCS 12791, pp. 144–155, 2021.
https://doi.org/10.1007/978-3-030-78358-7_9

Amount of vehicles in the city is growing, but the urban space of parking is limited, even we could have other ways to help, the best way to solve the problem is to reduce the amount of car using [5, 6]. With the increase of 5G network transmission speed and the popularity of self-driving car technology, carpooling systems and the social internet of the vehicle have become increasingly popular in cities [7]. In addition to public transportation, carpooling sharing system could improve the problem of insufficient parking spaces [9].

According to research findings, people who tend to ride-sharing mainly consider freight costs, and social and safety are also important indicators for determining whether to ride-sharing. Appropriate fare subsidies or higher-intensity social services may be helpful for the frequency of use of the urban carpool system.

This research purpose a design concept, establish a city car-sharing system by the internet of vehicle. Meanwhile, increasing subsidies for voluntary drivers and encourage the public to use the carpooling system in the city in order to reduce the need for urban parking spaces and ease traffic congestion. Besides, try to create a self-operating system model through the use of value-added ride-sharing data so that the carpooling system can continue to operate continuously through profit.

Blockchain has a particular characteristic of Decentralization. It has a significant value to preserve the public data, even could be used in save of decentralization data [10, 11]. If the system could collect the car-sharing data by blockchain, It can promise fairness without modification, also get the rely on advertiser and research institute to improve the willingness of using and paying [12]. In the end, If the blockchain can be used in conjunction with the Internet of Vehicles mechanism to create a decentralized system and reduce the system burden and traffic requirements of data obtained by the central host, it will also help the long-term operation of this system.

The research has two main goals: including how to design the carpooling system and how to measure the feasibility of this system.

## 2  Related Work

### 2.1  System Design

**Platform**
This idea is to design a shared ride platform that provides matching services between drivers and passengers. Then transform the data generated in the service into working capital for sustainable operations. Through the blockchain, The data is placed directly on the devices of all drivers and passengers who participate in this platform and is properly encrypted and stored. And all transactions on the platform will also be conducted through digital currencies similar to Bitcoin. The following is an explanation of the various roles in the service and its relationship with the platform:

**Passenger**
This service must provide passengers with an opportunity to find the car so that passengers can find drivers who drive in the same direction at the same time and travel together. Ridesharing allows you to meet more friends, and you can save the trouble of finding

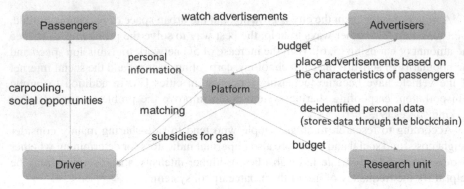

**Fig. 1.** System structure

parking spaces when you arrive at your destination. Not only reduces exhaust emissions, but it also lowers car expenses! To obtain the platform's services, passengers must agree to provide departure location and destination data and the time spent on the ride for free during the service period or assist in clicking and watching the advertisements provided by the platform so that the platform can continue to operate.

## Driver

The platform provides drivers with a service to subsidize fuel costs. In addition to giving drivers the opportunity to meet more friends, this shared ride platform will also offer appropriate fuel subsidies based on the length of the route they drive, and at the same time help the government solve the city problems of traffic congestion and insufficient parking spaces. If the government unit can actively encourage such high-capacity driving and give them rewards, I believe it will significantly improve urban traffic.

## Research Unit

This service provides research units with an opportunity to obtain a large amount of personal movement path data. This platform uses the mechanism of blockchain. The information provided by this service is credible and traceable. Moreover, the information of the starting point and ending point can be accurately grasped, so there is an opportunity to obtain cooperation and profit from the research unit through this service. This service can maintain long-term operation by assisting research units in conducting surveys of audiences to obtain higher compensation.

## Advertisers

Since this service can obtain the movement trajectory of drivers and passengers, the trajectory information can be further analyzed and used. The analyzed data can be used to create advertising products with more commercial potential. In the current advertising market, most of the advertising is to analyze the ethnicity or region and place the advertisement in the right place. Less based on the movement trajectory, if the movement trajectory can be provided and provide various advertising services closely related to life, It is believed that such applications will become a new trend in the future development of 5G.

## 2.2 Main Services

### Ride Matching

The operation of the platform must be profited mainly by obtaining the rider's trajectory data. Therefore, the more users use, the more data they use, which is crucial to the growth of the platform. Thus, the operator must try to match drivers and passengers through various methods to increase the matching rate between drivers and passengers. The methods that can be used include allowing potential drivers to see the needs of passengers or providing more passengers with different types of needs to obtain ride-sharing services. The service mechanism can provide passengers with an interface for making demands. After statistics, the system exposes all the routes that need to be provided with driving services and uses various reward mechanisms to make drivers willing to provide services. The feasible method includes allowing drivers to choose passengers or giving different levels of fuel supplements, and these supplements can also be provided by the passengers. At the same time, if the travel path can be arranged appropriately, it will be able to meet the travel needs of more passengers.

### Matchmaking

If ride-sharing data can be used, it may be able to create more value in the current standard social services. Shared riders must provide accurate destination and time in order to be matched to a suitable rideable vehicle. Therefore, users cannot provide fake data, so the data can be trusted, and the accuracy rate is high. After statistical analysis, it is possible to know when and where each passenger will be at, so it can accurately confirm the moment of encounter between any two passengers. If the meeting between people is cleverly packaged into a social service product, there will be a chance to profit from it. For example, the following types of social products may be mentioned: New friends who appear in the same place today, friends who often meet, friends who meet at a particular time, friends who meet in a specific location, friends who often go to the same place, and so on. As long as you use data analysis to get different lists of friends, you can provide various services to allow users to follow their personal preferences and pay the price they are willing to pay to expand their connections.

### Advertising Matching

The primary source of profit for the platform is advertising. Therefore, how to provide advertisers with various advertising needs is the core of this service. Standard advertising services in the market generally analyze user groups based on their behavior and provide advertising services. The service we hope to provide is not to judge user groups through their behavior. Still, the trajectory of movement, frequency, and location are directly used as the criteria for placement so that advertisers can directly contact the customer groups on a specific moving path. The advertising platform's design uses blockchain to create an advertising system that truly belongs to the Internet of Vehicles. The goal is to store all advertising content directly on each car networking node in the ecosystem. When a specific user needs information, each node will assist in providing content to reduce flow demand. Also, all the shared data is not stored in a particular location by a specific unit. After collection, information is placed on all participating nodes through algorithms and encryption mechanisms to ensure the data's authenticity (Fig. 2).

## 2.3  Data Analysis

**Carpooling Data**

Ride-sharing data is composed of the starting point of the ride to the end of the ride of a specific object, and it also contains the purpose of the journey and the time of movement. We can organize the item's data, starting point, ending point, purpose, and time, and cooperate with various statistical algorithms to gradually transform the ride-sharing data into different service needs. Including conversion to frequency, period, touchpoint, amount, trajectory, and then conversion to a group, key path, gathering point, meeting point, and finally, you can get the necessary information for social and advertising applications. The following table is a reference for some data calculation results (Tables 1, 2, 3, 4, 5, 6, 7, 8, 9 and 10).

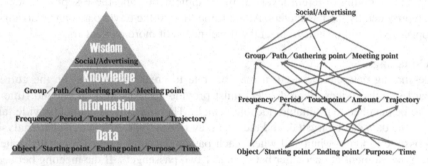

**Fig. 2.** Data structure

**Table 1.** Object/Starting point/Ending point/Purpose/Time

| Object | Starting point | Ending point | Purpose | Weekday | Time |
|--------|----------------|--------------|---------|---------|------|
| Person1 | Dongmen | Xihu | Working | Tuesday | 18 |
| Person1 | Dongmen | Kunyang | Traveling | Saturday | 10 |
| Person2 | Huilong | Xindian | Working | Monday | 9 |
| Person2 | Huilong | Xindian | Working | Wednesday | 9 |
| Person2 | Huilong | Xindian | Working | Friday | 9 |

**Table 2.** Frequency/Time

| Object | Location | Time | Frequency |
|--------|----------|------|-----------|
| Person1 | Xihu | 9 | 17 |
| Person2 | Xindian | 14 | 22 |
| Person2 | Tamsui | 18 | 30 |
| Person3 | Zhongshan | 8 | 30 |
| Person3 | Zhongshan | 18 | 22 |
| Person3 | Zhongshan | 21 | 14 |

**Table 3.** Touchpoint/Amount

| Touchpoint | Weekday | Time | Amount |
|------------|---------|------|--------|
| Shongshan | Monday | 9 | 146 |
| Shongshan | Monday | 10 | 22 |
| Zhongxiao sogo | Monday | 17 | 45 |
| Zhongxiao sogo | Friday | 19 | 144 |
| Guanghua | Friday | 12 | 65 |
| NTUT | Sunday | 12 | 11 |
| NTUT | Sunday | 15 | 33 |

**Table 4.** Trajectory

| Path | Weekday/Weekend | Time | Amount |
|------|-----------------|------|--------|
| Shongshan-Dongmen | Monday | 12 | 129 |
| Shongshan-Dongmen | Monday | 13 | 213 |
| Tamsui-Zhongshan | Monday | 8 | 131 |
| Tamsui-Zhongshan | Thursday | 8 | 153 |
| Tamsui-Zhongshan | Friday | 8 | 164 |
| Kunyang-Xindian | Saturday | 15 | 382 |
| Kunyang-Xindian | Sunday | 15 | 453 |

# 3 Feasibility of System

Because of the different conditions in each city, the way of lane planning and the dense working places are also different. For the convenience of explanation, we first use a relatively simple model to look at the possible effects and results of this whole mechanism. We can simulate a city with a commute distance of about one hour and observe the results

**Table 5.** Group

| Group | Monday | Tuesday | Wednesday | Thursday | -> |
|-------|--------|---------|-----------|----------|-----|
| Office worker | 22 | 21 | 18 | 20 | -> |
| Backpacker | 2 | 3 | 2 | 1 | -> |
| Retired people | 18 | 14 | 12 | 14 | -> |

| Friday | Saturday | Sunday | Working | Entertainment | Traveling |
|--------|----------|--------|---------|---------------|-----------|
| 23 | 1 | 1 | 22 | 1 | 4 |
| 7 | 12 | 12 | 1 | 1 | 6 |
| 18 | 16 | 20 | 0 | 8 | 10 |

**Table 6.** Path

| Path | Percent of weekday | Percent of weekend | Level |
|------|--------------------|--------------------|-------|
| Area1->Area2 | 20 | 80 | 1 |
| Area3->Area5 | 70 | 30 | 4 |
| Area7->Area12 | 40 | 60 | 3 |
| Area15->Area9 | 45 | 55 | 2 |
| Area6->Area13 | 10 | 90 | 6 |
| Area25->Area16 | 5 | 95 | 6 |

**Table 7.** Gathering point

| Path | Level of weekday | Level of weekend | Average amount per day |
|------|------------------|------------------|------------------------|
| Area1 | C | B | 2232 |
| Area2 | B | A | 1543 |
| Area4 | C | B | 1643 |
| Area5 | B | A | 2289 |
| Area7 | C | C | 889 |
| Area8 | A | A | 4388 |

of the changes in vehicle movement to determine whether the carpooling mechanism can operate smoothly in the city. Suppose there are some fixed main lanes in this city, and commuters must travel through these lanes when they want to go to work. Also, for office workers in a particular area, we assume that commuters can initially come from

**Table 8.** Meeting point

| Object A | Object B | Meeting place | Time |
|----------|----------|---------------|------|
| Person1 | Person2 | Area3 | 9 |
| Person3 | Person2 | Area7 | 22 |
| Person1 | Person4 | Area2 | 12 |
| Person5 | Person3 | Area6 | 15 |
| Person4 | Person3 | Area9 | 8 |
| Person2 | Person5 | Area4 | 16 |
| Person5 | Person4 | Area8 | 17 |
| Person3 | Person6 | Area1 | 4 |
| Person1 | Person8 | Area5 | 23 |

**Table 9.** Social

| Time | Weekday/Weekend | Location | Type | Amount |
|------|-----------------|----------|------|--------|
| Morning | Monday | Area3 | New friend | 336 |
| Morning | Tuesday | Area5 | New friend | 543 |
| Afternoon | Tuesday | Area7 | Meet each other sometimes | 888 |
| Evening | Wednesday | Area1 | Meet each other sometimes | 456 |
| Evening | Sunday | Area6 | Meet each other always | 292 |

**Table 10.** Advertising

| Group | Path | Frequency | Amount |
|-------|------|-----------|--------|
| Office worker | Area2->Area21 | High | 336 |
| Office worker | Area4->Area16 | High | 543 |
| Office worker | Area15->Area18 | Middle | 888 |
| Office worker | Area3->Area17 | Middle | 456 |
| Office worker | Area2->Area6 | Low | 292 |

lanes in various directions. Therefore, after simplifying the entire model, we can use the following path reference diagram to illustrate (Fig. 3):

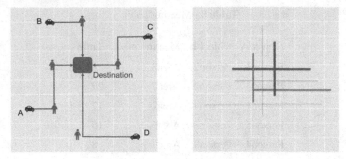

**Fig. 3.** Vehicle trajectory simulation

As shown in the picture above on the left, there are four ABCD vehicles in the picture. When these vehicles depart for a specific destination, they will take passengers to the destination on the way. Therefore, the use of traffic arteries in the city may be There will be changes as shown on the right. A road with fewer vehicles may not have a greater impact, but for certain traffic arteries, the number of vehicles is reduced, so it is bound to help ease the traffic on the path. This help will produce different levels of size according to different carrying capacity. There are two situations worth discussing here, one of which is the impact of commuting time to and from get off work, and the other part is the impact of non-working hours.

When the number of vehicles is reduced by half from the original number, it will theoretically affect the two projects, including the smoothness of the traffic path and the demand for parking spaces. Since the shared ride is to carry passengers along the road to a specific destination, it is not necessarily The ride-sharing service is used throughout the journey. Therefore, if the traffic conditions of the whole journey can be reduced by half, more people will use the ride-sharing service. However, this is related to the distribution of households and offices in the city, and this Part is not the core theme of this research, so I won't discuss it in depth for the time being.

**Table 11.** Trial calculation of the influence of carpooling system on the number of parking spaces

|  | Fixed parking space | Temporary parking space | Total |
|---|---|---|---|
| Passengers number | 2 |  |  |
| Original parking space | 1000 | 100 | 1100 |
| Original total number of vehicles | 1000 | 250 | 1250 |
| Original remaining parking spaces | 0 | −150 | −150 |
| Number of vehicles after sharing | 800 (Not 500) | 125 | 925 |
| Remaining parking spaces after sharing | 200 | −25 | 175 |
| Reduced number of vehicles | 200–500 | 125 | 325–625 |
| Impact ratio | 62% | 38% | 100% |

Let's take a look at the impact of shared rides on the number of parking spaces. Due to the limited parking spaces in the city, each workplace of course has its own parking spaces. If the number of parking spaces can be reduced, this part of the there is a chance that the usage can be converted into temporary parking spaces during normal business hours. Therefore, the use of parking spaces that can be affected by shared rides can not only reduce the usage of general temporary parking spaces during the day because of ride-sharing but also allow more regular parking spaces for work are released for temporary parking (Table 11).

As shown in the above table, we assume that there are 1,000 fixed parking spaces for work in an area, and there are 100 temporary parking spaces in the area. The area has 1,000 vehicles that must be parked for work. The number of temporary stops at the same time is 250. Obviously, the number of parking spaces in this area is not sufficient. Once the parking is full, there will be about 150 vehicles with no parking spaces.

When using the shared ride system, we assume that the number of people in each car is 2; that is, the total number of vehicles should be at least half. Therefore, for parking spaces that originally needed to be parked for work, it may save up to 50% of Vehicles, but considering human nature may not be willing to accept, especially some long-distance commuters, so we assume that only about 20% of the vehicles will be reduced in the end, that is, 200 vehicles. Generally speaking, it may be easier to accept shared rides due to the temporary parking needs of public utilities. Therefore, the same calculation is carried out by saving 50% of the vehicles, which is 125 vehicles. After calculation, we can finally get 175 available parking spaces that should have a chance.

In the above figures, the demand for fixed parking spaces and temporary parking spaces for work is estimated from observations. Generally speaking, the number of parking spaces is only about half of the peak period. From the above figures, it is not difficult to see that if the fixed parking spaces used above can be moved out and used, and the number of temporary parking spaces can be reduced by half, it is actually easy to solve the problem of insufficient parking spaces during peak periods, and this is just based Each car only carries one more passenger, not to mention the situation when it carries 3 or 4 people.

## 4  Discussion

According to the result simulating of the experiment, by reducing self-driving and using the carpooling system, the need for parking spaces in popular locations can be effectively reduced, and urban social relationships can be strengthened, giving the public the opportunity to meet more friends from the daily geographical life circle. However, there are two important conditions for the operation of this service. First of all, is expansion. There must be enough participants to maintain the operation of the system so that everyone is willing to continue to use the service and allow the service to continue to expand in size. Secondly, there must be a stable profit so that the service can be a sustainable operation, so how to build on this service and further enhance the value of the service, to enhance the profitability of the service is another important thing.

**How to Extend**

Since the number of users and driving may be insufficient at first, consideration could be given to working with specific operators to bring the volume of traffic to a certain level in the first were, thus enabling the service to continue to operate. If the self-driving system continues to develop in the future, it may also be considered to complete this ride-2 mechanism through self-driving cars, or even to achieve a full ride-driven system in smart cities, but due to different laws and regulations and different acceptances of new technology models, such innovative solutions still need to be officially supported before they can be gradually promoted.

User expansion at the beginning of the platform is very important. The fastest way is, of course, through word-of-mouth introduction, but to make the average user willing to go to another person's car is not so easy. Here we can consider several modes, such as allowing users to get used to it from vehicles with fewer co-occupants to vehicles with more co-occupancy or by pairing users with the same friends to increase familiarity and reduce their anxiety. If the number of users is enough, you can also use interests to categorize or even do a ride together for dinner or outings and constantly pull more new users in.

**How to Profit**
Because of the characteristics of co-multiplied data, which provides accurate geographic information, such data is of great use-value to a location-sensitive surveyor or advertiser. Therefore, if the research unit is willing to pay directly to commission the survey of geographic user data, this will be the system's best source of profit, only in the platform in accordance with the user's path to provide the relevant questionnaire, can be directly in exchange for income. Of course, because we have the frequency of user exchanges, so even if the survey unit to carry out user identification requirements, we can immediately provide a relevant reference. Another extension of the application is advertising. The same data can also be direct as a filter for ad delivery so that advertisers can search directly through the data to filter out the need for the object and advertising, will also be the main source of profit for the service. As for the novelty of the data, because the same place will continue to have new users to enter, and therefore the demand unit will continue to invest funds repeatedly.

Another project that has the opportunity to suck money is the social needs of human nature. The nature of co-location data depends on the location, and it is easy to connect people to a specific time and place, so by packaging the results into merchandise, you can continue to create a steady stream of new buying needs, both for advertisers and the general passenger, which is very attractive, as technology becomes more advanced, and may even achieve more immediate geographic interaction and charge higher service charges.

## 5  Conclusion

The study is designed to be based on social vehicle networking, where passengers exchange anonymous information in exchange for free or cheaper transport services, drivers use social vehicle networking to increase social opportunities, and platforms that mediate passengers and drivers through car networking and urban parking problems are addressed jointly by providing services to reduce the need for parking spaces during peak urban periods.

This research design will be mediated when the supply and demand side of the information provided by the calculation, into available geographical area user group data, and in accordance with the needs of service objects and advertisers, further design-related social services or advertising mechanism, so that the system can take advantage of this long-term profit, but can be the sustainable operation of the self-operation mechanism, but also enable the demand side of all walks of life to enjoy a long-term sustainable geographical area survey analysis and mass access system.

# References

1. Ali, Y., et al.: The impact of the connected environment on driving behavior and safety: a driving simulator study. Accid. Anal. Prev. **144**, 105643 (2020)
2. Roopa, M.S., et al.: DTCMS: Dynamic traffic congestion management in social internet of vehicles (SIoV). Internet Things 100311 (2020)
3. Butt, T.A., et al.: Privacy management in social internet of vehicles: review, challenges and blockchain based solutions. IEEE Access **7**, 79694–79713 (2019)
4. Zia, K., Shafi, M., Farooq, U.: Improving recommendation accuracy using social network of owners in social internet of vehicles. Future Internet **12**(4), 69 (2020)
5. Hamid, U.Z.A., Zamzuri, H., Limbu, D.K.: Internet of vehicle (IoV) applications in expediting the implementation of smart highway of autonomous vehicle: a survey. In: Al-Turjman, F. (ed.) Performability in Internet of Things. EICC, pp. 137–157. Springer, Cham (2019). https://doi.org/10.1007/978-3-319-93557-7_9
6. Hammoud, A., et al.: AI, blockchain, and vehicular edge computing for smart and secure IoV: challenges and directions. IEEE Internet Things Mag. **3**(2), 68–73 (2020)
7. Li, Y., et al.: Fog-computing-based approximate spatial keyword queries with numeric attributes in IoV. IEEE Internet Things J. **7**(5), 4304–4316 (2020)
8. Tao, M., Wei, W., Huang, S.: Location-based trustworthy services recommendation in cooperative-communication-enabled internet of vehicles. J. Netw. Comput. Appl. **126**, 1–11 (2019)
9. Hyland, M., Mahmassani, H.S.: Operational benefits and challenges of shared-ride automated mobility-on-demand services. Transp. Res. Part A: Policy Pract. **134**, 251–270 (2020)
10. Gurumurthy, K.M., Kockelman, K.M., Simoni, M.D.: Benefits and costs of ride-sharing in shared automated vehicles across Austin, Texas: Opportunities for congestion pricing. Transp. Res. Rec. J. Transp. Res. Board **2673**(6), 548–556 (2019)
11. Gerte, R., Konduri, K.C., Ravishanker, N., Mondal, A., Eluru, N.: Understanding the relationships between demand for shared ride modes: case study using open data from New York City. Transp. Res. Rec. J. Transp. Res. Board **2673**(12), 30–39 (2019)
12. Rathee, G., et al.: A blockchain framework for securing connected and autonomous vehicles. Sensors **19**(14), 3165 (2019)

# Design of Natural Human-Computer Interaction for Unmanned Delivery Vehicle Based on Kinect

Kaidi Wang and Lei Liu[✉]

Southeast University, Nanjing, China
liulei@seu.edu.cn

**Abstract.** There are more and more application scenarios for unmanned delivery vehicles, and traditional human-computer interaction methods can no longer meet different task scenarios and user needs. The primary purpose of this paper is to apply natural human-computer interaction technology to the field of unmanned delivery, changing the traditional mode that users can only operate unmanned delivery vehicles through the touch screen to complete tasks. The task scenarios of unmanned delivery vehicles are classified through the concept of context. Participatory design and heuristic research are used to allow users to define interactive gestures. Two groups of gesture interaction set that meet different task scenarios and can be accepted and understood by general users are designed. Based on Kinect's deep imaging and bone tracking technology, large number of preset gesture samples are collected, and the Adaboost algorithm is used for machine training to realize gesture interaction. Through recognition and detection, it is proved that the gesture recognition achieved by this method has a high recognition rate and responding speed. Finally, based on Unity3D, the task scene of the real unmanned delivery vehicle is simulated in the virtual scene. Through the usability test of the human-computer interaction system, it is concluded that this interaction mode guarantees the task efficiency to a certain extent and improves the user experience.

**Keywords:** Natural human-computer interaction · Unmanned delivery · Kinect · Gesture recognition

## 1 Introduction

Stimulated by the Internet economy, the logistics industry has developed rapidly. The addition of artificial intelligence technology can effectively reduce labor costs in the distribution process and improve the efficiency of logistics terminal personnel and equipment. Especially under the epidemic situation, unmanned delivery vehicles designed to solve the "last mile" terminal distribution problem can help solve various problems and have a wide range of application scenarios [1]. However, at present, the non-material delivery vehicle is still in the early stage of development, and the research focus remains on the level of unmanned driving technology, including vehicle positioning, environment perception, path planning, and vehicle control execution technologies. However, even autonomous delivery vehicles can only have partial autonomy, requiring manual

© Springer Nature Switzerland AG 2021
H. Krömker (Ed.): HCII 2021, LNCS 12791, pp. 156–169, 2021.
https://doi.org/10.1007/978-3-030-78358-7_10

intervention in certain scenarios. In addition, the primary goal of unmanned delivery vehicles is to serve users. With the gradual maturity of autonomous driving technology for unmanned delivery vehicles, the focus of the next phase of research on unmanned delivery vehicles needs to shift to the interaction between humans and vehicles, combining specific task scenarios and iteratively updating human-computer interaction technology to meet Various complex user requirements for terminal distribution. As shown in Fig. 1, during the epidemic, doctors are using unmanned delivery vehicles. Doctors wearing protective gloves cannot directly use the interactive interface, and touch in the public environment will also increase the risk of infection. Therefore, how to reduce the contact of people in the public environment under the epidemic situation also needs to be included in the thinking of human-computer interaction.

**Fig. 1.** The interaction scene between the user and the unmanned delivery vehicle.

As shown in the figure above, in the field of human-computer interaction research, the traditional mode is that the user uses the mouse or multi-touch to interact with the machine through pointing devices, windows, menus, icons, etc. [2]. With the development of technology, the limitations of these operations have become more and more obvious. People look forward to using a more natural and efficient way to interact with devices, that is, natural human-computer interaction. Natural human-computer interaction enables people to communicate with computer equipment in the simplest way, so that people can focus on tasks without having to perform tedious operation tasks, thereby improving user experience and task efficiency. Natural interaction methods include language, posture, gestures, etc. People usually use different ways to express different ideas and information. The characteristics of natural human-computer interaction include ambiguity, implicitness, imitation and metaphor of real behavior, etc. These characteristics reflect its powerful advantages in the rapid development of natural interaction. With the maturity of key technologies, one of the future development trends of human-computer interaction is the development of multi-modal natural human-computer interaction.

## 2 Related Work

### 2.1 Technical Support

Traditional human gesture recognition has two methods based on embedded sensors and computer vision. The former has the advantages of fast response speed and high accuracy, but the disadvantage is that the user needs to wear a specific device and the movement is limited; Based on the computer vision method, the user does not need to wear anything, and the depth camera can be used to directly obtain the depth image to realize the recognition of human posture and gestures. Kinect V2 is a depth sensor developed by Microsoft. The device analyzes the depth image itself through the collected data, determines the information contained, and obtains human skeleton information. It has been widely used in education, medical treatment, games and other fields. Relevant scholars have studied the usability of depth sensing equipment represented by Kinect from the perspective of human-computer interaction, and evaluated its ergonomics in typical interactive tasks, which has been well verified [3].

### 2.2 Task Scene and Gesture Classification

Human body gesture interaction can be divided into mapping gesture interaction, operation gesture interaction and dialogue gesture interaction according to different modes of action in different scenarios. Dialogue interactive gestures that are closely related to the user's daily operating experience are suitable for interactive scenarios in unmanned vehicle delivery. In conversational gestures, the user and the device establish an agreement through gestures, and the semantic meaning of the gestures corresponds to the control function [4]. To explore a more natural human-computer interaction relationship, the complete task path of the unmanned delivery vehicle from the cargo distribution center to the completion of the distribution has been tracked and studied, and the task path diagram obtained is shown in Fig. 2:

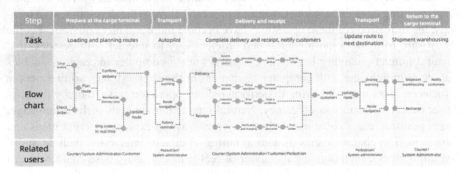

**Fig. 2.** The complete mission path of the unmanned delivery vehicle.

By analyzing the task path of the unmanned delivery vehicle, the scenarios involving human-vehicle interaction are refined. The interaction scenarios of the unmanned delivery vehicle are classified into two categories: the user controls the driving of the unmanned delivery vehicle and the user operates the unmanned delivery vehicle inter-active interface (Fig. 3). Two groups of interactive gestures will be designed to satisfy different scenarios.

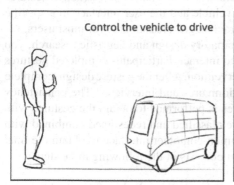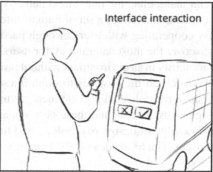

Control the vehicle to drive

Interface interaction

**Fig. 3.** Two mission scenarios for unmanned delivery vehicles.

## 2.3  Gesture Design Principles

Boulabiar et al. discussed the complete process of gesture interaction in detail, and believed that gesture interaction is a complete loop constructed by brain activity, body manipulation, detection, algorithm, object performance, mental model and feedback [5]. For conversational gesture interactions, after the interaction event begins, an agreement between the subject and the object is first established (for example, "stretching your hand to the right means turning right"), and using this agreement to guide the formation of a dialogue between the subject and the object. Positive feedback means that the role is generated, and a mental model will be established at the same time, which will be repeatedly strengthened in the subsequent interaction process. On the contrary, the subject will reconfirm the previous agreement and try the dialogue again. The state of the object and the recognition feedback are more critical. Based on the characteristics of the dialogue gesture interaction and its construction process, scholars from Tongji University proposed the design principles of reliability, fault tolerance, continuity and inheritance in the design process of gesture interaction [6]. On this basis, this article supplements the two principles of metaphor and universality, hoping that gestures are an imitation of real life and can be accepted by most people. In the human-computer interaction scene of the unmanned delivery vehicle, the application of the above principles and the subdivision of the task scene can make the interaction logic more smooth and clear.

# 3  User Research

## 3.1  Research Purpose

In the related research of gesture interaction, there are few problems to explore and research from the perspective of users. At this stage, most of the design of interactive gestures comes from technicians, lacking design thinking and design empathy. Accurately understand the cognitive differences of different users for the meaning of different gestures, which leads to the low recognition rate of some gestures. Therefore, this paper hopes to understand the user's cognitive understanding and habits of interactive gestures that can manipulate the unmanned delivery vehicle and the user interface in a specific task context, and design a set of natural interactive gestures suitable for most users.

By cooperating with users through participatory design and heuristic research, you can discover the most natural way for users to interact. Participants completed gestures autonomously in an environment without intervention. After the gesture design, complete the feedback on the gestures through questionnaires and interviews. The consistency analysis of the collected user-defined gestures is performed to obtain the gestures with the highest usage rate, and these gestures are classified and redesigned combined with the knowledge of design psychology and human-computer interface to obtain the final gesture set. The basic idea of the experiment consists of the following three steps:

1. Research the function of gestures. When gestures are applied to unmanned delivery vehicles, the design of gestures must be able to express the functions that users want to achieve, so before you start, you need to clarify which functions need to be implemented, such as click gestures in interface operations;
2. Collect user gestures. The goal of this step is to collect user gestures that can represent an application function in the previous step. In this step, through WoZ heuristic research [7], participant allowed to design their own gestures and record the experimental process through a camera. The user is in the gesture design There is no need to consider the limitations of technical factors, so it is best for the subjects to have no relevant experience.
3. Design gestures. Extract user gestures by analyzing the video, pay attention to the continuity and consistency of most people's gestures and experimental participants' gestures when analyzing the video, and then complete the gesture set design based on the user's interview feedback and application scenarios.

## 3.2 Experiment Preparation

The experiment is carried out in an indoor environment. The equipment required is: a computer to provide the task scene; a camera to record the experiment process; the experiment record sheet needs to record the user's gestures completed under each task and the interview feedback after completion; The task card prepared in advance describes the task to the user.

Based on the user intervention in the driverless car driving scene and the scene of the user operation interface summarized in the previous article (referred to as scene 1 and scene 2 respectively), two groups of ten tasks are summarized, including "Command the car to turn left", "Command the car to turn right", "Command the car to approach", "Command the car to move back", "Command the car to stop in place" and "Command the car to leave" in scene 1. The tasks in scenario 2 include "forward/next page", "return/previous page", "click to confirm" and "cancel operation". Write each task separately on the task card. On the computer, a dynamic video that simulates the realization of each task is prepared.

## 3.3  Participant

The experiment recruited 20 healthy participants who could clearly understand the experiment process and the tasks they need to complete. Participants are 24–28 years old (M = 25.2, SD = 4.60). These participants are all undergraduate or graduate students in school, so they have a relatively high frequency of receiving express delivery and have experience in unmanned vehicles. But none of them have relevant experience in gesture interaction.

## 3.4  Procedure

Each participant completes the experiment in turn, and needs to repeat the following steps:

(1) Introduce experiment purpose, equipment, process and experiment scene to experiment participants. After confirming that the participants are familiar, remind users to ignore the experimenters as much as possible after the experiment starts, do not need to consider the limitations of technical factors when doing gesture interactions, and operate in the most natural way that they think;

(2) Experiment participants face the simulated task scene in the computer, randomly select a task card, and make the corresponding interactive gestures according to their own ideas, and do it 3 times every time they are packed to avoid errors in the process of single data collection;

(3) After completing the ten tasks, the subjects are required to give a simple explanation of the gestures they make for each task, and then give participants a suggestion sheet for each interactive gesture;

(4) After giving certain operation suggestions, ask participants to repeat step two;

Finally, the participants were interviewed and asked if there were other operational suggestions or concerns.

## 3.5  Consistent Gesture Set

Through the above heuristic research, a total of 400 gestures from 20 users were collected, including the 200 before the suggestion and the 200 after the suggestion. Through statistical analysis, the gesture with the highest number of repetitions for each task can be found. For the actions used by the users, the same gestures that most people make for a certain task are defined as user-defined gestures. To evaluate the consistency of the gestures, the agreement rate is calculated for each gesture. The agreement rate can reflect the consensus of the participants on the gesture [8]. The calculation formula is:

$$Ar = \frac{\sum_{p_i \in p_t} \frac{1}{2}|p_i|(|p_i| - 1)}{\frac{1}{2}|p_t|(|p_t| - 1)} \tag{1}$$

The agreement rate $Ar$ is defined as the number of actions that participants use the most of the same number under a single task divided by the total number of participants. Where $t$ represents the label of the task, $p_t$ is the number of all gestures in task $t$, and $p_i$ is the number of the same gesture in task $t$. The range of the consistency score $Ar$ is

0–1. When the actions used by the participants in a certain task are the same, $Ar = 1$, and when the actions used by all people are not at the same time, $Ar = 0$. The higher the consistency score, the more participants use this posture for a specific person's task, which is more in line with the user's mind. Table 1 shows the consistency score statistics of the most frequently performed actions under different tasks.

**Table 1.** Action consistency statistics with the highest agreement rate under different tasks.

| Scene | Mission | User action description | $Ar$ |
|---|---|---|---|
| User intervene in driving and controlling the unmanned vehicle | Command to turn left | Raise left arm to the left | 0.56 |
| | Command to turn right | Raise right arm to the right | 0.55 |
| | Command the car to approach | Wave your arm to the inside of your body (any hand) | 0.59 |
| | Command the car to go back | Push the arm to the outside of the body (any hand) | 0.46 |
| | Order the car to leave | Wave of goodbye | 0.41 |
| | Let the car stop in place | Reach out hand and do stop motions | 0.82 |
| User operate the unmanned delivery vehicle interactive interface | Start operation instruction | Any hand up | 0.74 |
| | Left swipe operation/previous step | Swipe left with any hand | 0.52 |
| | Right swipe operation/next | Swipe right with any hand | 0.52 |
| | Click to confirm | Fist (any hand) | 0.52 |
| | Cancel | Hands folded | 0.74 |

Vatavvu has made a three-level delineation of the consistency gesture score. When the consistency score is higher than 0.5, it means that the gesture has strong representativeness. When the consistency score is higher than 0.25, it means that the gesture can be referred to. When the consistency score is less than 0.2, it means that the gesture requires expert participation in the design [9]. According to the results, most of the gesture consistency scores are higher than 0.5, which can meet the universal design requirements. The back command and the leave command have relatively low consistency, but they can also meet the design requirements. Some users worry that the back and stop gestures are easy to confuse, the gestures in the two different scenarios are easy to confuse, and it is easy to cause problems such as misoperation when turning the page left and right. The stop command gesture and the left and right page turning gestures have been optimized. Finally, two sets of gestures for different task scenarios are obtained, which can be summarized as the body gestures in the scene of controlling the car driving, using "raise hand to the left", "raise hand to the right", "push forward", and "to the body". The six basic instructions for turning left, turning right, going back, approaching, leaving, and stopping corresponding to the six gestures of "inside beckoning", "wave of hand", and "stop gesture" are shown in Fig. 4; local gestures are used when the user controls the operation interface, "Raise any hand", "Slide right hand to left", "Slide left hand to right", "Make a fist", "Cross hands", corresponding to the basic operations of start, slide left, slide right, click to confirm, and cancel, respectively, As shown in Fig. 5.

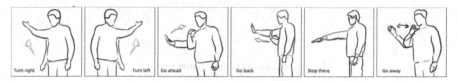

**Fig. 4.** Gestures used in scenario 1.

**Fig. 5.** Gestures used in scenario 2.

## 4   Technical Realization

### 4.1   Machine Training

Based on Kinect's deep imaging technology and skeleton tracking technology, this paper uses the iterative machine training method to realize gesture interaction, which can meet the recognition of discrete and continuous actions at the same time. The basic process is shown in Fig. 1. First, we need to collect the action samples for training through Kinect. This article invites 40 experimental subjects, each of whom completes the 6 gestures in Scene 1 described above, and each action is repeated 4 times, including three positive samples and one negative sample. The reason for adding negative samples is to effectively avoid the ambiguity of similar actions by the machine. It took about 5–6 min. for each person to complete the six sets of actions, and a total of 960 action clips were collected. Then create a training project in Visual Gesture Builder, the training samples need to remove the invalid action time, so 40 sets of samples of each action are edited and imported separately. All gestures only involve the upper body, only the upper body 16 skeleton points need to be called, as shown in Fig. 6:

Before training, we need to determine whether it is a discrete action or a continuous action. If it is a discrete action, we only need to mark the segment that conforms to the action as True, and the non-conformity is False, and output a bool value to describe whether the action occurs. If it is a continuous action, we need to add a progress parameter, which represents the degree of completion, and mark the start and end points of the

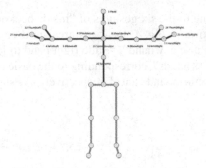

**Fig. 6.** Kinect human skeleton points

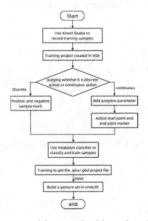

**Fig. 7.** Human gesture recognition method based on VGB machine training.

action, which are marked as 0 and 1, respectively. Then a True mark needed to add at the discrete action at the end-point to determine when to get effective feedback (Fig. 7).

After completing the above steps, use the Adaboost classifier for machine training. Adaboost is a framework that provides a series of classifiers, sub-classifiers can be customized according to their own needs, with the ability of adaptive learning [10]. The basic flow of the algorithm is shown in Fig. 8:

Adaboost does not require repeated sampling during the training process, the weight update of the sample is related to whether the sample is correctly classified. Initially, the weights of all samples are equal. During the training process, the weight of the correctly classified sample will be reduced, and vice versa. In the next round of training, increase the focus on misclassified samples, and each iteration will recalculate the weight value of the sample. To prevent the ambiguity of the training results, negative samples are added, so according to the original algorithm logic samples, the weight of each round of negative samples will increase, resulting in overtraining. Therefore, in each round of sample training, a weight threshold needs to be defined, combined with judging whether the sample is classified correctly in each round of training and whether the sample weight

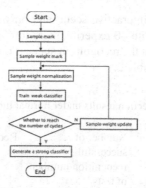

**Fig. 8.** Adaboost iterative training algorithm flow

in this round of training is greater than the threshold S, three different methods are used to update the weight, the weighting rules are shown in the following formula:

$$w_{t+1} = \begin{cases} w_t \beta, & h_t(x_i) = y_i \\ w_t \beta^{-1}, h_t(x_i) \neq y_i w_t < S_t \\ w_t, h_t(x_i) \neq y_i w_t < S_t \end{cases} \tag{2}$$

$w_t$ is the weight of the training sample in the $t$ round, $w_{t+1}$ is the weight of the $t + 1$ training round.

$$\beta_t = \frac{e_t}{1 - e_t} \tag{3}$$

If the sample is correctly classified, $e_i = 0$, otherwise, $e_i = 1$. $S_t$ is the weight threshold of the $t$ round iteration.

$$S_t = \frac{\sum_{i=1}^{n} w_{t,i}}{n} \tag{4}$$

After updating the weight rule, during the training process, if a sample is misclassified and the weight of the sample is less than the given weight threshold, the weight of the sample will continue to increase. Continue iterative training. If the sample is still not classified correctly, once the sample weight is greater than the weight threshold, the sample is judged to be a negative sample, and the weight stops increasing. The algorithm logic can effectively prevent the phenomenon of overtraining, and the training efficiency and results are good. In this paper, a total of 1200 weak classifiers are set up, and 95% of the CPU threads are called to run the Adaboost algorithm. Finally, six strong classifiers are obtained, that is, six gesture recognition engineering files.

## 4.2 Gesture Recognition Detection

After the training is completed, six classifiers corresponding to the actions are obtained. To verify the accuracy of the machine training to recognize the human body pose, a custom gesture library is established in Unity3D, and the six project files are imported separately. Establish a simple interactive scene and only provide text feedback that a certain gesture is detected. Invite 40 experimenters to complete 6 gestures, repeat 10 times for each person, and count the recognition success rate and average response time. The test results are shown in Table 2.

**Table 2.**  Gesture detection results under normal lighting environment.

| Gestures description | Gesture semantics | Number of successful recognition/number of tests | Recognition rate | Average response time |
|---|---|---|---|---|
| Raise left arm to the left | Command to turn left | 385/400 | 96.3% | 0.16 s |
| Raise right arm to the right | Command to turn right | 382/400 | 95.5% | 0.24 s |
| Wave your arm to the inside of your body (any hand) | Command the car to approach | 393/400 | 98.3% | 0.12 s |
| Push the arm to the outside of the body (any hand) | Command the car to go back | 374/400 | 93.5% | 0.28 s |
| Wave of goodbye | Command the car to leave | 386/400 | 96.5% | 0.12 s |
| Reach out hand and do stop motions | Command the car stop in place | 397/400 | 99.3% | 0.08 s |

The results show that both discrete actions and continuous actions have a higher recognition rate and faster response time. Considering that the use scene of the unmanned delivery vehicle is outdoors, to verify whether the depth image-based gesture recognition can reduce the influence of light, three groups of control groups are added to repeat the above experiment. The results are shown in Table 3.

The results show that based on Kinect's depth imaging technology, the illumination environment and complex background have little influence on the gesture recognition obtained by this training method, and it has a better detection and recognition rate under complex background, uneven illumination and weak illumination environment. The gesture set in scene 2 is also constructed through the same steps.

**Table 3.** Comparison of gesture recognition rate under complex background and lighting environment.

| Gesture semantics | Normal environment | Complex background | Uneven light | Weak light |
|---|---|---|---|---|
| Command to turn left | 96.3% | 94.5% | 94.3% | 95.0% |
| Command to turn right | 95.5% | 94.5% | 94.8% | 95.5% |
| Command the car to approach | 98.3% | 97.3% | 96.3% | 97.3% |
| Command the car to go back | 93.5% | 92.5% | 91.3% | 93.0% |
| Command the car to leave | 96.5% | 94.8% | 95.5% | 95.8% |
| Command the car stop in place | 99.3% | 97.8% | 98.0% | 98.5% |

## 5   Usability Testing

To check the rationality and interactive efficiency of the system design, a virtual scene was constructed in Unity 3D to simulate two real-life task scenarios, namely, to construct a delivery scene, import the model of the unmanned delivery vehicle created by Rhino, The user controls the task of unmanned vehicle driving; creates a UGUI interface that contains the entire interactive operation interface for the user from the beginning to the completion of the pickup, and simulates the task of user interface operation.

To check the rationality of the system design and the efficiency of interaction, a virtual scene was constructed in Unity 3D to simulate two task scenarios at reality. A task scene is constructed in Unity, and the model of the unmanned delivery vehicle created by Rhino is imported to simulate the user's task of controlling the driver of the unmanned vehicle; a UGUI interface is created, which contains the entire interactive operation interface for the user from the beginning to the completion of the pickup, Simulate user interface operation tasks.

The test invites 10 participants, randomly divided into group 1 and group 2, respectively, using the natural human-computer interaction system constructed in this article and the traditional human-computer interaction interface to complete a series of task scenarios. Two groups of users are required to complete a complete task process in two scenarios. Scenario 1 requires participants to complete six control instructions. Participants in group 1 complete them through gesture interaction, and participants in group two complete them through the remote control. Scenario 2 requires the user to start from the initial interface and complete an operation process of picking up goods. The user in group one completes it through gestures, and the user in group two completes it through screen touch (Fig. 9).

**Fig. 9.** Usability test experiments that simulate actual task scenarios in virtual scenarios.

After the task is completed, use the operation completion degree and time-consuming to evaluate the system efficiency, and complete the user's satisfaction and ease of use evaluation in the form of a scale questionnaire, focusing on the satisfaction related to the operation method. The test results show that the operation efficiency of the traditional human-computer interaction interface is higher than the natural human-computer interaction method constructed in this article at the beginning, but most users have a significant improvement in the operation efficiency in the natural human-computer interaction after learning, but it is not as good as the traditional, therefore, the user's learning cost and efficiency improvement still need to be considered. But in terms of user experience, the satisfaction and user experience of natural human-computer interaction systems are significantly better than traditional human-computer interaction interfaces.

The most natural interface is not necessarily the most effective interface. Although the interaction system constructed in this article can meet the operating efficiency of basic tasks, it is still not as good as the traditional touch interaction interface, but it can replace the traditional touch interaction in many special scenarios. Especially in scenarios that are difficult to achieve with traditional interaction methods.

## 6   Conclusion and Future Work

In this paper, a new unmanned delivery vehicle human-machine interaction system is constructed through the task situation classification of the unmanned delivery vehicle and depth sensing technology. Large number of user-defined gesture samples are collected through the Kinect depth sensor, and the gesture library obtained by the improved machine training algorithm has a high recognition rate of human posture detection, which provides a certain idea for the research and design of unmanned delivery vehicles and their interaction methods. Usability test results show that the operating efficiency of the natural human-computer interaction system is not as good as the traditional interaction method, but the user experience has been significantly improved, and it can replace the traditional interaction method in some specific scenarios. However, in the actual delivery process, more complicated situations will be encountered. To ensure that the natural human-computer interaction system can still maintain efficiency in complex scenarios, it is necessary to subdivide the scenes and improve the situational awareness of unmanned delivery vehicles. Modal parallel interaction, including how to better integrate interaction methods such as gestures, somatosensory, voice, and eye tracking, is a key research direction in the future.

# References

1. Lu, N., Ying, X.: Development status and application prospects of "last mile" unmanned vehicle distribution. Integr. Transp. **43**(1), 117–121 (2020)
2. Tao, X.: Research on the history and trend of human-computer interaction. Technol. Commun. **11**(22), 137–139 (2019)
3. Deng, R., Zhou, L., Ying, R.: Research on gesture extraction and recognition based on Kinect depth information. Appl. Res. Comput. **30**(4), 1263–1265 (2013)
4. Albrecht, S.: The Encyclopedia of Human-Computer Interaction. Liverpool John Moores University, UK (2016)
5. Boulabiar, M.-I., Coppin, G., Poirier, F.: the study of the full cycle of gesture interaction, the continuum between 2D and 3D. In: Kurosu, M. (ed.) Human-Computer Interaction. Advanced Interaction Modalities and Techniques. LNCS, vol. 8511, pp. 24–35. Springer, Cham (2014). https://doi.org/10.1007/978-3-319-07230-2_3
6. Sun, X., Zhou, B., Li, T.: Design key elements and principles of in-air gesture-based interaction. Packag. Eng. **36**(8), 10–13 (2015)
7. Jessica, R., Jane, L.: Drone & me: an exploration into natural human-drone interaction. UBICOMP **15**(9), 7–11 (2015)
8. Williams, A.S., Garcia, J., Ortega, F.: Understanding multimodal user gesture and speech behavior for object manipulation in augmented reality using elicitation. IEEE Trans. Visual. Comput. Graphics **26**(12), 12 (2020)
9. Vatavu, R.: User-defined gestures for free-hand TV control. In: Proceedings of the 10th European Conference on Interactive TV and Video 2012, pp. 45–48. ACM, Berlin (2012)
10. Rojas, R.: Adaboost and the super bowl of classifiers a tutorial introduction to adaptive boosting. Int. Conf. Image Process. **28**(12), 156–170 (2009)
11. Nielsen, J.: Non-command user interfaces. Commun. ACM **36**(4), 83–99 (1993)

# Research on the Function Design of 5G Intelligent Network Connected Cars Based on Kano Model

Zheyin Yu and Junnan Ye[✉]

East China University of Science and Technology, Shanghai 200237, China

**Abstract.** The fifth-generation mobile communication technology (5G) is the latest generation. Under 5G environment, the automatic driving technology will be upgraded, and the automobile networking will be developed, rendering the cars smarter and more closely connected with each other. However, the existing researches all focus on technologies such as autonomous driving, car networking, etc., and lack the researches concentrated on the evolution of car functions under 5G environment. In this paper, the innovative car function research method based on Kano model is constructed, and the promotion of 5G technology to car networking, auto-driving, VR/AR, mobile office and so on is deeply studied. The functional requirements of users are obtained through interviews and expert evaluation methods, and Kano questionnaire survey is carried out as well. By sorting and analyzing the data, the real functional requirements of intelligent network connected car are derived. The core functions are screened through Better-Worse coefficient analysis, and the innovative function model of intelligent network connected cars is thereafter formed. This study makes up for the lack of researches in this field, and constructs the Kano model research method for automobile functions, and simultaneously digs out the users' needs, which has immense theoretical and practical significance. By redesigning the functions of intelligent network connected car under 5G environment, the potential of 5G technology can be activated and the market's satisfaction of intelligent network connected cars can be improved. It also provides a model for the function design of intelligent network connected cars in the future.

**Keywords:** Intelligent network connected car · 5G · Function design · Kano model

## 1 Introduction

In recent years, the fifth-generation mobile communication technology has become the focus of research in the field of communication technology. In 2019, 5G was officially put into commercial use in China, and the global 5G infrastructure construction was also carried out simultaneously. According to the statistics of the Global Mobile Suppliers Association (GSA), by the end of 2020, a total of 140 operators from 59 countries or regions are providing 5G network services. The International Telecommunication

© Springer Nature Switzerland AG 2021
H. Krömker (Ed.): HCII 2021, LNCS 12791, pp. 170–183, 2021.
https://doi.org/10.1007/978-3-030-78358-7_11

Union (ITU) has defined a 'full' 5G system, which includes enhanced Mobile Broadband (eMBB), Ultra Reliable Low Latency Communications (URLLC) and massive Machine Type Communications (mMTC). EMBB can greatly improve the Internet speeds. The ITU claims that the upload rate of 5G network can peak at 10 Gbps and the download rate can peak at 20 Gbps. MMTC enables the connection number density to reach 1 million per square kilometer [1]. Data are obtained from terminals in the Internet of Things (IoT) and analyzed using cloud computing, AI, and other technologies to form big data and coordinate the nodes in a unified way. URLLC provides low delay and high reliability, thus the latency time of 5G network can be theoretically reduced to 1ms. This historic break-through can greatly improve the efficiency of network data transmission, enable seamless signal switching between vehicles, and greatly improve the stability of signal transmission [2].

EMBB, URLLC and mMTC usher in a breakthrough development for the automotive industry. In 5G network environment, cars will be connected to the Internet of Vehicles (IoV) and autonomous driving technology will be upgraded. Therefore, in the future, the traditional driving and riding functions of intelligent connected cars will be gradually weakened, and the driving activity will be gradually reduced, which result in the possible increase of car functions such as mobile office, entertainment and games. Researchers have already carried out certain researches on the future of automobile functions. For example, Guofang Zhang et al. built a hierarchical demand model for automobile customer customization based on AHP [3]. Wenyu Ji studied the functional design of automobile travel system and entertainment system [4]. Hui Wang deeply studied the development status of automobile entertainment system [5]. Anable J et al. investigated people's demands in work travel and leisure travel [6]. However, the application of 5G network is very insufficient in the research, and the advantages of 5G network technology are not fully brought into play in automobile function design. In addition, through the research on the existing intelligent connected cars in the market, it is found that the existing intelligent connected cars' function settings fail to give play to the advantages of 5G network environment, and the users are not very satisfied. Therefore, the research of 5G network technology and automobile functions as well as the exploration of auto-mobile core functional requirements have become the focus of this paper.

Kano model is a useful tool for user demand classification and prioritization in-vented by Professor Noriaki Kano of Tokyo University of Technology. Its goal is to classify different demands based on the analysis of the impact of user needs on user satisfaction. Kano model can reflect the nonlinear relationship between product performance and user satisfaction. Using Kano model to study the functional design of 5G intelligent connected cars can help researchers understand the functional needs of users at different levels, find out the key points to improve the satisfaction of car users, so that reference can be provided for future car function design. Firstly, this paper builds up the design method for car function based on Kano model. Secondly, this paper focuses on the impact of development of the 5G network technology in the field of automobile industry, in order to explore the potential changes in car functions. development may auto function potential. The third part introduces the process of user demand acquisition as well as the design and distribution of Kano questionnaire. Finally, according to the results of the questionnaire data analysis, different levels of functional requirements are classified, and the car function blueprint is carried out.

## 2   The Design Method for Car Function Based on Kano Model

### 2.1   A Theoretical Overview of Kano Model

Kano model is an effective and widely used method to reflect customers' needs in product design and concept generation of new products. It reflects the nonlinear relationship between products' attributes and customers' satisfaction level. Generally speaking, Kano model defines five types of requirements, which are: Must-be Quality, One-dimensional Quality, Attractive Quality, Indifferent Quality and Reversible Quality [7].

### 2.2   The Process of Car Function Design Based on Kano Model

**Car Functional Requirements Acquisition.** In the process of automobile design, it is of vital importance to obtain the users' requirements, which is related to whether the product can really solve the users' problems and meet the users' requirements. In addition, it is also the prerequisite for Kano questionnaire design and questionnaire survey. Methods of demand acquisition include literature survey, interview, questionnaire survey, expert evaluation, observation method and experience method. For the design of automobile function, we can use the combination of interview method and expert evaluation method: the number of automobile users is huge, and different people have different needs. Thus, it is more comprehensive and objective to conduct a large number of interviews than literature survey method or experience method. Due to the specialty and difficulty of automobile design, the users' requirements and conversation record materials collected in the interviews are summarized, and the expert evaluation is carried out. The list of users' requirements is sorted, screened and supplemented by means of expert meetings and brainstorming to ensure the effectiveness and realizability of the requirements.

**Design and Distribution of Kano Questionnaires.** The satisfaction questionnaire of Kano model shown in Fig. 1 divides a function into positive and negative questions, and each question has five levels, from 1 to 5, which are respectively 'I like it that way', 'It must be that way', 'I am neutral', 'I can live with it that way', and 'I dislike it that way', as shown in Fig. 1. Then, each function can get a two-dimensional satisfaction result. The Kano evaluation table can be used to help determine the classification of function attributes [8]. The Kano evaluation table is shown in Fig. 2. Q stands for questionable result, which usually does not occur unless the interviewee has a wrong understanding of the problem or the problem itself is wrong; A stands for Attractive Quality; I stands for Indifferent Quality; R stands for Reversal Quality; M stands for Must-be Quality; O stands for One-dimensional Quality.

**Statistics and Analysis of Questionnaire Results.** Statistic the data in all questionnaire samples by function, and each function will get a percentage statistics table. The highest percentage item in the table is regarded as the kano classification of this function.

| Function A$_1$ | I like it that way | It must be that way | I am neutral | I can live with it that way | I dislike it that way |
|---|---|---|---|---|---|
| If it has this function | | | | | |
| If it doesn't have this function | | | | | |

**Fig. 1.** Kano questionnaire

| Customer requirements | | Dysfunctional form of the question | | | | |
|---|---|---|---|---|---|---|
| | | I like it that way | It must be that way | I am neutral | I can live with it that way | I dislike it that way |
| Functional form of the question | I like it that way | Q | A | A | A | O |
| | It must be that way | R | I | I | I | M |
| | I am neutral | R | I | I | I | M |
| | I can live with it that way | R | I | I | I | M |
| | I dislike it that way | R | R | R | R | Q |

**Fig. 2.** Kano evaluation table

Because the answer with the highest percentage is selected as the final quality classification of a certain function in Kano questionnaire statistics, there exist practical problems. If a certain function gets 40% Attractive Quality answer and 41% Reversal Quality answer, according to the rules, the function is classified as Reversal Quality, but obviously the result is not accurate enough. Therefore, the Better-Worse coefficient is usually calculated to increase the accuracy of questionnaire statistics. The function of Better-Worse coefficient is to illustrate to what extent the function can increase satisfaction or eliminate dissatisfaction. According to the Better-Worse coefficient, the functions (or services) with higher absolute scores should be implemented prior to all.

**The Construction of the Innovative Function Model of Intelligent Network Connected Cars.** According to the data statistics and analysis results, an objective and scientific function requirement list can be obtained, which shows the Must-be Quality, One-dimensional Quality, Attractive Quality, Indifferent Quality and Reversal Quality in detail. According to the function list, the innovative function model of intelligent network connected cars is constructed, which provides theoretical basis for the subsequent design practice.

## 3    Relevant Research and Requirements Acquisition

### 3.1    Trends in Car Functions

**Internet of Vehicles and Autonomous Driving.** Vehicle networking and autonomous driving are the areas where 5G has the greatest impact on the automobile industry. Based on the agreed communication protocol and data interaction standard, the Internet of Vehicles (IoV) establishes connections between vehicles, vehicles and people, vehicles and roads, and vehicles and the cloud for wireless communication (V2X technology). For example, BAIC MOTOR ARCFOX αT can connect traffic light information and feed it back to the people in the car. The characteristics of large connections in 5G networks allow a large number of cars to interact with each other in the same network, and the 1 ms delay of 5G network ensures the real-time information exchange between vehicles. From this point of view, 5G network optimizes V2X technology and improves the cooperation among vehicles, people and environment, which is the foundation of automatic driving technology.

5G improves the working ability of the IoV, and further guarantees the safety and reliability of automatic driving. According to the five-level division of SAE automatic driving standard, automatic driving under 5G network environment is expected to reach L4 and L5 levels. L4 highly automated level means that the driving system can independently complete vehicle driving in a specific environment; L5 full automation level means that the system can complete all driving, monitoring, judging and decision-making tasks completely and autonomously, without the intervention and support of human drivers. At that time, the "driver" will completely cease to exist. It can be predicted that drivers will engage in more non-driving activities in the automatic driving environment [9]. Upgrading automatic driving to L4 means that the driver does not need to participate in driving operation most of the time. In this case, the difference between "driver" and "passenger" is blurred, and the concept of "driver" may gradually disappear. With the arrival of 5G and automatic driving, drivers can spend more time "riding" and finish some non-driving activities.

**Video/Audio Entertainment and VR/AR.** At present, the commonly used display output devices in automobiles are electronic screens, including electronic instrument panel screens, central control area screens, front headrest screens and so on. Thanks to the high speed of 5G eMBB, it takes only a few seconds to download a 4K or even 8K movie, and URLLC renders it possible to broadcast UHD live in real time without any delay. During the 2019 Boao Forum for Asia Annual Conference, a live webcast was conducted, and the transmission rate of 5G network reached 10Gb/s with the 1 ms network delay [10]. In the future, the screens in automobiles may change due to the increasing demand for audio and video entertainment.

In addition, VR technology has become increasingly mature in various industries in recent years, and VR devices allow users to become "witnesses" and "field observers" in immersive, three-dimensional and multi-sensory reception situations [10], and to have a better using experience. Users can retrieve programs or data from the cloud directly through the 5G mobile network by using VR terminal equipment in the car, and hard disk is needed no more. It is pointed out that 7 ms delay is ideal in the scene where VR devices participate, while 5G can reach 1 ms delay [11], which is far superior to the ideal standard. Soon, VR games, VR movies and other VR activities in the car will become a new leisure trend. Audi has released a set of VR equipment called Experience Ride, and the rear passengers can realize the close combination and linkage between the game plot and the driving state of the vehicle through VR glasses [12], which attests the realizability of VR equipment in automobiles.

AR equipment can also benefit from V2X technology based on 5G to play its role in vehicles. AR equipment provided to passengers will be capable to display the basic information of surrounding roads, buildings and vehicles as well as the vehicle condition information, and assist drivers in judging and making decisions. In addition to the assisting driving function, AR devices can also provide entertainment activities.

**Mobile Office.** After the COV19 pandemic outbreak in 2020, mobile office has changed from an option to a mandatory choice [13]. Office workers can handle anything related to business at anytime and anywhere [14]. Whether working in an office or not, employees need wireless network to support a more flexible way of working [15]. On the one hand, 5G provides a wireless network foundation for in-car office work. Studies have specifically put forward the research ideas of security architecture design and key security technology system under 5G mobile edge computing (MEC) conditions, and given two solutions: security control of sensitive places and mobile office system at the edge of cloud terminals [16]. On the other hand, 5G autonomous driving frees up spare time for drivers, which makes working in cars a potential demand. Correia, GHD and other researchers pointed out that if people choose autonomous vehicles with office functions, they are more willing to save time in the office (using travel time instead of time at home) rather than working overtime in the morning/evening rush hour [17]. Nissan cooperated with Studio Hardie to transform its e-NV200 car into a mobile office version E-NV200WORKSPACE, with foldable desk, mobile office chair, wireless charger and other equipment. Under the precondition of automatic driving, there is no need to hire a driver. In the future, the in-car office environment may be a complete office integrated with computers, keyboards and desk lamps, or it may only provide office desks and chairs and require users to bring their own personal computers. However, in-car office may be a new and growing demand for automotive functions.

**Intelligent Push.** The development of mobile edge computing (MEC) and artificial intelligence will automatically create, capture, distribute, receive and consume the contents and services required by users in different intelligent scenes [18], so that the server can accurately and punctually deliver the vehicle maintenance services and entertainment services required by users to each user's vehicle. With the goal of extracting customers' personalized demand for automobile products and customizing service experience, a system of customer entry management, e-commerce service, personalized demand management, vehicle configuration and intelligent service with customers as the core can be developed [19].

## 3.2 Design and Implementation of User Interview

The basic purpose of this interview is to further explore the users' demand for car function and understand the characteristics of people of different ages and identities when using automobiles (whether riding or driving) based on the existing research results. In order to get the interviewees' real thoughts, semi-structured interviews are adopted to obtain true, reliable and abundant information. The content of the interview roughly includes the interviewee's basic information, daily life, thoughts on existing cars, imagination of future cars, etc.

## 3.3 Analysis of User Needs

The interviewees in this interview ranged in age from 20 to 50, including experts, car owners, and students majoring in design. Experts have rich professional knowledge, car owners have rich experience, and students have good imagination and forward-looking power. Through in-depth communication, the explicit and implicit needs of the interviewees are collected and sorted out. According to the information obtained from the interview, we conducted the next expert meeting and evaluated the operability of the users' needs. And based on the users' needs, we conducted a brainstorming meeting to supplement the needs. After sorting out all the requirements, four space forms of new network connected cars are established, including office space, entertainment space, rest space and conversation space. The requirements are classified into five types: basic function requirements, office space requirements, entertainment space requirements, rest space requirements and conversation space requirements. All requirements are classified and integrated as shown in Table 1. It must be noted that the classification of functions is not completely accurate, and different automobile function spaces may share one function.

**Table 1.** The list of car function requirements

| Group | | Code | Element |
|---|---|---|---|
| Space function | Office space(A) | $A_1$ | Folding desk |
| | | $A_2$ | On-board computer |
| | | $A_3$ | On-board conference system |
| | | $A_4$ | Coffee machine |
| | | $A_5$ | Desk lamp |
| | Rest space(B) | $B_1$ | Screen of large size |
| | | $B_2$ | VR/AR devices |
| | | $B_3$ | Immersive sound system |
| | | $B_4$ | Video game system |
| | | $B_5$ | On-board refrigerator |
| | | $B_6$ | Karaoke |
| | | $B_7$ | Dressing table |
| | Entertainment space(C) | $C_1$ | Reclining seat |
| | | $C_2$ | Seat massage and heating |
| | | $C_3$ | Air purification |
| | | $C_4$ | Panoramic sunroof |
| | | $C_5$ | Space for pets |
| | Conversation space(D) | $D_1$ | Water dispenser |
| | | $D_2$ | Sofa seat |
| | | $D_3$ | Green plants |
| | | $D_4$ | Table of large size |
| Basic function(E) | | $E_1$ | Modular purchasing |
| | | $E_2$ | Active information push |
| | | $E_3$ | Car summon |
| | | $E_4$ | Automatic parking |
| | | $E_5$ | Remote boot |

# 4  Data and Result

## 4.1  Questionnaires and Semi-structured Interviews

We get experimental data from people from different areas. There are 132 people in the sample, who are generally between 20 and 50 years old. They are car owners, experts or students majoring in design, and their data is of more practical value.

After the questionnaire survey, we conducted semi-structured interviews with some interviewees, aiming at getting a deeper understanding of the satisfaction and opinions of potential users on the innovative functions of automobiles. The questions include but are not limited to "How often do you think you use this function?", "Will this function be the decisive factor in your choice of buying this car?", "Which of these two functions do you think is more important to you?". Through semi-structured interviews, we can have a richer understanding of the user satisfaction level of the listed innovative functions.

### 4.2  Data and Analysis of Kano Questionnaire

The Kano model defines five types of requirements.

- Must-be Quality. Without this attribute, users will be very disappointed. With this attribute, users' satisfaction level will not increase significantly. This is the quality that a product must contain, and it is an obligation to provide users with such a function.
- One-dimensional Quality. There is a linear relationship between the attributes and users' satisfaction level. The more satisfied this attribute is, the higher the user satisfaction is. This is the expected demand of users. When products need to further improve the satisfaction extent of users, this indicator should be improved.
- Attractive Quality. When this quality exists, users will feel joyful, and when it does not exist, it will not cause users to be disappointed.
- Indifferent Quality. Whether exists or not, it will not affect users' satisfaction level. This attribute needs to be excluded as much as possible, and it is meaningless to improve users' satisfaction degree.
- Reversal Quality. The attribute will reduce users' satisfaction and should be completely avoided [7].

According to the statistical results of Kano questionnaire, 9 of all 26 functional requirements are Attractive Quality, including coffee machine, screen of large size, on-board refrigerator, reclining seat, seat massage and heating, green plants, car summon, automatic parking and remote boot. Three items are One-dimensional Quality, including immersive sound system, air purification and panoramic sunroof. Two items are Reversal Quality, including space for pets and table of large size. Other functional requirements are Indifferent Quality. In addition, folding desk (36.55%), on-board computer (31.03%) and dressing table (37.93%) also received higher answer of Reversal Quality. See Table 2 for relevant data.

### 4.3  Better-Worse Coefficient Analysis

By bringing the statistical results summarized by Kano questionnaire into the calculation formula of Better-Worse coefficient, we can get the specific influence degree of the increase or decrease of a certain function demand on users' satisfaction level. The Better coefficient (also known as SI) and the Worse coefficient (also known as DSI) are formulated as follows.

$$Better/SI = \frac{A + O}{A + O + M + I} \tag{1}$$

$$Worse/DSI = -1 \times \frac{O + M}{A + O + M + I} \tag{2}$$

The value of Better coefficient is usually positive, which means that if a certain function is provided, users' satisfaction degree will increase. The closer the value is to 1, the easier it is to improve users' satisfaction degree. Worse coefficient is usually negative, which means that if a certain function is not provided, the users' satisfaction degree will decrease. The closer the value is to −1, the easier it is to reduce users' satisfaction. The data calculation results are summarized in Table 2.

**Table 2.** Results of the questionnaire data analysis

| Code | A | O | M | I | R | Q | Better | Worse |
|------|---|---|---|---|---|---|--------|-------|
| $A_1$ | 15.86% | 0.00% | 5.52% | 41.38%[a] | 36.55% | 0.69% | 25.27% | − 8.79% |
| $A_2$ | 16.55% | 1.38% | 0.00% | 48.28%[a] | 31.03% | 2.76% | 27.08% | − 2.08% |
| $A_3$ | 3.45% | 0.69% | 0.69% | 66.21%[a] | 26.90% | 2.07% | 5.83% | − 1.94% |
| $A_4$ | 49.66%[a] | 0.69% | 0.69% | 36.55% | 9.66% | 2.76% | 57.48% | − 1.57% |
| $A_5$ | 29.66% | 2.07% | 11.72% | 52.41%[a] | 1.38% | 2.76% | 33.09% | − 14.39% |
| $B_1$ | 34.48%[a] | 25.52% | 19.31% | 15.86% | 3.45% | 1.38% | 63.04% | − 47.10% |
| $B_2$ | 35.17% | 0.69% | 0.69% | 55.17%[a] | 5.52% | 2.76% | 39.10% | − 1.50% |
| $B_3$ | 31.72% | 35.17%[a] | 15.17% | 16.55% | 0.69% | 0.69% | 67.83% | − 51.05% |
| $B_4$ | 19.31% | 6.90% | 4.83% | 35.17%[a] | 33.10% | 0.69% | 39.58% | − 17.71% |
| $B_5$ | 53.10%[a] | 6.90% | 10.34% | 28.28% | 0.69% | 0.69% | 60.84% | − 17.48% |
| $B_6$ | 14.48% | 5.52% | 10.34% | 53.79%[a] | 13.79% | 2.07% | 23.77% | − 18.85% |
| $B_7$ | 15.17% | 1.38% | 0.69% | 43.45%[a] | 37.93% | 1.38% | 27.27% | − 3.41% |
| $C_1$ | 48.97%[a] | 15.86% | 11.03% | 23.45% | 0.00% | 0.69% | 65.28% | − 27.08% |
| $C_2$ | 29.66%[a] | 28.97% | 15.86% | 24.83% | 0.00% | 0.69% | 59.03% | − 45.14% |
| $C_3$ | 32.41% | 40.00%[a] | 12.41% | 14.48% | 0.00% | 0.69% | 72.92% | − 52.78% |
| $C_4$ | 25.52% | 38.62%[a] | 0.69% | 27.59% | 6.90% | 0.69% | 69.40% | − 42.54% |
| $C_5$ | 8.97% | 1.38% | 0.69% | 37.93% | 50.34%[a] | 0.69% | 21.13% | − 4.23% |
| $D_1$ | 15.17% | 0.69% | 0.69% | 69.66%[a] | 7.59% | 6.21% | 18.40% | − 1.60% |
| $D_2$ | 36.55% | 7.59% | 9.66% | 40.00%[a] | 5.52% | 0.69% | 47.06% | − 18.38% |
| $D_3$ | 59.31%[a] | 6.21% | 0.00% | 26.21% | 7.59% | 0.69% | 71.43% | − 6.77% |
| $D_4$ | 2.76% | 0.69% | 0.00% | 35.86% | 53.79%[a] | 6.90% | 8.77% | − 1.75% |
| $E_1$ | 31.72% | 17.24% | 2.76% | 45.52%[a] | 0.69% | 2.07% | 50.35% | − 20.57% |
| $E_2$ | 28.97% | 11.03% | 4.83% | 47.59%[a] | 6.21% | 1.38% | 43.28% | − 17.16% |
| $E_3$ | 66.21%[a] | 6.90% | 0.69% | 25.52% | 0.00% | 0.69% | 73.61% | − 7.64% |
| $E_4$ | 66.90%[a] | 24.14% | 0.69% | 7.59% | 0.00% | 0.69% | 91.67% | − 25.00% |
| $E_5$ | 57.24%[a] | 15.17% | 2.76% | 23.45% | 0.00% | 1.38% | 73.43% | − 18.18% |

[a]represents the category with the highest percentage

According to the calculation result of Better-Worse coefficient, the scatter plot can be drawn, as shown in Fig. 3. In the first quadrant is One-dimensional quality. In the second quadrant is Attractive Quality. In the third quadrant is Indifferent Quality. In the fourth quadrant is Must-be Quality. According to the position distribution of points of each function item, statistics can be derived: One-dimensional quality: 7 items, Attractive Quality: 5 items, Indifferent Quality: 14 items, Must-be Quality: 0 items. See Table 3 for final Kano classification of car function.

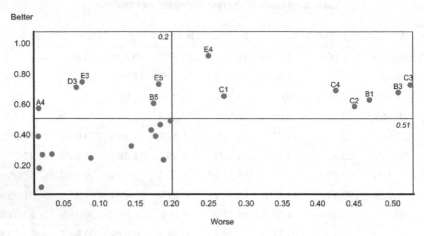

**Fig. 3.** Scatter plot of the Better-Worse coefficient analysis

**Table 3.** Kano classification of car function

| Elements classification | Code |
| --- | --- |
| One-dimensional quality (O) | $B_1$, $B_3$, $C_1$, $C_2$, $C_3$, $C_4$, $E_4$ |
| Attractive quality (A) | $A_4$, $B_5$, $D_3$, $E_3$, $E_5$ |
| Indifferent quality (I) | $A_1$, $A_2$, $A_3$, $A_5$, $B_2$, $B_4$, $B_6$, $B_7$, $C_5$, $D_1$, $D_2$, $D_4$, $E_1$, $E_2$ |

### 4.4 The Qualitative Analysis

**Office Space.** In the category of mobile office, five functional requirements are defined, among which only the coffee machine is considered as Attractive Quality. Many interviewees said that they did have the experience of working in the car, but only a few times, and the work can be done by mere mobile phone. The functions such as on-board computer, folding desk and desk lamp are not strongly needed, but they will not be disgusted if they do not occupy a large space in the car.

**Entertainment Space.** In the analysis of Better-Worse coefficient, immersive sound system and large-size screen show One-dimensional Quality. In interviews, respondents generally think that intelligent network connected cars lacking these two functions are "outdated" and "boring". In addition, the video game system is more attractive to male, but less attractive to female. On the contrary, the on-board dressing table attracts more female interviewees, therefore it needs careful consideration when adding these two functions. In addition, respondents are generally curious about VR/AR devices, but they don't know enough about them and lack experience in using them, which may be the reason why this function belongs to Indifferent Quality, but there are still more than 35% of the answers classify this function as Attractive Quality. When VR/AR devices are prevalent and powerful enough, this function may be more needed and attractive.

**Rest Space.** In the interview, we found that the interviewees pay more attention to the comfort of the rest space, which is reflected in the design of seats. The interviewees are very interested in the seats that can lie down, and think that the massage function can make them rest more comfortably. However, possibly due to the cost considerations, the absence of these two functions will not make users feel sad, as long as the price of cars is low enough. In addition, the design of pet space is not optimistic, because it is a burden for users without pets, and those who have pets prefer pets to stay beside themselves or other passengers, instead of staying alone in the space planned for them.

**Conversation Space.** Respondents are most interested in green plants, and think that green plants can improve mood and keep healthy. However, some respondents complain that green plants may be difficult to take care of, so when choosing green plants, designers should try to choose ones that are not easy to wither and do not require human intervention. There are also some respondents who are interested in sofas. Their main consideration is whether the existence of sofas will greatly increase the price of cars. In addition, tables are considered to occupy too much interior space and are repugnant.

**Basic Functions.** Car summon, automatic parking and remote start are newly introduced functions in recent years, which only exist in a few models. After we explained them in detail, the respondents were interested in these functions. However, the functions of modular purchasing and active information push are not easy to be understood by respondents, so respondents showed caution.

### 4.5  Innovative Function Model of Intelligent Network Connected Cars

According to the conventional prioritization criteria: Must-be Quality > One-dimensional Quality > Attractive Quality > Indifferent Quality, the innovation function model of intelligent networked vehicles is constructed. Due to the lack of necessary attributes, we set up a primary function layer with the One-dimensional Quality function as the core, in which the functions are provided first, so as to meet the users' expected needs under 5G environment and improve users' satisfaction degree. Attractive Quality functions make up a high-level function layer. Functions at this level should be provided as appropriate, before which factors such as cost, target population and target market are globally considered, so as to further improve users' satisfaction level and brand loyalty. The innovative function model of intelligent network connected cars is shown in Fig. 4.

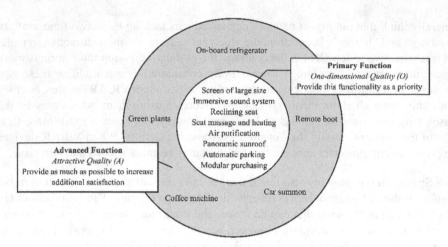

**Fig. 4.** Innovative function model of intelligent network connected cars

# 5    Conclusion and Prospect

In this paper, we construct a research method of car function based on Kano model. There is no doubt that people have realized the impact of 5G network technology on various fields, especially for the automobile industry. The evolution of car networking and autonomous driving technology enables drivers to spend most of their time on non-driving activities. However, how the 5G network technology will lead to the change of automobile functions remains to be explored. Therefore, based on the research method, we get the users' demand for car function. After a questionnaire survey of 26 different functions (including office space function, entertainment space function, rest space function, conversation space function and basic function), they are classified into demand attributes, and finally the car function model is constructed. Facts have proved that this research method can help enterprises and designers scientifically and accurately design the car functions under 5G network environment, giving full play to the advantages of 5G technology, and thus maintain or further improve users' satisfaction level and brand loyalty, which has theoretical significance. Secondly, the research results of 26 car functions and the innovative car function model constructed in this paper can directly provide experience for car function design and promote the upgrading of car functions under 5G network environment, which has the practical significance.

The limitation of this paper is that the interviewees and questionnaire respondents used to obtain the information are clustered between 20–50 years old, hence it is insufficient to attain and study the demand of children, teenagers and elderly people. Because the acceptance extent and understanding of new technologies are inferior to those of young people, interviews with children, middle-aged and elderly people may need other auxiliary means. In the future research, the demand of automobile users of all ages needs to be comprehensively acquired and studied, so as to make the car function more universally applicable.

# References

1. White paper on the economic and social impact of 5G. http://www.caict.ac.cn/kxyj/qwfb/bps/201804/t20180426_158438.htm. Accessed 12 Feb 2021. (in Chinese)
2. Zhou, P., Yang, Q.R., Yang, J., Wu, X.J.: Brief analysis of 5G technology and the development of intelligent connected vehicles. J. Kunming Metall. Coll. **35**(4), 90–94 (2019). (in Chinese)
3. Zhang, G.F., Xu, W.S., Song, J.F.: Automobile product customer demand analysis based on AHP-BP neural network. China Econ. Trade Herald **2020**(7), 136–140 (2020). (in Chinese)
4. Ji, W.Y.: Study on demand anlysis and system design of intelligent vehicle travel entertainment system. Inf. Commun. **2020**(8), 102–105 (2020). (in Chinese)
5. Wang, H.: Analysis of the development trend of automobile entertainment system design. SME Manage. Technol. **2019**(12), 185–186 (2019). (in Chinese)
6. Anable, J., Gatersleben, B.: All work and no play? The role of instrumental and affective factors in work and leisure journeys by different travel modes. Transp. Res. Part A-Policy Pract. **39**(2–3), 163–181 (2005). https://doi.org/10.1016/j.tra.2004.09.008
7. Yang, C.C.: The refined Kano's model and its application. Total Qual. Manag. Bus. Excell. **16**(10), 1127–1137 (2005). https://doi.org/10.1080/14783360500235850
8. Matzler, K., Hinterhuber, H.H.: How to make product development projects more successful by integrating Kano's model of customer satisfaction into quality function deployment. Technovation **18**(1), 25–38 (1998)
9. Kun, A.L., Boll, S., Schmidt, A.: Shifting gears: user interfaces in the age of autonomous driving. IEEE Pervasive Comput. **15**(1), 32–38 (2016)
10. Lu, F.: Enabling and transmutation: discussion on media convergence applications in 5G context. China Radio TV Acad. J. **2020**(07), 21–23 (2020)
11. Feng, D.G., Xu, J., Lan, X.: Study on 5G mobile communication network security. J. Softw. **29**(6), 1813–1825 (2018). https://doi.org/10.13328/j.cnki.jos.005547 (in Chinese)
12. Zhang, J.: Audi: build the interior of the car into an entertainment experience center. Light Veh. **2019**(7), 49–53+28 (2019). (in Chinese)
13. Zhu, D.: Telecommuting in the epidemic is forcing enterprises to upgrade their "bandwidth" from 2G to 5G. Sino Foreign Manage. **2020**(Z1), 211–214 (2020). (in Chinese)
14. Li, J.Q.: On the construction of enterprise mobile office platform. Hongshui River **36**(1), 121–124 (2017). (in Chinese)
15. Focus on 5G: Enterprises need indoor 5G technology -- the importance of 5G and Wi-Fi 6 for the future of the office. Inf. Commun. Technol. **14**(5), 69–76 (2020). (in Chinese)
16. Ren, F., Wen, X., Lyu, L.: Security application of 5G MEC architecture in mobile office system. Secrecy Sci. Technol. **2019**(4), 28–34 (2019). (in Chinese)
17. Correia, G.H.D., Looff, E., van Cranenburgh, S., Snelder, M., van Arem, B.: On the impact of vehicle automation on the value of travel time while performing work and leisure activities in a car: theoretical insights and results from a stated preference survey. Transp. Res. Part A-Policy Pract. **119**, 359–382 (2019). https://doi.org/10.1016/j.tra.2018.11.016
18. Yu, G.M., Li, F.P.: Media innovation in 5G era: a research on the applied paradigm of edge computing. J. Shanxi Univ. (Philosophy and Social Science Edition) **43**(1), 65–69 (2020). https://doi.org/10.13451/j.cnki.shanxi.univ(phil.soc.).2020.01.006 (in Chinese)
19. Zheng, J.F.: Research on the new intelligent manufacturing model of automotive mass personalization customization. China Collective Economy **2018**(16) (2018). (in Chinese)

# Designing a New Electric Vehicle Charging System: People's Preference and Willingness-To-Pay

Lanyun Zhang[1], Tracy Ross[2], and Rebecca Cain[2(✉)]

[1] Nanjing University of Aeronautics and Astronautics,
29 Yudao Street, Nanjing 210016, P. R. China
[2] Loughborough University, Epinal Way, Loughborough L11 3TU, UK
R.Cain@lboro.ac.uk

**Abstract.** On-street Vehicle-to-Grid (V2G) allows battery electric vehicles (BEVs) to park on the street, charge from the electricity grid, and give electricity back to the grid. It could encourage people without off-street parking at home to adopt electric vehicles. On-street V2G is technically feasible, but there is a lack of understanding of consumer perceptions of On-street V2G features and services. This study aims to quantitatively and empirically explore people's attitudes towards On-street V2G with a focus on preference and willingness-to-pay. An online survey was carried out and 495 successful responses were collected. In the survey, a video clip that explained On-street V2G was included to better sensitise participants to the futuristic scenario of On-street V2G. This study found that 'required plug-in hours per month' was viewed as more important regarding preference for On-street V2G, while 'minimum level of battery guaranteed' was more important regarding willingness-to-pay. People hold opposite views towards these two features concerning preference and willingness-to-pay. The theoretical contribution, practical implications, and future research directions are discussed.

**Keywords:** Electric vehicle · Vehicle-to-grid · On-street · Preference · Willingness-to-pay · Quantitative

## 1 Introduction

The UK government has decided to encourage uptake of battery electric vehicles (BEV), as a strategy to decarbonise transportation that now remains significantly dependent on fossil fuels [1, 2]. Transition to electric vehicles requires a resilient power system so that it can deal with the massive electricity burden on the electricity grid caused by BEV charging. To reduce the electricity burden, a new charging concept has been developed: Vehicle-to-Grid (V2G) [4]. It allows BEVs to charge from the electricity grid and discharge electricity stored in the batteries back to the grid. V2G users can choose suitable time slots to charge when electricity is cheap and sell it when electricity is expensive [3]. Based on average daily electricity consumption in the UK [5], V2G financially benefits users the most if they charge BEVs overnight when electricity is

© Springer Nature Switzerland AG 2021
H. Krömker (Ed.): HCII 2021, LNCS 12791, pp. 184–195, 2021.
https://doi.org/10.1007/978-3-030-78358-7_12

cheap and use it or sell it during the day when electricity is expensive. This pattern suits most people who have a normal and regular work shift. Overall, V2G allows users to take control of their BEV charging behaviour, according to their lifestyles, driving patterns, financial requirements, and personal preferences, and at the same time, minimise the overall consumption of electricity.

BEV as a relatively new technology compared to conventional vehicles, has received social justice concerns. It is argued that BEV currently benefit a limited number of citizens such as people able to afford chargers and off-street parking at home [1]. To address these equity issues, On-street V2G is proposed to encourage people without off-street parking at home to adopt BEV and V2G [6, 7, 8]. On-street V2G utilises street furniture such as lampposts and bollards for V2G, which allows BEV drivers to park on the street, connect to the electricity grid, charge and discharge. Only a few studies have mentioned the concept of addressing equality through supporting charging on the street.

At a technical level, On-street V2G is feasible with the collaboration of manufacturers, power suppliers, charging facility companies, and V2G operators. However, to design a system that meets users' needs, it remains unknown how people without off-street parking at home (as potential consumers) will perceive an On-street V2G system. Existing studies have summarised some key features in V2G packages that greatly influence people's decision-making, and two of these features were explored in this study: 'required plug-in hours per month' and 'minimum level of battery range guaranteed'. In most V2G packages, BEV owners are required to plug in a certain number of hours during a fixed period of time to enable the electricity grid to take battery from their BEVs. As a guarantee, V2G packages will not draw battery beneath a pre-set level to guarantee a minimum driving range. Researchers and designers lack an understanding of how people perceive these two features for On-street V2G.

This study aims to quantitatively and empirically explore people's perceptions of On-street V2G features with a focus on preference and willingness-to-pay (WTP), for the purposes of designing a new charging system to further accelerate electric vehicle revolution and adoption. The objectives of this study are:

- To explore the underlying factors that influence user preference for On-street V2G services.
- To explore the underlying factors that influence willingness-to-pay for On-street V2G services.

## 2 Background and Related Work

This section introduces related work in two parts: (1) features related to EV and V2G service (including driving range, charging time, parking locations, discount time slots, etc.) and (2) user preference and willingness-to-pay.

### 2.1 Features Related to EV and V2G Service

Existing studies have looked into several features related to EV and V2G services, in which some are explored in-depth, and some are briefly mentioned in context. This section introduces the main features that have been studied and focuses on the two features that are studied in this paper.

**Driving Range on Full Battery:** Existing works have reported that the driving range of an EV is regarded one of the most important features when consumers are considering adopting EVs. As discussed in [21], for people who take many long-distance trips, electric vehicles may not be able to compete with conventional vehicles. On the other hand, for people who mainly take local trips (e.g., serving commuting purposes and grocery trips), electric vehicles might be a better choice.

**Battery Charging Time:** Research has found that the time taken to charge an electric vehicle is another important factor when consumers are considering adopting EVs [22]. It is common that people tend to compare a new product with what they have been used to. In this case, most conventional vehicles only need a trip to a petrol station to refuel every 300 miles, while electric vehicles need at least 30 min for 80% battery capacity on fast charging, which only gives about 50 miles driving range.

**Acceleration of EV:** Another interesting fact about electric vehicles is that acceleration, as a main driving experience, has changed compared to conventional vehicles. Acceleration represents the smooth or rapid process when increasing a vehicle's speed. Electric vehicles are popular for their faster acceleration performance that *"EV driving experiences are gradually overtaking their petrol supercar equivalents in popularity, with faster acceleration a major draw"* [12]. [13] has taken acceleration performance into consideration when testing the preference of adopting electric vehicles.

**Pollution of EV on V2G Contract:** Since electric vehicles are known for their environment-friendly property, consumers are keen to know whether joining V2G would increase the benefit to the environment or not. [14] found that people would like to know if V2G uses 'greener' electricity, including the percentage and the type of sustainable sources.

**Guaranteed Minimum Driving Range on V2G Contract:** [13, 14] have explored people's opinions towards signing up to V2G contract and found that although people showed interest in the idea of selling electricity back to the grid, they did not want to be left without enough battery charge in case of emergencies. Therefore, a new feature is developed in V2G contracts – guaranteed minimum driving range. It represents the lowest amount of charge that the vehicle's battery would have 'all the time' and consumers can select the guaranteed range according to their needs. It is displayed to respondent as both percentage charge and electric range in mile.

**Required Plug-in Time Per Day to Sell on V2G Contract:** Overall, people see high inconvenience cost with signing a V2G contract, due to factors such as people's desire for flexibility of car use, and their lack of awareness of how many hours their cars can be parked to sell electricity [13]. This feature is developed to make sure that V2G users are committed to selling electricity back to the grid on daily basis. V2G users are free to select the length of plug-in time to sell per day, which is related to a contractual financial benefit.

**Monetary Value on V2G Contract:** Studies have explored the financial benefit of signing up to a V2G contract, that drivers are not certain about earning money from re-selling power back to power companies [13]. Monetary value on a V2G contract comprises the discount of the price of EVs on V2G [13] and the discount of electricity cost on V2G [14]. This paper focuses on two main features related to the use of EV on V2G contract: guaranteed minimum driving range and required plug-in hour per month.

## 2.2 Consumer Preference and Willingness-To-Pay

Preference and willingness-to-pay (WTP) have been used to explore consumers' attitudes towards a product. Preference refers to certain characteristics any consumer wants to have in a good or service to make it preferable to the consumer [19]. The preferred characteristics may lead to consumer happiness, satisfaction, etc. Preference is the main factor that influences consumer demand. On the other hand, WTP refers to the maximum price a consumer is willing to pay for a product. It is typically represented by a price range or a value figure. With proper data collection and analysis, WTP will inform the maximum amount that consumers will pay, and also reveal what influences consumers' WTP [20].

Consumer preference and WTP have been studied in different areas, including the food industry, such as meat substitutes [18] and infant milk formula [17]; transportation policy, such as renewable fuel standard policy [16] and controlled charging schemes [13]; and medical treatment, such as vaccination services [15]. Many studies explored preference and WTP among different user groups to compare whether different user segments hold different attitudes towards a product. Most studies investigate preference and WTP separately with different latent factors.

Preference and WTP are used to explore consumers' attitudes towards a product on two levels. The first level is to examine preference and WTP for a specific product or variations of products, while the second level is to examine the underlying factors (i.e., latent factors) that explain what drives user preference and WTP [18]. Different approaches are used to measure the drivers, such as binary models and linear models [18].

In business, consumer preference and WTP are two important metrics to help evaluate whether a product suits its target users and whether it is profitable for investors [19]. On the one hand, measuring these two metrics can be an effective approach to understand users and consumers. On the other hand, outcomes of preference and WTP will determine a company's pricing strategy or whether a product needs to go back to the research and design phase, to make sure that it is market ready.

This paper aims to explore people's perceptions of On-street V2G with a focus on preference and WTP, for the purposes of designing a new charging system to further accelerate electric vehicle adoption.

## 3 Method

### 3.1 Study Design and Procedure

A major challenge of this study is that it is difficult to obtain people's preference and WTP of a future technology that does not currently exist. Therefore, visualisation and sensitising concepts [9] were employed to make it easier for participants to understand On-street V2G. Secondly, it is hard to ask for people's preference and WTP for different features in V2G packages directly. Thus, mock-up V2G packages with different combinations of feature values were created. Stated preference and WTP were collected after presenting all hypothetical V2G packages to participants, so that participants could decide on their optimal choices to maximise their utility.

An online survey was developed, guided by the framework of a potential customer journey. It describes a customer's experience with a product from initial engagement and into hopefully a long-term relationship. This survey adopted the first two stages of the customer journey: awareness of a product and consideration of a product.

**Table 1.** Study procedure.

| Part | Task | Questions |
|------|------|-----------|
| 1 | Participants were asked to watch a short video clip that illustrated the concept of On-street V2G | |
| 2 | On-street V2G 'webpage' was presented that included more about its feature variations | |
| 3 | Nine mock-up On-street V2G packages were presented | Preference ranking |
| | | WTP amount |
| | | Demographic Info |

In the first part of the survey, a short video clip was displayed to visually explain the idea of On-street V2G, shown in Table 1. This video was an animation that clearly illustrated the features of V2G, and the benefits customers could get from it. The details illustrated in the video was derived from the results of a workshop [10], in which all stakeholders were invited to create a holistic vision of On-street V2G. This video aimed to mimic one of the potential routes to consumer 'awareness' of an On-street V2G service.

In the second part of the survey, to mimic the 'consideration' stage in the customer journey map, an On-street V2G information 'webpage' was created and shown to the participants. On this 'webpage', mock-up On-street V2G packages were given to participants to indicate the variations of 'required plug-in hours per month' and 'minimum level of battery range guaranteed'.

In the third part of the survey, nine options for On-street V2G packages were shown to participants (see Table 2), followed by two questions. The first question asked participants to rank the given packages according to their preference, while the second question asked participants to assign a value to each package, indicating their WTP for On-street V2G service per month from £5 to £22. The details included in the 'webpage' and the range of monthly WTP was derived from a workshop [10]. Finally, demographic information about participants were collected.

**Table 2.** Options for on-street V2G packages.

| Package | Required plug-in hours per month | Minimum level of battery range guaranteed |
|---|---|---|
| 1 | 120 h | 80 miles |
| 2 | 60 h | 60 miles |
| 3 | 180 h | 40 miles |
| 4 | 180 h | 60 miles |
| 5 | 60 h | 80 miles |
| 6 | 120 h | 60 miles |
| 7 | 60 h | 40 miles |
| 8 | 180 h | 80 miles |
| 9 | 120 h | 40 miles |

### 3.2  Participants and Data Collection

A purposive screening process was followed to get access to a certain range of participants: adult, owning a car, not having access to off-street parking at home, and living in England, UK. MRFGR [11], an online survey platform, was employed to implement the questionnaire and recruit participants. Incentives were given to participants through a points system. In total, 495 completed surveys were received. Table 3 shows the participant information.

### 3.3  Data Analysis

To explore people's preference and WTP in terms of 'required plug-in hours per month' and 'minimum level of battery range guaranteed', descriptive analysis and statistical analysis were employed via SPSS. Firstly, to describe preference and WTP, median values were used due to the skewed data that both datasets did not match normal distribution. Secondly, conjoint analysis was carried out to investigate the underlying perceptions of these two features: 'required plug-in hours per month' and 'minimum level of battery range guaranteed', concerning preference and WTP.

## 4  Results

This paper reports our findings in two sections: preference and willingness-to-pay for an On-street V2G service, with a focus on two features - 'required plug-in hours per month' and 'minimum level of driving range guaranteed'.

**Table 3.** Participant information.

| Age | 18–29 | 55 |
|---|---|---|
| | 30–39 | 241 |
| | 40–49 | 98 |
| | 50–59 | 53 |
| | 60+ | 48 |
| Gender | Male | 313 |
| | Female | 182 |
| Household gross income | <£18,999 | 51 |
| | £19,000 - £31,999 | 96 |
| | £32,000 - £63,999 | 175 |
| | >£64,000 | 173 |
| Residence type | Detached | 250 |
| | Semi-detached | 79 |
| | Terraced house | 112 |
| | Flat/apartment | 54 |
| Type of car driven most frequently | Conventional car (petrol/diesel) | 370 |
| | Electric vehicle (hybrid and pure battery) | 125 |
| Number of cars in the household | 1 | 398 |
| | 2 | 71 |
| | 3+ | 26 |
| Typical car usage over a week | Very regular | 330 |
| | Quite regular | 122 |
| | Neutral | 16 |
| | Less regular | 23 |
| | Not regular at all | 4 |
| Young children (age 0–12) in the household | Yes | 309 |
| | No | 186 |
| EV chargers at the workplace | Yes | 272 |
| | No | 223 |

## 4.1  Preference

As shown in Fig. 1, a descriptive analysis indicates that participants showed greater preferences for package 2, 5, and 7 (the smaller the value, the greater the preference). Package 2, 5, and 7 had the shortest required plug-in hours per month.

Conjoint analysis indicates that 'required plug-in hours per month' (*Importance value: 86*) was viewed significantly more important than 'minimum level of battery range guaranteed' (*Importance value: 14*) when preference for On-street V2G services was considered. *Importance value* is provided automatically when conjoint analysis is performed via SPSS. It represents the significance level that each factor impacts on overall user preference, the greater the value, the greater the impact of a specific factor. The values are normalised in the scale from 0 to 100 by SPSS.

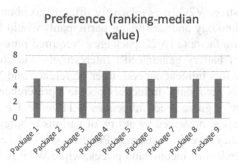

**Fig. 1.** Ranking-median value of preference (the smaller the value, the greater the preference).

### 4.2  Willingness-To-Pay

As shown in Fig. 2, descriptive analysis indicates that participants showed greater WTP for package 8, 5, and 1 (the greater the value, the greater the WTP). Package 8, 5, and 1 had the longest driving range guaranteed.

Conjoint analysis indicates that when WTP was considered, participants viewed 'minimum level of battery guaranteed' (*Importance value: 83*) as a significantly more important factor, compared with 'required plug-in hours' (*Importance value: 17*).

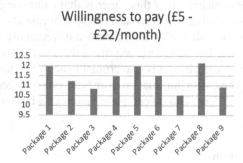

**Fig. 2.** Ranking-median value of willingness to pay (the greater the value, the greater the WTP).

## 5  Discussion

This section discusses findings from this study, including theoretical contributions and practical implications. Theoretical contributions contain consumer preference and WTP in the context of an On-street V2G service, and the importance of investigating preference and WTP in the design process. Practical implications cover a range of design considerations that might provide better On-street V2G services for potential consumers.

## 5.1 Theoretical Contributions

In the context of On-street V2G, it is relatively difficult to obtain people's opinions since it is a future technology and sensitising participants would be a challenge. This study explored two main factors in V2G packages: 'required plug-in hours per month' and 'minimum level of battery guaranteed'. Interestingly, via conjoint analysis, this study found that the factors influencing consumer preference and WTP were different. 'Required plug-in hours per month' influences people's preference significantly, while 'minimum level of battery guaranteed' influences WTP more. Consumer preference is defined as subjective tastes, measured by utility, of various bundles of goods according to the levels of utility they give the consumer. Preference is different from WTP in that preferences are independent of income and prices, while WTP is highly dependent on monetary factors, such as income, price, cost, and incentives. Findings from this study show that potential On-street V2G users place a higher priority on 'required plug-in hours per month' than 'minimum level of battery guaranteed', but only have a willingness to pay more for an increased 'minimum level of battery guaranteed'. It means that when the total amount of cost is limited, potential users would like to spend the limited amount for an increase in guaranteed driving range, over a decrease in required plug-in time. In other words, people would like to pay for more driving range, rather than having more flexible access to their vehicles. On the contrary, 'required plug-in hours per month' was viewed as more important when consumers were stating their preferences. Consumers' likes and dislikes do not determine their ability to purchase goods, and vice versa. Theoretically, this finding revealed how people weigh-up On-street V2G between its two main features when consumer preference and WTP are considered.

Another theoretical contribution of this paper is that examining consumer WTP at the early stage of user-centred design plays an as important role as examining consumer preference. This study found that WTP might show a different aspect of user response compared with preference. For designers, acquiring the knowledge of potential users' preference is important; meanwhile, it is also crucial to acquire what they would like to spend their money on. Thus, with information from both sides (i.e., consumer preference and WTP), designers will form a more complete vision of what target users need, and will pay for.

## 5.2 Practical Implications

This section will explain how the findings from this study can guide the design of On-street V2G services. 'Required plug-in hours per month' and 'minimum level of battery guaranteed' explored in this paper essentially represent potential user experiences of On-street V2G in two dimensions: 'driving time freedom' and 'driving distance freedom'.

From this study, the main underlying factor of preference is 'driving time freedom', while WTP is influenced by 'driving distance freedom'. The first design implication is to treat 'driving time freedom' and 'driving distance freedom' equally in the design phase. Designers and practitioners do not need to focus too much on whether consumers are keen on time-freedom or distance-freedom in the context of using On-street V2G. On-street V2G packages can be designed such that that all parameters are flexible to set up. It means that users can select their desired level of 'driving time freedom' and 'driving distance freedom' to create their personalised On-street V2G package. In this case, users are given the maximum freedom to choose and decide, while designers need to come up with a universal solution that can calculate how much money to charge for any personalised V2G package. This strategy will make using On-street V2G more like signing up to a 'pay as you go' package, that people pay for what they have consumed. The second design implication is to create levels between 'driving time freedom' and 'driving distance freedom'. The practical outcomes could be in various forms. For example, On-street V2G services could be packaged in groups for different user needs. 'Driving distance freedom' could be prioritised for people who would like to pay more for more driving distance.

# 6 Conclusions, Limitations and Future Work

## 6.1 Conclusions

This study contributes to the knowledge of how people without off-street parking at home perceive two key features of On-street V2G: 'required plug-in hours' and 'minimum level of battery guaranteed', focusing on preference and WTP. This study found that 'required plug-in hours' was viewed more important when preference was considered, while 'minimum level of battery guaranteed' was viewed more important when WTP was assessed. It means that people preferred 'driving time freedom' but were more willing to pay for 'driving distance freedom'.

Theoretically, this study found that people hold opposite views toward these two features when asked about preference and WTP. Preference is often investigated in the early stage of a user-centred design process, while WTP is often neglected in design research. Academics tend to approach users for design from many aspects, such as studying user perceptions and user cognition, but rarely consider WTP. WTP is usually related to the commercial or business side of developing products, which can sometimes be overlooked in design research. This paper found that WTP showed different sides of user response, which helped create a more complete vision of user segments and their view of products. This complete vision will help design and re-design. Therefore, exploring WTP is also very useful in design cycles. Practically, design implications for On-street V2G service development were discussed. This paper explained finding the balance between the two dimensions in user experience of On-street V2G: 'driving time freedom' and 'driving distance freedom'. 'Driving time freedom' and 'driving distance freedom' could be approached equally in design or have different priorities depending on consumer lifestyles.

## 6.2 Limitations and Future Work

Firstly, this study only focused on exploring drivers without off-street parking at home in England, UK, and future studies could investigate drivers from other regions to assess if there are cultural differences. Secondly, this study only explored two main features of On-street V2G packages and there are many other important features, such as the location of off-street charging, electricity rates in V2G packages, etc. Future work could further examine these features. Thirdly, this study used conjoint analysis to explore preference and WTP with full factorial analysis. Future work would employ partial factorial analysis to explore a greater number of features at once. Fourthly, this study used a vision of On-street V2G created among our project partners. Although these partners represented all the key stakeholder in V2G, we are aware that there might be different thoughts and ideas about how On-street V2G will be embodied in the future. Future work could focus on exploring new features and new visions of On-street V2G alongside the development of new technology.

**Acknowledgements.** This work was supported by the Innovate UK funded project V2Street [grant number 104224] and represents research carried out at Loughborough University. We thank the project partners Durham County Council, E-Car Club, EDF Energy, Imperial College London, Southend on Sea Borough Council, Ubitricity, UKPN, and Upside Energy for their contributions to the initial workshop and the development of the study reported in this paper. The authors are also appreciative to Nanjing University of Aeronautics and Astronautics for the Grant YAH20099, which enabled the continuing involvement of the primary author in the research beyond the initial project completion date.

# References

1. Sovacool, B.K., Noel, L., Axsen, J., Kempton, W.: The neglected social dimensions to a vehicle-to-grid (V2G) transition: a critical and systematic review. Environ. Res. Lett. **13**(1), 013001 (2018)
2. IEA 2015 Energy and Climate Change: World Energy Outlook Special Report (Paris: OECD/IEA), https://www.iea.org/reports/energy-and-climate-change, last accessed 2020/12/13.
3. Guille, C., Gross, G.: A conceptual framework for the vehicle-to-grid (V2G) implementation. Energy Policy **37**(11), 4379–4390 (2009)
4. Madawala, U.K., Thrimawithana, D.J.: A bidirectional inductive power interface for electric vehicles in V2G systems. IEEE Trans. Industr. Electron. **58**(10), 4789–4796 (2011)
5. Amber, K.P., Aslam, M.W., Hussain, S.K.: Electricity consumption forecasting models for administration buildings of the UK higher education sector. Energy and Buildings (2015)
6. Lam, A.Y.S., Yu, J.J.Q., Hou, Y., Li, V.O.K.: Coordinated autonomous vehicle parking for vehicle-to-grid services: formulation and distributed algorithm. IEEE Trans. Smart Grid **90**, 127–136 (2007)
7. Jiang, B., Fei, Y.: Decentralized scheduling of PEV on-street parking and charging for smart grid reactive power compensation. In: Innovative Smart Grid Technologies IEEE (2013)
8. Twahirwa, T., Kasakula, W., Rwigema, J., Datta, R.: Optimal exploitation of on-street parked vehicles as roadside gateways for social iov-a case of kigali city. J. Open Innov. Technol. Market Complex. **6**(73), 73 (2020)
9. Bowen, G.A.: Grounded theory and sensitizing concepts. Int. J. Qual. Methods **5**(3), 12–23 (2006)
10. Zhang, L., Cain, R., Ross, T.: Designing the user experience for new modes of electric vehicle charging: A shared vision, potential user issues and user attitudes. Int. Assoc. Soc. Design Res. **5**, 12–23 (2019)

11. MRFGR Homepage. https://mrfgr.com Accessed 13 Dec 2020
12. Lightning bolt…EV driving experiences overtake petrol supercars with faster acceleration a major draw. https://www.automobilesreview.com/auto-news/lightning-boltev-driving-experi ences-overtake-petrol-supercars-with-faster-acceleration-a-major-draw/127290/. Accessed 23 Dec 2020
13. Parsons, G.R., Hidrue, M.K., Kempton, W., Gardner, M.P.: Willingness to pay for vehicle-to-grid (v2g) electric vehicles and their contract terms. Energy Econ. **42**(1), 313–324 (2014)
14. Bailey, J., Axsen, J.: Anticipating PEV buyers' acceptance of utility-controlled charging. Transp. Res. Part A **82**, 29–46 (2015)
15. Nguyen, A.T.L., Le, X.T.T., Do, T.T.T., Nguyen, C.T., Hoang Nguyen, L., Tran, B.X.: Knowledge, preference, and willingness to pay for hepatitis b vaccination services among woman of reproductive age in Vietnam. Biomed Research International, 1–7 (2019)
16. Shin, J., Hwang, W.S.: Consumer preference and willingness to pay for a renewable fuel standard (RFS) policy: focusing on ex-ante market analysis and segmentation. Energy Policy **106**, 32–40 (2017)
17. Yin, S., Li, Y., Xu, Y., Chen, M., Wang, Y.: Consumer preference and willingness to pay for the traceability information attribute of infant milk formula. British Food J. **119**(6), BFJ-11-2016-0555 (2017)
18. Weinrich, R., Elshiewy, O.: Preference and willingness to pay for meat substitutes based on micro-algae. Appetite **142**, 104353 (2019)
19. Definition of Preference. https://economictimes.indiatimes.com/definition/preferences. Accessed 13 Jan 2021
20. Willingness to pay: What it is & how to calculate. https://online.hbs.edu/blog/post/willin gness-to-pay. Accessed 22 Jan 2021
21. Franke, T., Krems, J.F.: What drives range preferences in electric vehicle users? Transp. Policy **30**, 56–62 (2013)
22. Chéron, E., Zins, M.: Electric vehicle purchasing intentions: the concern over battery charge duration. Transp. Res. Part A: Policy Practice **31**(3), 235–243 (1997)

# Cooperative and Automated Mobility

# Users' Expectations, Fears, and Attributions Regarding Autonomous Driving – A Comparison of Traffic Scenarios

Hannah Biermann$^{(\boxtimes)}$ , Ralf Philipsen , Teresa Brell , Simon Himmel , and Martina Ziefle

Human-Computer Interaction Center, RWTH Aachen University,
Campus-Boulevard 57, 52074 Aachen, Germany
{biermann,philipsen,brell,himmel,ziefle}@comm.rwth-aachen.de
https://www.comm.rwth-aachen.de

**Abstract.** The development towards autonomous driving is about to change our future mobility. Though there is public consent on perceived benefits of autonomous driving, potential drawbacks are also considered. In this survey, we focused on the user-centered assessment of autonomous driving in comparison of two traffic scenarios ("city" vs "highway") using an online questionnaire. Participants were generally more affirmative in regard to using autonomous vehicles on the highway than in the urban area. Attributions measured on a semantic differential were more positive with respect to highway use. Motives regarding the use of autonomous vehicles differed between the scenarios, mainly in terms of the simplicity and complexity of the driving situation, with autonomous vehicles being considered feasible for highways but not for cities. However, some evaluation patterns were independent of the context, revealing controversial attitudes and perceived trade-offs in both scenarios. Participants indicated positive expectations regarding the novel transport experience, but also fears of use, mainly described in terms of control issues. Findings of this survey can be used in further research in the field of human-automation interaction and build the basis for education and communication concepts to increase public awareness.

**Keywords:** Autonomous driving · Benefits and barriers · Semantic differential · Traffic scenario · Highway · City

## 1 Introduction

Advanced driver assistance systems (adaptive cruise control, lane keeping, etc.) support human drivers in specific traffic situations to improve road safety and enhance driving comfort. The development towards autonomous mobility, i.e., vehicles that completely take over driving and parking, shall provide further

© Springer Nature Switzerland AG 2021
H. Krömker (Ed.): HCII 2021, LNCS 12791, pp. 199–212, 2021.
https://doi.org/10.1007/978-3-030-78358-7_13

advantages, e.g., transport efficiency, environmental benefits (especially if vehicles are electrified), and the potential to mobilize vulnerable user groups and people without a driver's license [6,7,9].

Research has revealed public consent on perceived advantages of using autonomous driving, though potential drawbacks are also considered. Schmidt et al. [12] identified increases in road safety as major benefit of use and concerns about data safety and privacy as potential disadvantages. König and Neumayr [8] discovered that autonomous vehicles were seen as a great mobility aid for elderly and disabled people and that multitasking while driving was highly valued, whereas legal issues and fears of hacker attacks were considered as barriers to use. A review of stated preference including a summary of benefits and barriers in autonomous driving is given in Gkartzonikas and Gkritza [5].

Certainly, user-centered requirements and needs for the use of autonomous driving are very diverse and individual, and the trade-off between perceived advantages and disadvantages may be determined by multifaceted factors (including human factors, vehicle type and applications, context of use, etc.). One research focus is on the use of autonomous vehicles in diverse traffic situations that require specific driving maneuvers due to varying on-road conditions (e.g., crossing pedestrians in urban areas, lane change at high speed on highways), which, apart from engineering challenges, may affect the perception and evaluation from the user's point of view.

Yet, there is empirical evidence that usage assessments vary with regard to different traffic environments [1,10]. However, the research focus has often been on other factors, such as contrasting automation levels or predicting the intention to use autonomous vehicles with regard to context as *one* explanatory factor. Studies that measure user-centered assessments of different traffic conditions and requirements in regard to using autonomous vehicles to identify common (possibly generic) as well as specific evaluation patterns are scarce, though such findings are essential to ensure that mobility services of the future get the chance to meet human needs in all their aspects.

In order to contribute to this research gap, we addressed the user-centered assessment of autonomous driving contrasting a city and highway traffic scenario. Both scenarios were created on the basis of the current state of technical development and previous studies in order to provide comparative research findings. For one thing, we focused on the participants' willingness to use autonomous vehicles in the respective traffic situation as well as perceived benefits and barriers. Secondly, we used a semantic differential to measure public perceptions, impressions, and connotations in regard to autonomous driving. This method allowed us to implicitly investigate user attitudes and has already been successfully applied before (cf. Brell et al. [2], for example).

Findings of this survey serve as a basis for follow-up research in the field of innovative mobility and human-automation interaction as well as for education and communication concepts to increase public awareness.

# 2 Questions Addressed and Empirical Research Approach

Based on qualitative data from preceding interviews, a quantitative online survey (i.e., questionnaire) was conducted in Germany. We addressed the following research questions comparing two traffic scenarios (city vs highway) from the perspective of future users:

1. How is the willingness to use autonomous driving?
2. Which attributions are associated with autonomous driving?
3. Which expectations and fears are there in regard to autonomous driving?

To answer our research questions, we conducted a two-step, consecutive empirical approach using qualitative and quantitative methods. Initially, guided interviews were conducted to identify expectations, perceived fears, and associations in regard to autonomous driving. In a second step, these findings were quantified in an online questionnaire survey, which is the focus on in this paper.

## 2.1 Questionnaire Design

The questionnaire was developed based on the interview results and consisted of two main parts: personal data as well as the scenario-based assessment of autonomous driving.

**Personal Data.** Demographic data were requested, including age, gender, and education. The participants were asked about their mobility behavior concerning the possession of a driver's license, car use, and driving style. For the latter, we used a semantic differential with attributes that distinguished between a restrained and brisk driving style, e.g., *slow – fast* or *anxious – bold* (range: 1–6 with lower numbers indicating a defensive and higher numbers indicating an offensive driving style).

**Assessment of Autonomous Driving.** Initially, the participants were asked whether they had heard of autonomous driving – as an introduction to the topic and to get an unbiased impression of the current public knowledge. Then, we provided all participants with the same definition of autonomous driving in terms of driverless vehicles that are capable of handling all tasks while driving in all situations and under all on-road conditions autonomously, i.e., a human driver is no longer required. As the survey was conducted in Germany using German language, the definition of autonomous driving based on a report of the German Association of the Automotive Industry (VDA) [14]. We referred to automation level 5 of the VDA, which is comparable to automation level 5 according to the standard of the SAE International [11].

Subsequently, the participants were randomly divided into two groups for the assessment of using autonomous driving in a respective traffic scenario: city and highway, respectively. For each traffic scenario we provided a textual description and a photo collage illustrating the scenario (available in the appendix). We ensured that the material was similar in regard to factors not under investigation (e.g., composition, wording) and different only in regard to the experimental factors (i.e., driving situation: city vs highway).

In both groups, the participants were asked the same questions about their perception and evaluation of autonomous driving. First, the willingness to use autonomous driving was evaluated (item: "Would you want to drive autonomously in this situation?"; answer options: "yes"/"no"). Second, perceived benefits and barriers were evaluated. Here, we focused on general usage motives, expectations, and fears (regarding safety, vehicle control, etc.) and aspects specific to the traffic situation (simplicity, complexity, speed). We used 6-point Likert scales (min $= 1$ "full disagreement" to max $= 6$ "full agreement"). Items based on preceding qualitative data and according to Trübswetter [13]. For deeper insights into the affective evaluation and motivation to adopt or reject autonomous driving, we used a semantic differential with contrasting attributes, e.g., *dangerous – safe* or *uncontrollable – controllable* (range: 1–6 with lower numbers indicating a negative and higher numbers indicating a positive evaluation) [13].

## 2.2 Data Collection and Analysis

As the survey addressed potential users of autonomous vehicles, which could be anyone in the future, ad-hoc participants were acquired using online distribution channels in social media (e.g., Facebook), messaging and communication services (e.g., mailing lists) to ensure an unbiased diversity of perspectives. Participation was on a voluntary basis, without payment or other gratification. People who participated in the survey indicated their interest in the topic.

The survey started with introductory information on the topic (i.e., user perceptions and attitudes towards autonomous driving) as well as data protection and privacy policies according to ethical research standards guaranteeing the participants' anonymity. The participants were made aware that there were no "right" or "wrong" answers and we encouraged them to answer intuitively, as we were interested in their personal opinions.

Descriptive and inferential statistics were used for data analysis. Cronbach's Alpha ($\alpha$) was measured for scale consistency and interpreted as good reliability for all scales according to $\alpha \geq .7$. The level of significance was set at 5%. For effect sizes, the partial eta-squared ($\eta^2$) was reported.

## 2.3 Participants

In total, 180 participants ($= N$) took part in the survey. Age ranged between 20 and 76 years ($M = 37.5$, $SD = 15.2$). The proportion of men (51.7%) was slightly higher than that of women (48.3%). With 52.7% university graduates, the level of education was above average (cf. Federal Statistical Office [3]).

Nearly all participants held a driving license (99.4%). The majority reported a daily car use (53.3%). According to self-reports (semantic differential), the participants considered their driving style, compared to other drivers, as rather bold ($M = 4.2$, $SD = 1.0$), fast ($M = 4.1$, $SD = 1.1$), sporty ($M = 3.8$, $SD = 1.2$), and offensive ($M = 3.6$, $SD = 1.2$), but also cautious and less risky ($M = 3.0$, $SD = 1.0$). Most of the participants (88.9%) reported that they had already heard of autonomous driving.

## 3  Results

For the assessment of autonomous driving in comparison of the two traffic scenarios "city" and "highway", the sample was divided into two equally sized groups, each with 90 group members. Next, we describe our obtained results.

### 3.1  City vs Highway: Usage Evaluation and Attributions

The willingness to use autonomous vehicles in the urban area amounted to 38.9%; the majority of the participants (61.1%) stated that they would not want to drive autonomously in this traffic scenario. In contrast, the willingness to use autonomous vehicles on the highway amounted to 60.0%; less than half of the participants (40.0%) stated that they would not want to drive autonomously in this traffic scenario.

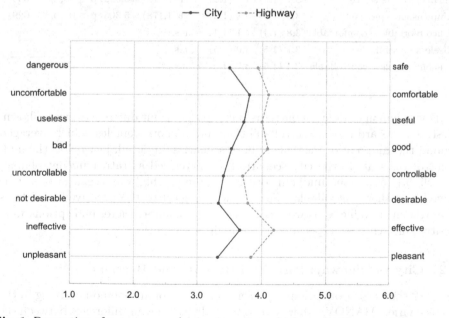

**Fig. 1.** Perception of autonomous driving in a city vs highway traffic scenario measured on a semantic differential (mean values).

H. Biermann et al.

To better understand the perception of autonomous driving, we further explored connotative meanings and affective evaluations, measured on a semantic differential, in comparison of the city and highway scenario with regard to functional and non-functional (e.g., emotional, social) aspects. Overall, evaluations were more positive in the highway scenario compared to the city scenario (see Fig. 1). A multivariate analysis of variance (MANOVA) showed a statistically significant difference between the traffic scenarios (as independent variable) on the combined dependent variables (i.e., all pairs of attributes) $(F(8, 171) = 2.064,$ $p < .05, \eta^2 = .088)$. Post-hoc univariate analysis of variance (ANOVA) were conducted for every dependent variable. Table 1 presents descriptive and inferential statistics. There were significant differences in the affective evaluation (e.g., in terms of "bad" and "good") and with regard to aspects of safety and effectiveness, provided that evaluations of using autonomous vehicles were more positive on highways than in the city.

Table 1. Descriptive (mean values, standard deviation) and inferential statistics ($F$-tests) of attributions measured on a semantic differential with regard to a city vs highway traffic scenario.

| Attributions | "City" | "Highway" | F-tests |
|---|---|---|---|
| Dangerous – safe | 3.5 (1.5) | 3.9 (1.4) | $(F(1,178) = 4.141, p < .05, \eta^2 = .023)$ |
| Bad – good | 3.5 (1.5) | 4.1 (1.4) | $(F(1,178) = 6.994, p < .01, \eta^2 = .038)$ |
| Not desirable – desirable | 3.3 (1.6) | 3.8 (1.5) | $(F(1,178) = 3.928, p < .05, \eta^2 = .022)$ |
| Ineffective – effective | 3.7 (1.5) | 4.2 (1.3) | $(F(1,178) = 6.699, p < .05, \eta^2 = .036)$ |
| Unpleasant – pleasant | 3.3 (1.6) | 3.8 (1.5) | $(F(1,178) = 5.305, p < .05, \eta^2 = .029)$ |
| Uncomfortable – comfortable | 3.8 (1.4) | 4.1 (1.4) | n.s. |
| Useless – useful | 3.7 (1.5) | 4.0 (1.5) | n.s. |
| Uncontrollable – controllable | 3.4 (1.5) | 3.7 (1.5) | n.s |

Few common evaluation patterns (i.e., no significant differences) were shown. First, with regard to the controllability of autonomous vehicles, which averaged around the midpoint of the scale in both scenarios: in detail, just below the midpoint of the scale in the city scenario (i.e., perceived as rather uncontrollable) and slightly above the midpoint of the scale in the highway scenario (i.e., perceived as rather controllable). Second, with regard to the perceived usefulness of autonomous vehicles as well as the expected comfort: here, perceptions were positive in each case, especially in the highway scenario.

## 3.2 City vs Highway: Perceived Benefits and Barriers

Figure 2 shows the evaluation of perceived benefits of autonomous driving in the two scenarios. MANOVA showed a statistically significant difference between the traffic scenarios (as independent variable) on the combined dependent variables (i.e., all perceived benefits) $(F(10, 169) = 3.313, p < .01, \eta^2 = .164)$. Post-hoc univariate ANOVA were conducted for every dependent variable. Results show a statistically significant difference between the traffic scenarios for the items "autonomous driving is suitable here due to the simplicity of the situation" $(F(1, 178) = 21.256, p < .001, \eta^2 = .107)$ and "autonomous driving is suitable here due to the complexity of the situation" $(F(1, 178) = 10.104, p < .01, \eta^2 = .054)$. In both cases, participants agreed that traffic conditions (simplicity and complexity) were appropriate for autonomous vehicle use on the highway, but not in the city center.

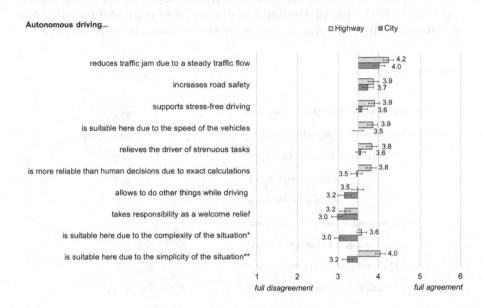

**Fig. 2.** Evaluation of perceived benefits of using autonomous driving (mean values and standard error; * corresponds to $p < .01$, ** corresponds to $p < .001$).

Besides, similar evaluation emerged in terms of whether or not an item was perceived as motive for use. In both scenarios, traffic optimization in the sense of congestion reduction through improved traffic flow was mainly considered beneficial, whereas the ability to do other things while driving and reduced responsibility were not considered as advantages of the use of autonomous vehicles, neither on the highway nor in the city. Increases in road safety, reliefs of the human driver, and the situational speed suitability were perceived as usage motives in both scenarios, though agreements were slightly higher with regard to highways. Notably, the perceived reliability of autonomous vehicles differed between the scenarios (although not significantly): participants tended to rely more on autonomous vehicle technologies on the highway than in the city.

Figure 3 shows the evaluation of perceived barriers of autonomous driving, differentiating between the two scenarios. MANOVA showed a statistically significant difference between the traffic scenarios (as independent variable) on the combined dependent variables (i.e., all perceived barriers) $(F(10, 169) = 3.155,$ $p < .01, \eta^2 = .157)$. Post-hoc univariate ANOVA were conducted for every dependent variable. Results show a statistically significant difference between the traffic scenarios for the items "autonomous driving cannot drive as well as human drivers" $(F(1, 173) = 4.176, p < .05, \eta^2 = .023)$, "autonomous driving is not suitable here due to the speed of the vehicles" $(F(1, 170) = 10.272, p < .01,$ $\eta^2 = .055)$, "autonomous driving is not suitable here due to the unclear traffic situation" $(F(1, 170) = 26.222, p < .001, \eta^2 = .128)$, and "autonomous driving is not suitable here due to the complexity of the situation" $(F(1, 166) = 19.969,$ $p < .001, \eta^2 = .101)$. Overall, the items mentioned were more likely perceived as usage barriers in cities, whereas they were rejected as usage barriers on highways.

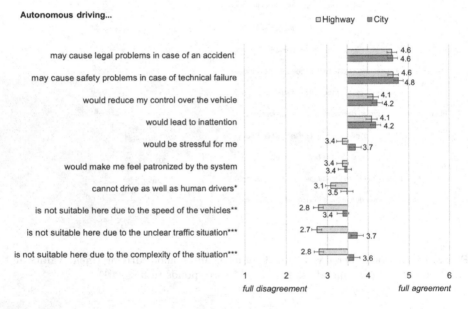

**Fig. 3.** Evaluation of perceived barriers of using autonomous driving (mean values and standard error; * corresponds to $p < .05$, ** corresponds to $p < .01$, *** corresponds to $p < .001$).

Evaluations were consistent in that safety and legal issues, reduced vehicle control, and possible inattention (of the human passenger) were perceived as potential drawbacks in both scenarios. Considering autonomous driving to be stressful was confirmed as disadvantage in the city scenario, but rather not in the highway scenario, whereas feelings of being patronized by the (vehicle) system were denied as potential barrier to use in both scenarios.

# 4 Discussion

This survey presented insights into users' perceptions and needs in relation to future autonomous driving in comparison of two traffic scenarios: "city" vs "highway". We discovered assessment differences in regard to the willingness to use autonomous vehicles, attributions, perceived benefits and barriers. In general, we found that the perception and evaluation of using autonomous vehicles from the users' perspective was comparatively more affirmative on the highway than in the city. Furthermore, we showed that the evaluation of situation-specific traffic characteristics (here: simplicity and complexity of the driving situation) as being suitable for autonomous vehicles differed between the scenarios and was confirmed for highways, but negated for cities. However, some evaluation patterns appeared independent of the context. Key findings revealed controversial attitudes in both scenarios (albeit to a different extent): Participants of both groups indicated positive expectations regarding the novel transport experience, but also fears of use, mainly described in terms of control issues.

Next, we discuss our research findings in detail, also with regard to limitations of this survey and opportunities for future work. We then reflect on our methodological approach.

## 4.1 Interpreting Results

In a first and general impression, autonomous driving tended to be considered useful and comfortable in both traffic scenarios, although uncertainties regarding its (un)controllability were indicated alike, similar to the findings of Brell et al. [2]. With further respect to individual connotations and associations (measured on a semantic differential), perceptions of autonomous driving were overall more positive in regard to highway use, especially in terms of "good", "safe", and "effective". In contrast, using autonomous vehicles in the city was described as rather "unpleasant" and "not desirable". This assessment trend was also shown in the willingness to use autonomous vehicles, which was higher for highways than cities.

Payre et al. [10] already revealed a comparatively high interest in the use of autonomous vehicles on highways from the perspective of French users, which we have now been able to confirm within a German sample. In addition to cross-national similarities in the evaluation of future autonomous driving, it remains exciting to explore culturally determined differences. Since the handling of (technology) innovation, related mind sets and connotations may differ – as further results suggest (cf. Freadrich and Lenz [4], for example) – broader and likewise deeper cross-national and cross-cultural comparisons in this context are welcome to enrich the variety of perspectives in the field of innovative mobility research and human-automation interaction.

Considering usage benefits and barriers, we saw general evaluation differences with regard to the level of agreement and disagreement: In the highway scenario, there was a strong agreement on usage advantages, while in the city scenario more consensus was reached on the disadvantages of using autonomous vehicles, again indicating higher levels of skepticism. There were significant evaluation differences concerning situational characteristics in terms of perceived simplicity and complexity of the two traffic scenarios, whereas speed did not seem to decisively guide users' evaluation of autonomous driving. With regard to highway use, traffic conditions were considered as manageable and feasible for autonomous vehicles. This perceived situational appropriateness was supported by the rejection of the assumption that autonomous vehicles cannot drive as well as human drivers as a perceived barrier to use on highways. In contrast, the use of autonomous vehicles in the city was not described as appropriate due to an unclear and complex nature of the (driving) situation, which was also reflected in the assessment of barriers to use, as autonomous driving was perceived as rather stressful here, in combination with strong safety concerns.

It may be assumed that the perception and evaluation of autonomous driving is affected by the type and variety of specific traffic conditions and requirements, which apparently were perceived as more complex in the urban area (e.g., due to crossing pedestrians, stop and go situations at traffic lights, parking) compared to driving tasks on the highway (e.g., lateral and longitudinal vehicle guidance). It remains the task of future studies to find out which driving maneuvers in particular are perceived as simple or complex. Furthermore, it is reasonable to include additional traffic scenarios in the investigation, such as rural traffic, as this may determines the perception and evaluation for the use of autonomous vehicles due to more difficult road conditions (e.g., uphill, downhill, curves, gravel) and in turn may reveal further requirements to be met in the technical development from the users' perspective. In this regard, individual conditions of use (e.g., manual control options) could also be examined to better understand the decision of diverse user groups to accept or reject autonomous driving.

Besides, we discovered trade-offs in the evaluation of general usage benefits and barriers that were independent of the traffic scenario. Participants of both groups expected benefits of using autonomous vehicles, especially in terms of improved traffic flow, less traffic jam, and increased road safety, similar to Schmidt et al. [12]. Interestingly, safety concerns, e.g., regarding technical failure, were likewise reported as a major barrier to use, presumably due to a strong fear of devastating consequences in the event of an accident. In this case, the trade-off between potential safety gains and drawbacks may be driven by a perceived loss of control – which was already indicated when neither a reduction of human responsibility nor the opportunity to do other things while driving (different from König and Neumayr [8]) were considered as usage benefits. Instead, reduced vehicle control and possible inattention of the passenger(s) on board were decisively determined as usage barriers.

Since control issues have also been discussed in other traffic situation-specific scenarios [1], investigations of personality and individual attitudes would be helpful in future work, e.g. with regard to people's innovation readiness and risk perception, and, of course, control dispositions when interacting with (innovative) technology. This would allow to see to what extent the evaluation of autonomous driving changes (or remains the same) in consideration of diverse user constellations, also in order to take this into account in technical development with regard to individual requirements. Only this way it will be possible to provide access to the new mobility for as many users as possible.

## 4.2 Reflection on Methods

In the following, we reflect on our empirical approach of this survey in regard to the sample and the methods used.

Considering the sample, slightly more men than women took part and the participants were, on average, rather young and highly educated. In addition to a more balanced gender and education distribution (to identify diverse information needs of the public), future studies should address elderly user groups in particular, as they may have special requirements for the design and use of autonomous vehicles due to physical or cognitive limitations. It is relevant to investigate more closely which user groups (in terms of socio-demographic data) would use autonomous vehicles, for what purpose, and thus in which traffic situations, in order to identify target-group-specific requirements.

With regard to the participants' mobility behavior, the majority was in possession of a driver's license, which may have influenced the evaluation of autonomous vehicles. In order to obtain valid results in this respect, a comparison to user groups without a driver's license would be very interesting, especially with regard to perceived control issues. Considering driving license holders, it would also be exciting to identify driver types and personalities to measure the extent to which these affect the evaluation of autonomous mobility. A first indication of characteristics to be measured may be provided by the semantic differential used in this study to explore driving behavior and style, which was self-reported to be both offensive and sporty, but also cautious compared to others and thus already indicates diverse and ambiguous points of view.

Considering our methods used, evaluations were scenario-based to first gain various insights into the perceptions of a broad, but random sample. Follow-up studies can build on the key findings of this study to validate them in empirical-experimental research designs with selected participants (e.g., according to demographic, mobility, or attitudinal data). If in-person field research settings are not practical in the short term, especially due to the Covid-19 pandemic and high risk of contagion, integrating short video clips into the scenario description (in addition to photo collages) could be helpful to more clearly depict and present various traffic situations, especially to novice users.

**Acknowledgments.** The authors thank all participants for their openness to share opinions on autonomous driving. Special thanks are given to Farina Heilen for research assistance. This work has been funded by the Federal Ministry of Transport and Digital Infrastructure (BMVI) within the funding guideline "Automated and Networked Driving" under the project APEROL with the funding code 16AVF2134B.

# A    Appendix

*City Traffic Scenario.* Please imagine that you are driving in an autonomous vehicle in the city. You have already scheduled the start and destination of your trip. The vehicle system takes over all driving tasks. You are driving at 50 km/h[1] on a moderately busy road, with another vehicle ahead of you at a sufficient distance. A traffic light turns red, your vehicle stops and starts again when the light changes back to green, then turns right into a side street. After some time, you reach a crosswalk where a pedestrian wants to cross the street. Your vehicle stops for the pedestrian to pass and then continues driving (Fig. 4).

**Fig. 4.** Photo collage illustrating a standard driving situation in the city.

*Highway Traffic Scenario.* Please imagine that you are driving in an autonomous vehicle on the highway. You have already scheduled the start and destination of your trip. The vehicle system takes over all driving tasks. You are driving at 130 km/h[2] on a moderately busy two-lane highway, with another vehicle ahead of you at a sufficient distance. Your vehicle starts an overtaking maneuver, it swerves and, after completing the maneuver, rejoins the lane. After some time, you reach a slip road, another vehicle is in the acceleration lane. Your vehicle swerves again and rejoins the original lane after having passed the acceleration lane (Fig. 5).

---

[1] Note: The scenario description was translated from German. 50 km/h is a maximum speed limit in German built up areas, corresponding to approx. 31 mph.

[2] Note: The scenario description was translated from German. 130 km/h is an advisory speed limit on German highways, corresponding to approx. 81 mph.

**Fig. 5.** Photo collage illustrating a standard driving situation on the highway.

# References

1. Bjørner, T.: A priori user acceptance and the perceived driving pleasure in semi-autonomous and autonomous vehicles. Paper presented at European Transport Conference 2015, Frankfurt, Germany, pp. 1–13 (2015)
2. Brell, T., Philipsen, R., Ziefle, M.: sCARy! Risk perceptions in autonomous driving: the influence of experience on perceived benefits and barriers. Risk Anal. **39**(2), 342–357 (2019)
3. Federal Statistical Office: Bildungsstand (2020). https://www.destatis.de/DE/Themen/Gesellschaft-Umwelt/Bildung-Forschung-Kultur/Bildungsstand/_inhalt.html
4. Fraedrich, E., Lenz, B.: Automated driving: individual and societal aspects. Transp. Res. Rec. **2416**(1), 64–72 (2014)
5. Gkartzonikas, C., Gkritza, K.: What have we learned? A review of stated preference and choice studies on autonomous vehicles. Transp. Res. Part C Emerg. Technol. **98**, 323–337 (2019)
6. Harper, C.D., Hendrickson, C.T., Mangones, S., Samaras, C.: Estimating potential increases in travel with autonomous vehicles for the non-driving, elderly and people with travel-restrictive medical conditions. Transp. Res. Part C Emerg. Technol. **72**, 1–9 (2016)
7. Igliński, H., Babiak, M.: Analysis of the potential of autonomous vehicles in reducing the emissions of greenhouse gases in road transport. Procedia Eng. **192**, 353–358 (2017)
8. König, M., Neumayr, L.: Users' resistance towards radical innovations: the case of the self-driving car. Transp. Res. Part F **44**, 42–52 (2017)
9. Olaverri-Monreal, C.: Autonomous vehicles and smart mobility related technologies. Infocommunications J. **8**(2), 17–24 (2016)

10. Payre, W., Cestac, J.J., Delhomme, P.: Intention to use a fully automated car: attitudes and a priori acceptability. Transp. Res. Part F: Traffic Psychol. Behav. **27**(Part B), 252–263 (2014)
11. SAE: Taxonomy and definitions for terms related to driving automation systems for on-road motor vehicles. J3016. Technical report, SAE International (2018). https:// www.sae.org/news/2019/01/sae-updates-j3016-automated-driving-graphic
12. Schmidt, T., Philipsen, R., Ziefle, M.: Safety first? V2X - Percived benefits, barriers and trade-offs of automated driving. In: Proceedings of the 1st International Conference on Vehicle Technology and Intelligent Transport Systems (VEHITS-2015), pp. 39–46 (2015)
13. Trübswetter, N.M.: Akzeptanzkriterien und Nutzungsbarrieren älterer Autofahrer im Umgang mit Fahrerassistenzsystemen. Ph.D. thesis, Technische Universität, München (2015)
14. VDA (Verband der Automobilindustrie): Automatisierung. Von Fahrerassistenzsystemen zum automatisierten Fahren. Technical report, Verband der Automobilindustrie (2015)

# Should Self-Driving Cars Mimic Human Driving Behaviors?

Jamie Craig and Mehrdad Nojoumian(⌗)

Department of Computer and Electrical Engineering and Computer Science,
Florida Atlantic University, Boca Raton, FL 33431, USA
{jcraig18,mnojoumian}@fau.edu

**Abstract.** Recent studies illustrate that people have negative attitudes towards utilizing autonomous systems due to lack of trust. Moreover, research shows a human-centered approach in autonomy is perceived as more trustworthy by users. In this paper, we scrutinize whether passengers expect self-driving cars (SDC) to mimic their personal driving behaviors or if they hold different expectations of how a SDC should drive. We developed a survey with 46 questions that asked 352 participants about their personal driving behaviors such as speed, lane changing, distance from a car in front, acceleration and deceleration, passing vehicles, etc. We further asked the same questions about their expectations of a SDC performing these tasks. Interestingly, we observed that most people prefer a SDC that drives like a *less aggressive version of their own driving behaviors*. Participants who reported they trust or somewhat trust AI, autonomous technologies, and SDCs expected a car with behaviors similar to their personal driving behaviors. We also found that the expectation of a SDC's level of attenuated aggressiveness witnessed among all other participants was *relative to their personal driving behavior aggressiveness*. For instance, male drivers showed to be more aggressive drivers than female drivers, and therefore, their expectations for a SDC was slightly more aggressive. These findings can be useful in developing certain profiles or settings for SDCs, and overall they can help in designing a SDC that is perceived as trustworthy by passengers.

**Keywords:** Self-driving cars' behavior · Mimicking human-driving cars' behavior · Trust in self-driving cars

## 1  Introduction

Self-driving cars are quickly becoming a reality and will have significant consequences on our society. A substantial amount of research is being conducted to make vehicles that are fully autonomous, where humans are no longer needed for operation. AI and machine learning have been instrumental in the development of these advanced systems and continue to improve at a rapid rate. While the academia, tech community and researchers in the field of AI and autonomy

© Springer Nature Switzerland AG 2021
H. Krömker (Ed.): HCII 2021, LNCS 12791, pp. 213–225, 2021.
https://doi.org/10.1007/978-3-030-78358-7_14

eagerly work towards delivering a fully autonomous vehicle, there is evidence showing that the general public is apprehensive towards using such systems.

Recent studies show that people have negative attitudes towards utilizing autonomous platforms [9,19]. A survey conducted by Continental AG found that 31% of respondents stated they were unnerved by the development of automated vehicles, and 54% claimed that they do not believe that such vehicles will function reliably[1]. Besides, researchers at the University of Michigan found that 46% of adult drivers preferred to retain full control while driving. Just under 16% of the 618 respondents said they would rather ride in a completely self-driving vehicle[2]. This hesitation to utilize SDCs is not unfounded [12,21,22]. In fact, Howard and Dai identified five challenges to the adoption of SDCs including: the lack of a robust legal and regulatory framework, cost of technology and it's result on economic equality, control and trust, privacy, and safety [7]. Driving autonomous vehicles with adaptive and personalized features is a key technological challenge that can improve some of the aforementioned problems [4,13,14,16].

## 1.1  Trust and Levels of Autonomy

Human trust in AI or autonomy is a major theme seen throughout the literature on autonomous system adoption [1,2,5,15,18,20]. The focus of our paper is to explore this trust/distrust and determine if there is a relationship between a person's driving behaviors and their expectations of the driving behaviors of a SDC. A definition of trust is needed in order to investigate trust in relation to driving behaviors and expectations of SDCs. Lee and See [10] define trust as "the attitude that an agent will help achieve an individual's goals in a situation characterized by uncertainty and vulnerability." Another more detailed definition of trust is an entity that will act with benevolence, integrity, competence, and predictability [11]; *Benevolence* means the entity will act in the subjects interest rather than acting opportunistically; *Integrity* means the entity will fulfill what it promises to do; *Competence* means the entity has the ability, expertise, or authority to do the task at hand; and *Predictability* means the actions of the entity, whether good or bad, are consistent enough that they can be forecasted in a given situation. Based on this definition, in order for a person to trust a SDC, the passenger will expect the SDC to act in the best interest of its passenger and keep them safe, i.e., to perform autonomous driving tasks successfully, and to operate consistently so that the passenger can predict the SDC's actions in most driving situations. The SAE International standard J3016 defines six levels of autonomy as follows:

- Level-0 - No Automation: The human controls the system 100% of the time.
- Level-1 - Driver Assistance: The vehicle must have at least one advanced driver-assistance feature, e.g., adaptive cruise control or lane-keeping assist.

---

[1] https://www.continental.com/en/press/initiatives-surveys/continental-mobility-studies/mobility-study-2013.

[2] https://news.umich.edu/vehicle-automation-most-drivers-still-want-to-retain-at-least-some-control/.

- Level-2 - Partial Automation: The vehicle has two or more advanced driver assistance systems that control speed and acceleration, steering, and braking.
- Level-3 - Conditional Automation: The vehicle is capable of taking over full control and can operate for selected parts of a journey. However, conditions must be ideal and human supervision is needed to take over in case of failures.
- Level-4 - High Automation: The vehicle can complete an entire journey without human intervention. In rare situations, a human may need to intervene.
- Level-5 - Full Automation: A vehicle does not require any human intervention and can operate under all circumstances.

Trust is essential from levels 1 through 5 as the car starts to take over operational tasks. SDCs at level 2 and 3 are at transition levels between the driver having full control over all the car's functions to the car having full control over all the car's functions. Levels 2 and 3 is where we see the most interaction between the driver and the autonomous system in terms of operation. At these levels, the human driver is expected to make decisions with the system, interact with the system, and at times take over the system. The trust at these levels depends on communication and mutual understanding. Because of this, one hypothesis for the distrust in SDCs is that the user does not understand why the car is making certain decisions, and therefore the user cannot predict what the car will do in various driving situations [8]. If the user is expected to interact with the system but cannot anticipate the car's actions, it is likely that the user's trust will decrease. According to Butakov and Ioannou [4], "The closer the automated vehicle dynamics are with those of a manually driven vehicle, the more likely that the comfort level of the automated vehicle user will improve."

## 1.2  Our Motivation and Contribution

There are two approaches to tackle the aforementioned problem. First is to design a SDC that mimics real human driving behaviors, more specifically, mimicking the driving behaviors of the actual passenger [13]. Second is to control a SDC to be responsive to the driver's emotional state [14,16,17] or to control it based on the driver's expectations of the SDC's driving behaviors. A question we hope to answer in this research is: *Do users expect the SDC to exhibit their personal driving behaviors or do they hold different expectations of how a SDC should drive?* Once this question is answered, it will be imperative to design a system that can communicate with the user through a user-friendly interface.

Previous studies that have used surveys and questionnaires to collect empirical data on trust between humans and SDCs have focused more on the users overall feelings towards SDCs and their potential to use them [3,7,9]. In contrast, our survey asks the users to report their own driving behaviors in various situations, and then it asks similar questions in the scenario that they are in a SDC and not in control of driving. With this approach, we expect to find certain profiles of drivers and explore the possibilities of there being a difference between users' own driving behaviors and the expectations of driving behaviors of SDCs.

The rest of this paper is organized as follows. In Sect. 2, we review some remarks and important observations in existing literature. In Sect. 3, we discuss our research methodology. In Sect. 4, we present our technical results. Finally, in Sect. 5, we conclude the paper with remarks and future directions.

## 2    Remarks in Existing Literature

In a 2015 study, Butakov and Ioannou [4] assert that the two major determining factors in the successful adoption of SDCs are for them to be perceived as safe and comfortable. There are four conditions that ultimately need to be satisfied in order for humans to perceive the technology as safe and useful.

1. The SDC should always perform better than the human driver.
2. While the SDC safely and reliably operates within its limitations, the driver should have a clear understanding of what those limitations are.
3. The driver should know when the SDC is in control and what it will do.
4. The SDC's behavior should be predictable and acceptable to the passenger.

The researchers collected experimental data from a twenty seven years old human driver who made daily trips over the course of four months in a customized vehicle equipped with side-facing radars and front and rear facing LiDAR (Light Detection and Ranging). This experiment illustrated how the autopilot personalization feature can make autopilot behavior more transparent and intuitive to the driver. Overall the driver is able to detect the boundaries and behavior of the autonomous vehicle because the autopilot personalization is mimicking their own driving behaviors. One point to mention is that the researchers assumed the human subject in their study would prefer a more conservative personalization of his own driving behaviors when setting up the autopilot features. This assumption deserves a more in-depth analysis and raises the question of whether humans feel safer in a SDC that drives like them or drives like a more conservative version of themselves. This assumption is explored in more depth in our research, as explained in Sect. 4.

Goodrich and Boer [6] laid out the same four conditions for safety and usefulness as mentioned in Butakov and Ioannou's research. However, their motivations came from a case study on an automated car following systems whose fundamental design principle was to use the human operator as a template for automation. Through their case study, they demonstrate the need for a human-centered approach in the design of automation and they were able to support the following hypotheses. A human-centered approach:

1. Can improve the users detection of nonsupport situations.
2. Improves the user's evaluation of the system's performance.
3. Facilitates the development of a proper level of trust within the user.
4. Improves the ability for the user to take over control.
5. Enhances safety of the automation theoretically.

They conclude that advanced vehicle system design can benefit from in-depth analysis of driver behavior by constructing a control system that can be perceptible to the human driver. It's worth mentioning that our research is an attempt to get a better understanding of driving behavior and driver's expectations of SDCs in a variety of driving scenarios. Ultimately, this can be used to construct driving models for SDCs.

Finally, in a recent study [20] Shahrdar et al. designed and implemented a VR-based self-driving car simulator using a realistic driving scenarios captured by 360-degree camera. They recorded the fluctuations of passengers' trust levels through various trust building and trust damaging driving scenarios. The simulation ran for approximately ten minutes and the users were asked to report their levels of trust through the VR simulation in increments of two minutes. Prior to the simulation, participants were also asked sixteen demographic and psychological questions through an anonymous questionnaire. The authors showed that trust levels of human subjects were directly correlated to the driving style of the SDC. They also found that certain demographic attributes such as gender, cultural background, and current attitude toward autonomous driving technology had an effect on the way people trusted the simulated SDC. They later repeated this experiment by using both subjective and objective data collections [16]. We therefore intend to take Shahrdar's et al. research a step further by asking detailed questions about driving behaviors and expectations of SDC driving behavior. Would the subjects in this research have had a higher trust level if the SDC simulation matched the driving style of the subject? If the subject's performance expectations of a SDC were known before and reflected in the simulation, would this also increase the level of trust in the human subject? Could a simulation model that is trustworthy be constructed based on the demographics of the subject? These are questions we intend to expand upon in this paper.

## 3   Research Methodology

We surveyed 352 participants that were recruited on social media platforms, PollPool.com, and through e-mails sent out to students and faculty.

### 3.1   Survey Procedure and Instruments

Participants were told that the purpose of the survey was to gain a better understanding of driving behaviors among the population and how these driving behaviors can be modeled by SDCs. In addition, they were told the survey was completely anonymous, voluntary, and that it would take fifteen minutes to complete. Participants were not offered compensation for responding.

This survey was created in Google Forms and consisted of 46 fixed-response and forced-choice questions. The 46 questions were broken down into three categories: (a) Demographic questions, (b) Personal driving behavior questions, and (c) Questions involving SDCs. With the demographic questions, we collected the following information: age, gender, ethnicity, education, employment status, income, and marital status. The personal driving behavior questions focused on various behaviors such as *speed, lane changing, breaking, acceleration, decel-eration, passing other vehicles, signaling, lane preference, and parking.* These behaviors were asked in regards to 4 driving situations such as highway and non-highway roads, driving in a less than perfect weather condition, and driving at night. The final section of the survey was focused on SDC-related questions.

First, the participants were asked about their trust levels to utilize AI or fully autonomous technologies on a fuzzy set of trust states, i.e., *distrust, somewhat distrust, neutral, somewhat trust, trust.* Next, the participants were asked what are their trust levels to utilize a SDC when this technology becomes available using the same fuzzy set of trust states. The rest of the SDC questions were similar to the personal driving behavior questions but from the perspective of the subject traveling in a SDC.

## 3.2   Quantitative Measurement

In order to examine driving behaviors and the expected driving behaviors of a SDC, we generated a score for each one. These two scores were normalized between 0 and 1 and represented the aggressiveness of the driver or the car. To generate this score, we assigned a numerical value to each answer of each question. The answer was assigned a 0 if the response represented a behavior of a *cautions/conservative* driver, a 0.5 if the response represented a behavior of a *moderate driver*, and 1 if the response represented a behavior of an *aggressive driver.* Table 1 illustrates an example of how each question was coded.

**Table 1.** Example of individual question coding.

| Question: Which best describes your behavior most of the time in terms of speed while driving on: THE HIGHWAY | Aggressiveness score |
| --- | --- |
| I typically drive under the speed limit (more than 5 mph UNDER the speed limit) | 0 |
| I typically drive the speed limit (with plus or minus 5 mph) | 0.5 |
| I typically drive over the speed limit (more than 5 mph OVER the speed limit) | 1 |

Once each question was coded using this approach, we summed the answers in each group, i.e., personal driving behaviors and expected driving behaviors of SDCs, and then divided the result by the total number of questions used in each group. To insure that the questions from each section were the same, we combined the numerical values from the highway and non-highway equivalent questions and took the average. After making this adjustment, there were 6 questions used to calculate the *Driving Behavior Aggressiveness* score (DBA) and 6 questions used to calculate the *Self-Driving Car Aggressiveness* score (SDCA). For instance, if participants had a DBA score of roughly 0.9, they would be considered aggressive drivers. Likewise, if participants had a SDCA score of 0, they would be considered conservative drivers, and so on.

# 4    Technical Analyses and Results

We compared the DBA scores against SDCA scores for various categories. We found that overall, most people prefer a SDC that is less aggressive than their personal driving behavior. We highlight specific categories that we compared. We report significant results found within these categories with $p-value<0.05$.

## 4.1    Trust and Driving Aggressiveness

The first significant comparison we found was between all DBA scores and all SDCA scores with $p-value = 0.000$ calculated using the Mann-Whitney U Test. The summary statistics of these two groups are shown in Table 2. This illustrates that the personal driving behaviors significantly differ from the expectations of a SDC's driving behaviors.

**Table 2.** Summary statistics for DBA scores and SDCA scores.

| Score | Median | Mean | Standard deviation | Min | Max |
|---|---|---|---|---|---|
| DBA scores | 0.492 | 0.482 | 0.147 | 0.083 | 0.925 |
| SDCA scores | 0.417 | 0.381 | 0.191 | 0.000 | 1.000 |

The two histograms illustrated in Fig. 1 and 2 show that while the DBA scores have a normal distribution, the SDCA scores were skewed to the left. Based on the summary statistics, the participants from our sample preferred a SDC that was less aggressive than their personal driving behaviors. This may also indicate that people do not trust SDCs compared to the trust that they have in their own driving. In other words, it confirms the apprehension of trust towards self-driving cars as stated earlier.

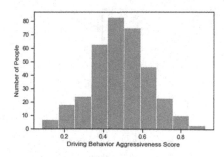

**Fig. 1.** Distribution of DBA scores.

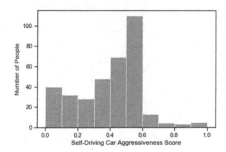

**Fig. 2.** Distribution of SDCA scores.

## 4.2   Gender and Driving Aggressiveness

When comparing Female DBA scores to Female SDCA scores, we found a significant difference between the two scores with $p - value = 0.000$ calculated using the Mann-Whitney U Test. Similarly, we also found a significant difference between male DBA scores and male SDCA scores with $p - value = 0.036$ calculated using the Mann- Whitney U Test. We can conclude that both male and female drivers have expectations of self-driving cars that differ from their personal driving behaviors.

We then compared the DBA scores of female drivers to the DBA scores of male drivers and found a significant difference, i.e., $p - value = 0.004$. Table 3 displays the summary statistics of male and female drivers. From this table, we can see that the average male DBA score and SDCA score are higher than the equivalent scores of an average female driver. We can see that male drivers tend to be more aggressive drivers than female drivers, and therefore, their expectations for a SDC is slightly more aggressive.

**Table 3.** Summary statistics for gender DBA scores and SDCA scores.

| Score | Median | Mean | Standard deviation | Min | Max |
|---|---|---|---|---|---|
| Female DBA scores | 0.475 | 0.470 | 0.136 | 0.083 | 0.808 |
| Female SDCA scores | 0.333 | 0.336 | 0.173 | 0.000 | 0.750 |
| Male DBA scores | 0.508 | 0.497 | 0.158 | 0.083 | 0.925 |
| Male SDCA scores | 0.417 | 0.439 | 0.198 | 0.000 | 1.000 |

To conclude, female drivers are less aggressive drivers than male drivers and while both male and female drivers prefer a SDC that is less aggressive than their personal driving behaviors, a female expects an SDC to be less aggressive than the equivalent male expectations of a SDC, as shown in Fig. 3.

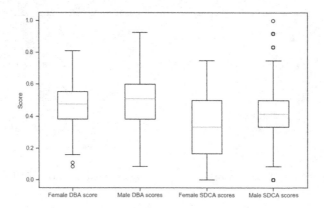

**Fig. 3.** Boxplot of gender DBA scores and SDCA scores.

## 4.3    Trust Levels and Driving Aggressiveness

Next, we examine the results from the following questions and separate the results into three trust groups, as shown in the pie charts in Figs. 4 and 5.

- What is your trust level to utilize AI or fully autonomous technologies?
- What is your trust level to utilize a SDC when it becomes available?

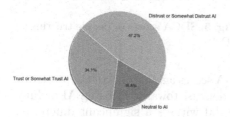

**Fig. 4.** Trust levels to utilize AI/ Autonomy.

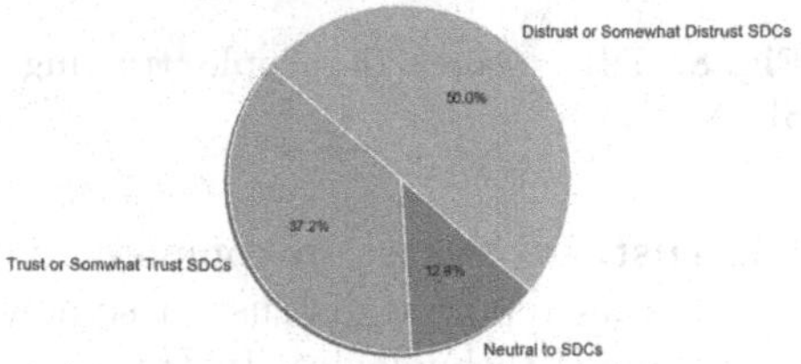

**Fig. 5.** Trust levels to utilize SDCs.

**Trust.** We compared the DBA scores against the SDCA scores for people that reported trust or somewhat trust towards utilizing artificial intelligence or fully autonomous technologies. We did not witness a significant difference between the two scores for this group. Likewise, we did not witness a significant difference between DBA scores and SDCA scores of people who said they trust or somewhat trust towards utilizing a SDC when this technology becomes available. We can conclude that people who report they trust or somewhat trust utilizing AI, autonomous technologies, and SDCs would want a SDC that exhibits the same driving behaviors as their own. The results are shown in Figs. 6, 7, 8 and 9.

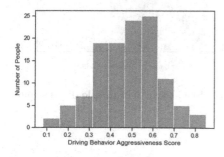

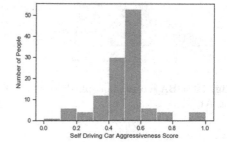

**Fig. 6.** DBA scores of people trusting AI.

**Fig. 7.** SDCA scores of people trusting AI.

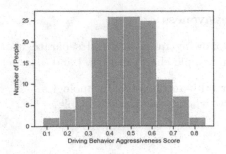 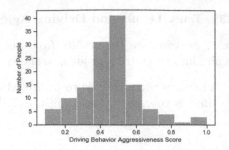

**Fig. 8.** DBA scores of people trusting SDCs.

**Fig. 9.** SDCA scores of people trusting SDCs.

**Distrust.** In contrast, we compared the DBA scores against the SDCA scores for people that reported distrust or somewhat distrust towards utilizing AI or fully autonomous technologies. In this case, we did witness a significant difference, i.e., $p - value = 0.000$ calculated using the Mann-Whitney U Test, between the DBA scores and SDCA scores of this group. Similarly, we found a significant difference, i.e., $p - value = 0.000$ calculated using the Mann-Whitney U Test, between the two scores for people who reported distrust or somewhat distrust towards utilizing SDCs. From this, we can conclude that people who report distrust or somewhat distrust towards AI, autonomous technologies, and SDCs are less trusting of these technologies, and therefore, they would expect an SDC that is less aggressive than their personal driving behaviors. The results are shown in Figs. 10, 11, 12 and 13.

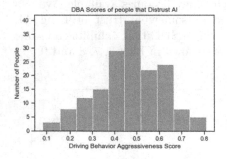

**Fig. 10.** DBA scores of people distrusting AI.

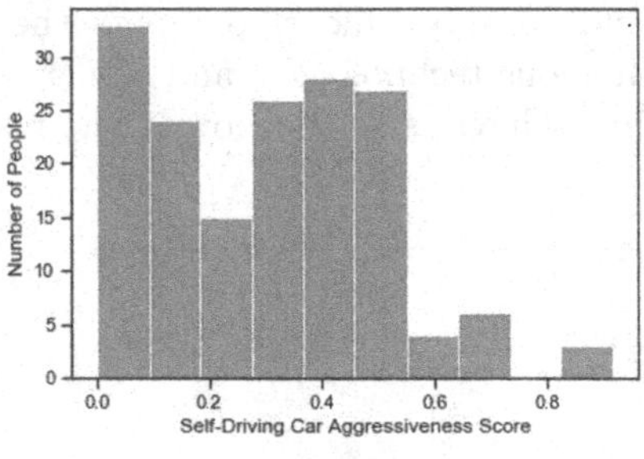

**Fig. 11.** SDCA scores of people distrusting AI.

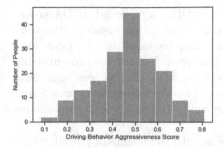

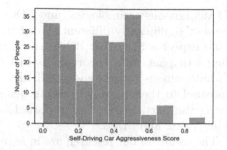

**Fig. 12.** DBA scores of people distrusting SDCs.

**Fig. 13.** SDCA scores of people distrusting SDCs.

**Neutral.** As a final comparison, we looked at the DBA scores against the SDCA scores for people who reported being neutral towards AI and autonomous technologies and found that the results confirmed the overall trend in our sample of people expecting a less aggressive SDC. We report a significant difference with $p-value = 0.00003$ calculated using the Mann-Whitney U Test. We also found a significant difference between these two scores for people who reported that they are neutral towards utilizing SDCs with $p-value = 0.0006$ calculate using the Mann-Whitney U Test. In summary, people who are neutral towards AI, autonomous technologies, and SDCs still prefer a SDC that is less aggressive than their personal driving behaviors.

## 5 Conclusion and Future Directions

Our technical results provide prominent insights into the driving behaviors and the expectations of drivers when it comes to SDCs. We sought to answer the question: Do passengers expect a SDC to exhibit their personal driving behaviors or do they hold different expectations of how a SDC should drive? When looking at the aggressiveness level of personal driving behaviors compared to the expected aggressiveness level of a SDC through the lens of gender, age, race/ethnicity, income and education, most people expect a SDC that is less aggressive than their personal driving behaviors. In other words, a SDC that drives in a way that is more conservative than their personal driving behaviors could be deemed more trustworthy. The SDC's level of aggressiveness is relative to each particular driver's DBA. Therefore, if the driver has a DBA score of 0.5, then an SDC should have an SCDA score that is lower, for instance, 0.4. When looking at male and female DBA and SDCA scores, we observed that female drivers would expect a less aggressive SDC compared to male drivers since female drivers were less aggressive drivers to begin with. This is an example of how the aggressiveness of the SDC is relative to each driver's driving behavior.

We found that current attitudes towards artificial intelligence, autonomous technologies, and SDCs had an effect on their expectations of a SDC. The one group that stood out were those that trust or somewhat trust towards utilizing

AI, autonomous technologies, and SDCs. Their DBA scores and SCDA scores were not significantly different. We therefore concluded that these participants would expect a SDC that drives the same as their personal driving behaviors. The participants that reported distrust or somewhat distrust towards utilizing AI, autonomous technologies, and SDCs prefer a SDC that is less aggressive compared to their personal driving behaviors. This same sentiment was found among the participants that reported being neutral towards AI, autonomous technologies, and SDCs.

The results of our research are in agreement with the results of [18,20] where the author found that gender and the current attitude toward autonomous driving technology had an effect on the way people trusted a SDC simulator. Our results also confirmed the assumption made by [4] that explored a human-centered approach. Overall, we can conclude that since the DBA scores were significantly different from the SDCA scores on the same driving tasks, and since the summary statistics showed that the average DBA scores were always greater than the SCDA scores, most people prefer a SDC that drives like a more conservative version of themselves. Only those who claimed they trust or somewhat trust SDCs or AI would want a car that matched their personal driving behavior. These results can be considered by engineers, computer scientist, and researchers to design a SDC or SDC simulator that is deemed trustworthy by the user.

The next steps are to create a driving simulator based on our results as well as the calculated DBA score and SDCA score. The simulation would include three different driving profiles for each type of drivers, i.e., conservative/cautious, moderate, and aggressive, and the profile would be set according to the DBA score. The experiment would measure trust levels before and after the simulation. In addition, an appended survey, which asks questions as why the participants trust or distrust these technologies and prior exposures to these technologies, is essential in understanding where this trust or distrust is originating from.

# References

1. Abd, M.A., Gonzalez, I., Ades, C., Nojoumian, M., Engeberg, E.D.: Simulated robotic device malfunctions resembling malicious cyberattacks impact human perception of trust, satisfaction, and frustration. Int. J. Adv. Rob. Syst **16**(5), 1–16 (2019)
2. Abd, M.A., Gonzalez, I., Nojoumian, M., Engeberg, E.D.: Trust, satisfaction and frustration measurements during human-robot interaction. In: 30th Florida Conference on Recent Advances in Robotics (FCRAR), pp. 89–93 (2017)
3. Abraham, H., et al.: Autonomous vehicles, trust, and driving alternatives: A survey of consumer preferences, pp. 1–16. Massachusetts Institute of Technology, AgeLab, Cambridge (2016)
4. Butakov, V., Ioannou, P.: Driving autopilot with personalization feature for improved safety and comfort. In: 2015 IEEE 18th International Conference on Intelligent Transportation Systems, pp. 387–393. IEEE (2015)
5. Choi, J.K., Ji, Y.G.: Investigating the importance of trust on adopting an autonomous vehicle. Int. J. Human-Comput. Interact. **31**(10), 692–702 (2015)

6. Goodrich, M.A., Boer, E.R.: Model-based human-centered task automation: a case study in ACC system design. IEEE Trans. Syst. Man Cybern.-Part A Syst. Humans **33**(3), 325–336 (2003)
7. Howard, D., Dai, D.: Public perceptions of self-driving cars: the case of Berkeley, California. In: Transportation Research Board 93rd Annual Meeting, vol. 14, pp. 1–16 (2014)
8. Koo, J., Kwac, J., Ju, W., Steinert, M., Leifer, L., Nass, C.: Why did my car just do that? explaining semi-autonomous driving actions to improve driver understanding, trust, and performance. Int. J. Interact. Design Manuf. (IJIDeM) **9**(4), 269–275 (2015)
9. Kyriakidis, M., Happee, R., de Winter, J.C.: Public opinion on automated driving: Results of an international questionnaire among 5000 respondents. Transp. Res. Part F Traffic Psychol. Behav. **32**, 127–140 (2015)
10. Lee, J.D., See, K.A.: Trust in automation: designing for appropriate reliance. Human Fact. **46**(1), 50–80 (2004)
11. Harrison McKnight, D., Chervany, N.L.: Trust and distrust definitions: one bite at a time. In: Falcone, R., Singh, M., Tan, Y.-H. (eds.) Trust in Cyber-societies. LNCS (LNAI), vol. 2246, pp. 27–54. Springer, Heidelberg (2001). https://doi.org/10.1007/3-540-45547-7_3
12. Merfeld, K., Wilhelms, M.P., Henkel, S., Kreutzer, K.: Carsharing with shared autonomous vehicles: uncovering drivers, barriers and future developments-a four-stage delphi study. Technol. Forecast. Social Change **144**, 66–81 (2019)
13. Nojoumian, M.: Adaptive driving mode in semi or fully autonomous vehicles (2019). US Patent App. 16/165,559
14. Nojoumian, M.: Adaptive mood control in semi or fully autonomous vehicles (2019). US Patent App. 16/165,509
15. Park, C.: Using EEG and Structured Data Collection Techniques to Measure Passenger Emotional Response in Human-Autonomous Vehicle Interactions. Master's thesis, Florida Atlantic University (2018)
16. Park, C., Nojoumian, M.: Social acceptability of autonomous vehicles: Unveiling correlation of passenger trust and emotional response. Florida Atlantic University, Technical report (2021)
17. Park, C., Shahrdar, S., Nojoumian, M.: EEG-based classification of emotional state using an autonomous vehicle simulator. In: 10th Sensor Array and Multichannel Signal Processing Workshop (SAM), pp. 297–300. IEEE (2018)
18. Shahrdar, S.: New Structured Data Collection Approach For Real-Time Trust Measurement in Human-Autonomous Vehicle Interactions. Master's thesis, Florida Atlantic University (2018)
19. Shahrdar, S., Menezes, L., Nojoumian, M.: A survey on trust in autonomous systems. In: Arai, K., Kapoor, S., Bhatia, R. (eds.) SAI 2018. AISC, vol. 857, pp. 368–386. Springer, Cham (2019). https://doi.org/10.1007/978-3-030-01177-2_27
20. Shahrdar, S., Park, C., Nojoumian, M.: Human trust measurement using an immersive virtual reality autonomous vehicle simulator. In: 2nd AAAI/ACM Conference on Artificial Intelligence, Ethics, and Society, pp. 515–520. ACM (2019)
21. Yan, C., Xu, W., Liu, J.: Can you trust autonomous vehicles: contactless attacks against sensors of self-driving vehicle. DEF CON **24**, 109 (2016)
22. Zang, S., Ding, M., Smith, D., Tyler, P., Rakotoarivelo, T., Kaafar, M.A.: The impact of adverse weather conditions on autonomous vehicles: how rain, snow, fog, and hail affect the performance of a self-driving car. IEEE Veh. Technol. Mag. **14**(2), 103–111 (2019)

# Autonomous Vehicles and Pedestrians: A Case Study of Human Computer Interaction

Subasish Das[1]([✉]) [iD] and Hamsa Zubaidi[2] [iD]

[1] Texas A&M University, 3135 TAMU, College Station, TX 77843, USA
s-das@tti.tamu.edu
[2] Oregon State University, Corvallis, OR 97331, USA

**Abstract.** Understanding public attitudes, sentiments, and perceptions is a significant first step to the widespread acceptance of autonomous vehicles (AVs). In recent years, many studies been conducted in recent years examining the general perception of AVs. The implementation of AVs has potential challenges involving pedestrians and bicyclists that need special attention. The current study analyzes survey data obtained from BikePGH, a non-profit organization in Pittsburgh, Pennsylvania . This study applied multiple correspondence analysis (MCA) to explore response patterns among the participants. The findings of the survey reveal that certain groups of people are much more receptive to AV technology than others who are vehemently opposed to the introduction of AVs to the roadways. The findings indicate that participants who have real world experience with human-AV interactions have more positive expectations and a higher level of interest than participants with no previous experience. The authors expect that these findings will contribute greatly to the development of safety policies related AV and pedestrian interactions.

**Keywords:** Autonomous vehicle · Human computer interaction · Pedestrian · Correspondence analysis

## 1 Introduction

The introduction of autonomous vehicles (AVs) has earned global attention in recent years. New technologies have thoroughly transformed the transportation infrastructure. Fully automated AVs can reduce the number of crashes caused by human error and handle various traffic conditions without human feedback. Traffic congestion and air pollution can also be minimized by the introduction of AVs. In order to ensure that customers are aware if these many advantages, is necessary to consider end-users' awareness and attitude towards AVs.

The rate of non-motorized traffic fatalities is currently increasing. In 2016, 16 percent (5,987) of traffic fatalities were pedestrians and 2.3 percent (852) were bicyclists in the U.S. [33, 34]. A non-motorized roadway user (i.e., bicyclist or pedestrian) is more likely to severely injured or killed in road traffic crash. Studies have shown that AVs can increase roadways safety, but recent news reported a crash incident that resulted in a

H. Krömker (Ed.): HCII 2021, LNCS 12791, pp. 226–239, 2021.
https://doi.org/10.1007/978-3-030-78358-7_15

pedestrian fatality, in which an AV was in full automation mode. This event attracted worldwide media coverage, and it brought attention to safety concerns and research areas that need to be studied further, such as AV and non-motorized user (i.e., pedestrians and bicyclists on the roadways) interaction.

To visually explore complex patterns among the responses, we applied multiple correspondence analysis (MCA) to public survey data gathered by BikePGH in Pittsburgh, Pennsylvania [1]. In a complex questionnaire survey, traditional survey analysis techniques restrict the comprehension of event phenomena, and the interpretation and findings are limited to significant generalizations. The findings suggest that limiting the assessment to smaller clusters greatly benefits the ability to understand and interpret the participants' response patterns towards AVs, based on their previous interaction level with AVs as pedestrians.

## 2 Literature Review

On public and private roads in multiple countries, large-scale testing of AVs is currently being introduced. However, these innovations are not necessarily accepted by all. technologies need to tackle technological challenges as well as social barriers in order to successfully join the marketplace. Several studies have previously analyzed consumers' views and public opinions about the potential implementation of AV technologies [17, 20, 22, 24, 25].

Schoettle and Sivak [38] used survey analysis to study public opinion about the use of AVs in the U.S., the U.K., and Australia. The findings suggested that women show more concern about AV technologies than men. Furthermore, a majority of the respondents indicated interest in possessing a vehicle with this technology, but a majority of them were not willing to pay extra costs. Howard and Dai [26] conducted a case study utilizing the responses of 107 AV users in Berkeley, California to assess public attitudes regarding AVs. An assessment was conducted to determine which vehicle features people liked and disliked, and the participants were asked to explain how they envisioned the implementation of the technology. The potential safety improvements, the ease of finding parking spots, and the ability to multitask while driving were all features that were associated with positive attitudes. In contrast, the traits for which people expressed concern included potential liability, the vehicle cost, and the possibility of losing control. The findings showed that women were less concerned with liability and more concerned with control than men. Choi et al. [9] studied the experience of users adopting AVs and explored which factors could promote people's trust in AVs. The study found that perceived trust and utility are significant factors that affect people's likelihood of using AVs. Furthermore, the findings showed that three structures had positive impacts on trust: technical competence, system transparency, and situation management. In another study, Kyriakidis et al. [30] analyzed user willingness to own AVs with different automation levels. An internet-based survey with 63 questions was used in this study; it produced 5,000 responses from 109 countries. Furthermore, participants were primarily concerned with hacking or software misuse, legal problems, and safety.

Bansal et al. [7] introduced a new simulation-driven fleet evolution system to forecast the long-term (year 2015–2045) adoption levels of AV technology in America based on

228 S. Das and H. Zubaidi

eight scenarios created to encompass 0%, 5%, and 10% annual increments in Americans' willingness to pay (WTP); 5% and 10% annual drops in technology prices; as well as new or changing government regulations. Krueger et al. [29] studied the travel behavior effects of AVs in order to advance future research. They identified features of consumers who are likely to adopt AV technology and assessed their willingness to pay additional costs for various service enhancements. The study findings revealed that service features such as travel cost, travel time, and waiting time are important factors of AV use and the users' acceptance of them. The adoption and acceptance levels may vary among people in different demographic groups. For example, young people and multimodal travelers are more likely to use AVs.

Daziano et al. [14] used data from a national online panel of 1,260 individuals. The study is designed in the setup of a discrete choice experiment for vehicle purchases based on energy efficiency and autonomous characteristics. The aim of their research was to find a proper estimate of the willingness of individuals to pay for AVs. The aggregate findings of their surveys show that the average household is prepared "to pay approximately $3,500 for partial automation and $4,900 for full automation." Menon [32] analyzed the future market segments of AV customers and disclosed the factors affecting AV adoption or non-adoption. The results revealed that the impact of giving up household vehicles varies among households with single or multiple vehicles as well as households with various triggers including socio demographics, vehicle purchase histories, and current travel characteristics.

The responses from a national survey of 1,765 adults in the U.S. were examined by Lee et al. [31] to classify the key factors of AV acceptance and to assess how age and other attributes are related to AV perceptions and attitudes. The findings of the study revealed age's negative effects on perceptions of AVs, interest in using them, and behavioral plans to use them when the technology is available. Furthermore, experiential characteristics related to age, such as knowledge of, experience level, and general trust toward technology, all had a significant impact on people's feelings towards AVs.

To assess their perspectives on AVs and related decisions, Bansal and Kockelman [5] surveyed 1,088 people in Texas and considered their demographics, travel patterns, crash histories, and built-environment attributes. The results demonstrate that older drivers and those with more driving experience are less willing to pay (WTP) for new vehicle technologies like AVs. Bansal and Kockelman [6] carried out a follow-up study using an internet-based survey that asked 347 Austinites about their opinions on smart-car technologies. The findings concluded that the participants consider less crashes to be the main benefit of self-driving cars and possible equipment failure to be their primary concern. Hulse et al. [27] conducted a survey of nearly 1,000 individuals to study their opinions of AVs regarding safety and acceptance. The results revealed that AVs are viewed as 'somewhat low risk' and there was not much resistance to their use on public roads. AVs were also perceived as having more risks than autonomous trains. It was found that gender, age, and risk-taking characteristics had various relationships with the perceived risk of various vehicle types and general opinions of AVs. For instance, men and younger adults showed a greater level of acceptance.

Canis [8] conducted a survey and found that 30 percent the respondents demonstrated unwillingness in purchasing an AV. The analysis also found that more than 50% of U.S.

drivers "feel less safe at the prospect of sharing the road with a self-driving vehicle." Das et al. [2] analyzed the attitudes of people towards AVs and the present polarities about content and automation levels used two natural language processing (NLP) tools to conduct knowledge discovery from a bag of approximately seven million words using YouTube video comments. The findings showed that the public is interacting with AV technologies more now than in the past; they also showed that as the degree of automation increases, so does the potential perception of safety. Finally, they concluded that there are more widespread positive feelings about AVs than negative, uncertain, or litigious feelings. Penmetsa et al. [37] assessed survey data obtained from BikePGH and found that respondents with previous experience of direct AV interaction reported significantly higher perceptions of their safety benefits than respondents without any experience of AV interaction.

The literature review shows several gaps. First, user attitudes are not reliant on the future safe implementation of AVs. Furthermore, the previous AV interaction level of participants is unknown in a majority of the studies. Another limitation is that not many studies have explored non-motorist perception of AVs. Additionally, traditional survey analysis approaches have limited interpretation abilities for co-occurrences of the responses. The present study provides an acceptable data analysis method that can recognize trends and correlations from a complex sample to clarify the attitude of roadway users towards the adoption of AVs.

## 3 Data Description

### 3.1 Survey Design and Participants

According to the first edition of the Highway Safety Manual (HSM), approximately 94 percent of collisions occur due to human errors [36]. AV companies expect that the AV technologies could eliminate the human error component of the driving tasks and therefore, greatly minimize these crashes. AVs offer many potential advantages independent of the automation modes (partial to full automation), such as minimizing roadway crashes and fatalities, reducing traffic congestion, and promoting mobility [35, 36]. However, AVs can also cause safety and infrastructure issues that state Department of Transportation (DOT) authorities must solve. DOTs must determine if a stable driving environment can be ensured with the current standards and methods of vehicle testing. Furthermore, AV interactions with other road users must be addressed. Infrastructure changes may be required for AV implementation, and policymakers will need to determine which changes they should pursue, while still catering to the needs of conventional vehicles which will continue to be used for decades. However, several regulating organizations do not have comprehensive plans that are readily available to set specific targets regarding the introduction of AVs to roadways [28].

AV firms have tested AVs in major cities since September 2016. In 2017, ten AV proving grounds were assigned by the United States Department of Transportation (USDOT) to promote testing of AV technologies; one of these testing grounds was located in Pittsburgh, PA. In early 2017, BikePGH developed a survey to gather information about how BikePGH donor-members as well as Pittsburgh residents at large feel about sharing the road with AVs as a non-motorized roadway user (i.e., bicyclist or pedestrian).

This survey aimed to help policymakers comprehend the complexity of futures involving AV technologies by demonstrating how AVs can bring introduce marketplace models to cities and town centers. The survey was conducted in two parts. In the first stage, the survey only included Bike PGH donor-members; this stage was conducted via email and it yielded 321 responses (out of 2,900). The second stage encouraged the public to participate by publishing it on the BikePGH website and promoting it via social media and several news articles. This stage produced 798 responses (mostly from Pittsburg residents for a combined total of 1,119 responses. The major questions from the survey are listed below.

- *InteractPedestrian:* Real-life interaction with AVs?
- *BikePghPosition:* Opinion on BikePGH's position on AVs?
- *CircumstancesCoded:* Six categories: AV Safer, Cautious about AV, Negative about AV, No Difference, No Experience, and Others.
- *SafetyHuman:* Feel safe with human driven vehicles on road?
- *SafetyAV:* Feel safe with AVs on road?
- *AVSafetyPotential:* AV safe enough?
- *RegulationTesting:* Regulation on AV testing?
- *RegulationSpeed:* Lower speed limit on roads that allow AVs?
- *RegulationSchoolZone:* Disallow AVs on school zones?
- *RegulationShareData:* Share AV movement data with public or regulatory agencies?
- *FeelingsProvingGround:* Use Pittsburgh's public streets as a proving ground for AVs?
- *AdvocacyIssues:* Should it be an advocacy issue for BikePGH?
- *PayingAttentionAV:* Pay attention AV news?
- *FamiliarityTechnoology:* Familiarity with AV technology?

### 3.2 Exploratory Data Analysis

The final dataset contains 321 responses from the BikePGH members (BPG) and 793 responses from the general public (Public) for a total of 1,114 respondents. The current analysis is limited to human-computer interaction framework for pedestrians. Table 1 lists attribute distribution by variables and results of chi-square test. The lower p-values ($p < 0.05$) from the chi-square tests indicate that some of the variables are significantly different for three types of participants based on their prior human computer interaction regarding AVs as pedestrians: yes (406 participants), no (599 participants), and not sure (109 participants).

The circumstances are coded into six categories: *AV Safer, Cautious about AV, Negative about AV, No Difference, No Experience, and Others.* Around 37% percent of the survey participants have prior experience of AV circumstances. the response patterns. Approximately 5.17% of the survey participants with prior interactions with AVs feel AV safer. This percentage is lower in other two groups with no (or not sure) prior experience with AVs. The response regarding *SafetyAV* (5 as very safe) is also higher in percentage in the responses of the participants with prior experience with AVs. This group also indicates they are mostly in favor of considering Pittsburgh as the proving ground of AVs. Two other information are important in this analysis: 1) familiarity with technology (*FamiliarityTechnoology*), and 2) paying attention to AVs (*PayingAttentionAV*).

**Table 1.** Compare groups

| Questions | Yes (N = 406) | No (N = 599) | Not sure (N = 109) | p-val |
|---|---|---|---|---|
| *BikePghPosition* | | | | < 0.001 |
| Actively oppose | 4.93% | 6.68% | 5.50% | |
| Actively support | 55.40% | 38.90% | 36.70% | |
| Neither support nor oppose | 31.00% | 42.90% | 45.90% | |
| No opinion | 8.62% | 11.50% | 11.90% | |
| *SafetyHuman* | | | | |
| 1 | 5.17% | 5.01% | 7.34% | |
| 2 | 30.30% | 26.90% | 30.30% | |
| 3 | 40.90% | 40.40% | 40.40% | |
| 4 | 21.40% | 22.40% | 16.50% | |
| 5 | 2.22% | 4.17% | 4.59% | |
| No experience | 0.00% | 1.17% | 0.92% | |
| *AVSafetyPotential* | | | | < 0.001 |
| 1: Yes | 73.60% | 60.30% | 56.90% | |
| 2: No | 4.43% | 7.68% | 8.26% | |
| 3: Not sure | 21.90% | 32.10% | 34.90% | |

| Questions | Yes (N = 406) | No (N = 599) | Not sure (N = 109) | p-val |
|---|---|---|---|---|
| *CircumstancesCoded* | | | | < 0.001 |
| AV Safer | 5.17% | 1.34% | 1.83% | |
| Cautious about AV | 11.60% | 2.50% | 6.42% | |
| Negative about AV | 5.91% | 2.00% | 0.00% | |
| No difference | 60.10% | 21.70% | 23.90% | |
| No experience | 14.30% | 69.60% | 65.10% | |
| Others | 2.96% | 2.84% | 2.75% | |
| *SafetyAV* | | | | |
| 1 | 3.94% | 5.84% | 6.42% | |
| 2 | 9.11% | 9.52% | 11.00% | |
| 3 | 21.40% | 24.20% | 25.70% | |
| 4 | 39.40% | 29.00% | 30.30% | |
| 5 | 25.10% | 14.50% | 16.50% | |
| No experience | 0.99% | 16.90% | 10.10% | |
| *RegulationTesting* | | | | 0.001 |
| 1: Yes | 64.50% | 75.10% | 69.70% | |

(continued)

**Table 1.** (*continued*)

| Questions | Yes | No | Not sure | p-val |
|---|---|---|---|---|
| *RegulationSpeed* | | | | < 0.001 |
| 1: Yes | 44.60% | 54.90% | 54.10% | |
| 2: No | 38.20% | 25.00% | 29.40% | |
| 3: Not sure | 17.20% | 20.00% | 16.50% | |
| *RegulationShareData* | | | | 0.873 |
| 1: Yes | 71.20% | 71.60% | 73.40% | |
| 2: No | 13.50% | 12.00% | 10.10% | |
| 3: Not sure | 15.30% | 16.40% | 16.50% | |
| *Advocacylssues* | | | | |
| 1: Not at all | 5.91% | 4.01% | 2.75% | |
| 2: Little | 16.30% | 13.70% | 13.80% | |
| 3: Some | 38.20% | 41.90% | 40.40% | |
| 4: Moderate | 33.50% | 34.90% | 36.70% | |
| 5: A lot | 6.16% | 5.51% | 6.42% | |
| *FamiliarityTechnoology* | | | | < 0.001 |
| 1: Not at all | 2.96% | 8.68% | 6.42% | |
| 2: Little | 20.20% | 20.00% | 30.30% | |
| 3: Some | 39.70% | 46.10% | 35.80% | |
| 4: Moderate | 23.60% | 17.90% | 20.20% | |
| 5: A lot | 13.50% | 7.35% | 7.34% | |
| 2: No | 16.50% | 8.68% | 11.00% | < 0.001 |
| 3: Not sure | 19.00% | 16.20% | 19.30% | |
| *RegulationSchoolZone* | | | | < 0.001 |
| 1: Yes | 18.50% | 27.90% | 30.30% | |
| 2: No | 56.90% | 41.60% | 42.20% | |
| 3: Not sure | 24.60% | 30.60% | 27.50% | |
| *FeelingsProvingGround* | | | | < 0.001 |
| 1: Not at all | 6.16% | 8.51% | 9.17% | |
| 2: Little | 7.39% | 9.02% | 12.80% | |
| 3: Some | 11.80% | 14.40% | 11.00% | |
| 4: Moderate | 14.30% | 23.90% | 22.90% | |
| 5: A lot | 60.30% | 44.20% | 44.00% | |
| *PayingAttentionAV* | | | | < 0.001 |
| 1: Not at all | 0.99% | 1.50% | 1.83% | |
| 2: Little | 6.40% | 10.50% | 11.90% | |
| 3: Some | 24.90% | 32.90% | 33.90% | |
| 4: Moderate | 33.50% | 33.10% | 33.00% | |
| 5: A lot | 34.20% | 22.00% | 19.30% | |
| *Group* | | | | 0.047 |
| BPG | 32.50% | 25.70% | 32.10% | |
| Public | 67.50% | 74.30% | 67.90% | |

The group with prior human-computer interaction with AVs shows that they are more familiar with technology and they pay attention to AVs than the other two groups.

## 4 Methodology

### 4.1 Multiple Correspondence Analysis

Multiple Correspondence Analysis (MCA), a dimension reduction method, mandates the formation of a matrix based on pairwise cross-tabulation of explanatory and response variables but does not differentiate between the two variable types. For instance, $1114 \times 10$ would be the dimension of the used data if the number of total survey respondents was $1114$ and there were 10 questions analyzed in the survey. For a categorical or qualitative variable table with the dimensions $1114 \times 10$, MCA can be described by taking an individual record (in row), i [i $= 1$ to $1114$], where the 10 categorical variables are represented by 24 columns that have different sizes for the categories.

Consider $P$ as the total variables (i.e., columns), and $I$ as the total responses (i.e., rows). This will produce a matrix of $I \times P$. If $L_p$ is the total attributes for variable $p$, then the sum of the categories for all variables is, $L = \int_{p=1}^{p} L_p$. This will produce another matrix $I \times L$, in which each variable will comprise multiple columns to display all of their possible categorical values. The cloud of categories is known as a weighted combination of $J$ points. Category $j$ is shown by a point denoted by $Cj$ with the weight of $nj$. For each of variables, the total of the weights of category points is $n$. Following this, for the entire set $J$, the sum is $nP$. The relative weight $wj$ for point $Cj$ is $wj = nj/(nP) = fj/P$. The total of the relative weights for category points is $1/P$; thus, the sum of the entire set is 1 [36]. $w_j = \frac{n_j}{nP} = \frac{f_j}{P}$ with $\int j \in J_q w_j = \frac{1}{P}$ and $\int j \in J w_j = 1$. Here, $n_j n_j$ is the number of individual points that have both categories $k$ and $k'$. The squared distance from $Cj$ to $Cj'$ is shown in Eq. 1 [36]:

$$\left( C^j C^{j'} \right)^2 = \frac{n_j + n_{j'} - 2n_{jj'}}{n_j n_{j'}/n} \tag{1}$$

In Eq. 1, $n_j + n_{j'} - 2n_{jj'}$ is the number of individual records associated with either j or j'. For variables $p$ and $p'$, the denominator is the familiar "theoretical frequency" for the cell $(j, j')$ of the $Jp \times Jp'$ two-way table. Other MCA frameworks have been created with similar goals [21, 23]. Several recent studies have applied MCA in transportation research [3, 4, 10–13, 15, 16, 18, 19].

## 5 Results and Discussion

The variables with low p-values (less than 0.10, see Table 1) indicate that the groups (pedestrian-AV interaction $=$ yes vs. pedestrian-AV interaction $=$ no) are significantly different in attitude towards AVs for the associated attributes. The variables that show significant differences among these three types of participants based on their prior human computer interactions regarding AVs include *BikePghPoisition, SafetyHuman, AVSafetyPotential, RegulationSpeed, AdvocacyIssues, FamiliarityTechnology, CircumstancesCoded, SafetyAV, RegulationTesting, RegulationSchool-Zone, FeelingsProvingGround, and PayingAttentionAV*. To conduct the MCA analysis, we considered seven variables (*SafetyHuman, AVSafetyPotential, FamiliarityTechnology, CircumstancesCoded, SafetyAV, PayingAttentionAV, and InteractPedestrian*) for analysis to identify the clusters based on their prior interactions with AVs as pedestrians.

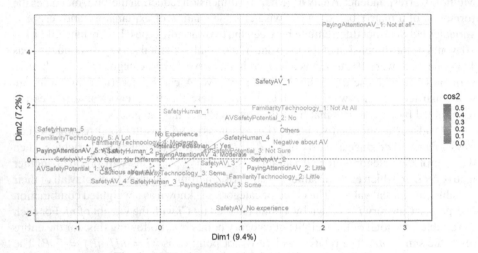

**Fig. 1.** MCA plot (pedestrian interaction group: yes)

Figure 1 to Fig. 3 illustrate the MCA plots by three groups: group 1 (yes), group 2 (no), and group 3 (not sure). The first two axes of the MCA plot explain nearly 20% of total inertia. Due to the presence of large number of attributes, MCA plots may sometimes become noisy. The closer the locations of the attributes in the MCA plot, the closer the associations between these attributes. Figure 1 (MCA plot for group 1) shows that the responses 'a lot' for two variable familiarity with technology (*FamiliarityTechnology*) and paying attention to AV (*PayingAttentionAV*) are closer to other relevant responses. The negative responses regarding AVs (*SafetyAV* as 1 or very low, *PayingAttentionAV* as 'not at all', *FamiliarityTechnology* as 'not at all') are not in the close proximity of other responses that are more positive towards AVs. On the other hand, Fig. 2 and Fig. 3 show MCA plots for other two groups (group 2 and group 3). The negative responses towards AVs are clustered in a group for group 3. Similarly, *PayingAttentionAV* as 'a lot', and *FamiliarityTechnology* as 'a lot' are not in the close proximity of other responses that are positive towards AVs in group 2.

The overall finding shows that participants of similar mindset are clustered in groups. For example, participants with negative views are in closer proximity for a cluster and same for the participants with positive views. The parameter squared cosine ($cos^2$) indicates the degree of association between variable attributes and a dimension or axis. If a variable attribute is well represented by two dimensions in a plane, the sum of the $cos^2$ will be approximately one. Sometimes, n-dimensional display is required if the

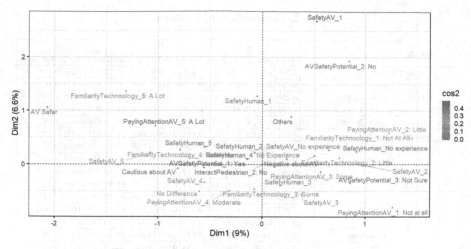

**Fig. 2.** MCA plot (pedestrian interaction group: no)

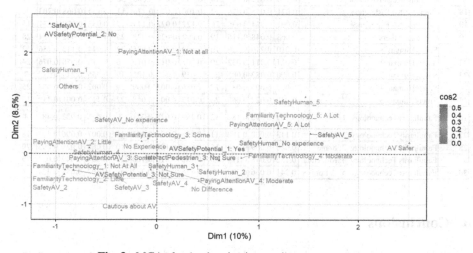

**Fig. 3.** MCA plot (pedestrian interaction group: not sure)

inertia or variance of the first two dimensions are inadequate. Table 2 lists the $\cos^2$ values for axis 1 (dimension 1) and axis 2 (dimension 2) for three groups. For each attribute, ranks are provided based on total $\cos^2$ values for both axes. The top 10 ranks for each group are colored in red. For group 1, the attribute that can explain more in both axes is *PayingAttentionAV* as 'a lot'. For group 2, *AVSafetyPotential* as 'yes', and *SafetyAV* as '1 (very low)' are the top two attributes. *SafetyAV* as '1 (very low)' is also a top attribute for group 3. The rank distribution also indicates that group 1 participants are more positive towards AVs compared to other two groups.

**Table 2.** Contributions by attributes

| ID | Attribute | Yes | | | | No | | | | Not Sure | | | |
|---|---|---|---|---|---|---|---|---|---|---|---|---|---|
| | | Dim 1 | Dim 2 | Tot | Rank | Dim 1 | Dim 2 | Tot | Rank | Dim 1 | Dim 2 | Tot | Rank |
| 1 | AVSafetyPotential: Yes | 0.366 | 0.013 | 0.379 | 3 | 0.475 | 0.006 | 0.480 | 1 | 0.336 | 0.063 | 0.399 | 6 |
| 2 | AVSafetyPotential: No | 0.084 | 0.137 | 0.221 | 9 | 0.058 | 0.310 | 0.368 | 5 | 0.124 | 0.039 | 0.162 | 11 |
| 3 | AVSafetyPotential: Not Sure | 0.250 | 0.004 | 0.254 | 6 | 0.342 | 0.057 | 0.399 | 4 | 0.000 | 0.085 | 0.085 | 23 |
| 4 | PayingAttentionAV_1: Not at all | 0.096 | 0.251 | 0.347 | 4 | 0.024 | 0.010 | 0.034 | 24 | 0.003 | 0.066 | 0.069 | 26 |
| 5 | PayingAttentionAV_2: Little | 0.076 | 0.002 | 0.078 | 21 | 0.177 | 0.040 | 0.217 | 8 | 0.076 | 0.011 | 0.087 | 22 |
| 6 | PayingAttentionAV_3: Some | 0.145 | 0.088 | 0.233 | 8 | 0.152 | 0.009 | 0.162 | 14 | 0.104 | 0.021 | 0.125 | 16 |
| 7 | PayingAttentionAV_4: Moderate | 0.006 | 0.032 | 0.038 | 25 | 0.050 | 0.116 | 0.167 | 13 | 0.172 | 0.002 | 0.174 | 9 |
| 8 | PayingAttentionAV_5: A Lot | 0.401 | 0.136 | 0.536 | 1 | 0.298 | 0.144 | 0.441 | 3 | 0.484 | 0.023 | 0.507 | 3 |
| 9 | SafetyAV_1 | 0.092 | 0.385 | 0.476 | 2 | 0.017 | 0.447 | 0.463 | 2 | 0.078 | 0.456 | 0.534 | 1 |
| 10 | SafetyAV_2 | 0.151 | 0.003 | 0.154 | 14 | 0.091 | 0.000 | 0.091 | 21 | 0.145 | 0.003 | 0.148 | 13 |
| 11 | SafetyAV_3 | 0.107 | 0.025 | 0.132 | 17 | 0.044 | 0.101 | 0.145 | 16 | 0.053 | 0.094 | 0.147 | 14 |
| 12 | SafetyAV_4 | 0.057 | 0.062 | 0.119 | 18 | 0.125 | 0.045 | 0.170 | 12 | 0.057 | 0.007 | 0.064 | 27 |
| 13 | SafetyAV_5 | 0.211 | 0.026 | 0.237 | 7 | 0.197 | 0.008 | 0.205 | 9 | 0.041 | 0.474 | 0.515 | 2 |
| 14 | SafetyAV_No experience | 0.011 | 0.038 | 0.049 | 24 | 0.083 | 0.014 | 0.097 | 20 | 0.089 | 0.002 | 0.092 | 21 |
| 15 | Circums: AV Safer | 0.027 | 0.006 | 0.034 | 27 | 0.058 | 0.015 | 0.073 | 23 | 0.009 | 0.001 | 0.010 | 30 |
| 16 | Circums: Cautious about AV | 0.001 | 0.010 | 0.010 | 30 | 0.017 | 0.000 | 0.017 | 28 | 0.102 | 0.061 | 0.163 | 10 |
| 17 | Circums: Negative about AV | 0.120 | 0.013 | 0.132 | 16 | 0.005 | 0.001 | 0.006 | 30 | 0.009 | 0.086 | 0.095 | 19 |
| 18 | Circums: No Difference | 0.021 | 0.055 | 0.076 | 22 | 0.100 | 0.070 | 0.170 | 11 | 0.134 | 0.068 | 0.203 | 8 |
| 19 | Circums: No Experience | 0.001 | 0.034 | 0.035 | 26 | 0.122 | 0.022 | 0.144 | 17 | 0.053 | 0.078 | 0.131 | 15 |
| 20 | Circums: Others | 0.068 | 0.048 | 0.116 | 19 | 0.002 | 0.023 | 0.025 | 25 | 0.003 | 0.052 | 0.054 | 28 |
| 21 | SafetyHuman_1 | 0.010 | 0.210 | 0.220 | 10 | 0.000 | 0.085 | 0.085 | 22 | 0.044 | 0.112 | 0.156 | 12 |
| 22 | SafetyHuman_2 | 0.021 | 0.011 | 0.032 | 29 | 0.002 | 0.017 | 0.019 | 26 | 0.018 | 0.058 | 0.076 | 24 |
| 23 | SafetyHuman_3 | 0.000 | 0.082 | 0.083 | 20 | 0.019 | 0.105 | 0.125 | 19 | 0.055 | 0.244 | 0.299 | 7 |
| 24 | SafetyHuman_4 | 0.027 | 0.005 | 0.032 | 28 | 0.006 | 0.003 | 0.009 | 29 | 0.002 | 0.070 | 0.072 | 25 |
| 25 | SafetyHuman_5 | 0.038 | 0.022 | 0.060 | 23 | 0.013 | 0.004 | 0.018 | 27 | 0.117 | 0.001 | 0.118 | 17 |
| 26 | FamiliarityTechnoology_1: Not at All | 0.085 | 0.095 | 0.180 | 12 | 0.199 | 0.022 | 0.221 | 7 | 0.451 | 0.032 | 0.482 | 4 |
| 27 | FamiliarityTechnoology_2: Little | 0.113 | 0.068 | 0.180 | 11 | 0.138 | 0.003 | 0.141 | 18 | 0.007 | 0.096 | 0.103 | 18 |
| 28 | FamiliarityTechnoology_3: Some | 0.080 | 0.069 | 0.149 | 15 | 0.006 | 0.181 | 0.186 | 10 | 0.068 | 0.025 | 0.093 | 20 |
| 29 | FamiliarityTechnoology_4: Moderate | 0.132 | 0.048 | 0.179 | 13 | 0.135 | 0.015 | 0.150 | 15 | 0.004 | 0.049 | 0.053 | 29 |
| 30 | FamiliarityTechnoology_5: A Lot | 0.241 | 0.066 | 0.307 | 5 | 0.136 | 0.146 | 0.282 | 6 | 0.367 | 0.058 | 0.424 | 5 |

## 6 Conclusions

For the comprehensive implementation of AVs, it is crucial to gather response from end-users and understand acceptance tolerances. Furthermore, understanding public perception of emerging technologies like AVs is critical for public policy. Major AV companies expect that AVs will be sufficiently trustworthy and economical to replace a majority of human driving by 2030, giving roadway users independent mobility [3]. As AVs continue to develop rapidly, there is an increasing need to control the adoption of these technologies in the real-world roadway environment.

Due to their health and environmental benefits, non-motorized modes of transport such as walking has become popular. Unfortunately, the high exposure of pedestrians in urban areas might have contributed to the sharp increase in pedestrian crashes in the recent years. Due to the high exposure of pedestrians and rapid growth of AV testbeds, it is critical to comprehend the public perception towards pedestrian and AV interactions.

We analyzed participants' attitudes towards AVs and considered their human AV interaction and gathered findings to inform policy development for AV adoption. We found that pedestrians with previous AV interactions think AVs are safer than human drivers and they recognize the potential safety benefits of AVs. The survey also found that participants without previous AV experience believe that Pittsburgh streets are safe with human drivers. The study's findings show that prior experience and knowledge of AVs are associated with more positive opinions and perceptions. This provides evidence to the value of AV demonstration projects that offer non-motorized roadway users with the opportunity to interact with AVs.

Our study has several limitations. First, only a limited region was included in the survey. Additionally, detailed information about the AV interactions within this survey were not publicly available for use. Further studies could perform text mining on these narratives to help identify the missing link between the AV perception and acceptance among non-motorist roadway users.

# References

1. Website of Bike PGH. Accessed July 2019, https://www.bikepgh.org/
2. Das, S., Dutta, A., Lindheimer, T., Jalayer, M., Elgart, Z.: YouTube as a source of information in understanding autonomous vehicle consumers: natural language processing study. Transp. Res. Rec. **2673**(8), 242–253 (2019)
3. Ali, F., Dissanayake, D., Bell, M., Farrow, M.: Investigating car users' attitudes to climate change using multiple correspondence analysis. J. Transp. Geogr. **72**, 237–247 (2018)
4. Baireddy, R., Zhou, H., Jalayer, M.: Multiple correspondence analysis of pedestrian crashes in rural Illinois. Transp. Res. Rec. **2672**(38), 116–127 (2018)
5. Bansal, P., Kockelman, K.M.: Forecasting Americans long-term adoption of connected and autonomous vehicle technologies. Transp. Res. Part A Policy Pract. **95**, 49–63 (2017)
6. Bansal, P., Kockelman, K.: Are we ready to embrace connected and self-driving vehicles? a case study of Texans. Transportation **45**(2), 641–675 (2018)
7. Bansal, P., Kockelman, K.M., Singh, A.: Assessing public opinions of and interest in new vehicle technologies An Austin perspective. Transp. Res. Part C Emerg. Technol. **67**, 1–14 (2016)
8. Canis, B.: Issues in Autonomous Vehicle Deployment. Technical Report (2018)
9. Choi, J.K., Ji, Y.G.: Investigating the importance of trust on adopting an autonomous vehicle. Int. J. Human-Comput. Interact. **31**(10), 692–702 (2015)
10. Das, S., Jha, K., Fitzpatrick, K., Brewer, M., Shimu, T.H.: Pattern identification from older bicyclist fatal crashes. Transp. Res. Rec. **267**, 638–649 (2019)
11. Das, S., Minjares-Kyle, L., Lingtao, W., Henk, R.H.: Understanding crash potential associated with teen driving: survey analysis using multivariate graphical method. J. Safety Res. **70**, 213–222 (2019)
12. Das, S., Sun, X.: Factor association with multiple corre-spondence analysis in vehicle–pedestrian crashes. Transp. Res. Rec. **2519**(1), 95–103 (2015)
13. Das, S., Sun, X.: Association knowledge for fatal run-off- road crashes by Multiple Correspondence Analysis. IATSS Res. **39**(2), 146–155 (2016)
14. Daziano, R.A., Sarrias, M., Leard, B.: Are consumers willing to pay to let cars drive for them? analyzing response to autonomous vehicles. Transp. Res. Part C Emerg. Technol. **78**, 150–164 (2017)
15. Degraeve, B., Granie, M.A., Pravossoudovitch, K.: Men and women drivers: a study of social representations through prototypical and correspondence analysis (2014)
16. Diana, M., Pronello, C.: Traveler segmentation strategy with nominal variables through correspondence analysis. Transp. Policy **17**(3), 183–190 (2010)

17. Fagnant, D.J., Kockelman, K.M.: Dynamic ride-sharing and fleet sizing for a system of shared autonomous vehicles in Austin, Texas. Transportation 45(1), 143–158 (2018)

18. Fontaine, H.: A typological analysis of pedestrian accidents. In: 7th workshop of ICTCT, pp. 26–27 (1995)

19. Fort, E., Gadegbeku, B., Gat, E., Pelissier, C., Hours, M., Charbotel, B.: Working conditions and risk exposure of employees whose occupations require driving on public roads – factorial analysis and classification. Accid. Anal. Prev. 131, 254–267 (2019)

20. Gold, C., Korber, M., Hohenberger, C., Lechner, D., Bengler, K.: Trust in automation–before and after the experience of take-over scenarios in a highly automated vehicle. Procedia Manuf. 3, 3025–3032 (2015)

21. Greenacre, M., Blasius, J.: Multiple Correspondence Analysis and Related Methods. Chapman and Hall/CRC, Boca Raton (2006)

22. Heide, A., Henning, K.: The "Cognitive Car" a roadmap for research issues in the automotive sector. IFAC Proc. Vol. 39(4), 44–50 (2006)

23. Hoffman, D.L., De Leeuw, J.: Interpreting multiple correspondence analysis as a multidimensional scaling method. Mark. Lett. 3(3), 259–272 (1992)

24. Hohenberger, C., Sporrle, M., Welpe, I.M.: Not fearless, but self-enhanced: the effects of anxiety on the willingness to use autonomous cars depend on individual levels of self-enhancement. Technol. Forecast. Soc. Chang. 116, 40–52 (2017)

25. Hou, G., Chen, S., Chen, F.: Framework of simulation- based vehicle safety performance assessment of highway system under hazardous driving conditions. Transp. Res. Part C Emerg. Technol. 105, 23–36 (2019)

26. Howard, D., Dai, D.: Public perceptions of self-driving cars: the case of Berkeley, California. In: Transportation Research Board 93rd Annual Meeting, vol. 14, pp. 1–16 (2014)

27. Hulse, L.M., Xie, H., Galea, E.R.: Perceptions of autonomous vehicles relationships with road users, risk, gender and age. Saf. Sci. 102, 1–13 (2018)

28. Keeney, T.: Mobility-as-a-service: Why self-driving cars could change everything. Ark Invest 1, 3 (2017)

29. Krueger, R., Rashidi, T.H., Rose, J.M.: Preferences for shared autonomous vehicles. Transp. Res. Part C Emerg. Technol. 69, 343–355 (2016)

30. Kyriakidis, M., Happee, R., de Winter, J.C.F.: Public opinion on automated driving: results of an international questionnaire among 5000 respondents. Transport. Res. F Traffic Psychol. Behav. 32, 127–140 (2015)

31. Lee, C., Ward, C., Raue, M., D'Ambrosio, L., Coughlin, J.: Age differences in acceptance of self-driving cars: a survey of perceptions and attitudes. In: Zhou, J., Salvendy, G. (eds.) ITAP 2017. LNCS, vol. 10297, pp. 3–13. Springer, Cham (2017). https://doi.org/10.1007/978-3-319-58530-7_1

32. Menon, N.: Autonomous vehicles an empirical assessment of consumers' perceptions, intended adoption, and impacts on household vehicle ownership (2017)

33. National Highway Traffic Safety Administration (NHTSA). Traffic Safety Facts: 2017. Bicyclists and Other Cyclists (2019)

34. National Highway Traffic Safety Administration (NHTSA). Traffic Safety Facts: 2017. Pedestrians (2019)

35. U.S. Government Accountability Office. Automated Vehicle: Comprehensive Plan Could Help DOT Address Challenges (2017)

36. National Research Council (US). Transportation Research Board. Task Force on Development of the Highway Safety Manual and Transportation Officials. Joint Task Force on the Highway Safety Manual. Highway safety manual, vol. 1. AASHTO (2010)

37. Penmetsa, P., Adanu, E.K., Wood, D., Wang, T., Jones, S.L.: Perceptions and expectations of autonomous vehicles–a snapshot of vulnerable road user opinion. Technol. Forecast. Soc. Chang. **143**, 9–13 (2019)
38. Schoettle, B., Sivak, M.: A survey of public opinion about autonomous and self-driving vehicles in the US, the UK, and Australia. Technical Report. University of Michigan, Ann Arbor, Transportation Research Institute (2014)

# Great Expectations: On the Design of Predictive Motion Cues to Alleviate Carsickness

Cyriel Diels[1](✉) and Jelte Bos[2,3]

[1] Intelligent Mobility Design Centre, Royal College of Art, 4 Hester Road,
London SW11 4AN, UK
cyriel.diels@rca.ac.uk
[2] TNO Perceptual and Cognitive Systems, Soesterberg, Netherlands
[3] Faculty of Behavioural and Movement Sciences, VU University, Amsterdam, Netherlands

**Abstract.** Motion sickness has gained renewed interested in the context of the developments in vehicle automation in which we are witnessing a transition from a driver-centric to passenger-centric design philosophy. As a corollary, motion sickness can be expected to become considerably more prevalent which creates a hurdle towards the successful introduction of vehicle automation and its ultimate socio-economic and environmental benefits. We here review early proof-of-concept studies into the beneficial effects of providing passenger with predictive motion cues as an elegant and effective method to reduce motion sickness in future vehicles. Future design parameters are discussed to finetune such cues not only for optimum effectiveness but, importantly, also for acceptance including sensory modality, timing, information detailing, and personalization.

**Keywords:** Motion sickness · Automation · Interface design

## 1 Introduction

*"It's Wednesday morning, 8am, 2035, Martin is walking his kids to school. After dropping them off at the school gate, he walks another 5 minutes to his local Mobility Hub where he hops on his pre-booked commuter pod. Martin greets his fellow passengers, settles in, and gets on with his work for the day. His travel time has become valuable office time and, given his 45 mins commute each way, he is now able to pick Rosa and Rudy up from afterschool club at 4.30pm, take them to the park before heading home for the evening."*

While perhaps somewhat utopian, at the time of writing, the above scenario is increasingly starting to feel within the realms of possibility. Indeed, being able to use our travel time more enjoyable or productive is arguably one of the main benefits that vehicle automation may bring to our everyday lives. However, the ability to do so in comfort is far from trivial [1], not in the least due to the fact that a sizeable proportion of passengers may feel queasy and experience signs and symptoms of motion sickness while engaging in so-called Non-Driving Related Activities (NDRA) [2, 3]. Thus, to realize the full

© Springer Nature Switzerland AG 2021
H. Krömker (Ed.): HCII 2021, LNCS 12791, pp. 240–251, 2021.
https://doi.org/10.1007/978-3-030-78358-7_16

potential of vehicle automation, we need to understand not only the causes of motion sickness, but also the effectiveness and acceptance of design solutions that may prevent or at least reduce the likelihood of motion sickness [4]. While no silver bullet, we argue here that predictive motion cues may go some way towards achieving this goal. Before discussing motion cueing in more detail, we will provide a brief introduction to the topic of motion sickness in general and anticipation in particular.

Motion sickness is a natural response to an unnatural motion environment and is commonly reported aboard ships, in space, virtual reality, simulators, and cars. Although the ultimate manifestation of motion sickness is vomiting, this is typically preceded initially by signs and symptoms such as (cold) sweating, pallor, flatulence, burping, salivation, apathy, and finally by nausea and retching [5, 6]. These symptoms may vary considerably between people regarding their (order of) occurrence, and degree.

A mismatch between sensed and expected motion is widely regarded as the root cause of motion sickness [5, 7]. It occurs under conditions in which the actual sensory information following motion is sufficiently at odds with the expected bodily sensory state as based on prior experiences [8]. Motion sickness is experienced when we are exposed to motion that, from an evolutionary perspective, we are not used to, such as low frequency oscillating motion [9]. Whereas sea and airsickness are mainly caused by slowly oscillating vertical motion, carsickness, on the other hand, is mainly caused by horizontal accelerations due to accelerating, braking, and cornering [10–12]. Hence, an aggressive driving style involving plenty of these actions is therefore more likely to result in carsickness.

In addition to the motion of the vehicle per se, there are several modulating factors that have the potential to aggravate carsickness [13]. These modulating factors are becoming increasingly important in the design of automated vehicles in which we are witnessing a transition from a driver-centric to a passenger-centric design philosophy. Whereas automation creates a new set of design opportunities with respect to the vehicles' interiors, exteriors, and passenger experiences, there are a number of reasons why these may inadvertently lead to an increased prevalence of car sickness [2, 3]. For the successful introduction and acceptability of vehicle automation, it is imperative that we understand not only the fundamental mechanisms and relevant parameters of these modulating factors, but also how they can be integrated within the design process and the design of the overall passenger experience.

We can identify the following four major future scenarios that will impact the occurrence of carsickness: First, unlike drivers, passengers are not in control of the vehicle and are less able to predict the future motion trajectory with sufficient accuracy and more likely to suffer from motion sickness [14]. Secondly, vehicle automation opens up the opportunity to engage in leisurely or economically-productive so-called Non-Driving Related Activities (NDRA). Where NDRA preclude a view of the outside world, such as using in-vehicle displays or reading a book, this may lead to conflicting motion information provided by the visual and vestibular system and a reduced ability to predict future motion [15]. Thirdly, future vehicles may involve flexible seating arrangements including rearward facing seats. Depending on the design of the vehicle, this may preclude an out of the window view but invariably prevent the ability to anticipate future motion and lead to increases motion sickness levels [16]. Fourthly, automation also offers the possibility to optimize the control of vehicle motion for comfort, provided the sensitivity of humans to specific motion characteristics and personal factors would be known.

## 1.1 The Role of Anticipation in Motion Sickness

A common denominator in all the above scenarios is the passengers' difficulty or inability to anticipate future motion. Anticipation plays a key role in the development of motion sickness. This can be understood by considering that our Central Nervous System (CNS) not only reckons sensed motion, but also makes a prediction about self-motion based on previous experiences [8, 17]. The necessity of such a feedforward system can be understood by the sensory imperfections, neural delays, and the fact that our organs of balance cannot make a distinction between inertial and gravitational accelerations that would prevent our CNS to adequately control body motion and attitude [18–20]. Here, attitude refers to our orientation with respect to gravity, which seems of particular interest with respect to motion sickness [17].

A discrepancy or conflict between integrated sensory afferents indicative for specifically attitude, and a prediction thereof by a so-called internal model or neural store, is assumed responsible for generating motion sickness [5, 8, 17]. A mathematical model of this concept has been able to explain the origin of the peak in sickness incidence about 0.16 Hz [17, 18]. This suggests that our CNS does apply a kind of feedforward mechanism.

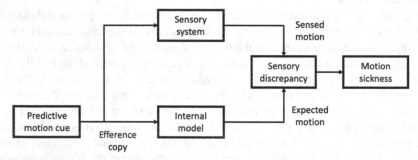

**Fig. 1.** Simplified motion sickness model illustrating the principle of the impact of predictive motion cues, activation of an internal model and subsequent impact on sensory discrepancies and associated motion sickness

In the context of carsickness, it becomes apparent that unlike passengers, drivers are able to anticipate the future motion due to the tight coupling between the control of pedals and steering wheel and subsequent known (learnt) vehicle motion, and thus minimising the likelihood of motion sickness (see Fig. 1 for a simplified representation of the proposed underlying principle). Further, whereas a forward-looking passenger will be able to see a curve ahead, only the driver knows when the vehicle will decelerate and whether this curve will be taken wide or sharp, thus having optimal information about upcoming self-motion, resulting in the smallest possible conflict. Likewise, braking and accelerating will cause a difference in conflict and hence a difference in sickness between drivers and passengers.

Importantly, this anticipatory mechanism concerns two major factors. First the motion itself can be more or less predictable. Kuiper et al. [21], for example showed that a series of equal motions following each other repeatedly in exactly the same way, does lead to less sickness as compared to the same motions following each other in a more random way, the two series yet showing the same RMS acceleration and peak frequency. Secondly, a view on the visual environment may also provide the passenger with cues about upcoming events, such as signs indicative for accelerating and braking (e.g., traffic lights), and curves. Note that this factor differs from the fact that instantaneous visual cues may result in a sense of self-motion, also referred to as vection [22].

The importance of visual information per se is demonstrated by the fact that rear seat passengers are particularly prone to carsickness under conditions where external visual views are limited [11–13]. The importance of anticipatory visual information is also suggested by the findings that backward looking passengers suffer more from carsickness than forward looking passengers, the former only seeing the trajectory that has been followed, the latter seeing the trajectory that will be followed [23]. The beneficial effect of anticipation on the basis of visual information was furthermore clearly demonstrated by [24], who showed a fourfold reduction in motion sickness when a visual track to be travelled was presented in a motion simulator. Assuming cognition to play a role in anticipation, even if that would be unconsciously, this would also imply that cues of a different modality could be helpful. Using an audio cue preceding certain events, Kuiper et al. [25], indeed did find a beneficial effect. Another observation on the role of cognition comes from Perrin et al. [26], who assessed motion sickness in rally co-drivers during the actual rally and "reconnaissance" drives. The reconnaissance drive allows the co-driver to write down shorthand notes (the pacenotes) on how to best drive the stage. Perhaps surprisingly given the differences in motion input, sickness was lower during the actual rally. The authors hypothesised this effect, at least in part, to be related to co-driver's ability to cognitively process and anticipate upcoming motion via the pacenotes.

From the above it follows that anticipation may be achieved in various ways. By extension, the method of delivery may be secondary to its effectiveness, which means that the information related to upcoming information may be provided via a range of sensory channels. This opens up the potential to use predictive motion cues as motion sickness countermeasures in automated vehicles. Whereas control cues are by definition not available since the passenger is not in control of the vehicle, the reported benefits of alternative predictive cues may prove promising. Importantly though, effectiveness in itself is not sufficient and the design and interaction of such cues with NDRAs need to provide an enjoyable, or at the very minimum, acceptable passenger experience. As suggested previously [27], the use of peripheral ambient displays may be particularly suitable in this context, providing effective yet unobtrusive and intuitive passenger information. In the following section we review recent studies that investigated the potential of such cues.

## 2 Effectiveness and Acceptance of Predictive Motion Cues

Several studies have recently explored the potential of predictive cueing to reduce motion sickness in passengers and, by extension, riders in future automated or shared vehicles.

In the absence or limited availability of automated vehicles, these studies have adopted either Wizard of Oz approaches [28], passenger positions in conventional vehicles [29], or motion simulators [25].

## 2.1 Karjanto et al. (2018)

Karjanto et al. [28] developed a "peripheral visual feedforward system (PVFS)" installed in an instrumented vehicle (Renault Espace) modified to provide an automated driving experience. Participants were asked to watch videos on a television display placed on a wall partition separating the driver of the vehicle from the participants which were seated in the rear of the vehicle. The windows of the vehicle were made opaque to prevent passengers from being able to see any upcoming corners and junctions.

Predictive motion cues were provided via the peripheral visual feedforward system consisting of vertical arrays of 32 blue LED lights placed on the left and right of the television display. To indicate the upcoming motion, i.e. a left or right turn or righthand corner, the lights would start to move 3 s ahead of the actual manoeuvre from the bottom to the top on the left or right side, respectively.

The vehicle was driven on a pre-defined route on the University campus for a period of 9 min. The driving speed was set at 30 kph with lateral accelerations being generated during turning and cornering to about $0.29$ g $(2.84 \text{ ms}^2)$, while longitudinal accelerations were kept to the minimum. Using a within-subjects design, 20 participants experienced the same drive with and without the predictive cues. In addition to motion sickness as assessed by the MSAQ and heart rate measurements (BPM), acceptance and mental workload was measured by a User Experience Questionnaire (UEQ) and the Rating Scale Mental Effort (RSME), respectively.

The study results showed a significant beneficial effect of the predictive cues as measured by the Motion Sickness Assessment Questionnaire (MSAQ) [30]. The median MSAQ difference score (post minus pre MSAQ score) was reduced by 90% (Factor 7.4) in the test condition with predictive cues present. The heart rate data, on the other hand, failed to show any significant difference between the conditions. In terms of mental workload, there was no indication that the predictive cues resulted in elevated workload levels. Finally, system acceptance was relatively good as indicated by positive scores on the UEQ.

In conclusion, the study suggests that the presentation of predictive visual cues can have a sizeable and beneficial effect while also enjoying relatively high levels of acceptability. On the one hand, this is despite the fact that the exposure duration was relatively short and larger effects may have been observed over time. Also, the cues were relatively non-specific in that they only indicated the direction of the corner or junction but not its intensity, radius or position relative to corners [28]. On the other hand, however, it cannot be ruled out that some of the beneficial effects observed could, at least partly, be explained by the fact that participants were asked to press a button as soon as the cue was perceived. It has previously been shown that engagement in such mentally engaging tasks can divert the attention away from the stomach and lead to lower levels of reported sickness [31]. Lastly, the question remains to what extent such cues may be effective when applied to longitudinal vehicle motions, i.e. braking and accelerations in stop-start traffic or urban driving.

## 2.2 Diels et al. (2018)

Adopting a similar approach to Karjanto et al. [28] study, Diels et al. [29] explored the impact of auditory as opposed to visual predictive cues. In this this study, a total of 24 participants sat in the front passenger seat of a conventional vehicle (Ford Mondeo) and were driven around a city circuit (test track) at speeds up to 64 kph. The route was representative of urban driving, navigating roundabouts, junctions, corners, and including several stop-and-start manoeuvres and took approximately 18 min in total.

During each of the two drives, participants were engaged in a visual search task, the Surrogate Reference Task (SuRT). The SURT was presented on a 10-inch tablet placed in a head down position at the glove compartment. This head down location has previously been shown to lead to considerably more motion sickness than a head up display allowing for more peripheral vision and assumingly less sensory conflict [15].

Each participant was driven around on two occasions. In the "no cueing" condition, participants performed the visual task only whereas in the "cueing" condition, participants received auditory cues informing them of upcoming vehicle manoeuvres (e.g. *left hand corner ahead, slowing down to a stop, turning right*). The driver pressed the "next" button located on the steering wheel to trigger the pre-recorded motion cue for each upcoming manoeuvre which were provided approximately 1 s ahead of the actual manoeuvre.

Motion sickness was assessed using both subjective responses using the MISery sCale (MISC) [31] and the MSAQ and objective physiological measures (heart rate variability and electrodermal activity) while vehicle and occupant head accelerations were measured to ensure inter drive consistency and to explore potential effects of cueing on participant head movements.

The study findings vividly demonstrated the provocative nature of engaging in NDRA whereby all but one (96%) participant reported motion sickness, 50% had to terminate trials prematurely due to sickness levels MISC score of 6, while two incidences of emesis were reported, during a drive and following a drive. Several participants commented that they did not did not anticipate to suffer so much given the apparent innocuous nature of the task. It further shows that motion sickness is not a luxury problem and requires to be a fundamental consideration in the design of future automated vehicles and user interactions.

Returning to the main objective of the study, the results also showed that auditory predictive cues led to a significant 17% reduction in motion sickness as measured by the MISC. Similar to Karjanto et al. [28], none of the physiological measures were able to detect a difference between the conditions. In contrast, however, unlike the ambient visual cues, the auditory cues used in this study were perceived to be mentally demanding. Some participants experienced the cues as distracting and annoying with some reporting that they stopped paying attention to the cues and "tuned out". In turn, this may have suppressed the effectiveness of the motion cues and to some extent led to a conservative estimate of their effectiveness.

In conclusion, as for visual cues, auditory predictive cues can significantly reduce motion sickness levels. They can however also be perceived as rather annoying and distracting which points towards a design challenge and develop not only effective but also acceptable predictive motion cues.

## 2.3 Kuiper et al. (2019)

Whereas anticipation appears to have an effect on motion sickness as implied by the motion sickness literature as well as the studies explicitly addressing the role of anticipation, the effect size of anticipation as such is difficult to gauge due to the presence of potentially confounding factors such as mental engagement or the effectiveness of predictive cues in conveying anticipatory information, as discussed above.

In an attempt to avoid some of these pitfalls and to get a better grip on the exact importance of anticipation, Kuiper et al. [21] assessed motion sickness by exposing people to repeated fore-aft motion on a sled on a 40-m rail. 17 participants were asked to sit in an enclosed cabin positioned on top of the sled which did not allow for an external view.

In each of the three 15-min conditions, each participant was exposed to the repeated fore-aft motion at 1) constant intervals and consistent motion direction (i.e. predictable: condition P); 2) at constant intervals but varied motion direction (i.e. directionally unpredictable: condition dU); and 3) varied intervals but consistent motion direction (i.e. temporally unpredictable: condition tU).

Each single displacement lasted for 8 s and had an amplitude of 9.0 m, corresponding to a peak acceleration of 2.49 $m/s^2$. In conditions P and dU, there was a fixed 8-s pause between each displacement, resulting in a regular 16-s cyclic motion. In condition dU, half of the displacements had their sign inverted semi-randomly, that is, motion was backward-then-forward instead of forward- then-backward. In condition tU, the pauses in between the displacements were varied semi-randomly between 4 and 12 s, still averaging 8 s over the 15-min experiment. The conditions were otherwise identical in motion intensity and displacement, as they were composed of the same repetitions of identical blocks of motion. Illness ratings were recorded at 1-min intervals using the MISC scale.

As expected, the average illness ratings after exposure were significantly lower for the predictable condition, compared to both the directionally and temporally unpredictable condition. With regard to the relative size of the effect of anticipation, the unpredictable conditions led to 52% higher illness ratings compared to the predictable condition.

## 2.4 Kuiper et al. (2020)

Following on from the previous study, Kuiper et al. [25] explored the use of auditory predictive cues. Using the same experimental setup and metrics, 20 participants were exposed on a sled on a rail track to two 15-min conditions. In terms of motion, the two conditions were identical being composed of the same repeated 9 m fore-aft displacements, with a semi- random timing of pauses and direction.

The auditory cues were either 1) informative on the timing and direction of the upcoming motion, or 2) non-informative. In the anticipatory condition (A), the auditory cues informed both of timing and of direction, by occurring consistently 1 s before the motion started and with the actual direction of upcoming motion. A sound clip was played over headphones communicating either "forward" or "backward". In the control condition (C), the auditory cues were presented at semi-random timings, 2–6 s after a motion was already initiated and were therefore non-informative, not aiding in the participants ability to anticipate the upcoming motion. The auditory cues in the control condition were included to ensure that the level of stimulation (i.e. hearing an auditory cue) was identical in both conditions.

The results showed that the average illness ratings were significantly lower for the condition that contained informative auditory cues, as compared to the condition without informative cues. The effect of the anticipatory cues averaged to a difference of 17%, similar to the effects observed by [29]. The fact that no such reduction in motion sickness was observed when presenting false cues with no predictive value suggests that the alleviating effects were not the result of "stimulation" per se.

## 3   Design Considerations

Together, the above findings indicate that anticipatory information provided by predictive motion cues might be an elegant and effective method to reduce motion sickness in future vehicles in particular when engaging in Non-Driving Related Activities (NDRA), and able to reduce sickness levels by 17% or more. In fact, this estimate may be considered conservative in the light of the limited time of exposure used in these studies. It is widely known that sickness increases for longer exposure durations and as such the differences between conditions can be expected to become more pronounced over longer periods of time. At the same time, it is apparent that the cues were not sufficiently effective to *eliminate* sickness altogether, at least under the conditions studied. Also in real car driving, even with a perfect view on the road ahead, passengers still can get sick.

This means that motion cueing by itself may not be sufficient and able to provide a single solution and raises the question as to the relative effectiveness of motion cueing which may be a function of the nature of the provocative environment. For example, are cues more effective in less provocative environments such as the use of displays that allow for more peripheral vision and motion profiles involving fewer accelerations (highway driving)? These questions would benefit from future studies.

Furthermore, their real potential is yet to be determined as the above proof-of-concept studies did not consider the exact nature of the predictive cues. We here discuss several design parameters (see Table 1) that should be considered to finetune the cues to enhance both their effectiveness and acceptance.

Sensory modality is one of the key design parameters under consideration. Are visual cues more effective than auditory or tactile cues, or should we consider multisensory cues? The above review shows that both visual and auditory cues can be effective. However, perhaps the single most disadvantage of visual motion cues is the fact that occupants have to direct their attention and/or gaze towards these visual cues in order for these to be effective. This may be appropriate for passive occupants looking out the window or at a display showing such visual motion cues. However, once passengers engage in non-driving tasks that involve redirecting their attention away from such visual cues, their effectiveness loses its potential. In conditions in which occupants are using in-vehicle displays, these visual cues may be presented co-located with the media content of interest. Where this may prove to be effective, additional concerns here would relate to interference with the task at hand and ultimately user acceptance. Perhaps even more importantly, when artificial visual cues are used, these should be (near) perfect, bad

**Table 1.** Overview of parameters for the design of predictive motion cues

| Parameter | Description |
| --- | --- |
| Sensory modality | Motor, visual, auditor, tactile, vestibular, multisensory cues |
| Timing | Time at which the cue is presented relative to the upcoming vehicle maneuver (e.g. 1 vs. 3 s) |
| Discrete vs continuous | Presented at discrete moments (i.e. upcoming change in velocity) vs. available at all times |
| Information sensitivity | True positive rate, should be high |
| Information specificity | True negative rate, should be high |
| Information detailing | Level of detail to describe the upcoming motion (e.g. announcing "change in velocity" vs. "degree and direction of change in velocity") |
| Attentional demand | Centrally (intrusive) vs. peripherally (ambient) presented information (i.e. low vs. high level cognitive processing) |
| Customisation | Design for all vs. personalised and adaptive approach |

cues likely causing more sickness. Also, when based on predictive mechanisms, both sensitivity (positive response rate for sickening events) and specificity (negative response rate for non-sickening events) should be high. Even if 1 out of 100 events would give a false alarm, it could jeopardize the passengers' system trust, while false alarms are known to be annoying irrespectively.

Alternatively, visual cues in the form of ambient displays may be considered to avoid to some extent this issue. In the context of automated vehicles and the ability to engage in NDRA, there would be significant benefit in using anticipatory motion cues that are less contingent on the occupants' direction of gaze and attention. Such ambient displays may provide a valuable direction for automated vehicles if they prove to be not only effective in reducing motion sickness by also enjoy a high level of user acceptance and, for example, result in limited interference with other tasks. The results from Karjanto et al. [28] do indeed indicate that ambient visual cues are not only effective in reducing motion sickness but also enjoy a high level of user acceptance.

As mentioned in the previous section, auditory cues are equally able to provide beneficial effects. However, an important consideration in this context is that such auditory cues may become more distracting and demanding for occupants to process. In the study by Diels et al. [29] some participants found it difficult to perform the visual search task while also attending to the motion cues at the same time. This then raises the question whether motion cues of reduced complexity may be less demanding but possibly similarly effective in reducing motion sickness.

Furthermore, the auditory cues reviewed here were language based. Alternatively, the use of auditory cues may be based on sounds. As for ambient visual displays, these auditory sounds could be more abstract and could involve increasing/decreasing pitches to indicate vehicle acceleration/deceleration, while direction (left, right) may be indicated using 3D audio signals. More specific cues may require changes to the auditory signal including pitch, loudness, and 3D location. Yet, and apart from an undesired learning process, it seems to make sense that the more intuitive the cue would be the more effective it is.

Of particular interest in the context of alternative cueing mechanisms is the demanding nature of the cues. Some participants in the study of Diels et al. [29] experienced the cues as distracting and annoying with some reporting that they stopped paying attention to the cues. This may have suppressed the effectiveness of the motion cues and to some extent led to a conservative estimate of its effectiveness and highlights the need to consider attentional demands of anticipatory motion cues.

The actual information detailing is a further variable. The studies reviewed here, the level of detail was low and future research would benefit from exploring motion cues with an increased or decreased level of detail. An increase in detail may allow the occupant to predict with a higher level of accuracy the upcoming motion and thereby reducing the discrepancy between the sensed motion.

Timing, the temporal characteristics of the predictive cues, is a further parameter to consider in future. In the studies by Diels et al. [29] and Kuiper et al. [25] cues were presented at approximately 1 s ahead of the motion manoeuvres where as a time window of 3 s was used by Karjanto et al. [28]. A longer period could allow for more time to cognitively process the cue, while, conversely, a shorter time could enable participants to estimate more accurately the time when the motion will occur. Future research would benefit from exploring if and to what extend different timings affect the level of motion sickness.

Finally, one of the most consistent findings in the field of motion sickness is that individuals show immense variability in their susceptibility to motion sickness. This provides a real design challenge in that a solution for all may not be desirable from an acceptance point of view. Personalised solutions may be desirable in particular in the context of future shared mobility.

## 4 Conclusions

The experience of motion sickness in automated vehicles, no matter how slight, is one of the main barriers to the successful introduction of this technology. This is particularly relevant under conditions in which passengers engage in Non-Driving Related Activities such as reading and thus jeopardizes the perceived benefit of future automated or shared mobility. The provision of predictive motion cues has the potential to considerably alleviate the severity of motion sickness in such circumstances. However, our understanding of the design of such predictive cues is still immature and their real potential is yet to be determined.

## References

1. Diels, C., et al.: Designing for comfort in shared and automated vehicles (SAV): a conceptual framework. In: Proceedings of the 1st International Comfort Congress (ICC), 7–8 June 2017, Salerno, Italy (2017)
2. Diels, C.: Will autonomous vehicles make us sick? Contemp. Ergon. Hum. Factors 301–307 (2014)

3. Diels, C., Bos, J.E.: Self-driving carsickness. Appl. Ergon. **53**, 374–382 (2016)
4. Diels, C., Bos, J.E.: User interface considerations to prevent self-driving carsickness. In: Adjunct Proceedings of the 7th International Conference on Automotive User Interfaces and Interactive Vehicular Applications (2015)
5. Reason, J.T., Brand, J.J.: Motion Sickness. Academic Press, London, New York, San Francisco (1975)
6. Bos, J.E., MacKinnon, S.N., Patterson, A.: Motion sickness symptoms in a ship motion simulator: effects of inside, outside, and no view. Aviat. Space Environ. Med. **76**, 1111–1118 (2005)
7. Money, K.E.: Motion sickness. Physiol. Rev. **50**, 1–39 (1970)
8. Oman, C.M.: A heuristic mathematical model for the dynamics of sensory conflict and motion sickness. Acta Otolaryngol. Suppl. **392**, 1–44 (1982)
9. O'Hanlon, J.F., McCauley, M.E.: Motion sickness incidence as a function of the frequency and acceleration of vertical sinusoidal motion. AeroSpace Med. **45**, 366–369 (1974)
10. Guignard, J.C., McCauley, M.E.: The accelerative stimulus for motion sickness. In: Crampton, G.H. (ed.) Motion and Space Sickness. CRC Press, Boca Raton (1990)
11. Turner, M., Griffin, M.J.: Motion sickness in public road transport: the relative importance of motion, vision and individual differences. Br. J. Psychol. **90**(Pt 4), 519–530 (1999)
12. Turner, M., Griffin, M.J.: Motion sickness in public road transport: passenger behaviour and susceptibility. Ergonomics **42**, 444–461 (1999)
13. Schmidt, E.A., Kuiper, O.X., Wolter, S., Diels, C., Bos, J.E.: An international survey on the incidence and modulating factors of carsickness. Transp. Res. F: Traffic Psychol. Behav. **71**, 76–87 (2020). https://doi.org/10.1016/j.trf.2020.03.012
14. Rolnick, A., Lubow, R.E.: Why is the driver rarely sick? the role of controllability in motion sickness. Ergonomics **34**(7), 867–879 (1991)
15. Kuiper, O., Bos, J., Diels, C.: Looking forward: In-vehicle auxiliary display positioning affects carsickness. Appl. Ergon. **68**, 169–175 (2018)
16. Salter, S., Diels, C., Herriotts, P., Kanarachos, S., Thake, D.: Motion sickness in automated vehicles with forward and rearward facing seating orientations. Appl. Ergon. **78**, 54–61 (2019). https://doi.org/10.1016/J.APERGO.2019.02.001
17. Bles, W., Bos, J.E., de Graaf, B., Groen, E., Wertheim, A.H.: Motion sickness: only one provocative conflict? Brain Res. Bull. **47**(5), 481–487 (1998)
18. Bos, J.E., Bles, W.: Theoretical considerations on canal-otolith interaction and an observer model. Biol. Cybern. **86**, 191–207 (2002)
19. Einstein, A.: On the relativity principle and the conclusions drawn fromit. Jahrb. Radioakt. Elektron. **4**, 411–462 (1907)
20. Mayne, R.: A systems concept of the vestibular organs. In: Kornhuber, H.H. (ed.) Handbook of Sensory Physiology: Vestibular System, pp. 493–580. Springer-Verlag, New York (1974)
21. Kuiper, O.X., Bos, J.E., Schmidt, E.A., Diels, C., Wolter, S.: Knowing what's coming: unpredictable motion causes more motion sickness. Hum. Factors (2019). https://doi.org/10.1177/0018720819876139
22. Kuiper, O.X., Bos, J.E., Diels, C.: Vection does not necessitate visually induced motion sickness. Displays **58**, 82–87 (2019). https://doi.org/10.1016/j.displa.2018.10.001
23. Salter, S., Diels, C., Herriotts, P., Kanarachos, S., Thake, D.: Motion sickness in automated vehicles with forward and rearward facing seating orientations. Appl. Ergon. **78**, 54–61 (2019). https://doi.org/10.1016/J.APERGO.2019.02.001
24. Feenstra, P.J., Bos, J.E., Van Gent, R.N.H.W.: A visual display enhancing comfort by counteracting airsickness. Displays **32**, 194–200 (2011)
25. Kuiper, O.X., Bos, J.E., Diels, C., Schmidt, E.A.: Knowing what's coming: Anticipatory audio cues can mitigate motion sickness. Appl. Ergon. **85**(March), 103068 (2020). https://doi.org/10.1016/j.apergo.2020.103068

26. Perrin, P., Lion, A., Bosser, G., Gauchard, G., Meistelman, C.: Motion sickness in rally car co-drivers. Aviat. Space Environ. Med. **84**(5), 473–477 (2013). https://doi.org/10.3357/ASEM.3523.2013

27. Löcken, A., et al.: Towards adaptive ambient in-vehicle displays and interactions: insights and design guidelines from the 2015 automotiveui dedicated workshop. In: Meixner, G., Müller, C. (eds.) Automotive User Interfaces. HIS, pp. 325–348. Springer, Cham (2017). https://doi.org/10.1007/978-3-319-49448-7_12

28. Karjanto, J., Yusof, N.M., Wang, C., Terken, J., Delbressine, F., Rauterberg, M.: The effect of peripheral visual feedforward system in enhancing situation awareness and mitigating motion sickness in fully automated driving. Transp. Res. Part F Traffic Psychol. Behav. **58**, 678–692 (2018). https://doi.org/10.1016/j.trf.2018.06.046

29. Diels, C., Cieslak, M., Schmidt, E., Chadowitz, R.: The effect of auditory anticipatory motion cues on motion sickness. Client project report, Coventry University (2018)

30. Gianaros, P.J., Muth, E.R., Mordkoff, J.T., Levine, M.E., Stern, R.M.: A questionnaire for the assessment of the multiple dimensions of motion sickness. Aviat. Space Environ. Med. **72**(2), 115–119 (2001)

31. Bos, J.E.: Less sickness with more motion and/or mental distraction. J. Vestib. Res. **25**, 23–33 (2015)

# Communication of Intentions in Automated Driving – the Importance of Implicit Cues and Contextual Information on Freeway Situations

Konstantin Felbel[✉], André Dettmann, Marco Lindner, and Angelika C. Bullinger

Chair for Ergonomics and Innovation, Chemnitz University of Technology, Chemnitz, Germany
konstantin.felbel@mb.tu-chemnitz.de

**Abstract.** In manual driving, implicit cues play an important role in the communication of intention and anticipation of upcoming driving situations. Considering mixed traffic situations, all interaction partners need to be able to detect and interpret implicit cues, as they are central to design smooth, efficient, and safe driving styles. However, most current automated driving functions do not incorporate the communication and anticipation of implicit cues. The lack of anticipation of upcoming events in automated driving increases the probability of inadequate actions (e.g., sudden breaking maneuver). This concerns especially freeway situations as the driving speeds are considerable high, requiring more anticipation. To show the importance of implicit cues, a study on German freeway was conducted where over 1000 km of 360° video material was recorded. The video material was then annotated with the focus on the identification of situations where implicit cues can be observed. Beside the situations, implicit and explicit cues as well as contextual information were annotated, too. The results show i) that specific situations can be categorized where ii) implicit and explicit as well as contextual information can be identified. Beside the findings of the study, the article provides an outlook on a naturalistic driving study design to examine implicit cues during the drive.

**Keywords:** Highly automated driving · Driving styles · Implicit cues · Anticipation

## 1 Theoretical Background

In the near future, a mixed traffic scenario of manually driven as well as highly automated vehicles (HAVs) is expected [1, 2]. This situation will be present until the full transition where only a small amount of manual driven vehicles are in use, which could take decades [3]. Till then, HAVs must be able to handle situations requiring interactions with other road users [4, 5]. As road traffic is a social system [5, 6] these interactions should be efficient, smooth, and safe [7]. Current automated driving systems, such as Lane Keep Assist or Active Lane Change Assist, attempt to take these principles into account, but due to their less anticipative trajectory, interactions with other road users are not optimal.

© Springer Nature Switzerland AG 2021
H. Krömker (Ed.): HCII 2021, LNCS 12791, pp. 252–261, 2021.
https://doi.org/10.1007/978-3-030-78358-7_17

These driving systems for example guide the vehicle in the center of the lane as much as possible, which results in an trajectory that is contrary to a natural, manual trajectory, which is in an case of oncoming traffic laterally shifted [8, 9]. This unnatural driving behavior makes it difficult for other road users to for example anticipate an upcoming lane change. Additionally, only a few current automated systems can detect the lane change intention of manually driven vehicles to some extent. Situations where these systems fail to detect a lane change at an early stage can lead to an abrupt braking reaction of the HAV. Commonly, first HAVs will be able to drive on freeways as they are less complex than urban environment. But, driving on the freeway is by no means uncritical. Due to the high speeds, wrong decisions can lead to serious accidents. Therefore, it is essential that HAVs function not only in urban scenarios but also on freeways robustly and safely [10], are able to anticipate future events [11] and then react accordingly [4]. A HAV which meets these requirements, is perceived as comfortable and safe and is more accepted [12–14]. These three factors affect the use of the HAV [15]. To increase these factors, the anticipation performance and the communication of the respective driving intention is of significant importance [16]. Driving intentions should be communicated in an understandable way and should be easy to anticipate [17]. In addition to explicit cues (e.g., indicator, horn) implicit cues, where information is not spoken aloud but communicated, for example through longitudinal and lateral vehicle movements, play in particular an important role [5, 18]. The HAV could use implicit cues to detect the lane change intentions of other road users early on and proactively adjusts its trajectory. In this context, understanding implicit cues could serve as a basis for a proactive driving style [13, 19]. In current literature, neural networks are used to interpret the driving environment and anticipate future events [20, 21]. Another research approach is to investigate the anticipation performance of human driver from naturalistic driving [22, 23]. It shows that human driver are generally good at anticipating future driving events [24]. They are able to name implicit and explicit cues which could indicate lane changes of road users ahead [22, 25]. This research approach holds a lot of potential but, previous implementation of this explorative design did not focus specifically on implicit cues. Additionally, it is uncertain if specific driving situations can be categorized to find implicit and explicit cues. Another factor which could affect the anticipation of upcoming lane changes are contextual information (e.g., road structure, signs). These information are not only important in the characterization of maneuver types [26], they also important for the development of a mental model of the situation. This might facilitate anticipation of lane changes by focusing the drivers' attention on relevant points in the driving scene [27]. Till this day, it is unclear to what extent driver use implicit cues and contextual information to anticipate lane changes of other road users and if any implicit cue can be observed in any driving situation. As a first step, an on-road study on German freeway is conducted.

## 2 Methodology

The aim of the study was the observation of the behavior and the reaction of road users in common freeway situations in order to extract implicit and explicit cues and contextual information. Therefore, 360-degree video data was recorded from on top of the car driven by the investigator. The data recording took place on German freeway.

The chosen routes were the BAB 4, 9, 14 and 72. These freeways consisted of roadway sections with two and three lanes including several freeway entrances and exits. In total, four individual drives were recorded. To record regular commuter traffic a period from 7 a.m. to 11 a.m. was chosen. This increased the probability of capturing as many traffic situations as possible in which drivers needed to communicate and cooperate. Drives in rain or on wet road surfaces were avoided. To record the 360-degree video data a GoPro Max camera with a 5.6 K pixel resolution and framerate of 30 fps was used. To achieve identical views between the drives, a fixed camera position was marked on the roof. Additional marking tape was used to set-up the correct perspective orientation of the views in post-production (i.e., converting 360-degree video into 2D-video; see Figs. 1 and 2). Furthermore, a smartphone navigation app (HERE WeGo) was used to record the position of the vehicle and, in a later step, to map the vehicle speed to the camera recordings. The simultaneous display of the front and rear view and the GPS data allowed a better overall assessment of the recorded situations and an estimation of the relative speeds between road users.

**Fig. 1.** Camera positioning and markers (left: rear view; right: front view)

**Fig. 2.** Fixed views of the 360-degree camera (left rear view; right front view) with GPS-data.

Regarding data analysis and categorization of the video footage both views (rear view and front view) were considered. All four recordings were viewed and categorized by three scientists with 4+ years of expertise in the field of driver-vehicle-interaction. The focus was on situations where lane changes and overtaking maneuvers took place. Subsequently, all other occurring situations were filtered out.

# 3 Results

In total, 12 different driving situations can be derived. Except for situation 12, every situation can be described through the combination of two scenarios shown in Table 1. A scenario is an observable rear view or front view scene which is featured through different characteristics. In Table 2 all situation described through the respective combinations are shown.

**Table 1.** Identified scenarios

| Scenario | Characteristics |
| --- | --- |
| 1 | — No active braking by observed vehicle<br>— End of acceleration lane<br>— Observed vehicle is moving into another freeway lane |
| 2 | — Active braking by observed vehicle<br>— End of acceleration lane<br>— Observed vehicle is moving into another freeway lane |
| 3 | — First an implicit cue (e.g., lateral shift) is seen by the observed vehicle then an explicit cue (e.g., flashing) |
| 4 | — Observed vehicle allows lag traffic to pass and then changes lanes<br>— First an explicit cue (e.g., flashing) is seen by the observed vehicle then an implicit cue (e.g., lateral shift) |
| 5 | — Observed vehicle changes lane in front of lag traffic<br>— First an explicit cue (e.g., flashing) is seen by the observed vehicle then an implicit cue (e.g., lateral shift) |
| 6 | — Observed vehicle accelerates (to change lane in front of lag traffic or to avoid obstruction of lag traffic) |
| 7 | — Enabling another vehicle to merge onto the freeway<br>— Enable overtaking or lane change for another vehicle |

All situations are typical freeway situation (e.g., passing slow vehicles or merging on two or three lanes). It shows that some situations occurred more frequently. The four most frequently observed situations account for 79% of all observed situations (situation 2, 3, 6 and 12). Commonly, these situations require higher attention from the driver. For a valid interpretation of the prevailing cues and contextual information only these four situations are considered. The other situations occurred less frequent ($N \leq 5$) and thus

**Table 2.** Combination of the scenarios to situations and their frequency of occurrence. The four most frequently observed situations are highlighted

| Driving situation | | Combination | | | Frequency of occurrence |
|---|---|---|---|---|---|
| Situation 1 | = | Scenario 1 | + | Scenario 3 | 2 |
| Situation 2 | = | Scenario 1 | + | Scenario 4 | 27 |
| Situation 3 | = | Scenario 1 | + | Scenario 5 | 20 |
| Situation 4 | = | Scenario 1 | + | Scenario 6 | 1 |
| Situation 5 | = | Scenario 1 | + | Scenario 7 | 2 |
| Situation 6 | = | Scenario 2 | + | Scenario 4 | 8 |
| Situation 7 | = | Scenario 2 | + | Scenario 7 | 3 |
| Situation 8 | = | Scenario 3 | + | Scenario 4 | 1 |
| Situation 9 | = | Scenario 3 | + | Scenario 5 | 3 |
| Situation 10 | = | Scenario 4 | + | Scenario 7 | 2 |
| Situation 11 | = | Scenario 5 | + | Scenario 6 | 5 |
| Situation 12 | = | Scenario 7 | | | 17 |

are not represented often enough to generalize implicit and explicit cues and contextual information. In Table 3, the four selected situations are described in further detail. For easier interpretation of the four selected situations the different road users are defined in Table 4.

The analysis of the recorded situations indicate that the behavior of road users and the associated implicit cues emitted depend on the contextual information (i.e. prevailing infrastructure) and the different road users involved. Depending on how many lanes the freeway has, different reactions and thus implicit and explicit cues can be expected. For example, a lane change across two lanes can only occur, if more than two lanes are available. The same is also valid for the type of vehicle involved. A truck driving at a low average speed shows different behavior than a passenger car traveling at higher speeds (i.e. lane changes are slower and less dynamically and therefore less likely at the same gap sizes). Based on the four selected driving situations, there are implicit and explicit cues as well as contextual information which speak likely for or against an upcoming lane change of the EV. An Overview of cues and contextual information which speak likely **for** and **against** an upcoming lane change are shown in Table 5.

**Table 3.** Selected situations characterization with schematic, brief situation description and special remarks

| Situation schematic | Maneuver description |
|---|---|
| Lane change by EV before SV | The EV approaches a POV. No speed reduction through active braking by EV. Despite the approaching SV, EV changes lanes before SV.<br><br>**Special remarks**<br>— This situation occurred mainly when traffic density was low<br>— Here, typical slow truck overtakes another slow truck occurred more frequent during uphill-drive |
| Lane change by EV after SV | EV approaches a POV. No speed reduction through active braking by EV. After the SV has been able to pass, the lane change/overtaking maneuver is initiated by the EV.<br><br>**Special remarks**<br>— This situation could be observed more often when there was a steady traffic flow on the adjacent left lane of the EV |
| Active breaking by EV | The EV approaches a POV. Speed is reduced by active braking to allow the SV to pass. After the SV has been able to pass, the lane change/overtaking maneuver is initiated by EV.<br><br>**Special remarks**<br>— This situation occurred mostly when traffic density was high |
| Cooperative driving by SV | The SV enables EV to change lanes or to merge.<br><br>**Special remarks**<br>— More cooperative driving behavior could be observed when the SV approached an infrastructural change:<br>• Exit<br>• On-ramp<br>• Lane narrowing<br>• Added extra lane<br>— A free adjacent lane increased the possible for cooperative driving behavior of SV |

**Table 4.** Road user definition referring to [26]

| Description | Abbreviation | Corresponding icon |
|---|---|---|
| **Subject Vehicle** The vehicle that approaches the examined vehicle | SV | |
| **Examined Vehicle** The vehicle that does or does not carry out the lane change | EV | |
| **Principal Other Vehicles** Surrounding vehicles that are likely influential on the EV, such as a slow lead vehicle | POV | |

**Table 5.** Overview of generalized implicit/explicit cues and contextual information which speak likely for and against an upcoming lane change of an EV

| | explicit cue | implicit cue | contextual information |
|---|---|---|---|
| Speak likely **for** an upcoming lane change | — Indicators turned on | — Large longitudinal distance between SV an EV<br>— Slow change in longitudinal distance between SV and EV<br>— Low speed difference between SV an EV<br>— Acceleration by EV<br>— Lateral offset on lane by EV | — Free adjacent left lane for SV or EV<br>— Overall low traffic density<br>— EV approaching infrastructural changes (exit, on-ramp, lane narrowing, extra lane) |
| Speak likely **against** an upcoming lane change | — Breaking lights turned on | — Small longitudinal distance between SV an EV<br>— Fast change in longitudinal distance between SV and EV<br>— Slow POV<br>— High speed difference between SV and EV<br>— Deceleration without active braking by EV | — Blocked adjacent left lane for SV or EV<br>— Steady traffic flow on adjacent left lane of EV<br>— Overall high traffic density<br>— EV driving between solid lines |

# 4  Conclusion and Outlook

By analysing the recorded video data, four frequently occurring situations could be categorizing (lane change by EV before SV; lane change by EV after SV; active breaking by EV and cooperative driving by SV). To these four situations, implicit and explicit cues as well as contextual information could be observed which speak likely for (e.g., indicator turned on, low speed differences,) or against (e.g., small longitudinal distance, overall high traffic density) an upcoming lane change. It shows that drivers use the indicator regularly to indicate their intention to change lanes but this is sometimes done after the vehicle has already shifted its lateral position on the lane (i.e., implicit cues). This implicit cue could be used by HAVs to detect the driving intentions of other traffic at an early stage. Certainly, it must be ensured that misinterpretations are avoided. Therefore, several implicit cues in combination with contextual information should be used for intention detection. If, for example, an on-ramp is in sight and a vehicle is on the on-ramp, the probability increases that another road user will change his lane in order to allow the other road user to enter the freeway. Accordingly, the HAV could adapt to this situation and proactively adjust its trajectory by earlier breaking which could lead to a more accepted driving style by the driven person.

An underlying disadvantage of the presented study approach is the factor that investigators (i.e., experts) analyzed and categorized the situations. This procedure could result in loss of information as non-experts (i.e. everyday driver) could have categorized these situations differently. Furthermore, these drivers could have named additional cues and contextual information while experiencing the situations first hand. Therefore, a modified approach is needed. Under these circumstances, a naturalistic driving study should be conducted where participants record their day to day drives over a period of time with a recording device installed in their car. The recording device records a video from the perspective of the driver as well as a voice commentary. The participants comment on every relevant driving situation on their trips through short speech protocols. Considered as relevant for recording is every situation, in which the participants realize that they actively observe another car in order to anticipate what it is going to do next, for example, when assessing whether the car will merge into the lane of the observer. After the trial period, recorded situations are reviewed and included in a questionnaire taking place after the trial. The aim of the questionnaire is to understand (uncertain) situations and the drivers' experience when interacting with other road users.

**Acknowledgements.** The research was funded by the Deutsche Forschungsgemeinschaft (DFG, German Research Foundation) – Project-ID 416228727 – SFB 1410.

# References

1. Ghiasi, A., Hussain, O., Qian, Z., Li, X.: A mixed traffic capacity analysis and lane management model for connected automated vehicles: a Markov chain method. Transp. Res. Part B Methodol. **106**, 266–292 (2017). https://doi.org/10.1016/j.trb.2017.09.022

2. Patel, R.H., Härri, J., Bonnet, C.: Braking strategy for an autonomous vehicle in a mixed traffic scenario. In: 3rd International Conference on Vehicle Technology and Intelligent Transport Systems, Porto, Portugal: SCITEPRESS - Science and Technology Publications. pp. 268–275 (2017). https://doi.org/10.5220/0006307702680275
3. Sven, A., Hans-Paul, K., Alex, A.D.M.: Einführung von Automatisierungsfunktionen in der Pkw-Flotte: Auswirkungen auf Bestand und Sicherheit (2018)
4. Schwarting, W., Pierson, A., Alonso-Mora, J., Karaman, S., Rus, D.: Social behavior for autonomous vehicles. Proc. Natl. Acad. Sci. USA **116**, 24972–24978 (2019). https://doi.org/10.1073/pnas.1820676116
5. Rasouli, A., Kotseruba, I., Tsotsos, J.K.: Agreeing to cross: how drivers and pedestrians communicate, pp. 264–269 (2017). https://doi.org/10.1109/IVS.2017.7995730
6. Müller, L., Risto, M., Emmenegger, C.: The social behavior of autonomous vehicles. In: Lukowicz, P., Krüger, A., Bulling, A., Lim, Y.-K., Patel, S.N. (eds.) UbiComp 2016: The 2016 ACM International Joint Conference on Pervasive and Ubiquitous Computing, 12 Sept 2016 – 16 Sept 2016, Heidelberg, Germany. ACM, New York, pp. 686–689, 09122016. https://doi.org/10.1145/2968219.2968561
7. European Road Transport Research Advisory Council (ERTRAC). Automated driving roadmap (2019). https://www.ertrac.org/uploads/documentsearch/id57/ERTRAC-CAD-Roadmap-2019.pdf
8. Voß, G.M., Keck, C.M., Schwalm, M.: Investigation of drivers' thresholds of a subjectively accepted driving performance with a focus on automated driving. Transp. Res. F: Traffic Psychol. Behav. **56**, 280–292 (2018). https://doi.org/10.1016/j.trf.2018.04.024
9. Schick, B., Seidler, C., Aydogdu, S., Kuo, Y.-J.: Fahrerlebnis versus mentaler Stress bei der assistierten Querführung. ATZ Automobiltech Z **121**, 70–75 (2019). https://doi.org/10.1007/s35148-018-0219-9
10. Schüller, H.: Modelle zur Beschreibung des Geschwindigkeitsverhaltens auf Stadtstraßen und dessen Auswirkungen auf die Verkehrssicherheit auf Grundlage der Straßengestaltung. Technische Universität Dresden, Dresden (2010)
11. Grahn, H., Kujala, T., Silvennoinen, J., Leppänen, A., Saariluoma, P.: Expert drivers' prospective thinking-aloud to enhance automated driving technologies - investigating uncertainty and anticipation in traffic. Accid. Anal. Prev. **146**, 105717 (2020). https://doi.org/10.1016/j.aap.2020.105717
12. Hartwich, F., Beggiato, M., Krems, J.F.: Driving comfort, enjoyment and acceptance of automated driving - effects of drivers' age and driving style familiarity. Ergonomics **61**, 1017–1032 (2018). https://doi.org/10.1080/00140139.2018.1441448
13. Rossner, P., Bullinger, A.C.: Do you shift or not? influence of trajectory behaviour on perceived safety during automated driving on rural roads. In: Krömker, H. (ed.) HCII 2019. LNCS, vol. 11596, pp. 245–254. Springer, Cham (2019). https://doi.org/10.1007/978-3-030-22666-4_18
14. Hartwich, F., Pech, T., Schubert, D., Scherer, S., Dettmann, A., Beggiato, M., et al.: Fahrstilmodellierung im hochautomatisierten Fahren auf Basis der Fahrer-Fahrzeuginteraktion: Abschlussbericht "DriveMe" (2016). https://doi.org/10.2314/GBV:870302329
15. Alawadhi, M., Almazrouie, J., Kamil, M., Khalil, K.A.: A systematic literature review of the factors influencing the adoption of autonomous driving. Int. J. Syst. Assur. Eng. Manag. **11**(6), 1065–1082 (2020). https://doi.org/10.1007/s13198-020-00961-4
16. Bengler, K., Rettenmaier, M., Fritz, N., Feierle, A.: From HMI to HMIs: towards an HMI framework for automated driving. Information **11**, 61 (2020). https://doi.org/10.3390/info11020061
17. Zwicker, L., Petzoldt, T., Schade, J., Schaarschmidt, E.: Kommunikation zwischen automatisierten Kraftfahrzeugen und anderen Verkehrsteilnehmern – Was brauchen wir überhaupt?

In: Bruder, R., Winner, H. (eds.) Hands off, Human Factors off? Welche Rolle spielen Human Factors in der Fahrzeugautomation? Darmstadt, pp. 47–57 (2019)

18. Dey, D., Terken, J.: Pedestrian interaction with vehicles: roles of explicit and implicit communication. In: Proceedings of the 9th ACM International Conference on Automotive User Interfaces and Interactive Vehicular Applications (AutomotiveUI 2017), pp. 109–113 (2017). https://doi.org/10.1145/3122986.3123009

19. Dettmann, A., et al.: Comfort or not? automated driving style and user characteristics causing human discomfort in automated driving. Int. J. Hum. Comput. Interact. 331–339 (2021). https://doi.org/10.1080/10447318.2020.1860518

20. Schmuedderich, J., Rebhan, S., Weisswange, T., Kleinehagenbrock, M., Kastner, R., Nishigaki, M., et al. (eds.): A novel approach to driver behavior prediction using scene context and physical evidence for intelligent Adaptive Cruise Control (i-ACC) (2015)

21. Tomar, R.S., Verma, S., Tomar, G.S.: Prediction of lane change trajectories through neural network. In: 2010 International Conference on Computational Intelligence and Communication Networks (CICN 2010), 26 Nov 2010 – 28 Nov 2010, pp. 249–253. IEEE, Bhopal. https://doi.org/10.1109/CICN.2010.59

22. Simon, K., Bullinger, A.C.: Was stresst, ärgert und beunruhigt Fahrer? Emotionale Reaktionen auf alltägliche Fahrsituationen bei jüngeren und älteren Fahrern, Braunschweig (2017)

23. Lietz, H., Petzoldt, T., Henning, M., Haupt, J., Wanielik, G., Krems, J., et al.: Methodische und technische Aspekte einer Naturalistic Driving Study: BASt FE 82.0351/2008. VDA, Berlin (2010)

24. Sommer, K.C.: Vorausschauendes Fahren: Erfassung Beschreibung und Bewertung von Antizipationsleistung im Straßenverkehr. Universität Regensburg (2012)

25. Simon, K., Bullinger, A.C.: To change or not to change – that is the question. In: Detecting Lane Change Signals for Anticipatory Highly Automated Driving, Berlin, 08 Oct 2018 bis 10 Oct 2018

26. Lee, S.E., Olsen, E.C., Wierwille, W.W.: A Comprehensive Examination of Naturalistic Lane-Changes (2004)

27. Mühl, K., Stoll, T., Baumann, M.: Look ahead: understanding cognitive anticipatory processes based on situational characteristics in dynamic traffic situations. IET Intel. Transp. Syst. **14**, 233–240 (2020). https://doi.org/10.1049/iet-its.2018.5557

# Talking Automated Vehicles

## Exploring Users' Understanding of an Automated Vehicle During Initial Usage

Mikael Johansson(✉) ⓘ, Fredrick Ekman ⓘ, MariAnne Karlsson ⓘ,
Helena Strömberg ⓘ, and Lars-Ola Bligård ⓘ

Division Design and Human Factors, Chalmers University of Technology,
SE-412 96, Gothenburg, Sweden
johamik@chalmers.se

**Abstract.** With the introduction of automation, vehicles have become increasingly complicated and difficult for users to understand. Users' understanding of Automated Vehicles (AVs) is a key aspect for safe and successful implementation of AVs. However, more research is needed into how users understand AVs based on actual use experience. In this paper, users' understanding of AVs is explored by investigating how they refer to and describe an AV, during and after initial usage. 18 participants experienced a seemingly fully automated vehicle, being driven with two distinctly different driving styles on a test course. The findings show that the participants had specific preconceptions of what they regarded as machine-like versus human-like driving characteristics of the AV. The participants also referred to the AV with gender pronouns and used human similes to describe the different driving styles. The different driving styles evoked different associations that influenced the participants' perceptions of AV behaviour as a result of individual preconceptions and previous experiences. The results imply that the participants initially used a human-like mental model of the AV. However, further investigations are necessary into users' initial comprehension of AVs, to better understand how they will experience and interact with future AVs.

**Keywords:** Automated Vehicles · Understanding · Mental models · Driving behaviour · User study

## 1 Introduction

As vehicles get increasingly automated, they also become more complicated, making it harder for users to fully understand the vehicle and its behaviour. This can create problems in that insufficient understanding of the functionality of an automated vehicle (AV) may lead to confusion regarding system limitations [1]. Studies suggest that for AVs to fully work in the intended way the user needs to understand the capabilities of the technology [e.g. 2]. In addition, users' understanding of the AV may not only impact system usage directly but also indirectly by affecting their trust in it, in turn influencing over-use or under-use [3, 4], as well as their acceptance [5] of the AV. Research has shown for example that users' levels of trust in an AV increased with their understanding of

© Springer Nature Switzerland AG 2021
H. Krömker (Ed.): HCII 2021, LNCS 12791, pp. 262–272, 2021.
https://doi.org/10.1007/978-3-030-78358-7_18

its capabilities [6] and studies by Beggiato and Krems [7] as well as Nees [8] have illustrated that the portrayal of an AV and therefore the users' initial understanding of its functionality affects user acceptance. However, even though the importance of understanding users' comprehension of AVs, especially of system capabilities, has been emphasised, there is (to the authors' knowledge) limited research on how users actually understand an AV based on experience of use.

The purpose of this paper is to present the results of a study in which users' understanding of an AV was explored by investigating how they refer to and describe the AV, during and after initial usage.

## 2  Method

A study involving 18 participants was conducted in a simulated fully automated vehicle on a test track. Each participant experienced two different driving styles during two different test runs. The aim of the study was twofold; one aim was to investigate the effect of driving styles on trust (Ekman et al. [9]) and the second aim was to explore how users understand the AV which is the focus of this paper.

### 2.1  Participants

The study involved 18 participants, 10 male and 8 female, between 20 and 50 years old (mean age 36.7, SD = 11.1) and with different backgrounds and education. The participants were recruited via a newspaper advertisement. The sole criterion for the selection of the participants was for them to have a valid driver's license. The participants' driving experience ranged from 3 to 35 years. As for driving frequency, half the participants drove almost every day and the other half drove from a couple of times per week to a couple of times per year. Most participants (16/18) had previous experience of driver assistance systems in terms of cruise control and some (5/18) had experience of combinations of adaptive cruise control and steering assist. Only two participants had no previous experience of any kind of advanced driver assistance system.

### 2.2  Study Design and Procedure

A Wizard of Oz approach was used to simulate an AV system, using a modified Volvo XC90 with a hidden driver in the back seat of the vehicle (see Fig. 1). Previous studies have shown that the driving style of the AV affects the perceived character of the AV [10, 11]. Hence, two distinctly different driving styles, here referred to as 'Aggressive' and 'Defensive', were simulated with the help of the hidden test driver, in order to investigate if the driving style affected the way users referred to it. The styles differed regarding acceleration/deceleration, distance maintained to other vehicles, lane positioning, and starting/stopping behaviour. A within-subject design was applied, where all participants experienced both driving styles in a counterbalanced order.

Each test run consisted of an approximately 15 min long drive on a test course, including both a rural road (with normal road standard for bi-directional traffic, indicated by blue line in Fig. 2) and a city area (with buildings and intersections, indicated by

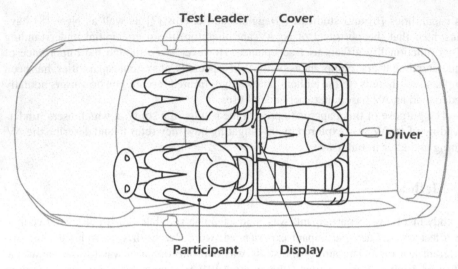

**Fig. 1.** Woz setup.

brown line in Fig. 2). During each test run the participants first experienced one test lap on the rural road to get acquainted with the driving style and the AV experience. Next, each run continued with two more laps of the test course where the participants encountered situations that are common in everyday driving (see Fig. 2 for overview), such as stopping at a red light in an intersection, overtaking a moving car, and stopping for a pedestrian at a pedestrian crossing, and where the difference in driving style and hence vehicle behaviour became evident.

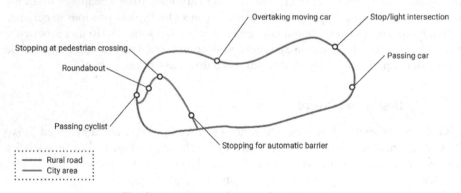

**Fig. 2.** Test course with predefined situations.

### 2.3  Data Collection and Analysis

Data was collected during and after each test run in order to extract the participants' descriptions of the AV. A think-aloud procedure [cf. 12] was used during the test runs, where the participants were asked to explain their experiences in connection with each of the seven situations encountered. A final interview was conducted after the participants had experienced both driving styles, where they were asked to further elaborate and compare their experiences. All verbalised data was recorded and later transcribed and analysed.

An initial analysis revealed that the participants referred to the AV in three recurring ways: using pronouns, using similes, and describing different behaviours as human-like or machine-like. A second analysis was then initiated and divided into three parts, each of which focused on one of the previously identified types of referrals. Part one consisted of identifying which specific pronouns the participants used to refer to the AV (e.g. I, we) and then quantifying these pronouns. The focus of the second part of the analysis was to identify what in the vehicle's behaviour was perceived as 'human'-like and what was perceived as 'machine'-like. All segments of the interviews referring to the vehicle or a specific behaviour as being either human-like or machine-like were coded, for example a statement such as *"the acceleration felt more like when a human is driving"* was considered human-like whereas *"This is just how I expected a self-driving car would behave"* was coded as machine-like. The aim of the third and last part of the analysis was to investigate, in more detail, the similes used by the participants to describe the AV. All segments mentioning a simile were identified, and then coded according to the comparison used. All parts of the analysis were conducted by two researchers (1st and 2nd authors) who individually analysed the transcriptions and subsequently compared their coding in order to resolve any discrepancies and reach full consensus.

## 3 Findings

The analysis shows that the participants referred to the AV in three recurring ways: using pronouns, describing different behaviours as human- or machine-like, and by using similes.

### 3.1 Pronouns

Using pronouns, the participants referred to the AV as 'the car' but mostly they referred to the AV in terms of the neutral pronoun 'it' (approximately 3/4 of the times) (see Fig. 3). The pronoun 'it' was used by all participants to describe several aspects, such as what the AV was doing: *"<u>It</u>drove a bit faster and positioned itself closer to the edge"* (P8) or by describing the reasoning behind the AV's actions: *"<u>It</u>takes the decision when <u>it</u>sees the cyclist... and changes lane after taking the decision that it is safe"* (P15). On a few occasions, the AV was referred to as a human with the gender 'he' (but never 'she'): *"It seems like <u>he</u>saw the person"* (P9). Furthermore, the participants referred to the AV as 'the technology' or 'the system', often when describing the inner workings of the AV: *"I got no feed-forward about the overtaking, so it was hard to know what the <u>system</u> knew"* (P8) or when referring to the capability of the AV: *"I think the <u>system</u>will do it better than I can"* (P14). However, the participants also used terms that indicated a more collaborative view, referring to themselves and the AV as 'we'. The participants frequently did so when there was a perceived risk involved in the situation, for example when the AV performed a manoeuvre that was perceived as dangerous to the driver: "*We were driving quite fast when we got to the crest, it was unsafe but we pulled out into the other lane anyway*" (P14) or to other road users: "*It felt like we started [overtaking] earlier, which makes it clearer for me sitting here as well as for the cyclist*" (P7).

The only pronoun used by all participants was 'it'. However, most participants used several different ways of referring to the AV, and different pronouns were used interchangeably (e.g. one participant sometimes used 'it' and sometimes 'he'). Not all ways of referring to the AV were used by all participants, those least commonly used (i.e. 'we', 'he', 'technology' and 'system') were only used by a few. When the participants referred to the AV in these ways, this does not seem to have been affected by the driving style but rather by the situation itself or certain aspects of the AV (e.g. using 'we' in hazardous situations or 'system' when describing the inner workings of the AV).

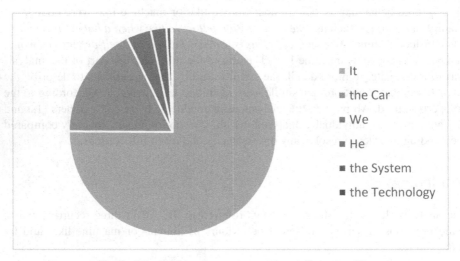

**Fig. 3.** Distribution of pronouns used to refer to the AV.

### 3.2 Human-Like or Machine-Like

Different characteristics of the AV that were regarded as human- or machine-like were identified. The characteristics that the participants considered to be human- or machine-like were in some respects similar among the participants but differed in others. The participants did not appear to connect what was regarded as human- or machine-like to a specific AV driving style, that is to say 'defensive' or 'aggressive', but rather to specific driving properties (e.g. acceleration or frequency of acceleration) or a combination of several properties.

When referring to what was regarded as machine-like, the participants sometimes talked about these as generic machine characteristics (e.g. "*a machine drives like …*") but also as specific AV characteristics (e.g. "*a self-driving car drive like …*"). Human-like characteristics were often referred to as general human qualities (e.g. "*a human drives like ...*") but sometimes also as a particular person (e.g. *my friend drives like ...*") or in terms of the participant's own characteristics.

The characteristics that were considered to be human- or machine-like seemed to be strongly affected by individual preconceptions, since the same characteristics were regarded as human-like by one participant and as machine-like by another. For example, 'smooth driving', that is to say steady acceleration patterns, was one of the most commonly mentioned characteristics when participants referred to what was perceived as machine-like behaviour: "*The driving is calm and smooth. That is just how I expect a self-driving car to drive*" (P10) but the same characteristics were also regarded as human-like: "*Another person [human driver] would not have been driving that jerkily*" (P7). On the other hand, while different participants considered calm and careful actions to be machine-like or human-like, aggressive actions with hard acceleration and braking were generally regarded as more human-like. One participant commented for example the ride as: "*It feels strange to try to reproduce human driving in terms of hard acceleration and deceleration*" (P3).

In addition, human-like driving was to a greater extent considered as prescient, i.e. involving early detection of situations and execution of actions: "*...it noticed the person crossing and therefore kept its distance and did not start to turn until the person had crossed the road. It felt just like a human [driving]*" (P10). Machine-like driving was to a greater extent believed to be more aware of itself (e.g. the position) and its surroundings. Some participants expressed high expectations of the AV's driving capabilities regarding these aspects: "*I think it drives calmly and carefully but I also believe that it is so much more aware [than a human driver] of where it is. I probably expect more from a car [than a human driver]*" ('P11). Another participant compared their expectations about the AV's awareness and driving to a human by stating: "*It was quite jerky, and I do not expect that from a self-driving car in the same way. You expect it to be smooth and pleasant, and that the vehicle will be aware of everything. I do not expect it to drive this way, because this is something that I would associate with a human driver*" (P12).

### 3.3 Similes

The participants used different kinds of similes, i.e. comparisons of two unlike things, to describe and compare the 'aggressive' and 'defensive' AV, in relation to the whole test run as well as in relation to the specific situations. Altogether five categories of similes were identified: Driving student, Taxi driver, Senior citizen, Father, and Friend (see Table 1 for overview). Hence, all similes included comparisons with different human drivers.

The participants used the different similes when referring to one or the other driving style. The most common comparison used to describe the 'defensive' AV was that of a 'driving student', also referred to as a 'beginner'. Participants using this comparison described the AV as being careful and correct, sometimes using exaggerated movements, but at the same time it was not perceived as having full control of the vehicle's movement or its positioning in the lane. One participant described their impressions as follows: "*I have a son at home who is learning to drive and there are similarities with his way of driving. It is a bit jerky, and positioning in the lane is not perfect. I do not feel unsafe, but it is a bit off, and sometimes needs extra adjustments in the steering*" (P8). Another participant commented: "*I think the other one felt almost too kind [referring to the*

**Table 1.** Summary of similes used by the participants during and after the test runs.

| Simile | Summary | Driving style |
|---|---|---|
| Driving student (n = 6) | Drives carefully and according to the rules but does not have full control of lanes and placement | Defensive |
| Senior citizen (n = 4) | Drives slowly and tries to drive safely but does not have full control. Make bad turns and has bad positioning | Defensive |
| Taxi driver (n = 3) | Drives aggressively but at the same time comfortably | Aggressive |
| Father (n = 2) | Drives fast, with hard acceleration and deceleration, but at the same time safely | Aggressive |
| Friend who likes to drive fast (n = 2) | Drives fast, with hard acceleration and deceleration, but may not have control | Aggressive |

*Defensive AV]. The drive felt good or very respectful but to me it was a bit exaggerated. A bit like when you just have got your licence"* (P4).

The second most commonly used comparison when referring to 'defensive' AVs was that of a 'senior citizen', also referred to as an 'old lady' or a 'grandmother'. The AV was described as being slow and careful but at the same time as not having full control, for example in terms of lane positioning and when turning: *"It did not feel like the car was aware of its size when we turned. It felt like we would get to our destination, but it did not feel so good... You may think that the fact that it drives slowly would give you a feeling of safety, but it also made it feel like a grandma"* (P11). Thus, the comparison with 'senior citizen' and 'driving student' shared several characteristics, such as being careful, but they were at the same time evaluated differently. Whereas the same characteristics were considered as something positive in case of the 'driving student', they carried negative associations for the 'senior citizen'.

Just as the similes that referred to the 'defensive' AV shared similarities, so too did the similes used for the 'aggressive' AV. In this case, the similes differed in how much control the AV was perceived to have. The most common simile used to describe the 'aggressive' AV was 'taxi driver'. This AV was regarded as rather aggressive, especially when accelerating and braking: *"The first one [referring to the aggressive AV] was more like a taxi driver driving in a city, who is a bit aggressive at crossings and drives first, checks later"* (P11). However, some participants described the same style of driving as comfortable: *"It generally drove like an experienced taxi driver this time. It was more stressful the first time"* (P2).

The second simile used for the 'aggressive' AV was that of a 'father'. The characteristics were similar to the 'taxi driver' in that the AV was described as fast, with hard acceleration and braking, but was at the same time considered to feel safe: *"It drove quite fast, braking, overtaking. It felt like driving with your father who was in a hurry but still drove at a speed that allowed you to see everything. The road was flat and there were no cars around, so it felt quite safe*" (P5). The 'friend who likes to drive fast' simile, also referred to as a 'crazy guy', had similarities with both 'father' and 'taxi driver'. The explanations in this case too were hard acceleration and braking, but the association was with a driver not having full control: "*The other one was more aggressive [referring to 'aggressive'], it feels like a friend who likes driving fast and enjoys twisting and turning. I became more inclined to look out myself. I probably trusted the system, but I also wanted to be aware*" (P14).

## 4   Discussion

The purpose of the paper is to describe how users refer to and describe the AV, during and after initial usage. The analysis shows that the participants used mostly neutral pronouns when referring to the AV, similar to any other car. Interestingly, however, some also referred to it in terms of 'the technology' or 'the system' or as a human (as 'he' but never as 'she').

Using gender pronouns when referring to technology or technical systems may indicate some level of personification. For example, gender pronouns are often used as indicators of personification when investigating people's social relationships with robots [13–15]. Many inanimate products today, ranging from computers to cars, are seen at least as partly human and users tend to attribute human-like mental capacities of intentions, beliefs, attitudes, and knowledge to them [16, 17]. This may often occur over a prolonged use of the product. However, in the study, some degree of personification was evident already after experiencing the AV for the first time, which may indicate that the personification will be more commonly occurring with AVs than in less advanced products, such as manually driven cars. This can possibly result from the more advanced functionality of the AV and users decoupling from the dynamic driving task, as more tasks are performed by the vehicle, leading users to have a harder time understanding the AV. It has been suggested that personification of non-humans may occur in order for a person to predict or comprehend an unfamiliar situation by projecting knowledge about other humans or oneself onto the other agent [18] as well as a way to establish a relation as if they were humans [19]. Hence, for the participants in this study, using knowledge about their own or other humans' driving behaviour may have been a way for them to make sense of the AV's behaviour. However, the effect of personification has previously been suggested to also lead to either long-term attachment or frustration if the object does not live up to expectations [14]. One relevant question is therefore whether personification of AVs is something to strive for or something to try to counteract in AV design.

In the same way as some participants used human pronouns, they also used human similes and information about human drivers to describe the AV's behaviour. Almost all the similes that the participants used to explain and characterise the AVs with different driving styles referred to human drivers ('driving student', 'father', etc.). One explanation is that although most participants had experience of some driver assistance system none had previous experience of highly automated vehicle systems. Without this experience they may not have been able to provide any technical explanations and therefore had to use their knowledge about human drivers to infer their understanding about the AV.

Another interpretation is that although most participants described the AV as a technical artefact and had some understanding of the underlying technical principles, they lacked the appropriate terminology and therefore needed to use human similes to describe their experiences of the AV. It is not possible to conclude which of the explanations is the most likely based on available data. However, interpretations of vehicle functionality and behaviour based on a reference to human drivers and driving, and not on how autonomous systems work, could lead to potential misunderstandings regarding the capabilities and limitations of the AV. For example, AVs which seem to have a human-like mind are believed to perform its intended functions better [20]. The consequence could be under- as well as overreliance as well as misinterpretations of cues provided by the vehicle.

Another finding regarding the similes is that most of them referred to one of the two driving styles: 'driving student' and 'senior citizen' were mentioned with reference to the 'defensive' driving style whereas 'taxi driver', 'father' and 'friend who likes to drive fast' were mentioned in connection with the 'aggressive' driving style. This shows that the different driving styles evoked different associations. Furthermore, even though the participants used similes with similar characteristics when describing the respective driving styles (e.g. using similes such as hard acceleration when referring to the 'aggressive' driving style), there were differences regarding how capable they were perceived as being. The 'father' and 'friend who likes to drive fast' were both described as aggressive with hard acceleration and deceleration but while the 'father' was perceived as having control the 'friend who likes to drive fast' was not. This indicates that the similes were not only used to describe or possibly explain a specific driving style but were also imbued with a value, probably affected by previous experiences with human drivers. For example, if a trusted person drives with an aggressive driving style this may trigger positive associations with that particular driving style, affecting also the way a user perceives and evaluates an AV with similar driving behaviour.

Even though the participants had limited previous experience of higher levels of AV systems, most expressed what they perceived as machine-like or human-like driving characteristics. Exactly which characteristics were regarded as human- or machine-like seemed to be affected by individual preconceptions regarding AV behaviour and capabilities. It is not possible from the data obtained to determine what has shaped these preconceptions, but it is probable that experience of other technology as well as descriptions of AVs in media have been crucial influences, as has been shown in for example Raue, D'Ambrosio [21]. These preconceptions could have a negative effect on users' trust and acceptance if there is a mismatch between users' expectations and the experienced behaviour or capabilities of the AV [22]. It is therefore of great importance to further investigate users' initial understanding of AVs to better understand how they will experience an AV, in particular since preconceptions and expectations seem to differ between individuals.

The combined findings show some level of personification of the AV and information about human drivers was used to understand and/or describe AV behaviour. Additionally, the human similes used indicate that the driving style affected the associations made. In addition, the data shows that individual preconceptions affected participants' expectations of the AV and its behaviour and the way it was perceived. It is possible that these expressions may be an effect of the participants' mental models. It is suggested that mental models are an internal representation of the environment [23] that allows users to describe, explain and predict current and future states of systems [24]. Furthermore, it has been proposed that mental models may develop as analogical and metaphorical reasoning, that is to say early versions of mental models are based on previous models of similar objects or concepts [25–27]. It is therefore possible that the participants' previous experiences of human drivers have shaped their initial mental model of the AV, creating a human-like model which in turn affects their reasoning. Thus, in this case, when the participants lacked any experience of highly automated systems, they instead used their models of human drivers. This may in turn affect how users believe the vehicle will behave, it may shape the attributes they assign to the AV and, further, determine how users will experience their interaction with the AV. Just how well the AV corresponds to expectations can lead to increased trust and a positive user experience whereas a lack of compatibility can have negative effects on acceptance [28, 29]. However, based on the findings, it is also possible to assume that individual users vary in how human-like their mental model is, and this is something that needs to be investigated further.

# References

1. Rajaonah, B., Anceaux, F., Vienne, F.: Trust and the use of adaptive cruise control: a study of a cut-in situation. Cogn. Technol. Work. **8**(2), 146–155 (2006)
2. Seppelt, B.D., Lee, J.D.: Making adaptive cruise control (ACC) limits visible. Int. J. Hum Comput Stud. **65**(3), 192–205 (2007)
3. Parasuraman, R., Riley, V.: Humans and Automation: Use, Misuse, Disuse. Abuse. Human Factors **39**(2), 230–253 (1997)
4. Seppelt, B.D., Lee, J.D.: Keeping the driver in the loop: dynamic feedback to support appropriate use of imperfect vehicle control automation. Int. J. Hum Comput Stud. **125**, 66–80 (2019)
5. Ghazizadeh, M., Lee, J.D., Boyle, L.N.: Extending the technology acceptance model to assess automation. Cogn. Technol. Work **14**(1), 39–49 (2012)
6. Khastgir, S., Birrell, S., Dhadyalla, G., Jennings, P.: Effect of knowledge of automation capability on trust and workload in an automated vehicle: a driving simulator study. In: Stanton, N. (ed.) AHFE 2018. AISC, vol. 786, pp. 410–420. Springer, Cham (2019). https://doi.org/10.1007/978-3-319-93885-1_37
7. Beggiato, M., Krems, J.F.: The evolution of mental model, trust and acceptance of adaptive cruise control in relation to initial information. Transport. Res. F: Traffic Psychol. Behav. **18**, 47–57 (2013)
8. Nees, M.A.: Acceptance of self-driving cars: an examination of idealized versus realistic portrayals with a self-driving car acceptance scale. In: Proceedings of the Human Factors and Ergonomics Society Annual Meeting. SAGE Publications Sage CA, Los Angeles, CA (2016)
9. Ekman, F., Johansson, M., Bligård, L.-O., Karlsson, M., Strömberg, H.: Exploring automated vehicle driving styles as a source of trust information. Transport. Res. F: Traffic Psychol. Behav. **65**, 268–279 (2019)
10. Kauffmann, N., Naujoks, F., Winkler, F., Kunde, W.: Learning the "Language" of Road Users - How Shall a Self-driving Car Convey Its Intention to Cooperate to Other Human Drivers? Springer, Cham (2018)
11. Oliveira, L., Proctor, K., Burns, C.G., Birrell, S.: Driving style: how should an automated vehicle behave? Information **10**(6), 219 (2019)

12. Charters, E.: The use of think-aloud methods in qualitative research an introduction to think-aloud methods. Brock Edu. J. **12**(2) (2003)
13. Purington, A., Taft, J.G., Sannon, S., Bazarova, N.N., Taylor, S.H.: Alexa is my new BFF: social roles, user satisfaction, and personification of the amazon echo. In: Proceedings of the 2017 CHI Conference Extended Abstracts on Human Factors in Computing Systems. ACM (2017)
14. Lopatovska, I., Williams, H.: Personification of the Amazon Alexa: BFF or a mindless companion. In: Proceedings of the 2018 Conference on Human Information Interaction & Retrieval, pp. 265–268, ACM, New Brunswick (2018)
15. Cooke, S.F.: Are Robots Animate or Inanimate? Children's pronoun use provides insight to categorization challenge. In: 2018 University of Montana Conference on Undergraduate Research (2018)
16. Epley, N.: A mind like mine: the exceptionally ordinary underpinnings of anthropomorphism. J. Assoc. Consum. Res. **3**(4), 591–598 (2018)
17. Aggarwal, P., McGill, A.L.: Is that car smiling at me? schema congruity as a basis for evaluating anthropomorphized products. J. Consum. Res. **34**(4), 468–479 (2007)
18. Epley, N., Waytz, A., Cacioppo, J.T.: On seeing human: a three-factor theory of anthropomorphism. Psychol. Rev. **114**(4), 864–886 (2007)
19. Airenti, G.: The development of anthropomorphism in interaction: intersubjectivity, imagination, and theory of mind. Front. Psychol. **9**(2136) (2018)
20. Waytz, A., Heafner, J., Epley, N.: The mind in the machine: anthropomorphism increases trust in an autonomous vehicle. J. Exp. Soc. Psychol. **52**, 113–117 (2014)
21. Raue, M., D'Ambrosio, L.A., Ward, C., Lee, C., Jacquillat, C., Coughlin, J.F.: The influence of feelings while driving regular cars on the perception and acceptance of self-driving cars. Risk Anal. **39**(2), 358–374 (2019)
22. Lee, J.D., See, K.A.: Trust in automation: designing for appropriate reliance. Hum. Factors **46**(1), 50–80 (2004)
23. Johnson-Laird, P.N.: Mental Models. Cambridge University Press, Cambridge (1983)
24. Rouse, W.B., Morris, N.M.: On looking into the black-box - prospects and limits in the search for mental models. Psychol. Bull. **100**(3), 349–363 (1986)
25. Gentner, D., Gentner, D.R.: Flowing Waters or Teeming Crowds: Mental Models of Electricity. In: Gentner, D.S, Albert L. (eds.) Mental Models, pp. 99–127, Erlbaum, Hillsdale (1983)
26. Staggers, N., Norcio, A.F.: Mental models: concepts for human-computer interaction research. Int. J. Man Mach. Stud. **38**(4), 587–605 (1993)
27. Cool, C., Park, S., Belkin, N., Koenemann, J., Ng, B.K.: Information seeking behavior in new searching environments. In: Royal School of Librarianship (1996)
28. d'Apollonia, S.T., Charles, E.S., Boyd, G.M.: Acquisition of complex systemic thinking: mental models of evolution. Educ. Res. Eval. **10**(4–6), 499–521 (2004)
29. Wolf, I.: The Interaction Between Humans and Autonomous agents. In: Maurer, M., Gerdes, J.C., Lenz, B., Winner, H. (eds.) Autonomous Driving, pp. 103–124. Springer, Heidelberg (2016). https://doi.org/10.1007/978-3-662-48847-8_6

# How is the Automation System Controlling My Vehicle? The Impact of the Haptic Feedback of the Joystick on the Driver's Behavior and Acceptance

Cho Kiu Leung[1](✉) [iD] and Toshihisa Sato[2] [iD]

[1] University of Tsukuba, Tsukuba, Ibaraki 305-8573, Japan
liang@css.risk.tsukuba.ac.jp
[2] National Institute of Advanced Industrial Science and Technology, Tsukuba,
Ibaraki 305-8560, Japan

**Abstract.** Vehicle's ADAS technologies are expected to provide benefits such as reducing accidents caused by human error while driving and improving the driving experience by supporting the driver in some driving tasks. This study summarizes how the haptic feedback support from the joystick affects the driver's visual actions, mental workload and user feeling when using autonomous driving. 24 elderly drivers ($M = 67/SD = 4.3$) took part in a driving simulation experiment. The result showed that there was a significant reduce in driver's visual actions with the haptic support, as the haptic feedback from the joystick provided a sense of reassurance and reduced the visual burden on the monitoring task when using ADAS. Also, the subjective evaluation NASA-TLX results showed that the haptic feedback of the joystick should not be a high mental and physical burden to the user.

**Keywords:** Advanced Driver-Assistance System · Human-machine interaction · Driver performance · Joystick

## 1 Introduction

### 1.1 Motivation

Vehicle's Advanced Driver-Assistance Systems (ADAS) technologies (e.g., Adaptive Cruise Control and Lane Keep Assist System) are expected to provide benefits such as reducing accidents caused by human error while driving and improving the driving experience by supporting the driver in some driving tasks. In terms of how the user interacts with the ADAS, it is generally conceivable to select, intervene and command the ADAS through the human machine interface (HMI) touch screen or steering wheel, buttons, etc. Furthermore, the ADAS interacts with the user by providing text messages, images, warning sound or/and animations through the visually supported or auditorily supported HMI. The interaction of text messages, images, sounds or animations is important to

© Springer Nature Switzerland AG 2021
H. Krömker (Ed.): HCII 2021, LNCS 12791, pp. 273–280, 2021.
https://doi.org/10.1007/978-3-030-78358-7_19

the user because the user is able to aware the situation through the interaction what the system is processing or will not be able to process, etc. This helps the user to better understand the operation of the ADAS because drivers might confuse that the ADAS does not work as they expect [1]. In addition to visual or auditory support methods, haptic feedback support is also a potential way to let the user more aware of the operation of the ADAS. In relation to haptic feedback support, there has been some research into the use of devices that transmit vibrations as support to let users know that they need to intervene and drive manually. There is still room for discussion on how haptic feedback can support the user's understanding of ADAS. This is because in the case of SAE level 2 autonomous driving for example, there are vehicles that do not require the user to hold the steering wheel when the adaptive cruise control (ACC) or lane keeping assist system (LKAS) of the autonomous system is activated simultaneously. This means that the user's hand is free during normal status of ADAS. The hand, in proportion to the foot or other body parts, are reliable receivers of tactile information. Also, we believe that all information the user needed should not be received exclusively visually. By distributing some information to other body parts through auditory sounds or tactile vibrations, the burden on the user's vision should be reduced. Therefore, more discussion is needed to make the operation of ADAS more understandable to the user visual behavior via haptic feedback from the device to the hand.

## 1.2  Related Knowledge of Joystick Control

It also should be mention that joystick is a traditional interaction device and are common in the field of air and marine transportation. The joystick is characterized by its ability to provide haptic feedback. In the case of aircraft, the automated system control of the aircraft is fed back via joystick in the cockpit. Furthermore, the control from two pilots and their control are fed back respective joysticks.

Related to the research of joystick control, it was regarded as a universal design for vehicle control and an alternative to steering wheel [2–4]. For example, in Japan, drivers can obtain a legal license permit to use joystick-control vehicle on public roads after being trained [4, 5]. Although some studies have been mentioned the issue of joystick control in manual vehicle such as the stability. For example, the joystick's sensitivity during different vehicle speed, might affect driver control input when changing lanes and lead to human error [2, 5]. Therefore, system for support the driver to adjust the sensitive and perform the part of driving task depending on different road conditions is a solution for preventing human error during use of joystick control [5]. Also, recent studies related to automated driving also have indicated that joystick control for short periods of time is possible [6].

Since the possibility of using a joystick as a control for an autonomous vehicle is still up for discussion, we focus on investigate the scenario where control via joystick is not required. The first objective of this study is to investigate whether the feedback support of the joystick can reduce the visual burden of the monitoring task of the user who is using ADAS as a support for the user to understand how the system controls the vehicle.

## 1.3  Research Question

This study investigates the user vision behavior and mental workload when using ADAS with the support of haptic feedback on the joystick through driving simulations and question to summarize the functional usability and user acceptance of the haptic feedback of the joystick.

1. Investigate whether the haptic feedback support of the joystick assists users in their monitoring tasks or reduce their burden of the vision when using the SAE level 2 automation system [7].
2. Whether the providing of haptic feedback on the joystick would have an impact on the mental or physical workload of the user?
3. How receptive users would be to haptic feedback on joystick devices?

This driving simulation experiment had totally 3 days scenario. It should be noted that this study only discusses the results of the first day of the experiment. Because the second day and third-day experiments included the consideration of joystick manipulation experiments, this study did not focus on the content of manipulability of the joystick, but rather on how the haptic feedback of the joystick assist users understand the automation system.

## 2  Introduction

**Fig. 1.** The appearance of the driving simulator and the recorded capture of the experiment

### 2.1  Participants and Apparent

This study was conducted with a joystick device connected to a driving simulator. 24 elder driver who were over 60-year-old (M = 67/SD = 4.3) with valid driving license took part in the driving simulation experiment. The joystick device used in experiment set on the left side of the driver's seat of the simulator (Fig. 1). The size of joystick device is 15 cm tall, 10 cm long and 10 cm wide. The simulation system for the driving simulator called D3-Sim produced by Mitsubishi Precision, Co. Ltd.

## 2.2 Experiment Design

The study considered single factor design with the 2 conditions called the support method. The detail shown in Table 1 and Fig. 2.

**Table 1.** Description of each condition

| Condition | Description |
|---|---|
| Baseline (B-condition) | The participants generally use the vehicle with a level 2 driving automation without any additional support |
| Joystick Haptic feedback Support (JHS-condition) | The participant holds the joystick when using the vehicle with level 2 driving automation. The joystick transfers the vehicle motion as haptic feedback to their hand, for example, when the vehicle moves horizontally to the right, the joystick will also be tilted to the right and the participant can feel it |

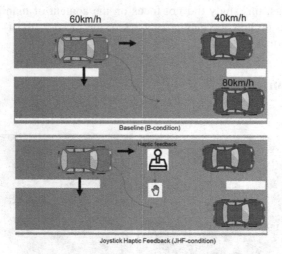

**Fig. 2.** Scenario descriptions for each situation

## 2.3 Scenario

To reduce the effect of learning, the order of the driving test was randomized. In day-1 experiment, each driving test was approximately 10 min in length (Fig. 3). Before the driving test, they had a 1 min for trying the haptic feedback function. Also, all participants receive an explanation of the functions of the ADAS prior to each experiment condition. Their task in the experiment is to monitor and keep focus on safely on the highway. After the driving test, all participants were asked about the user feeling. It included questions about the difficulty related to monitoring the traffic and the enjoyment of driving in the drive test.

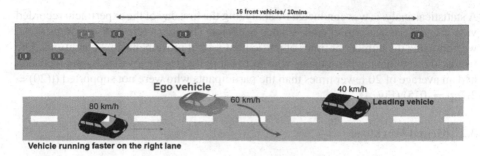

**Fig. 3.** Scenario of the experiment.

## 2.4 Measurements

To investigate their burden of the vision when using the SAE level 2 automation system. We confirmed their vision and head movement in two conditions. During the experiment, if the participants performed a confirmatory action, such as eye gaze towards the rear mirror/side mirror or direct gaze towards the side windows, it was considered as a one visual action. Also, this study evaluates their subjective feeling via using the NASA-TLX and question of driving difficulty and user feeling of the haptic feedback support of the joystick when using ADAS.

## 3    Results and Discussion

### 3.1    Vision Behavior

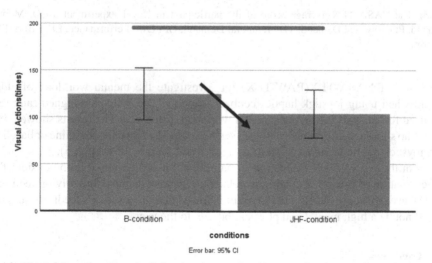

**Fig. 4.** The total number (times) of visual actions to the side, rear via mirror or direct sight during the 10 min of the experiment

A statistical analysis was conducted by analyzing the first day of the experiment recorded videos to observe the participants' glance behavior. Through a repeated-measures t-test, we compared the number of times of participant's sight movements in 2 conditions. The results showed that the participants who were supported by the joystick haptic feedback had an average of 20 fewer times than the participants who were not supported ($t(20) = 2.7$, $p = .015$) (Fig. 4).

### 3.2 Mental Workload

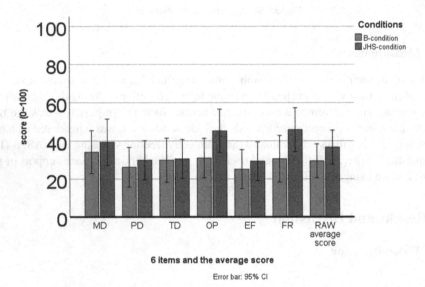

Fig. 5. The NASA-TLX average score of the participant in Day-1 experiment. (MD: Mental Demand, PD: Physical Demand, TD: Temporal Demand, OP: Own Performance, EF: Effort, FR: Frustration.)

The result of NASA-TLX(RAW-TLX) for investigate the mental workload of elder drivers when using joystick haptic feedback support. The Wilcoxon signed-rank test was used for analysis the driver's mental workload difference. The results showed that there is no significant difference of the driver's workload when using baseline or baseline with joystick haptic feedback support design ($Z = 1.96$ $p = .051$) (Fig. 5).

From the NASA-TLX evaluation 6 scale items, it is also possible to consider the self-evaluation of users in 2 conditions related to the task demand and driving feeling.

The average score of MD and PD results showed that the joystick's feedback support should not be a high mental and physical burden to the user.

### 3.3 User Feeling

After the participants have experienced the condition, they were asked to evaluate whether they feel enjoyment and difficulty of the driving experience. The evaluation

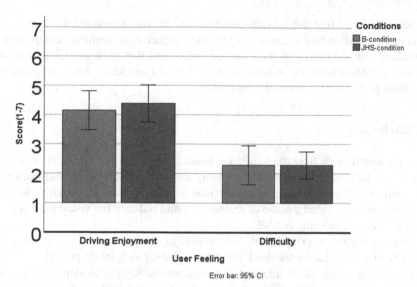

**Fig. 6.** Evaluation of the user's perceived driving enjoyment and perceived difficulty for each situation

was about whether the joystick as a new support method would bring greater enjoyment of automation driving or difficulty during using automation driving. The evaluation used a 7-point Likert scale approach, with 1 point: not felt and 7 point: very much felt. The results of the two conditions were compared using the Wilcoxon signed-rank test (Fig. 6). The results regarding the haptic feedback support of the joystick showed no significant difference in the ratings of enjoyment and difficulty between the two conditions. It was found that the evaluation scores of the two conditions were similar in this experiment, but there was no significant difference between the two conditions.

## 4   Discussion

The results showed no significant effect of the joystick on physical demand and mental demand, but there might be an effect on mental workload. In addition, the results of the visual action showed that the number of visual actions was less when supported by the joystick's feedback than when there was no support. This result implies that the potential of joystick feedback support to reduce the visual burden on the driver. They use the joystick to aware and understand the situation and increase their attention to the front. In addition, past study also has shown that users who feel uneasy their glance movement more than those who feel at ease during automation driving with no support [8]. A significant difference has been found between the number of sight movements of users who felt uneasy about the automation system and those who felt at ease. The past study result supports the findings of this study and therefore this result find that joystick haptic support might let the user more recognize what the vehicle is doing could increase understanding and peace of mind for the ADAS. Also, through haptic feedback support of the joystick, they can perceive how the system controls the vehicle, understanding more that the vehicle's motion may be able to induce the user's visual actions to the position that needs to be confirmed, rather than randomly confirming the surroundings. In addition, rom the recorded video data that most participants showed restlessness in their body movements such as they did not know where to place their hands when performing without the support.

## 5  Conclusion

This study summarizes how the haptic feedback from the joystick affects the driver's visual actions and mental workload when using autonomous driving. In summary, there was a significant reduce in driver's visual actions with the support, as the haptic feedback from the joystick provided a sense of reassurance and reduced the visual burden on the monitoring task when using ADAS.

Regarding mental workload, there was no significant difference found in this experiment. Therefore, the haptic feedback function of the joystick might provide users with more peace of mind about the automated system and reduce the burden on their vision required for monitoring tasks. This study mainly shares the findings from the first day experiment and our goal is to provide the joystick with effective support benefits in addition to a higher level of driving experience and enjoyment through joystick control.

## References

1. Crump, C., et al.: Differing perceptions of Advanced Driver Assistance Systems (ADAS). Proc. Hum. Factors Ergon. Soc. Annual Meeting **60**(1), 861–865 (2016)
2. Wada, M., Kameda, F., Saito, Y.: A study on steering control for a Joystick car drive system. Trans. Soc. Instr. Control Eng. **49**(4), 417–424 (2013)
3. Andonian, B., Rauch, W., Bhise, V.: Driver steering performance using joystick vs. steering wheel controls.: Journal of Passenger Car: Mechanical System Journal, 112(6), pp. 1–12 (2003)
4. Wada, M.: Development of a joystick drive car. Bull. Japan. Soc. Prosthetics Orthotics **33**(4), 232–238 (2017)
5. Wada, M., Kameda, F.: A Joystick car drive system with seating in a wheelchair. In: 35th Annual Conference of IEEE Industrial Electronics, pp. 2163–2168. IEEE Press, Porto (2009)
6. Large, D.R., Banks, V., Burnett, G., Margaritis, N.: Putting the joy in driving: investigating the use of a joystick as an alternative to traditional controls within future autonomous vehicles. In: Proceedings of the 9th International Conference on Automotive User Interfaces and Interactive Vehicular Applications. pp. 31–39. Association for Computing Machinery, New York (2017)
7. SAE.: Taxonomy and Definitions for Terms Related to Driving Automation Systems for On-Road Motor Vehicles (J3016_201806) (2018)
8. Ryota, T., Kenji, T.: Effect of suppressing driver's security feeling on improving attention in automated driving. In: Proceedings of the 46th Society of Instrument and Control Engineers (SICE) Symposium on Intelligent Systems, pp. 1–4, Tokyo (2018)

# Understanding Take-Over in Automated Driving: A Human Error Analysis

Jue Li, Long Liu(✉), and Liwen Gu

College of Design and Innovation, Tongji University, Fuxin Road 281, Shanghai 200092, China
liulong@tongji.edu.cn

**Abstract.** Automation offers a new way of driving, but often the human error (HE) in the process of take-over results in adverse effects of unrecognized risks. Hence, the impact of HE in safety of automated driving remains a major problem. This paper proposed a Human error analysis method based on analysis of the root cause of HEs events to understand the process of take-over and identify root cause of take-over failure in automated driving. Simulated driving practice with videos and questionnaire were conducted to identify the main factors leading to HEs in take-over. Human factors events diagram was used to better understand take-over as a human factor event and to provide information for root cause of take-over failure recognition. The results reveal that the most common failure mode in take-over is cognition error caused by driver poor mental state such as driver fatigue and reaction ability, followed by control error caused by inappropriate take-over request (TOR). Determination of these failure modes provide evidence for increasing or repairing barriers in the process of take-over. The suggested cognition-corresponding model of take-over showed that take-over is a complex human-machine interaction process, thus the causes of HEs should be discussed from a multi-dimensional perspective, and explored through empirical research.

**Keywords:** Automated driving · Take-over · Human error analysis · Human factor event

## 1 Introduction

Over the past few decades, an obvious trend toward increasing automation has characterized the auto industry [1]. Technology companies and Research institutes are working on different types and levels of driving automation systems and carrying out road tests and deployment within a certain range. Until automated driving systems are able to perform all driving tasks under all scenarios, human and vehicle systems will share responsibility for driving tasks and driving safety. Partially automated driving, which is already deployed in road, requires drivers to monitor the road and to be prepared for immediate intervention in case of unexpected events or conditions. At conditionally or highly automated driving, drivers are allowed to engage in non-driving-related tasks, while the automation executes the monitoring task and issues a TOR when the driver has to intervene [2].

H. Krömker (Ed.): HCII 2021, LNCS 12791, pp. 281–295, 2021.
https://doi.org/10.1007/978-3-030-78358-7_20

According to the above description of automation level and views of many scholars, the transition from automated driving to manual driving, referred to as the take-over, is the key to a successful deployment of conditional automation [3, 4]. Take-over is exactly the transition process in which automated system gives a danger warning to the driver when the car approaches the limit of system failure or the automated system fails to match a specific driving situation, and the driver restarted driving the car manually. For example, the vehicle leaves the highway that supports automated driving, and enters an undefined road. If such a system boundary is detected, a TOR is prompted and the driver should redirect their gaze from secondary tasks to the road to establish motor readiness, and to take over vehicle control within a limited time budget. At this stage, the safety of automated driving is mainly ensured by these systems by warning the driver about potentially dangerous conditions and also by manual active control.

Take-over can be divided into three categories according to the Initiator and compulsory [5]: a) optional human-initiated takeover, usually a transition generated by the driver's initiative to shut down the vehicle in a non-emergency state; b) compulsory human-initiated take-over, usually a transition generated by the driver actively shutting down the automation system after finding that the system is not up to the current task; c) compulsory vehicle-initiated take-over, usually a transition generated by the driver passively take-over the vehicle after the vehicle system finding that it is unable to perform the current task and issuing a TOR. Among the take-over types above, Passive take-over have attracted considerable research interest as it is not initiated subjectively by the driver and is more likely to cause an accident due to the lack of situational awareness during driving [6].

In fact, in the process of automated driving, the higher the degree of automation, the higher the number of non-driving tasks can be done, the less the driver will focus on environmental monitoring and system operation, and the worse his ability to take over the driving system [7, 8]. The new alarm receiving channel, continuous change of working environments and unpredictable non-driving related task during monitoring present further make the task of take-over excessively complex. While the existing research on take-over focuses mostly on the time limit or operation mode of take-over, TOR or vehicle HMI design strategy [9] and how well the driving performance after taking over [10]. Compared with those considerable studies on operations or performance of take-over, there are still few studies about the type and frequency of HEs that drivers make in take-over and how this, in turn, affect the driver's behaviours in a take-over. Given these situations, it is necessary to examine HEs in failed take-over and inappropriate take-over behavior by analyses incident mode and failure root cause which may poses serious safety implications for driving safety.

## 2   Background

There are many definitions of HE, and it is hard to say which is more authoritative, but it is generally believed that in the interaction between human and machine, HE is usually considered derogatory, and it is primarily attributed to a variety of adverse contextual conditions or the interaction among whole system instead of the failure of individual components in the system [11]. And with the continuous improvement of the automation level of human-machine systems, the number and types of system components have greatly increased, and the types and its impact of system failures related to HEs have also increased. Therefore, the role of HE in accidents become more and more important in some safety critical industry accidents, such as mobility industry and Nuclear Power industry. And after numerous observations and investigations of a series of historical catastrophic accidents in the last century (e.g. the Three Mile Island accident in 1979), theoretical research on HE is also becoming more abundant and scientific.

When HE was considered to be the cause of certain industrial accidents, some scholars put forward HE analysis methods mainly focused on applying structural models and calculation methods to solve mathematical problems. The best-known of these methods are The Technique of Human Error Rate Prediction (THERP) and the Human Error Assessment and Reduction Technique (HEART); Then the second-generation human error analysis methods mainly identify cognitive models of human behavior based on behavioral science and psychology to explain the mechanism of human error formation, such as General Error Modelling System, the Cognitive Reliability and Error Analysis Method (CREAM), and Cognitive Simulation Model (COSIMO). With the occurrence of various types of HE accidents and the in-depth investigation of these accidents by human factors experts, it believed that the focus of Human error analysis should shift from single HEs to the underlying organizational factors or management failures to find the hidden root causes that may be related to those serious consequences. Therefore, several "system approaches" have emerged. Wiegmann and Shappell [12] proposed a system method called Human Factors Analysis and Classification System (HFACS) to given a detailed classification to investigate accident causes. Mosleh and Groth (2010) established a method to develop a data-informed Bayesian Belief Network that can be used to improve HE probability prediction by reducing overlap among performance shaping factors (PSFs), then developed a performance shaping factors causal model for nuclear power plant Human error analyse on the basis of obtained data [13]. Kim et al. developed Korean-version HPES (human performance enhancement system) developed on the basis of INPO-HPES to reduce HE and enhance the personnel performance of nuclear power plants. The system draws on the ideas and procedures of the fault tree analysis method, starting from the characterization of human factor events, and looking for the root cause of the failure of the event [14].

The role of HE in accidents in most safety critical systems is already well known. However, there has been only limited attention focusing on the types of HEs of driving system at present [15]. Due to the human factor characteristics of the automated driving system itself: Human and machine share system control, support humans in various non-driving related task, and take human as a fallback option etc., HEs cannot be completely eliminated, but they should also be minimized and improved. It is acknowledged that driving is an extremely dynamic activity, and the process of take-over in automated driving takes only a few seconds [4]. Moreover, Automated vehicles have only been deployed on the road for a short time, which has results in insufficient field data on HE in automated driving available for scholars to analyze. These make the analysis of HE in automated driving take-over different from those in other industries, which bring greater challenges to this area.

This study used E&CF diagram [16] to construct the event timing diagram of the process of take-over, so as to provide useful information to clarification of the root cause of take-over failure. The whole human error analysis of this study is based on Analysis of The Root Cause of Human factor events, which is divided into 4 steps: Investigation and Determination of The Incident, Failure Mode Determination and Barrier Analysis, Analysis of The Root Cause of Human Factors, and Improvement Measures and Recommendations. A brief summary of the approach is given in Table 1. Before constructing the human factors events diagram of take-over, this article put forward the main reasons that affect the take-over performance during automated driving based on a simulated

**Table 1.** Analysis of the root cause of human errors events.

| Step of analysis | | Description |
|---|---|---|
| Investigation and Determination of the Incident | Incident investigation and data collection | Clarify the available information<br>Make clear the context of HE or improper action<br>Determine the content of further investigation, make a list of questions for interviews<br>Obtain the result of problem by interviews |
| | Chronological description of the event | Describe the entire process of the event in chronological order |
| | Construction of event timing diagram | Determine the start and end points of the event<br>Identify the order of events and failures |
| Failure Mode Determination and Barrier Analysis | Determination of failure mode | Match the failure mode of each stage from the failure mode database |
| | Barrier analysis | Identify barrier types and add or modify barriers |
| Analysis of The Root Cause of Human Factors | Construction of human factors events diagram (see Fig. 1) | Add the cause factor and failure barrier to event timing diagram<br>Establish its relationship with the event |
| | Root cause analysis of human factors events | Map the cause factor or contribution factor of the failure mode according to the cause factor table in the event analysis |
| Improvement Measures and Recommendations | | Obtained the cause of the event and the specific status of the failure barrier<br>Take appropriate measures to eliminate the root cause of incident |

driving practice and a questionnaire survey, and then analyzed possible HEs due to different driving situations. These studies and analyses can provide strong evidence for the subsequent proposal of Improvement Measures.

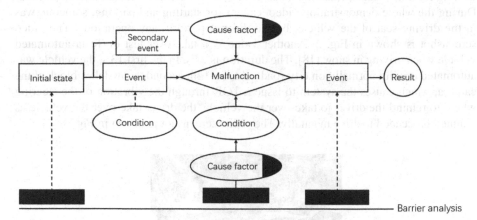

**Fig. 1.** Human factors events diagram.

# 3   Method

In the process of Analysis of The Root Cause of Human Errors Events, it is necessary to investigate the factors that affect take-over performance during automated driving so as to provide evidence for subsequent analysis. Therefore, simulated driving practice with two automated driving videos and a questionnaire survey utilizing five-point Likert Scale were conducted to identify the factors that affect take-over performance during automated driving. A short interview after that was conducted to obtain users' perceptions of these factors.

## 3.1   Participants

The simulated driving practice and questionnaire survey included 12 participants aged from 23 and 56 years. The sample consisted of more men (8) than women (4). All of them held a valid driver's license, but had no prior experience with fully automated driving. Each participant gave written informed consent at the beginning of the study and received a monetary compensation for their participation.

## 3.2   Materials

The simulated driving practice material is two videos about automated driving. They are intended to deepen the understanding of the automated functions of the vehicle for participants who have been used to manual driving, as well as raise their awareness of the take-over issues. The first one is a demo video from GeekCar of a Tesla car driving fully automated [17]. The vehicle interior in the video still maintains the basic layout of the manual driving car. The video takes 3 min and 46 s and includes demonstrations of operations such as starting a vehicle, crossing rural roads, driving on urban roads, waiting for traffic lights, changing lanes, courteous pedestrians, and automated parking. During the whole demonstration video, except for starting and parking, someone was in the driving seat of the vehicle, but he did not perform any operations. The video screenshot is shown in Fig. 2. Another video is a take-over test of Tesla automated vehicle when driving in snow [18]. The duration is 24 s. In the first 13 s, the vehicle was automated driving normally on city roads. At the 14th second, snowflakes blocked the camera, which causes the system to issue a TOR through the vibration of the steering wheel to remind the driver to take-over the vehicle. the driver take-over the vehicle as planned proceeded to drive manually. The video screenshot is shown in Fig. 3.

**Fig. 2.** Demonstration video fully automated driving.

**Fig. 3.** Take-over test video.

In the questionnaire, 24 factors that may affect take-over performance were proposed, and the participants were asked to measure the importance of these factors in take-over. These 24 factors are refined from the related factors of main elements in human-vehicle interaction: (1) driver factors, including driver gender, driver fatigue, driving experience (manual/automated), etc., a total of 10 items; (2) vehicle factors, including instant speed, dashboard and display interface, TOR time, etc., a total of 8 items; (3) scenario factors, including traffic flow, non-motor vehicles and pedestrians, road infrastructure, etc., a total of 6 items. The following empirical research is based on this classification. The questionnaire was presented in the form of 5-point Likert scale. Participants are asked to rate the impact degree of the above-categorized and refined factors from 0 (not important) to 4 (very important). At the end of the questionnaire, participants were asked to fill in three personal information on gender, age, and driving experience.

### 3.3 Procedure

The first step is to take take-over failure as a human factor event and analyse it according to the sequence of events, so as to establish the event timing diagram of take-over. In the second step, participants were asked to conduct simulated driving practice through video viewing. Then participants were accurately informed of the definition and characteristics of conditional automated driving and filled out the questionnaire. Finally, they were asked about perceived risks of automated vehicles and how they believe various errors in automated driving will affect road safety. The third step is to conduct barrier analysis and determination of failure mode of the events based on the defined factors, so as to construct the human factor event diagram. The fourth step is to find out the root cause of take-over failure through observation and analysis of the human factor event diagram, and propose improvement measures to reduce take-over failure. Figure 4 shows the procedure.

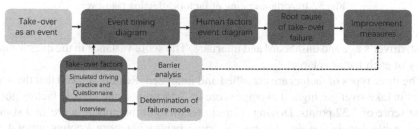

**Fig. 4.** Procedure of Human error analysis (light gray boxes represent operations, dark gray boxes represent the output of each step; solid arrows represent direct associations and dashed arrows represent weak associations).

## 4  Results

To assess the surveys and find significant results of take-over factors, this study used SPSS statistical software, version 23 to perform descriptive statistics and analysis of variance test (ANOVA) on the data sets. Twelve questionnaires were distributed, and simulation practice was carried out as required and 12 copies were collected. After statistical analysis, the reliability of the questionnaire data is 0.721, which indicates that the reliability of the data is good and has a high value to research and analyze.

### 4.1  Take-Over Factors: Overview Results

As shown in Fig. 5, participants believe that the biggest factor affecting take-over is driver fatigue, which is one of the most important causes of HE in general industrial processes. Besides, both the driver's reaction ability and the TOR time scored higher in the survey, indicating that they were both considered to have a greater impact on take-over performance. While seat comfort was considered the least important factor for take-over, and the factors considered also unimportant are driver's gender, interior

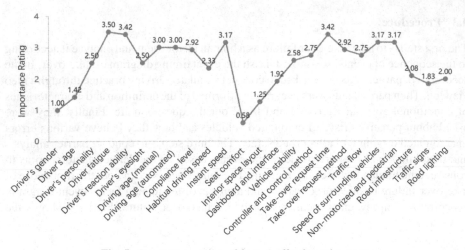

**Fig. 5.** Importance rating of factors affecting take-over.

layout, driver's age and dashboard and interface. The score variance in the questionnaire survey of each factor is large.

The three types of factors are classified and analyzed. Results showed that the driver factors in take-over got highest average score of 2.56 points, and vehicle factors got the lowest score of 2.32 points. Driving scenario factors got the middle score, and showed a more stable distribution. In addition, the driver factors got more 4 points out of 4, and the vehicle factors got less 4 points, which is consistent with the average score. Scenario factors got no 0 points in the survey, which indicates that the driving scenario factors do not have the greatest impact on take-over performance, but in the process of automated driving, all the factors in the driving scenario will act on the take-over behavior. As shown in Table 2.

**Table 2.** Comparison of three types of factors affecting take-over.

| Factor category | Number of factors | Proportion of full score factors | Proportion of 0 points factors | Average | Standard deviation |
|---|---|---|---|---|---|
| Driver factor | 10 | 80% | 30% | 2.56 | 0.86 |
| Vehicle factor | 8 | 63% | 38% | 2.32 | 0.84 |
| Scenario factor | 6 | 67% | 0% | 2.5 | 0.76 |

Analysis within groups within each factor category reveals more specific factors in take-over. Among driver factors, the average score of driver's fatigue and driver's reaction ability are the highest, which indicates that participants believed that these two factors in driver factors had the most impact on take-over. Among vehicle factors, the most important item for take-over is TOR time, followed by instant vehicle speed and the way to issue TOR. Among scenario factors, the average score of speed of surrounding vehicles, together with non-motor vehicles and pedestrians got the highest scores.

## 4.2 Take-Over Factors: Correlation Analysis

The data obtained from the questionnaire survey is mainly used to identify the factors affecting take-over. Therefore, correlation analysis on human factors was conducted to obtain different types of users' understanding of take-over. Human factors (gender, age and driving experience of the participants) were used as independent variables, and the factors affecting take-over were used as dependent variables. Significance was accepted at the α-level of $p < 0.05$.

Correlation analysis was performed, and only two of the 24 factors were found to be statistically correlated with the independent variables. Both of these two factors are vehicle factors. The results of analysis of variance show a significant main effects of vehicle stability in participant's age ($F = 0.683, p = 0.014 < 0.05$), and older participants believe that the impact of vehicle stability on automated driving take-over is smaller than that of smaller participants, as shown in Fig. 3. Similarly, the results showed a significant main effects of interior space layout in participant's driving experience ($F = 0.645, p = 0.024 < 0.05$), and experienced drivers believe that the impact of interior space layout on automated driving take-over is smaller than the participants who lack driving experience. The Correlations are summarized in Fig. 6 and Fig. 7.

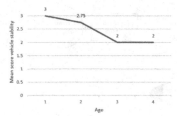
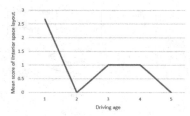

**Fig. 6.** Correlation of age and vehicle stability score.

**Fig. 7.** Correlation of driving experience and interior space layout score.

## 4.3 Analysis of the Root Cause of Take-Over Failure

In the event timing analysis of take-over in automated driving, the initial test state of take-over is that the automated driving system runs stably, so that the driver may participate in some Non-driving related tasks, such as sending message and chatting. And then the automation system fails for some unknown reasons, and the TOR is issued but the driver cannot quickly stop the non-driving related task and take over the vehicle, then, the vehicle out of control, the traffic accident with different severity will be caused according to the degree of failure of the take-over. An event timing diagram can be used to illustrate this whole process. As shown in Fig. 8.

In the whole root cause analysis, the determination of failure mode is to find the point where the accident occurred, and the barrier analysis is to clarify the defects that may lead to failure at this point. According to the analysis of the above survey, it can

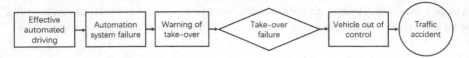

**Fig. 8.** Event timing diagram of take-over.

be found that the most common failure mode in take-over is cognition error of driver's poor mental state, and the second is control error caused by the inappropriate time of TOR. Therefore, these two kinds of HEs can be regarded as the cause factor of the human factor events, which together lead to the malfunction, that is, the take-over failure. While based on the identification of cause factor and interviews with the participants, it can be found that the biggest hole in take-over is the interface that information exchange between vehicle and human. Possible barriers like system inspection and TOR design can also be drawn along with the discovery of holes. All of the elements above can be abstracted into a diagram, which can help experts to identify failure mode of take-over, and to strengthen or supplement the barrier in the event, so as to prevent the repeated occurrence of similar failures. As shown in Fig. 9, the Human Factors Events Diagram provides a quick snapshot of take-over, including time sequence, event condition, failure cause factors, possible barriers, and so on.

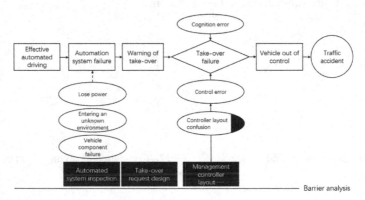

**Fig. 9.** Human factors events diagram for take-over.

## 5   Discussion

### 5.1   Take-Over Factors

The first step of human error analysis in take-over is investigation and data collection. In the study of traffic accidents, western researchers started research on the driver's emergency behavior earlier, and most of the research is through field research or analysis of accident causes or human error models [19, 20]. However, the construction of a traffic accident statistics system is still in its infancy in China. The investigation of special vehicles and accidents relies more on traffic police reports, insurance company reports and professional academic institutions. The amount and type of data are rich but not sufficiently clear and precise. Considering the time for development and application of automated driving is short and the field accident data is difficult to obtain, this study used simulated driving practice and questionnaire to obtain data of take-over factors. This method can ensure that ordinary users have a basic understanding of take-over in automated driving, and efficiently collect users' attitudes towards these factors and their understanding of HEs in take-over.

The results of questionnaire showed that driver factors are most relevant to take-over performance, and followed by system factors directly related to take-over. Among the driver factors, the average score of driver's fatigue and driver's reaction ability are the highest, which indicates that participants believed that these two factors had the most impact on take-over. This is similar to the conclusion of human error analysis of traditional manual driving [21]. According to the interview, some participants believed that while the automated system was working, they still needed to concentrate on monitoring the status of the system for a long time. Since humans are not good at monitoring, such long-term monitoring may cause higher fatigue than manual driving, thereby leading to HE and affecting take-over performance. Among system factors, the main focus of the participants was on the TOR, and according to the interview they believe that the vehicle prompts the driver to take over when the automation system is on the verge of failure, which is an important factor in driving safety and should get more accurate and empirical exploration. And the instant vehicle speed is directly related to the accident severity after a take-over failure. The way to issue TOR is the same as TOR time, which affects whether take-over can proceed smoothly.

In addition, the questionnaire and interview also revealed that driving scenario factors are generally considered to be less relevant to take-over performance, which is contrary to the fact in manual driving that drivers often complain about poor road conditions and confusion of pedestrians and non-motorized vehicles. According to the interviews, participants believed that after the vehicle was automated, the system should undertake the most of monitoring and analysis tasks of the complex road conditions outside the vehicle, and they have confidence in the system's ability to accomplish these tasks. In the correlation analysis, it also found that older drivers don't care much about the stability of vehicle system in take-over, and those with rich driving experience don't care much about the interior space layout of the vehicle. This may indicate that mature drivers are more familiar with different types of vehicle interior, and can quickly understand the layout of the display and control components in the automated vehicle. Through the interviews, it was learned that when the automated vehicle is in danger, they may be more willing to rely on their own control capabilities rather than the vehicle system and its components. This attitude may lead to the disuse of some automation functions during automated driving, which is a waste of automation, and may reduce the efficiency or safety of human-vehicle cooperation.

## 5.2  Analysis of the Root Cause

The status of the drivers in the simulated driving video and opinions of the participants in this study can actually reveal a possible HE in take-over of automated driving: Long-term monitoring leads to driving fatigue, resulting in Interactive errors in emergency situations, leading to take-over failure. In addition to excessive monitoring by the user, these errors can also be attributed to the display and controller design confusion. Therefore, this study puts forward cognitive errors and control errors as the conditions for the point of take-over failure, and puts forward the possible cause factors, as shown in Fig. 9.

Following this logic, more possible errors and the cause factors that led to take-over failure can be proposed. However, similar to the factors proposed in this paper, they can eventually fall into the human machine interface (HMI) design of automated vehicles in some way. In addition, it is the interface that information exchange between vehicle and human that is the biggest hole proposed in the barrier analysis of this study. A qualified HMI enables people to recognize the intentions and limitations of automation, and can improve driver performance and reduce the human-out-of-loop problems [22, 23]. Therefore, in automated driving where the drivers' mental workload and situation awareness generally decrease, thus requiring more reaction time [24], the barriers that can be repaired and perfected may be may be each interface where human-computer interaction occurs, such as presentation of system performance, interior layout and information design.

Returning to the purpose of human error analysis itself, it can be found HEs are usually caused by inherent defects in human cognition and bad system interaction. In order to eliminate HEs, human factor experts will review the accident cases to discover potential problems that conceal various bad operations or interactions through human error analysis. As a result, organization managers, system developers, interaction designers and everyone related to the operation of the system can propose effective improvement measures from their own perspectives. Through analysis of HEs in this study, the cause of the take-over failure and the specific information of the failure barrier can be obtained, also appropriate improvement measures and suggestions can be proposed to ensure the safe operation of the system and to avoid the occurrence of similar HE events: (1) Establish a complete take-over training and safety specifications to reduce the output of drivers' false long-term memory; (2) Optimize vehicle man-machine interface. improve driver's perception of vehicle status and road environment status to ensure reasonable matching of human, machines and environmental systems; (3) Standardize the scope and mode of non-motorized vehicles and pedestrian activities on roads that support automated driving, and improve road safety protection facilities to accommodate automated driving.

## 5.3  Human-Machine Model in Take-Over

Through the analysis of factors and root cause of take-over, it can be found that road traffic system is a complex system that integrates elements of people, vehicles and road. The driver has the ability to think and sum up experience and can continuously improve his subjectivity, whether it is automated driving or manual driving. Drivers, as the core factors of the system, play a leading role in the coordination and control of these elements of the system. Therefore, the human cognition-response model in take-over of automated driving should be human-centred and based on the general human-machine cognition model.

When take-over in automated vehicle occurs, the driver's cognitive-response behaviour can be divided into three stages: (1) perception: probe TORs and identify vehicle conditions; (2) judgment: filter, analyse and process the obtained information, and make the corresponding decision; (3) response: implement the corresponding response behaviour. Any error in this process will lead to HE and traffic accident. Through the analysis of the driver's cognitive response stage and the influencing factors of the automated driving take-over, cognition-corresponding model of automated driving take-over can be established, as shown in Fig. 10. This model is divided into three parts. During the perception stage, external stimuli are received through the eyes, ears and other senses. The external stimuli include vehicle factors and scenario factors. During the judgment stage, perceptual memory is transmitted to the central nervous system through sensory, converted into working memory, supplemented by long-term memory generated by automated driving training, so that drivers can identify, judge, and decide. In the operation stage, the information after judgment and decision is transmitted to the human operating organ, and the actions such as adjusting take-over posture, controlling the direction, and controlling the speed are generated. The model shows that take-over is a cycle of constant adjustment and correction, in which the causes of human error should not be analysed in isolation, but discussed from a multi-dimensional perspective of space and time, and explored through empirical research.

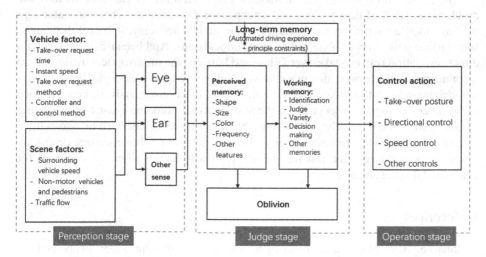

**Fig. 10.** Cognition-corresponding model of automated driving take-over.

## 6  Conclusion

In the transition from automated driving to manual driving, the driver switches from passenger to driver, and the cognitive and operational loads are all abrupt, which may cause driver's perceived failure of external stimulus. Together with the driver's long Non-driving related tasks, it could inevitably lead to HEs and possibly accidents. This study tried to map this general process into a new assessment for take-over in automated driving through Analysis of The Root Cause of take-over. Studies conducted using this methodology produce plausible and valuable results for further field take-over studies. It was indicated that human-related factors are the most important factors in take-over control, but users who are accustomed to manual driving do not seem to fully understand the characteristics of automated driving and overestimate the workload of people during automated driving. The reported correlation analysis moreover reveals differences in the attitudes of different types of drivers towards automated driving systems, and emphasizes the need to take the user's willingness or confidence to use the automation system into greater account in future research. According to the human factors events diagram established in this study, we recommend that increase the barrier of the system to reduce errors rather than increase the driver's operational responsibility, because the quality of take-over is largely independent of the execution speed of actions, but is determined by the drivers' cognitive processing of the situation. This study further proposes a cognitive-corresponding model of take-over by comparing human cognitive processes, which includes three continuous cognitive stages of perception, judgment, and operation. This makes it easier to understand the fact that take-over can be seen as a cycle of constant adjustment and correction. Human error may occur at any link in this cycle, and improvement measures can also be involved in any link, thus affecting the entire process of take-over.

The present study has several limitations. The overall survey and analysis have the problems of Lack of data volume and unobjective data sources. Therefore, although we have used a user interview to calibrate the study's accuracy, it is still difficult to summarized with very absolute and accurate conclusions. And because of the lack of a direct correlation between take-over failure and barriers, the improvement measures and recommendations are presented in a relatively subjective and broad style. Furthermore, non-driving-related tasks should be considered in human error analysis, because once the event chronological order changes, the possible conditions and cause factors leading to failure will change greatly. At this stage, however, it seems clear that human-vehicle information exchanges, rather than single human factors or vehicle factors, determine take-over performance, and this insight is expected to have design consequences for automated driving research.

# References

1. Stanton, N.A., Marsden, P.: From fly-by-wire to drive-by-wire: safety implications of automation in vehicles. Saf. Sci. **24**(1), 35–49 (1996)
2. Standard, SAE International. Taxonomy and Definitions for Terms Related to Driving Automation Systems for On-Road Motor Vehicles. J3016 (2018)
3. Zhang, B., De Winter, J., Varotto, S., Happee, R., Martens, M.: Determinants of take-over time from automated driving: a meta-analysis of 129 studies. Transp. Res. **64F**, 285–307 (2019)
4. Gold, C., Happee, R., Bengler, K.: Modeling take-over performance in level 3 conditionally automated vehicles. Accident Analysis & Prevention, 116. (2017)

5. Mccall, R., Mcgee, F., Meschtcherjakov, A., Louveton, N., Engel, T.: Towards a taxonomy of autonomous vehicle handover situations. In: Automotive'UI 16. ACM (2016)
6. Winter, J.C.F.D., Happee, R., Martens, M.H., Stanton, N.A.: Effects of adaptive cruise control and highly automated driving on workload and situation awareness: a review of the empirical evidence. Transp. Res. Part F: Psychol. Behav. **27**, 196–217 (2014)
7. Onnasch, L., Wickens, C.D., Li, H., et al.: Human performance consequences of stages and levels of automation an integrated meta-analysis. Hum. Factors **56**(3), 476 (2014)
8. Flemisch, F., Kelsch, J., Lo¨per, C., Schieben, A., Schindler, J., Heesen, M.: Cooperative control and active interfaces for vehicle assistance and automation. In: Proceedings of FISITA Automotive World Congress, Munich, F2008-02-045, VDI-FVT (2008)
9. Choi, D., Sato, T., Ando, T., Abe, T., Kitazaki, S.: Effects of cognitive and visual loads on driving performance after take-over request (tor) in automated driving. Appl. Ergon. **85**, 103074 (2020)
10. Merat, N., Jamson, A.H., Lai, F.C.H., Daly, M., Carsten, O.M.J.: Transition to manual: driver behaviour when resuming control from a highly automated vehicle - sciencedirect. Transport. Res. F: Traffic Psychol. Behav. **27**(26), 274–282 (2014)
11. Norman, D.A.: Design rules based on analyses of human error. Commun. ACM **26**(4), 254–258 (1983)
12. Mussulman, L., White, D.: The human factors analysis and classification system (hfacs). Approach the Naval Safety Centers Aviation Magazine (2004)
13. Mosleh, A., Groth, K.: A performance shaping factors causal model for nuclear power plant human reliability analysis. In: Proceedings of the 10th International Conference on Probabilistic Safety Assessment and Management (PSAM-10) (2010)
14. Kim, J.N.: The development of K-HPES: a Korean-version human performance enhancement system [for nuclear power plant control]. In: IEEE Sixth Conference on Human Factors & Power Plants, Global Perspectives of Human Factors in Power Generation. IEEE (1997)
15. Stanton, N.A., Salmon, P.M.: Human error taxonomies applied to driving: a generic driver error taxonomy and its implications for intelligent transport systems. Saf. Sci. **47**(2), 227–237 (2009)
16. Swain, A.D.: Handbook of human reliability analysis with emphasis on nuclear power plant applications. NUREG/CR-1278 (1980)
17. https://www.youtube.com/watch?v=C3DbrYx-SN4. Accessed 07 Oct 2020
18. https://v.qq.com/x/page/k3047km0b4r.html. Accessed 21 Dec 2020
19. Johnson, V., White, H.R.: An investigation of factors related to intoxicated driving behaviors among youth. J. Stud. Alcohol **50**(4), 320–330 (1989)
20. Atchley, P., Chan, M.: Potential benefits and costs of concurrent task engagement to maintain vigilance: a driving simulator investigation. Hum. Factors J. Hum. Factors Ergon. Soc. **53**(1), 3–12 (2011)
21. Crawford, A.: Fatigue and driving. Ergonomics **4**(2), 143–154 (1961)
22. Kaber, D.B., Perry, C.M., Segall, N., Mcclernon, C.K., Iii, L.J.P.: Situation awareness implications of adaptive automation for information processing in an air traffic control-related task. Int. J. Ind. Ergon. **36**(5), 447–462 (2006)
23. Inagaki, T.: Design of human–machine interactions in light of domain-dependence of human-centered automation. Cogn. Technol. Work **8**(3), 161–167 (2006)
24. Wu, Y., Kihara, K., Hasegawa, K., Takeda, Y., Kitazaki, S.: Age-related differences in effects of non-driving related tasks on takeover performance in automated driving. J. Saf. Res. **72**, 231–238 (2020)

# Human-Computer Collaborative Interaction Design of Intelligent Vehicle—A Case Study of HMI of Adaptive Cruise Control

Yujia Liu[1,2], Jun Zhang[1,2], Yang Li[3], Preben Hansen[4], and Jianmin Wang[2(✉)]

[1] College of Design and Innovation, Tongji University, Shanghai, China
{liuyujia,alexmaya}@tongji.edu.cn
[2] Car Interaction Design Lab, College of Art and Media, Tongji University, Shanghai, China
wangjianmian@tongji.edu.cn
[3] Karlsruher Institut of Technology, Baden-Württemberg, Germany
yang.li@kit.edu
[4] The Department of Computer and Systems Sciences,
Stockholm University, Stockholm, Sweden
preben@dsv.su.se

**Abstract.** With the rapid development of the intelligent vehicle industry, in order to improve the efficiency and usability of systems, the interface design of the Human-Computer Interaction of intelligent vehicles has received great attention. In this article, firstly, combined with the domestic and foreign taxonomy of automated driving systems and collaborative design concepts, a framework of human-computer collaborative interaction (Human-Engaged Automated Driving, HEAD) is proposed. Secondly, we discussed the engagement of people and intelligent vehicle at different levels. We analyzed the interaction flow chart under the Human-Computer Collaboration, established the information architecture of Human-Computer Engagement, and conducted the design practice on the Human-Machine Interface. Finally, a case study of the HMI design of Adaptive Cruise Control was performed to validate the HEAD framework. SUS usability test was performed together with an experiment evaluating interface elements in two interface designs: Original UI design and Iterative UI design. This also including a Likert scale questionnaire. The results show that HEAD can guide and improve the Human-Machine Interface design to enhance the efficiency of Human-Computer Interaction and user experience.

**Keywords:** Human-Computer Collaboration · Interaction Design · Human-Machine Interface · Adaptive Cruise Control · Intelligent vehicle

## 1 Introduction

With the rise of technologies such as Deep Learning, Machine Learning, and Cloud Computing based on big data, Artificial Intelligence (AI) [1] has promoted automated driving as an essential area of research, technology and economic development. The

H. Krömker (Ed.): HCII 2021, LNCS 12791, pp. 296–314, 2021.
https://doi.org/10.1007/978-3-030-78358-7_21

rapid development of intelligent vehicle has not only subverted the market pattern of the traditional automotive industry in the past but has also profoundly changed the relationship between people and vehicles. People think and behave in their own way. After computers predict people's thoughts and behaviors [2], they need to feedback to people so that people can make correct predictions about the next intentions and behaviors of the intelligent vehicle. The relationship between human and vehicle has moved from "supply and demand" to "collaboration", so how to better collaborate and interact with human and vehicle has become an urgent problem to be solved.

In the contemporary era of the rapid development of intelligent vehicle, this article focuses on how to better handle the interaction between human and intelligent vehicle, while improving the safety and work efficiency of intelligent driving, while also improving user experience. For the classification of automated driving technology, we proposed a Human-Computer interaction (HCI) framework for intelligent vehicle to guide the design of Human-Machine Interface (HMI); secondly, we discussed the engagement of human and intelligent vehicle at different stages; then, take the Adaptive Cruise Control (ACC) [3] function as an example, we analyze the cut-in scenario design, and establish an information architecture. In addition, based on the analysis of information exchange, we redesigned the Human-Machine Interface. Finally, usability test and interface elements related questionnaire (Likert scale from 1–5) were did in the experiment.

## 2 Related Work

### 2.1 Taxonomy of Automated Driving Systems

Intelligent vehicles are equipped with advanced onboard sensors, controllers, actuators, and other devices, as well as modern communication and network technology, to realize the exchange and sharing of intelligent information between the car and X (cars, roads, people, clouds, etc.). And its function of environmental perception, intelligent decision-making, collaborative control, and other functions realize people's desire for safe, efficient, comfortable, and Energy-Saving driving. In this context, SAE International announced in 2014 the Taxonomy of automated driving technology (SAE J3016 standard). From the perspective of the industry, Auto-pilot: In fact, automated driving is a gradual process that will eventually develop into autonomous driving. At present, the automated driving of major car companies is more to add a variety of intelligent assistance systems, so that the vehicle can complete a certain single "Automated Driving" actions such as changing lanes and overtaking, thereby becoming more intelligent. In the SAE J3016 standard, the process of the vehicle from full manual operation to full automation has mainly gone through six stages, including No Automation, Driver Assistance, Partial Automation, Conditional Automation, Highly Automation, and fully Automation.

The development of domestic intelligent vehicles technology takes into account the two paths of intelligence and networking. The integrated development of "Intelligence + Networking", with the ultimate goal of replacing humans with systems to achieve all driving tasks. The Chinese Society of Automotive Engineering has formulated level 5 intelligence in intelligent vehicle. In terms of intelligence, based on the current generally accepted definition of the SAE classification in the US, and considering the complexity of China's road traffic conditions [4], it is divided into Driving Assistance (DA), Partially Automated Driving (PA), and Conditional Automated Driving (CA), Highly Automated Driving (HA), Fully Automated Driving (FA) five levels, the specific content corresponding to each level can be seen in Table 1. The intelligent classification guides the specific technological path for the development of intelligent vehicle.

**Table 1.** Taxonomy of automated driving in China

| Intelligent level | Definition | Control | Supervision | Responding to failure |
|---|---|---|---|---|
| Driving Assistance (DA) | Provide support for one operation of direction and acceleration and deceleration through environmental information, and all other driving operations are operated by humans | Human and System | Human | Human |
| Partially Automated Driving (PA) | Provide support for one or more operations in direction and acceleration and deceleration through environmental information, and all other operations are operated by humans | Human and System | Human | Human |
| Conditional Automated Driving (CA) | All driving operations are completed by the unmanned driving system, and the driver needs to provide appropriate intervention according to the system request | System | System | Human |
| Highly Automated Driving (HA) | All driving operations are completed by the unmanned driving system. Under certain circumstances, the system will make a response request to the driver, and the driver can not respond to the system request | System | System | System |

*(continued)*

**Table 1.** (*continued*)

| Intelligent level | Definition | Control | Supervision | Responding to failure |
|---|---|---|---|---|
| Fully Automated Driving (FA) | The unmanned driving system can complete all the operations in the road environment that the driver can complete without the driver's intervention | System | System | System |

## 2.2 Human-Computer Collaborate Interaction in Automated Driving

Interaction refers to the communication between humans and things in the natural world [5, 6]. Its essence is the interaction and communication between the subject and the object. The subject of interaction is human-beings, and the object can be various things that can generate feedback. In the environment of intelligent vehicle, more and more scholars and R&D institutions in the field of Human-Computer Interaction have involved to study of intelligent collaborative interaction between humans and machines.

Ranney [7] proposed a Human-Machine cooperation strategy model for intelligent vehicle systems in 1994, which was decomposed from three levels of operation, tactics, and strategy. Salas [8] believe that the key to situational awareness of teamwork lies in the information interaction process of team members. Team members must have situational awareness related to their goals and surroundings, as well as other team members. Parasuraman [9] described the four stages of human information processing in the interaction between humans and automated driving: information acquisition, information analysis, decision making, and action implementation. Largillier, Ma Jun, and others applied the Analytical Hierarchy Process (AHP) to the design of the automotive Human-Machine Interface (HMI) and formulated the standards for constructing levels. Ren Xiangshi [10] integrated the organic collaborative interaction ideas of the Eastern philosophical concept of "the golden mean", and proposed the framework of Human-Engaged Computing (HEC). Engagement is "a state of consciousness in which people are fully immersed in it and aligned with the activity at hand", and it is a "process in which the interactors begin, maintain, and terminate their perceptual connection with each other during the interaction" [11]. In the scope of automated driving research, engagement is a communication process of human and vehicle. They are formed in the process of achieving (shared) goals. The relationship of mutual cooperation, engagement requires the interaction between humans and intelligent systems. Ben Shneiderman [12] proposed a Human-Oriented artificial intelligence framework, clarifying that by designing High-Level human control methods and increasing the level of computer automation, human performance can be improved and the danger of excessive human or computer control can be avoided.

# 3 The framework: Human-Engaged Automated Driving (HEAD)

Many types of research [13] on Human-Computer Interaction of intelligent vehicle are based on the interface between humans and machines (Fig. 1, part a). The human driver and the intelligent vehicle are regarded as two separate roles (the lines or faces of contact). This way of thinking is actually considered from the perspectives (interests) of both human and intelligent vehicle, and there is no co-development relationship between these. The engagement concept can be substituted into the role of the user, so that the user will not be overly dependent on the system in the process of interaction, and complete the loss of the perception of the surrounding environment. The Human-Engaged Automated Driving (HEAD) framework (Fig. 1, part b) is based on the intelligent level of human and vehicle, which hierarchically integrated into the engagement design concept, revealing the new interactive methods that are changing, and finally forming a collaborative interaction design framework.

Among them, different levels of automated driving intelligence are matched with human drivers who have different levels of engagement [14] in artificial intelligence, resulting in five stages. They are (Fig. 1, part b): Full Human, Full Automation, Driver Assistance, Human Supervision, and Collaboration Driving.

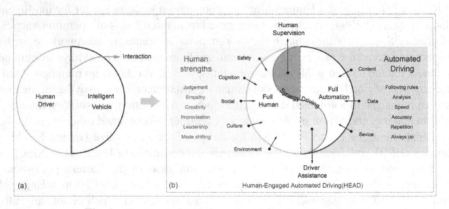

**Fig. 1.** The theoretical inference and framework of HEAD

## 3.1 Full Human

Full Human focuses on driving issues entirely from the driver-human perspective. The human driver treats the car as just a mechanical product and deals with driving issues based on their cognition in the current environment. Human is still an advanced complex creature. When dealing with driving issues, it will consider the impact of many factors such as "safety", "cognition", "society", "culture" and "environment". Human intelligence is not the product of informatization and automation. Humans are wise. Because human beings have the above elements as one of their considerations, this become the core and driver of intelligence. Such as society, it is a normative relationship of thoughts and behaviors that are commonly recognized between human and human, and between human and society. It involves the classification, analysis, and decomposition of facts in the group, and relates issues of value such as security, privacy, prejudice, substitution, and inequality.

The reason why human wisdom is smarter than intelligence [10] is that human beings may foresee things, reversals, and something unexpected that these intelligences cannot tell. Under the wisdom of man, man has their own unique abilities: judgment, empathy, creativity, improvisation, leadership, and mode shifting are all reasons why humans become the spirit of all things.

## 3.2 Full Automation

Full Automation focuses on the driving problem from the intelligence of automated vehicle. The machine has its own outstanding capabilities: data processing capabilities, data processing speed, data repeatability, and data analysis capabilities. To become intelligent, the three aspects of "content", "data" and "service" are inseparable [15]. The content is produced by a single person. The content of a single person is beneficial to accurately meet the personalized needs, but it does not contribute much to the needs of most people. Therefore, data should be collected from groups of people. Furthermore, the accumulation data result in patterns and tendencies of a group, which then may determine what kind of content and services will be generated in the intelligent future.

The inherent capabilities of automated driving are: compliance, analysis, speed, accuracy, repetition and consistency [16]. Because it uses mathematical logic as the basic thinking logic, it can certainly be used as an extension of some human abilities. For example, accuracy. automated driving can make analysis and behavioral responses in an extremely short time, avoiding the risks brought by many human carelessness or fluke.

## 3.3 Driver Assistance

Driver Assistance focuses on taking the intelligent system engaged with conventional vehicles. Automated driving system acts as an assistant for human drivers, reducing the information-intensive, repetitive or tedious tasks. Due to sensors and cameras on vehicles can make up the limitations of human perception organs, and become an extension of human physiological structure capabilities. The intelligent system of vehicle can detect the surrounding environment and communicates risks and deliver the unknown factors to the driver. At the same time, it helps humans handle the conversion of some information, but ultimately humans make decisions. In addition, in scenarios where the machine cannot be discriminated against or when people do not trust in the machine, they can be taken over by the person, who has the absolute control right of driving the vehicle.

## 3.4 Human Supervision

Human Supervision focuses on engaging human supervision with intelligent vehicle in order to accomplish the driving tasks. Intelligent vehicle of artificial intelligence have replaced human driving work, and information systems in the network constantly perform tasks and feed them back for review and supervision. Humans can be liberated from manual labour.

But the judgment of artificial intelligence sometimes conflicts with the human beings', because artificial intelligence considers more to achieve the "optimal solution" to complete the task, and even to reach the best path, try to figure out human intentions and make choices for humans. Moreover, machines do not have the same deep moral and ethical considerations as humans, so when artificial intelligence and human opinions are not unified, humans retain the highest decision-making power [16].

### 3.5 Collaboration Driving

Collaboration Driving represents an optimal balance of the desired Human-Computing (Vehicle) interaction. The essence of artificial intelligence is human-like, and Synergy driving is to put human wisdom and human thinking logic into artificial intelligence algorithms [17]. This is not to recreate the intelligence of a machine, but to construct the intelligence of autonomous driving in accordance with human logic. In the future, intelligent communication needs to construct data that serves human values and contents in the human-machine medium and the environment. These data can meet the ethics and judgment of the intelligent medium in society.

Then the wisdom of humans and intelligence of automated driving are engaged in driving activities, which is mutually beneficial. In this balanced state, the intelligent system of autonomous driving is not only used to complete certain envisioned tasks, but also to improve the boundary of human collective cognition and humanity by improving human capabilities and potential.

## 4 Case

How to make the interaction between human and vehicle more efficient and harmonious with the help of intelligent driving assistance systems has always been an issue being discussed in the industry [16]. This article proposes a framework for Human-Machine collaborative driving, discussing the engagement of human and intelligent vehicle in different driving statuses, and analyzes the design requirements of HMI. In this case, the driving scenario of ACC function (there is other vehicle cutting into the main lane from other lanes and driving in front of it) as an example to analyze the interaction flow chart under the Human-Machine collaboration and establish an information architecture, and finally carry out the prototype and interface design, expounding how to practice engagement Human-Machine collaborative interactive interface design.

### 4.1 Application Scenario Design

Adaptive Cruise Control (ACC) [2] is a driver assistance system that controls the longitudinal movement of the vehicle. It is one of the main functions required for automated driving vehicle and can reduce the workload for the driver. However, many interface designs for the ACC function cause people to be confused about the automated driving vehicle when they are used, and it will increase the workload effort and will not achieve cooperative driving. Good interface design can improve driving efficiency and at the same time promote the mutual symbiosis of both human drivers and automated driving. This case hopes to use the driving scenario under the ACC function as an example to establish an information architecture with Human-Computer engagement to help designer proto-type and interface design and iterate, and to verify that HEAD has a guiding role in the interaction of human and automated driving.

In the research on the ACC function [2], the vehicle cut-in is a typical application scenario. In this case, the driving scene is divided into two stages, with the front wheel on the right side of the auxiliary test vehicle (F) passing the lane line as the dividing point, the first stage before passing, and the second stage after passing. The first stage includes two statuses: (1) cruise, original set up and (2) cut-in A, and the second stage also includes two statuses: (1) cut-in B and (2) Following. In Cruise status, the driver sets the ACC to set the vehicle speed; in cut-in A status, the F car starts to overtake and starts preparing to cut into the main test lane; in cut-in B status, the F car starts to cut into the current lane; in Following status, the S car follows the F car at the speed of the F car. The status description of two vehicles in two phases is shown in Fig. 2.

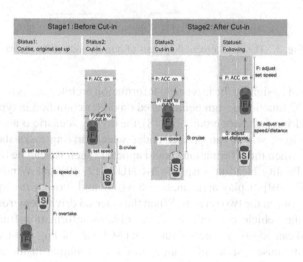

**Fig. 2.** The cut-in scenario design of ACC function

## 4.2 Information Architecture Design

In the cut-in scenario of ACC function, a flowchart of the human driver and the ACC system engagement parts are established as shown in Fig. 3. It analyzes the driver's use of ACC in different road scenarios involving physical operations and cognitive states in order to clarify the parts that human drivers and automated driving are engaged in. This may reduce the potential dangers caused by the driver's ignorance of the driving environment and the vehicle's ACC function itself ability and also increase the trust of humans for automated driving, and even improve the drivers' cognition abilities.

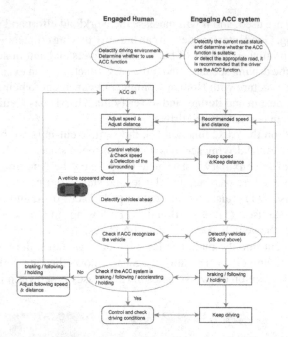

**Fig. 3.** The engagement flowchart of human and ACC system

The purpose of designing the in-vehicle information architecture is to find more deep and accurate ACC function design points based on driver cognition in typical scenarios. Combined with driver behavior prediction [18], the driving scenario is analyzed in stages and role dimensions, which can predict under which driving tasks the driver needs interface's help (when the information should appear). Then we can be displayed based on Augmented Reality Head-up display (AR-HUD) [2, 19] and Windshield head-up display (W-HUD) [19] display area, analyze the time and form of the information and the interaction between the two screens. When the external driving environment changes (such as when the vehicle cuts in), the driver obtains information from the external environment and can adjust the internal function (ACC) according to the current driving scene, which will cause a series of changes in the information inside and outside the vehicle. This case is to analyze the information architecture design displayed from the time sequence and the dimensions of different engagement roles. In terms of design, it is presented as the relationship between the AR-HUD and WHUD information elements [19].

Combining the driver's visual glance analysis with the vehicle's driving trajectory [20, 21], the driving trajectory of the vehicle under monitoring is obtained (Fig. 4). The approximate order of this trajectory [21] is left rear mirror/interior mirror—left window—front left windshield lane—Lane lane—the current lane, as shown in Fig. 7. The origin in the figure indicates the node of the vehicle's driving trajectory.

1–4 is the driver's identification range, and 4–6 is the common identification range between the driver and the vehicle. When the vehicle is in the 4–5 area, the driver and the vehicle need to make a judgment. Whether the state of the vehicle will affect the use of the

ACC function of the vehicle, and whether the driver needs to adjust the ACC set speed and set distance. Therefore, in this case, we used 4–5 as an area for design assistance to help the driver determine the track movement of the front car, enhance situational awareness, reduce driver workload, and make effective judgments and actions in a timely manner.

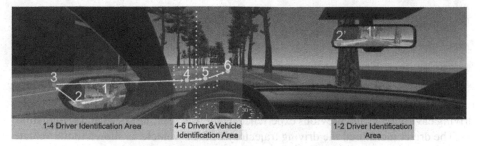

1-4 Driver Identification Area      4-6 Driver&Vehicle Identification Area      1-2 Driver Identification Area

**Fig. 4.** Driving trajectory of synergy driving for engagement

In order to more intuitively represent the changes in information architecture, icons [22] and symbols are used to express various types of information as shown in the Table 2.

**Table 2.** Correspondence table of information and graphic symbols

| Information | External Information Icon | Internal Information Icon |
|---|---|---|
| Set Speed | □ | □ |
| Set Distance | ○ | ○ |
| Detection Information | △ | △ |

Based on the four statuses of this scenario, the flowchart of the in-car information architecture design is shown in Fig. 5. It includes four parts:

(1)  driving scene analysis: analysis of the trajectory of the auxiliary vehicle (representing the timeline of the information structure in the vehicle);
(2)  four stages analysis diagram in cut-in scenario;
(3)  Driver's vision diagram;
(4)  HUD information layout and linkage information design prototype on W-HUD and AR-HUD.

**Status 1: Cruise, Original Set Up.** When the F car is driving behind the S car, this status is the first stage of the cut-in driving scenario. After the test vehicle reached a certain speed, the ACC function was turned on. At this time, the S vehicle was in cruise

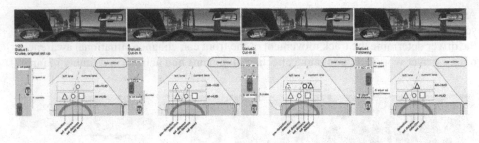

**Fig. 5.** The flowchart of information architecture design

mode. The driver made the initial settings of the ACC function based on the current driving environment and previous experience using ACC: setting speed and setting time.

The driver monitored the driving trajectory of the F vehicle as shown in the vehicle driving trajectory point 1–3. Dotted lines indicate related changes in the same information. In status one, AR-HUD only displays the time interval information, which indicates to the subjects the actual distance of the time interval bar on the road; W-HUD displays the set speed, time interval and identification information at this time.

**Status 2: Cut-in A.** At this status, the front car is driving on the left lane. The driver needs to determine whether the left side vehicle is overtaking or changing lanes, whether it will affect the use of the ACC function, and whether it is necessary to adjust the set speed and set time interval. At this stage, the driver monitors the driving trajectory of the F vehicle as shown in the vehicle driving trajectory point 4. At this status, the S car is still in cruise mode. The vehicle needs to assist the driver to predict the driving trajectory of the vehicle in front, to inform the driver of the vehicle's working status, whether the front car is recognized, and so on. The recognition at this status is a pre-detection state, which only informs the driver that the vehicle has recognized the F vehicle, but has not switched to the following mode.

**Status 3: Cut-in B.** At this status, the F car travels across the lane line to the front of the S car, and the driver monitors the driving trajectory of the F car as shown by the vehicle trajectory point 5. At this time, the S car has switched from the cruise status to the following status, and the AR-HUD will continue to display information of identification, and the identification information changes from pre-detection to detection, which enhances the prompt of ACC status change.

**Status 4: Following.** At this status, the F car is driving in front of the S car. The driver monitors the driving trajectory of the F car as shown in the vehicle trajectory node 6. The driver can adjust the set speed and set time interval according to the driving scene at this time, and the detection information is continuously displayed to indicate that the ACC function is in a continuous working state.

### 4.3  Interface Design

Many vehicles with automated driving functions are equipped with the ACC function and have a corresponding design interface. The basic information includes the ACC set speed, set time interval, and vehicle identification information. The ACC information is mainly displayed on the instrument panel. Although the design performance of ACC varies among different car manufacturers, a relatively stable design plan has been formed. Based on the previous work of the CarLxD laboratory at Tongji University, China [3, 17], we integrated the same ACC information on the driving simulator, which was displayed on the instrument and the head-up display, as shown in part a of Fig. 6.

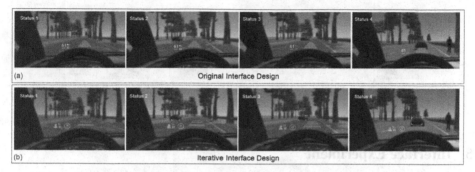

**Fig. 6.** Original interface design and iterative interface design

Based on the combing of the information architecture in the previous introduction, the original interface design is iterated (Fig. 6, part b). In the iterative design scheme, the real-time speed information is outlined by hollow circles, which makes the speed information more visible, and distinguishes the real-time speed from the ACC set speed. For the "spot the front vehicle" element, when the front vehicle has recognized the front vehicle, the vehicle icon changes from white to blue; in AR-HUD, a "pre-detection" state for the front vehicle is added, and the "blue semi-curved base" was iterated into a more obvious "marquee" form. When the vehicle recognizes the vehicle in the front or next lane, the detection information is white, and a white checkbox appears. When the vehicle changes from cruise mode to following mode and the detection information is blue, a blue checkbox appears. For the "set distance" element, reducing the five-time intervals to three-time intervals can more clearly visually determine the distance change caused by the time interval increase and decrease. In the AR-HUD, the judgment of the ACC function mode is added. When the vehicle is in cruise mode, the time interval rectangle is hollow; when the vehicle is in follow mode, the time interval rectangle is solid. The rectangular display changes the distance following the car in front of it in real-time. Visually, compared to the full solid design of the original design, the iterative design weakens the rectangular display, enhances transparency, and increases the gradient effect. At the same time, the judgment of the ACC function mode has been added to the "detection information" and "time interval" design elements. The design elements comparison of the two schemes is shown in Fig. 7.

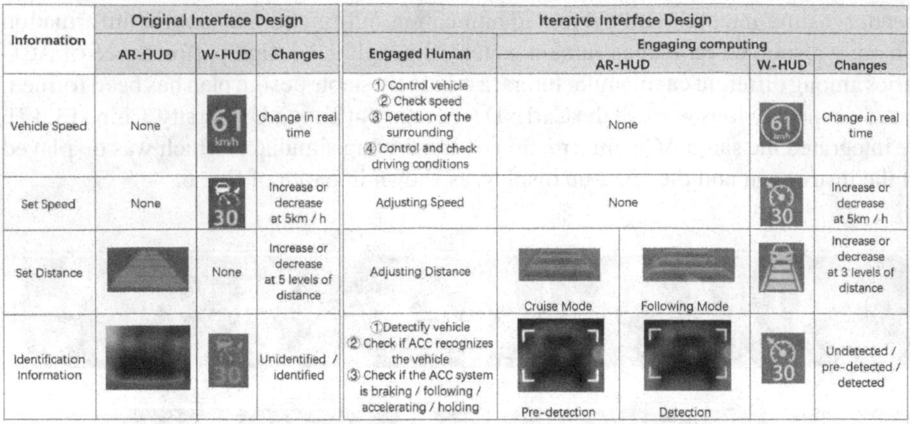

| Information | Original Interface Design | | | Iterative Interface Design | | | |
|---|---|---|---|---|---|---|---|
| | AR-HUD | W-HUD | Changes | Engaged Human | Engaging computing | | |
| | | | | | AR-HUD | W-HUD | Changes |
| Vehicle Speed | None | 61 km/h | Change in real time | ① Control vehicle ② Check speed ③ Detection of the surrounding ④ Control and check driving conditions | None | 61 km/h | Change in real time |
| Set Speed | None | 30 | Increase or decrease at 5km / h | Adjusting Speed | None | 30 | Increase or decrease at 5km / h |
| Set Distance | | None | Increase or decrease at 5 levels of distance | Adjusting Distance | Cruise Mode | Following Mode | Increase or decrease at 3 levels of distance |
| Identification Information | | 30 | Unidentified / identified | ① Detectify vehicle ② Check if ACC recognizes the vehicle ③ Check if the ACC system is braking / following / accelerating / holding | Pre-detection | Detection | 30 | Undetected / pre-detected / detected |

Fig. 7. The design elements comparison of original and iterative

# 5 Interface Experiment

In this experiment, the head-up display design interface of the ACC function was used as the Original Interface Design group. It was compared with the Iterative Interface Design group. The experiment was performed on a driving simulator to analyze the results. Design element questionnaires was used to analyze whether the information architecture and interface design can guide design and improve driver safety when using ACC functions.

## 5.1 Experimental Design

The experiment was conducted in a within-subjects design with one independent variable comparing the Original and Iterative interface design. Participants were all trained in two kinds of conditions above. Conditions were counterbalanced to minimize the sequence effect.

## 5.2 Participants

Ten university students and staff members (6 females) aged 22–30 (M = 25.9, SD = 2.47) were recruited. Most of the participants have automotive user interface design experience, they can be considered to have a certain ability to judge the interface in the experiment.

## 5.3 Apparatus

The experiment was completed on a driving simulator. The driving simulator provides a very realistic driving environment. The driver can fully realize vehicle driving control through the steering wheel and pedals, and can interact with other vehicles on the road. The hardware part of the experimental device includes two sets of Logitech G29 (including a steering wheel and three-foot pedals), two computers and six monitors, and so on. The experimental device software part contains Unity 5.0 and Unity related plug-ins that can simulate the real driving environment. The test measurement tool contains the participant information sheet, the introduction of experiment purpose and process, AR-HUD and W-HUD introductions and example pictures, ACC function introduction videos and user manuals, and questionnaires about relevant interface design elements.

### 5.4 Task and Procedure

Participants were asked to sign a letter of consent and information sheet. Later they were introduced to the procedure of the study. Participants were randomly assigned to either the Original or Iterative groups. Conditions were counterbalanced to minimize the sequence effect. Participants were instructed in the use of ACC. Participants were then allowed to perform auto-driving and manual driving exercises on the driving simulator (to avoid learning effects, the driving scenario during the practice was different from the driving scenario used in the actual experiments), and each mode was tested for at least 15 min. After getting familiar with the use of the simulator and the use of the ACC function settings, the formal experiment began. Test scenario were not revealed for the participants prior to the experimental tasks. When the subjects were ready to test, the tester issued mission instructions to the subjects. At the end of the experiment, the tester dictated the Usability Scale (SUS) [23], design element scoring and interview scales. After completing the filling, they continued to the next experiment. The test drive time is about 2 to 5 min, the time to perform each driving task is about 15 to 20 s, and the entire experimental process lasts about 50 to 60 min.

### 5.5 Measures

Questionnaires [24] are considered a simple and economical way to obtain and analyze data. In this experiment, this questionnaire combining closed questions (rated on a 5 points Likert-scale from 1 to 5) and open questions to obtain both quantitative data and qualitative data. The questionnaire is based on the design elements of "Set Speed", "Detection Information", "Set Distance", and "Mode Identification". It lists a total of 16 questions to test drivers' evaluation for Original and Iterative interface design.

### 5.6 Results

Data were checked using the Kolmogorov-Smirnov Test [25] and analyzed using the T-test. The comparison of SUS between the original group and the iteration group is shown in Fig. 8.

The average score of the original group was 62.3 points (SD = 6.1), which was lower than the score of the iteration group of 74.3 points (SD = 8.8), and the difference was significant (P = 0.016 < 0.05). Figure 9 summarizes quantitative data results.

**Set Speed.** In the Original group, two participants indicated that they did not see the symbol of set speed. In the acquisition of the set speed element information, compared with the Original group (M = 1.90, SD = 0.57), the mean value of the set speed in the

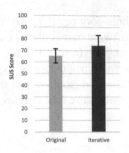

**Fig. 8.** Comparison chart of SUS system availability scores under different groups

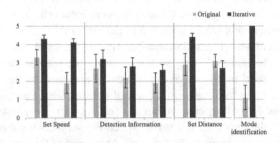

**Fig. 9.** Quantitative score results in questionnaire

iterative design scheme is significantly higher (M = 4.11, SD = 0.20). Simple main effects analysis showed a significant difference (P = 0.012 < 0.05).

**Detection Information.** The mean value of the Iterative group (M = 3.20, SD = 0.49; M = 2.80, SD = 0.47; M = 2.60, SD = 0.31) for the "three closed questions of "whether the front car was recognized", "the influence of the front car on the use of the ACC", and "the degree of attracting attention" are higher than the original group (M = 2.7, SD = 0.76; M = 2.20, SD = 0.57; M = 1.90, SD = 0.55), but it is not very significant (P = 0.557; P = 0.279; P = 0.354). In the Original group, the scores of the participants were much more discrete than Iterative, indicating that the subjects had a large difference in their cognition of Original. Regarding the question "Impact on the attention of driving information", the Iterative group may give participants provided more information and thus increased their impact on subjects' attention.

The results of the open question indicate that in the Original group, 3 of the 10 participants did not perceive the identification information. In the Iterative group, all the participants felt the detection symbol. [P3]: "In the process of judging the track of the preceding car, the decision time is not enough. The white box icon can increase the driver's decision time." [P8]: "The white box indicates that the front car and the own car are in different lanes; the blue box indicates that the front car and the own car are in the same lane and it can lock with the car target." However, there are 2 participants tended to the Original. [P5]: "The blue color is integrated with the car, will not cause too much distraction for the driver, but feel that the difference of detection status is very necessary."

**Set Distance.** Regarding the question "Judge the distance from the front car according to the blue rectangle", the mean value of Iterative (M = 4.40, SD = 0.2) is much higher than Original (M = 2.90, SD = 0.6). But the difference is not significant (P = 0.153 > 0.05). Regarding the question "The influence of the blue rectangle on the attention", The average score of Original (M = 3.10, SD = 0.35) and iteration (M = 2.70, SD = 0.4) is basically the same (P = 0.42), which shows that although the iterative design scheme increases the recognition status of the time rectangle, it does not cause much attention of the driver.

In the open question about "hollow rectangle and solid rectangle", one participant did not notice the change from "hollow rectangle" to "solid rectangle". Regarding the question "Design scheme that tends to display synchronously with time-distance rectangles or different display", 4 participants agreed design should have a differentiated display. [P1]: "Different from AR-HUD's dynamic time-distance display method, W-HUD gave the driver an iconic information prompt driver know their original settings." [P9]: "W-HUD and AR-HUD are better expressed in a consistent manner, and it is not easy to cause misunderstanding and confusion."; the remaining participants indicated that they have no special influence.

**Mode Identification.** The results of the Iterative (M = 5.00, SD = 0.00) are significantly better than the Original (M = 1.10, SD = 0.66), and the difference is significant (P = 0.0002 < 0.05), all participants can judge the mode of ACC function at that time by Iterative interface design.

The results of the open question indicate that in the original design scheme, the participants' perception of the mode of the ACC function in the driving scene at that time was quite confusing. During the car, the following target was lost due to lane departure, and it returned to the cruise mode.

## 5.7 Discussion

The SUS results show that the participants in the experiment believe that the interface design of the iteration group (74.3 points > 62.3 points) is more usable.

From the questionnaire of design elements, the average score of "**set speed**" in Iterative group is higher than 3 points, which illustrates that this design can provide the driver the information more effectively than in original group. The "**detection information**" in Iterative group promotes the driver with evaluating the forward vehicle state more effectively. Besides that, it has a positive impact on using the ACC function. However, except the question "whether the front car was recognized", the average scores of rest questions are lower than 3 points, so this design element needs to be further improved.

**Set Distance.** The reduction of five-time intervals to three-time intervals increases the difference in each time interval, which is more conducive to the driver's judgment of the time interval. The use of solid rectangles and hollow rectangles enhances the current mode of ACC, it helps driver to estimate ACC's status. On the other hand, this design does not affect the driver's attention greatly. The Iterative interface design is superior to the original design scheme in terms of "ACC function time display", but the design method of AR-HUD and W-HUD asynchronous display still needs further design and testing.

**Mode Identification.** Regarding the judgment of the ACC functional mode, the test results of the iterative design scheme are better than the original interface design. Compared to original design, the results show that the internal design improves driver's usability and help the driver clarify the status of ACC.

In conclusion, in the design element scoring and driver interview questionnaires, the results indicate that the Iterative interface design is superior to the Original interface design in four aspects: Set Speed, Detection Information, Set Distance, and Mode identification. This experiment can prove that the interaction concept in the HEAD framework is used as a guide to build an information architecture that can be jointly engaged by humans and machines, helping designers to perform better prototypes and interface designs and iterations.

## 6   Conclusions

Based on the research of intelligent vehicle, from a Human-Computer collaborative interaction design perspective, this paper proposes a theoretical framework based on engaged Human-Computer collaborative driving. The combination of the theoretical framework and the ACC example, we analyzed the interaction flow chart under the Human-Machine Collaboration. Then we established the information architecture, discussed the interface design practices, and usability test and interface elements related questionnaire (Likert scale from 1–5) were did in the experiment. Experiment results shows that the interactive interface design guided by the Human-Machine Collaborative Driving Framework (HMCD Framework) can improve the usability of the system, allow users to understand the system more clearly through more appropriate information on the interface, and improve the collaboration efficiency of Human-Vehicle interaction.

The work of this paper focuses on proposing a new design framework for human-computer interaction (the HMCD framework), and using this to explain how to guide design practice, future work will further explore specific practices and guidelines for design methods, and improve experimental evaluation methods. The research in this paper can bring new thinking to the research on the interaction between humans and automated driving, and it also has certain inspiration and guiding significance for the design of interactive applications of automated driving.

**Funding.** This work was supported by The National Key Research and Development Program of China (No. 2018YFB1004903), The Projects funded by the National Social Science Fund (No. 19FYSB040), shanghai Automotive Industry Science and Technology Development Fundation (No. 1717).

## References

1. Liao, J., Hansen, P., Chai, C.: A framework of artificial intelligence augmented design support. Hum. Comput. Inter. **35**(5–6), 511–544 (2020)

2. Ohn-Bar, E., Trivedi, M.M.: Looking at humans in the age of self-driving and highly automated vehicles. IEEE Trans. Intell. Veh. **1**(1), 90–104 (2016)
3. Wang, J., Wang, W., Hansen, P., Li, Y., You, F.: The situation awareness and usability research of different HUD HMI design in driving while using adaptive cruise control. In: Stephanidis, C., Duffy, V.G., Streitz, N., Konomi, S., Krömker, H. (eds.) HCI International 2020 – Late Breaking Papers: Digital Human Modeling and Ergonomics, Mobility and Intelligent Environments: 22nd HCI International Conference, HCII 2020, Copenhagen, Denmark, July 19–24, 2020, Proceedings, pp. 236–248. Springer, Cham (2020). https://doi.org/10.1007/978-3-030-59987-4_17
4. Wang, J., Cai, Z., Hansen, P., Lin, Z.: Design exploration for driver in traffic conflicts between car and motorcycle. In: Stephanidis, C. (ed.) HCII 2019. CCIS, vol. 1034, pp. 404–411. Springer, Cham (2019). https://doi.org/10.1007/978-3-030-23525-3_54
5. Harper, R.H.R.: The role of HCI in the age of AI. Int. J. Hum. Comput. Inter **35**(15), 1331–1344 (2019)
6. Stephanidis, C., Salvendy, G., Antona, M., et al.: Seven HCI Grand Challenges. International Journal of Human-Computer Interaction **35**(14), 1229–1269 (2019)
7. Ranney, T.A.: Models of driving behavior: a review of their evolution. Accid. Anal. Prev. **26**(6), 733–750 (1994)
8. Eduardo, S., Dana, E., Sims, C.: Shawn, Burke, is there a "Big Five" in Teamwork? Small Group Res. **36**(5), 555–599 (2005)
9. Parasuraman, R., Sheridan, T.B., Wickens, C.D.: A model for types and levels of human interaction with automation. IEEE Trans. Syst. Man Cybern. Part A: Syst. Hum. **30**(3), 286–297 (2000)
10. Ren, X.: Rethinking the relationship between humans and computers. Computer **49**(8), 104–108 (2016)
11. Ma, X.: Towards human-engaged AI. In: Twenty-Seventh International Joint Conference on Artificial Intelligence, pp. 5682–5686 (2018)
12. Shneiderman, B.: Human-centered artificial intelligence: reliable, safe & trustworthy. Int. J. Hum. Comput. Inter. **36**(6), 495–504 (2020)
13. Biondi, F., Alvarez, I., Jeong, K.A.: Human–vehicle cooperation in automated driving: a multidisciplinary review and appraisal. Int. J. Hum. Comput. Inter. **35**(11–15), 1–15 (2019)
14. Ren, X., Silpasuwanchai, C., Cahill, J.: Human-engaged computing: the future of human-computer interaction. CCF Trans. Pervasive Comput. Inter. **1**(1), 47–68 (2019)
15. Towards a new era of big data content services. http://www.cbbr.com.cn/article/117835.html
16. Redesigning work in an age of automation. https://www.slideshare.net/planstrategic/redesigning-work-in-an-age-of-automation-125030010
17. Naujoks, F., Forster, Y., Wiedemann, K., Neukum, A.: A human-machine interface for cooperative highly automated driving. In: Stanton, N.A., Landry, S., Di Bucchianico, G., Vallicelli, A. (eds.) Advances in Human Aspects of Transportation, pp. 585–595. Springer, Cham (2017). https://doi.org/10.1007/978-3-319-41682-3_49
18. You, F., Li, Y., Schroeter, R., Friedrich, J., Wang J.: Using eye-tracking to help design HUD-based safety indicators for lane changes. In: Proceedings of the 9th International Conference (2017)
19. Cheng, C., You, F., Hansen, P., Wang, J.: Design methodologies for human-artificial systems design: an automotive AR-HUD design case study. In: Karwowski, W., Ahram, T. (eds.) IHSI 2019. AISC, vol. 903, pp. 570–575. Springer, Cham (2019). https://doi.org/10.1007/978-3-030-11051-2_86
20. You, F., Zhang, J., Wang, J., Mengting, F., Lin, Z.: The research on basic visual design of head-up display of automobile based on driving cognition. In: Stephanidis, C. (ed.) HCII 2019. CCIS, vol. 1034, pp. 412–420. Springer, Cham (2019). https://doi.org/10.1007/978-3-030-23525-3_55

21. You, F., Wang, Y., Wang, J., Zhu, X., Hansen, P.: Take-over requests analysis in conditional automated driving and driver visual research under encountering road hazard of highway. In: Nunes, I.L. (ed.) Advances in Human Factors and Systems Interaction, pp. 230–240. Springer, Cham (2018). https://doi.org/10.1007/978-3-319-60366-7_22
22. You, F., et al.: Icon design recommendations for central consoles of intelligent vehicles. In: Ahram, T., Taiar, R., Gremeaux-Bader, V., Aminian, K. (eds.) Human Interaction, Emerging Technologies and Future Applications II. IHIET 2020. Advances in Intelligent Systems and Computing, vol. 1152, pp. 285–291 Springer, Cham 2020
23. Brooke, J.: SUS - A Quick and Dirty Usability Scale. Usability Evaluation in Industry, p. 189 (1996)
24. Matell, M., Jacoby, J.: Is there an optimal number of alternatives for likert-scale items? effects of testing time and scale properties. J. Appl. Psychol. **56**, 506–509 (1972)
25. Massey, F.J.: The Kolmogorov-Smirnov test for goodness of fit. J. Am. Stat. Assoc. **46**(253), 68–78 (1951)

# Multimodal Takeover Request Displays for Semi-automated Vehicles: Focused on Spatiality and Lead Time

Harsh Sanghavi[1], Myounghoon Jeon[1], Chihab Nadri[1], Sangjin Ko[1], Jaka Sodnik[2], and Kristina Stojmenova[2(✉)]

[1] Mind Music Machine Lab, Department of Industrial and Systems Engineering, Virginia Tech, Blacksburg, VA, USA
{harshks,myounghoonjeon,cnadri,sangjinko}@vt.edu

[2] Laboratory of Information Technologies, ICT Department, Faculty of Electrical Engineering, University of Ljubljana, Ljubljana, Slovenia
{jaka.sodnik,kristina.stojmenova}@uni-lj.si

**Abstract.** To investigate the full potential of non-speech sounds, this study explored the effects of different multimodal takeover request displays in semi-automated vehicles. It used a mixed design - the visual and auditory notification lead time was within-subjects, whereas the auditory notification spatiality was between-subjects. The study was conducted in a motion-based driving simulator with 24 participants. All participants were engaged in four 9-min driving tasks in level 3 automated vehicle and simultaneously performed a non-driving related task (NDRT, online game). Each driving session contained three hazardous events with takeover request (in total 12 requests per user). The results showed that 3-s lead time evoked the fastest reaction time but caused high perceived workload and resulted in unsafe and non-comfortable maneuver. In terms of workload and maneuver, 7-s lead time showed better results than others. Auditory displays with directional information provided significantly better reaction times and reaction types. Subjective evaluation, on the other hand, did not show any significant differences between non-directional and directional displays. Additionally, the results showed that braking is a more common first reaction than steering, and that the NDRT did not influence the takeover request.

**Keywords:** Takeover · Multimodal displays · Semi-automated vehicles · Sound spatiality · Lead time

## 1 Introduction

As the autonomous vehicle technology advances through the different levels of automation, it is important to ensure a safe transition from autonomous to manual mode. Much research has been conducted in studying driver performance for takeover requests in highly autonomous vehicles [1–4]. Gold et al. [2], for example, measured driver performance for different levels of lead time.

© Springer Nature Switzerland AG 2021
H. Krömker (Ed.): HCII 2021, LNCS 12791, pp. 315–334, 2021.
https://doi.org/10.1007/978-3-030-78358-7_22

Studies have also focused on takeover and the effect of non-driving related tasks (NDRT) performed in the vehicle during the autonomous mode [5–7]. The distraction from completing NDRT can impair performance [7]. Moreover, higher levels of automation correspond with higher levels of engagement in non-driving related tasks and lower situation awareness [1]. Therefore, to increase driver situation awareness and improve driver response, it is important for the takeover request to be as salient as possible to the driver. Ho et al. [8] found that multimodal warning signals effectively captured driver attention in demanding situations. Politis et al. [9] showed that multimodal displays performed better in takeover situations than unimodal visual displays. Petermeijer et al. [10] confirmed that multimodal displays are preferable over unimodal displays for takeover requests in autonomous vehicles.

## 2 Related Work

### 2.1 Selection of Lead Time

The lead time plays an important role in the effectiveness of takeover requests, especially in collision avoidance situations. The lead time is defined as the time from takeover notification offset until the time of collision into the obstacle if the participant would not perform any action to avoid the collision. Eriksson and Stanton [11] tested takeover requests in non-critical situations and reviewed other studies of takeover requests in critical situations. Different autonomous vehicle takeover related papers were reviewed and their details such as lead time and take over time were tabulated. They found that mean lead time between all 25 papers was found to be 6.37 ± 5.36 s having a mean reaction time of about 2.96 ± 1.96 s. They also noted that 3, 4, 6 and 7 s were the most frequently used lead times with 1.14, 2.05, 2.69 and 3.04 s as their corresponding reaction times [11].

Damböck et al. [4] tested lead times of 4, 6 and 8 s and found that for simpler conditions of takeover, 4 s lead time was good enough, but drivers required longer lead times for complex situations. In another study by Gold et al. [12] two lead times were used, 5 s and 7 s. The 5 s lead time led to harder stops, swerving and failure to check blind spots while switching lanes. Although 5 s had better reaction times, 7 s led to better braking. A caveat of this better braking performance for a lead of seconds was that the car was still in accelerating motion, and thus, it was not fully in control, suggesting even 7 s lead time being not entirely safe for takeover requests. Zeeb et al. [2] showed that the reaction times (1.14 s) were substantially lower than other existing studies but the faster reaction times also coincided with a higher number of collisions.

Wan et al. [13] used different lead times to measure the optimum lead times for collision warnings in connected vehicle settings. Although it is not a highly autonomous vehicle setting, it extensively tested different lead times between 0 and 30 s and used measures indicative of the safe avoidance of obstacles. The study used three major dependent measures: kinetic energy reduction, minimum time to collision and collision rate. Kinetic energy reduction was considered most important as it determined if a vehicle can slow down enough to avoid a collision. A major conclusion of this paper was that a lead time of 4–8 s was best for optimal safety benefit and a lead time between 5–8 s gave the best reaction time. It was also concluded that below 2 s, even at 45 mph the kinetic energy of the vehicle is too high for collision avoidance [13]. Mok et al. [14] used 2, 5 and 8 s lead times for their study and showed that 2 s lead time was not enough for drivers to regain sufficient control. Moreover, drivers rated the 2 s lead time as significantly less trustworthy. This suggests lead times below 2 s may be unsafe for takeover from highly autonomous vehicles. Mok et al. [14] showed that the 5 s lead time was found to be sufficiently long enough for drivers to regain control. The 8 s lead time also yielded good performance in takeover with higher levels of trust in drivers [14]. Wan et al. [13] found that lead time above 8 s showed a higher collision rate among drivers.

To summarize, the minimum value for lead time has been suggested as 2 s and maximum has been suggested as 8 s [13]. Many papers have used 7 s as their lead time [7, 9, 11, 15]. Optimum lead time in a connected vehicle setting has been suggested as between 5 and 8 s [13]. Based on this data, we used the following lead times, 3 s, 5 s, 7 s and 9 s; two falling within the "optimum range" and two outside. Moreover, with previous studies already testing 5 and 7 s, it was of interest to see if the present study could get similar reaction time values for those levels of lead times. Testing the two outside the optimum range of lead times would also help compare the results with the study by Wan et al. [13] and help define the optimum lead times within the autonomous vehicle setting in a collision avoidance setting.

## 2.2 Spatiality of Auditory Warnings

Research has shown that the presentation of relevant spatial cues can help facilitate responses to target events, whether it be as the driver or pedestrians on the road [8, 16].

Driving research on collision spatial warning signals found out that the recognition of auditory warnings for threats from the front and back of the vehicle was more effective than vibrotactile ones, and provided a meaningful facilitatory effect on response time to events [8]. A closer look to spatial auditory warning cues also indicated how meaningful such warnings can be at providing locational data and capturing attention [17]. Research on auditory perception for pedestrians also indicated significant effects from spatiality, with users detecting the direction of traffic coming at a speed greater than 12 mph at around 90% accuracy, indicating the importance and information provided from spatiality [16].

As drivers make similar decisions when evaluating obstacles on the road, research on directional warnings for drivers at intersection has also proven that driver behavior is improved by directional information in auditory warnings, significantly reducing drivers' brake time and deceleration rate [18]. Additionally, research on spatiality and display modality showed that spatially congruent auditory displays significantly improved drivers' performance as a unimodal display, with the hybrid visual-auditory performing better [19].

## 2.3 Selections of Subjective Characteristics

Marshall et al. [20] stated that it was important to consider the annoyance of an alert as annoyance can undermine the influence of warning systems. They also noted that the alert characteristics of warning signals affect perceived urgency and annoyance in different degrees. They also state that people judge highly urgent results as more appropriate while also being annoying. Auditory alerts having a high level of urgency may help drivers respond more effectively [20]. Studies have also shown that people respond to auditory alerts more quickly if they sound urgent [21, 22]. Therefore, annoyance and urgency were considered important parameters to measure the effectiveness of an auditory alert. Intuitiveness, the immediacy of recognizing cues and their relation to the users' mental model [23], was also considered an important characteristic as it would allow us to measure the effectiveness of the spatiality of the auditory warning.

### 2.4 Our Research Contribution

The main goal of our research is to revisit and extend some open questions related to suggested lead time in auditory displays for semi-automated driving mainly as previous reported results are not entirely coherent with each other. Additionally, we also studied the directionality of sound in this type of warning systems as it has shown to be very efficient in other HMI domains, but has not been extensively studied for takeover request displays.

Our main research questions are, therefore, the following:

- What is the optimal lead time in visual and auditory displays for takeover request?
- What are the main benefits and drawbacks of directional information in auditory displays for takeover request?

Both properties of auditory displays are studied through a wide combination of dependent variables, which reflect drivers' performance, safety related issues, mental workload and subjective preferences. This enabled us to draw comprehensive conclusions and provide some concrete guidelines for auditory takeover displays for semi-automated driving.

## 3 Experimental Setup

The experiment was conducted in a motion-based driving simulator by Nervtech [24]. It consisted of a racing car seat, a steering system and sport pedals. The visuals were displayed on a triple-screen configuration, which covered 120° horizontal field of view and consisted of three equal curved 48' HD TVs. The driving scenarios were developed using AV Simulations software [25], which run on a computer with an i7 - 8086K CPU and Nvidia GTX 1080 graphics card. The driving scenarios involved driving on a three-lane highway with 110 km/h (70 miles/h) speed limit. The visibility was lowered to approximately 100 m using fog (see Fig. 1).

**Fig. 1.** Reduced visibility using fog.

## 4 Methodology

A simulation of a SAE Level 3 [26] automated vehicle was used for the study. The automated driving system (ADS) was engaged at all times, except during unexpected situations (for example, car accident, deer on the road, construction work, etc.). When the ADS was on, the vehicle would always drive in the middle lane of the road with a fixed speed of 110 km/h. The ADS system could be disengaged and reengaged at any time with a simple pull up of the left lever on the steering wheel. The unexpected situation always resulted in closing two out of the three high-way lanes – the center and either the left or the right lane – so that the driver would have to perform a lane-change in order to avoid an accident.

The experiment involved a test drive and four 9-min-long trials. The test drive lasted 3 min and was intended for the participants to get familiar with the driving simulator, the ADS system, the auditory display and the NDRT. In each of the four trials, the participants were exposed to 3 critical situations in which they had to take over control of the vehicle. The event time and location of the free lane was randomized to try to avoid learning and anticipation effects.

To ensure consistency of the use of the ADS system, approximately 150 m after the obstacle on the road, a pre-recorded sound notification asked the participant to re-engage the ADS system and continue with the NDRT.

### 4.1 Participants

Twenty-four participants (20 male), aged from 20 to 42 years old ($M = 26.5$, $SD = 5.13$) participated in the study. Only participants with valid driving licenses were asked to participate in the study.

## 4.2 Tasks

During the experiment, the participants were asked to perform two tasks with the same level of priority:

- Non-driving-related task (NDRT),
- Take control of the vehicle in case of a takeover request.

The non-driving-related task was *2048* - a single-player sliding block puzzle game, whose objective is to slide and combine numbers on a grid with the purpose of achieving a sum of 2048.

Since the vehicle had a high level of automation, participants had to perform the takeover task only in case of disabled automation of the vehicle due to an unexpected situation or obstacle. When an obstacle appeared, the driver was given an auditory and visual takeover request, consisting of text and warning icon, as shown in Fig. 2. The visual notification was presented at the same with the auditory notification, on the dashboard display in the simulator (see Fig. 2). Upon the takeover request, the autonomous vehicle switched over to the non-autonomous mode. This allowed the driver to take control of the vehicle. When operated manually, the vehicle had an automatic transmission system.

**Fig. 2.** Visual notification used for takeover request.

## 4.3 Lead Times and Spatiality

Lead times and spatiality were the two independent variables primarily observed in the study. Lead times were assigned within-subject, whereas spatiality was observed between-subjects.

Based on previous research, lead times were set as 3 s, 5 s, 7 s and 9 s (LT3, LT5, LT7 and LT9). The lead time was defined as the time from takeover notification offset until the time of collision into the obstacle if the participant would not perform any action (brake or steer). Each lead time was observed in a separate trial, where the participant had to take over control of the vehicle three times.

Two types of auditory displays for takeover notifications were used in the study: non-directional and directional sound. The non-directional auditory display played the takeover notification from the front center of the vehicle. For the trials with the directional sound, the notification always came from the direction of the free lane, where the participant should drive in order to avoid collision. Using the Nervtech simulator surround sound system, the sound would be played from direction of the left and right front side of the simulated vehicle.

A non-speech auditory display was designed, including two dominant frequencies (880, 1760 Hz) repeated four times using sine wave, following ISO [27] guidelines. The sound was played at 65 dB, whereas the ambient environment sound was played at 60 dB.

## 4.4 Dependent Variables

We observed three groups of dependent variables:

- takeover reaction time and reaction type (steering or braking),
- Takeover Success Rate and Driving Performance, and
- self-reported data (perceived workload and a sound questionnaire).

Reaction times and reaction types were recorded for all takeover situations varying in lead time and sound direction. Reaction time was measured from the moment of the auditory-visual takeover request until the driver regained control by breaking or steering the wheel. Breaking was initiated by pressing the brake pedal for at least 10% while steering was detected as turning the steering wheel for more than $2°$, based on similar experimental data reported by Gold et al. [2]. As reaction type, we detected which of these two operations the participants would perform first, steering or braking.

Furthermore, the takeover success rate, and logs of speed, acceleration and jerk were observed. Each of these variables was observed from the auditory takeover request until the pre-recorded notification asking the participant to reengage the ADS.

After completing each trial (with a different lead time), participants were asked to evaluate the takeover task using the NASA TLX questionnaire. For each trial, the score of the 2048 (task success rate) was also recorded. At the end of the experiment, participants were further asked to evaluate their user experience with the auditory display using a sound questionnaire.

# 5 Results

Levene's test was used to assess the equality of variances and Shapiro-Wilk test for exhibiting the normal distribution of each group of data. Based on these results, appropriate parametric and non-parametric tests were used for analyzing the data.

## 5.1 Reaction Time and Reaction Type

All groups (lead time and sound display type) did not have a normal distribution. The between-subjects design results (sound direction) were analyzed with the independent-samples Mann-Whitney U test 4. The within-subject design results (lead time) were analyzed with the related-samples Friedman test and post-hoc Wilcoxon-signed ranks test. For all pairwise comparisons a Bonferroni adjustment was applied to control for Type-I error, resulting in adjusted alpha levels (critical alpha level = .0083 (0.05/6)).

**Reaction Time.** Figure 3 shows the mean reaction times for each of the tested lead time for trials with directional and non-directional sound display. Reaction times for the directional sound (M = 1857.24 ms, SD = 1161.94 ms) were significantly faster than the reaction times for the non-directional sound display (M = 1977.83 ms, SD = 1079.06 ms), U = 7843, p = .030.

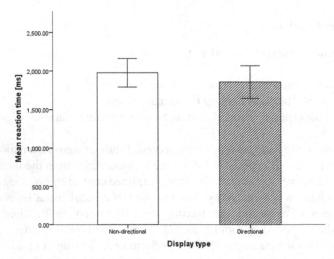

**Fig. 3.** Mean reaction times and SE to takeover requests with different lead times for non-directional and directional display

The results for different lead times showed that there was a statistically significant difference in reaction times to the takeover request, $\chi 2(3) = 50.961$, p < .001 (Fig. 4). The reaction times to takeover requests with LT3 were significantly faster than reaction times for takeover requests with LT5 (p < .001), LT7 (p < .001) and LT9 (p < .001). Furthermore, the reaction times to takeover request with lead time of LT5 were significantly faster than LT7 (p = .002) and LT9 (p = .004). There were no statistical differences between reaction times lead LT7 and LT9 (p = .305). Table 1 reports on the mean reaction times for each display type takeover request for all four observed lead times.

We also observed the reaction times for different types of reaction, steering or braking. These results showed that the reaction times for breaking (*M* = 1857.24 ms, *SD* = 1240.37 ms) were significantly faster compared to steering (*M* = 1977.84 ms, *SD* = 1079.06 ms), *U* = 11 960, *p* < .001 (see Fig. 5).

**Reaction Type.** In 68.1% of takeovers, the participants' first reaction to the takeover notification was breaking, and in 31.9% it was steering. These ratios are similar for all lead times as shown in Table 2 and Fig. 6.

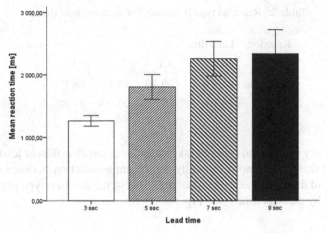

**Fig. 4.** Mean reaction times and SE to takeover requests for different lead times

**Table 1.** Mean reaction times to takeover requests for different lead times

| Display type | Lead time | | | |
|---|---|---|---|---|
| | 3 s | 5 s | 7 s | 9 s |
| Non-directional | 1296.71 ms | 2006.27 ms | 2165.18 ms | 2430.71 ms |
| Directional | 1233.82 ms | 1604.29 ms | 2348.21 ms | 2245.66 ms |

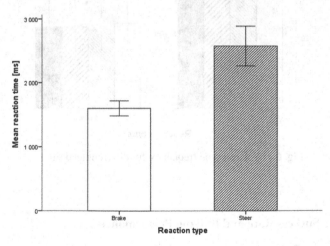

**Fig. 5.** Mean reaction times and SE for different types of takeover

**Table 2.** Reaction type frequency for different lead times

| Reaction type | Lead time | | | |
|---|---|---|---|---|
| | 3 s | 5 s | 7 s | 9 s |
| Brake | 66.7% | 69.4% | 68.1% | 68.1% |
| Steer | 33.3% | 30.6% | 31.9% | 31.9% |

The frequency of reaction types to takeover requests with different lead time, however, indicated that there is no statistically significant association between the reaction type and the lead time, $\chi(1) = 0.064$, $p = .800$; that is, the reaction type preference was not influenced by the lead times (see Fig. 6).

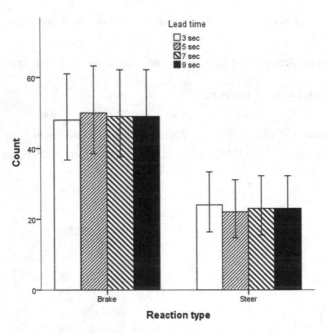

**Fig. 6.** Reaction type frequency for different lead time

## 5.2 Takeover Success Rate and Driving Performance

**Takeover Success Rate.** Each participant performed 12 takeover requests (3 with all four lead times). Out of all of the 288 takeover requests in this study, 99.986% were successful.

**Speed.** The results on speed showed that the average speed during the takeover (manual driving of the vehicle) was higher in trials with the takeover request with non-directional sound display (M = 92.26, SD = 18.02) compared to trials with the directional sound display (M = 88.31 km/h, SD = 17.69 km/h), U = 8979, p = .049 (Fig. 7). When observing the average speed for trials with different lead times, the analysis did not reveal any statistically significant differences, $\chi 2(3)$ = 2.050, p = .562 (Fig. 8).

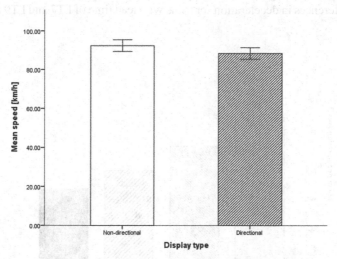

**Fig. 7.** Mean speed and SE during takeover for trials with non-directional and directional display

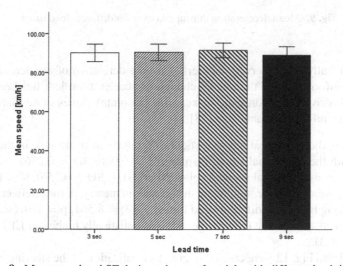

**Fig. 8.** Mean speed and SE during takeover for trials with different lead times

**Deceleration (Braking).** The results for deceleration (braking) did not show any statistical differences in the takeover situations with directional sound display (M = 0.65 km/h2, SD = 0.59 km/h2) compared to trials with non-directional sound display, U = 10674, p = .665.

Significant differences in deceleration were, however, revealed for different lead times (see Fig. 9), $\chi2(3) = 67.223$, p < .001. The pairwise comparisons showed that the average deceleration for trials with LT3 was higher compared to LT5 (p < .001), LT7 (p < .001) and LT9 (p < .001). Furthermore, the average deceleration for trials with LT5 was significantly higher than LT7 (p < .001) and LT9 (p < .001). There were no statistical differences in deceleration for trials with lead time of LT7 and LT9 (p = .053).

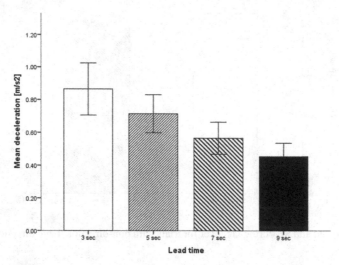

**Fig. 9.** Mean deceleration during takeover for different lead times

**Jerk.** Additionally, we observed also jerk – a time derivative of acceleration, which gives us the information of the rate of change of acceleration. jerk has been used to determine the driving comfort experience, with (absolute) values more than 0.6 m/s3 indicating uncomfortable maneuvers [28].

The results did not reveal any statistical differences in mean jerk in the takeover situations with the directional sound display (M = 0.086, SD = 0.236) compared to trials with the non-directional sound display (M = 0.031, SD = 0.256), U = 11577.5, p = .068. There were, however, significant differences in mean jerk for takeover situations when comparing trials with different lead times, $\chi2(3) = 8.561$, p = .036 (see Fig. 10). The post-hoc test showed that the average jerk was statistically higher for LT3 compared to LT5 (p = .003).

Figure 11 and Fig. 12 were created to give a visualization of the steering maneuvers performed after the request to takeover. The first steering directions, swerving and faster maneuvers that resulted into higher jerk can be seen for each lead time for both non-directional (see Fig. 11) and directional sound displays (see Fig. 12) to avoid the road obstacle.

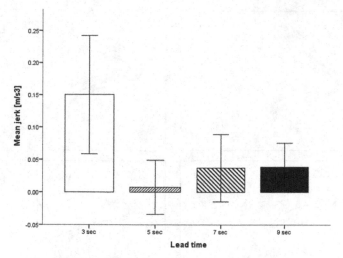

**Fig. 10.** Mean jerk during takeover for different lead times

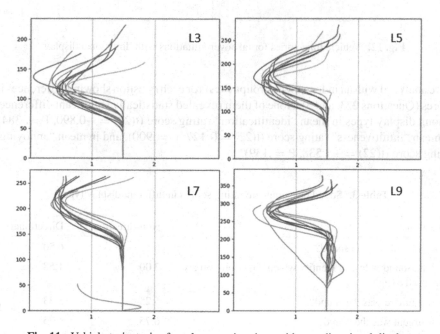

**Fig. 11.** Vehicle trajectories for takeover situations with non-directional display

## 5.3 Questionnaires

**Sound Questionnaire.** Table 3 and Fig. 13 show the results of subjective rating scores on a 0–100 scale for the non-directional display and the directional sound display. The directional sound display received a numerically higher score in "intuitiveness" and lower scores in both "identification" and "annoying" than the non-directional display. results

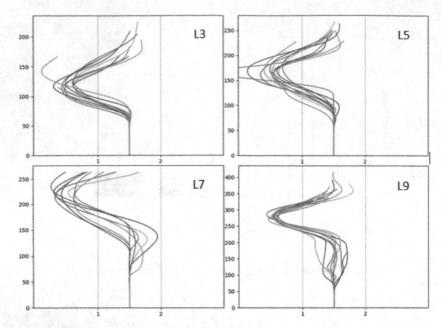

**Fig. 12.** Vehicle trajectories for takeover situations with directional display

were analyzed with an independent-groups t-test for each question showing differences in scores (Questions 2, 3, and 5). None of them revealed statistically significant differences among display types in mean "identification" rating score (t(22) = −0.890, P = .384), in mean "intuitiveness" rating score (t(22) = 0.127, p = .900), and in mean "annoying" rating score (t(22) = −1.538, P = .139).

**Table 3.** Sound questionnaire rating scores in different display types

| Questions | Non-directional | Directional |
|---|---|---|
| How helpful was the sound? | 6.50 | 6.50 |
| Did the sound help you identify which direction you are supposed to go? | 2.00 | 1.58 |
| How intuitive was the sound? | 5.25 | 5.33 |
| How urgent was the sound? | 6.75 | 6.75 |
| How annoying was the sound? | 2.92 | 2.00 |

**NASA-TLX Rating Score.** The results were analyzed with a 4 (Lead time) X 2 (Display) repeated measures analysis of variance (ANOVA) for each subscale including mental, physical and temporal demand, performance, effort, and frustration.

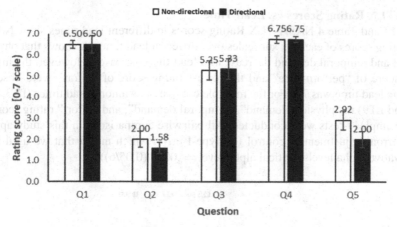

**Fig. 13.** Sound questionnaire rating scores of post-questionnaires

*Mental Demand.* The analysis showed that there were no statistically significant differences among display types in mean "mental demand" rating score, $F(1, 22) = 1.309$, p $= .265$ and lead times, $F(1, 22) = 2.368$, p $= .079$. There was no significant interaction between Display and Lead time, $F(3, 66) = 0.080$, p $= .970$.

*Physical Demand.* The analysis showed that there were significant differences among lead times in mean "physical demand" rating score, $F(3, 66) = 3.802$, p $= .014$. However, there were no statistically significant differences found among display types, $F(1, 22) = 1.309$, p $= .191$. There was no significant interaction between Display and Lead time, $F(3, 66) = 0.366$, p $= .778$.

*Temporal Demand.* The analysis showed that there were significant differences among lead times in mean "temporal demand" rating score, $F(3, 66) = 8.255$, P $< .0001$. However, there were no statistically significant differences found among display types, $F(1, 22) = 2.344$, P $= .140$. There was no significant interaction between Display and Lead time, $F(3, 66) = 2.143$, p $= .103$.

*Performance.* The analysis showed that there were no significant differences among lead times in mean "performance" rating score, $F(3, 66) = 0.334$, P $= .801$ and display types, $F(1, 22) = 0.037$, P $= .849$. There was no significant interaction between Display and Lead time, $F(3, 66) = 1.1820$, p $= .323$.

*Effort.* The analysis showed that there were significant differences among lead times in mean "effort" rating score, $F(3, 66) = 4.113$, p $= .010$. However, there were no statistically significant differences found among display types, $F(1, 22) = 1.746$, p $= .200$ or between Display and Lead time, $F(3, 66) = 0.308$, p $= .819$.

*Frustration.* The analysis showed that there were no significant differences among lead times in mean "frustration" rating score, $F(3, 66) = 2.488$, p $= .068$ and display types, $F(1, 22) = 0.035$, p $= .853$. There was no significant interaction between Display and Lead time, $F(3, 66) = 1.423$, p $= .244$.

**NASA-TLX Rating Scores vs. Lead Time**

Figure 14 and Table 4 NASA-TLX Rating scores in different lead times show NASA-TLX rating scores of each six subscales over different lead times. It shows that physical demand and temporal demand decrease as a lead time increases. However, the highest rating score of "performance" and the lowest rating score of "effort" were observed when the lead time was 7 s. For the multiple comparisons among lead times (LT3, LT5, LT7, and LT9) in "physical demand", "temporal demand", and "effort" rating scores, a paired-samples t-tests were conducted. All pairwise comparisons in this study applied a Bonferroni adjustment to control for Type-I error, which meant that we used more conservative alpha levels (critical alpha level = .0083 (0.05/6)).

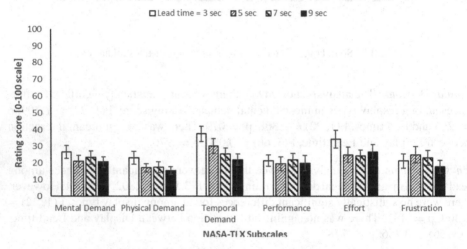

**Fig. 14.** NASA-TLX Rating scores of subscales

In "physical demand", participants showed significantly higher temporal demand in LT3 ($M = 23.123$, $SD = 18.928$) than LT9 ($M = 15.417$, $SD = 11.508$), $t(66) = -3.170$, $p = .002$. In "temporal demand", participants showed significantly higher temporal demand in LT3 ($M = 37.500$, $SD = 22.554$) than both LT7 ($M = 25.417$, $SD = 18.053$), $t(66) = -3.670$, $p = .005$ and LT9 ($M = 22.083$, $SD = 17.126$), $t(66) = -4.682$, $p < .0001$. In "effort", participants showed significantly a higher temporal demand in LT3 ($M = 34.375$, $SD = 26.716$) than LT5 ($M = 25.000$, $SD = 22.458$), $t(66) = -2.881$, $p = .005$. Also, it was significantly higher than LT7 ($M = 24.167$, $SD = 18.278$), $t(66) = -3.138$, $p = .003$.

## 5.4  NDRT Performance

The results for the average score of 2048 were analyzed with a 4 (Lead time) x 2 (Display) repeated measures analysis of variance (ANOVA). The analysis showed that there were no significant differences among lead times in average score of 2048, $F(3, 66) = 0.346$, p = .792 and display types, $F(1, 22) = 2.618$, p = .120. There was no significant interaction between display and lead time, $F(3, 66) = 2.313$, p = .084. There was numerically higher

**Table 4.** NASA-TLX Rating scores in different lead times

| Subscale | Lead time | | | |
|---|---|---|---|---|
| | 3 s | 5 s | 7 s | 9 s |
| Mental demand | 26.67 | 21.04 | 23.33 | 20.63 |
| Physical demand | 23.13 | 17.08 | 17.50 | 15.42 |
| Temporal demand | 37.50 | 30.21 | 25.42 | 22.08 |
| Performance | 21.46 | 19.58 | 21.88 | 20.00 |
| Effort | 34.38 | 25.00 | 24.17 | 26.67 |
| Frustration | 21.25 | 25.00 | 23.13 | 17.92 |

average score of 2048 with the directional sound display (675 M = 7321.396, SD = 4758.245) than the non-directional display type (M = 5068.063, SD = 3143). However, there were no statistically significant differences between display types.

## 6 Discussion and Conclusions

To design more optimized takeover request displays in automated vehicles, we evaluated takeover time and type, driving performance, perceived workload, and subjective experience measures for two spatial sounds with four lead times, while our participants performed a non-driving related task. The results for all of the observed variables did not show a consistent pattern for one specific lead time or display type. However, when observed together, it is possible to differentiate which lead time and display type offer safest takeover and to some extent better takeover experience.

The shortest (3 s) lead time evoked fastest reaction times, which is in line with previous research [13, 14]. However, based on the jerk results, this lead time resulted in the takeover maneuver surpassing the threshold of 6 m/s3, which suggests unpleasant maneuvers in driving. Figure 11 and Fig. 12 further support these results as it can be seen that the L3 resulted in most sharp swearing when avoiding the road obstacle. Furthermore, the results from the NASA-TLX showed that participants reported high perceived workload in terms of physical demand, temporal demand and effort. By extending the lead time to 5 s and 7 s, the reaction time decreased but the perceived workload and subjective preferences of users increased significantly. The avoiding maneuver also improved suggesting the 7 s lead time as the optimal choice among the tested signals. This is coherent with previous results reported in [13].

The reaction times for braking were significantly faster than steering, and in almost two thirds of all takeover request situations, the participant's first reaction was braking. This is probably due to the fact that the NDRT was manually operated. However, when explored the frequencies, the results revealed that the length of the lead time did not affect the first reaction type. Although not statistically significant, it is interesting to notice that in the 3-s lead time condition majority of the participants decide to steer instead of break. Due to the proximity of the road obstacle, which at this point was already visible in comparison to the situations with longer lead times, participants could have found braking as not efficient (not enough time to stop). In real life situations, such decisions for sudden lane changes without reducing the speed can be dangerous and less comfortable.

Furthermore, we observed the influence of spatiality and directionality of auditory signals. Regardless of the reaction type, the directional display evoked faster reaction times. Interestingly, the directional display showed lower average vehicle speed during the takeover. We can cautiously infer that the participants might be slower with the directional sound because they recognized the direction and so, prepared for changing the direction of the vehicle.

When observing the self-reported ratings, there was no significant difference in the sound experience questionnaire. However, at least, the annoyance question favored the directional sound (directional: 2.00 vs. non-direction: 2.92), which bodes well for further investigation of display directionality in this domain. Annoying sounds have been shown to have a positive correlation with alertness and urgency, but research has also suggested that too much annoying and startling sound will discourage users to use auditory displays. Given that our directional sound performed better and received lower annoyance, it seems desirable. Research with more participants will clarify whether there is different preference for either display type.

Finally, when observing the NDRT results, it can be noticed that all of the participants achieved high scores suggesting a high engagement in the task. Although previous research has indicated that higher engagement in NDRT can affect the takeover success rate, we could not find any evidence of such effect in this study. This could be further explored with different NDRT tasks and different durations of the trials.

As for multimodality, in this study we tested only auditory variables with a fixed visual display. Future research can include variations of other modalities (visual and tactile) and spatiality of visual and tactile displays. Planned research includes making a mathematical model and validating it with empirical research outcomes to formalize takeover procedure in automated vehicles.

**Acknowledgment.** This work was partially supported by a grant (number BI-US/19–21-008) from Slovenian Research Agency and grant (code 17TLRP-B131486–01) from Transportation and Logistics R&D Program funded by Ministry of Land, Infrastructure and Transport of Korean government.

# References

1. Merat, N., Jamson, A.H., Lai, F.C., Carsten, O.: Highly automated driving, secondary task performance, and driver state. Human Factors 54(5), 762–771 (2012)
2. Gold, C., Damböck, D., Lorenz, L., Bengler, K.: Take over!" How long does it take to get the driver back into the loop? In: Proceedings of the Human Factors and Ergonomics Society Annual Meeting 2013 Sep, vol. 57, no. 1, pp. 1938–1942. Sage Publications, Sage CA, Los Angeles (2013)

3. Zeeb, K., Buchner, A., Schrauf, M.: What determines the take-over time? an integrated model approach of driver take-over after automated driving. Accid. Anal. Prev. **1**(78), 212–221 (2015)
4. Damböck, D., Bengler, K.: Übernahmezeiten beim hochautomatisierten Fahren. In5. Tagung Fahrerassistenz (2012)
5. Feldhütter, A., Gold, C., Schneider, S., Bengler, K.: How the duration of automated driving influences take-over performance and gaze behavior. In: Schlick, C.M., et al. (eds.) Advances in Ergonomic Design of sysTEMS, PROducts and Processes 2017, pp. 309–318. Springer, Heidelberg (2017). https://doi.org/10.1007/978-3-662-53305-5_22
6. Neubauer, C., Matthews, G., Saxby, D.: The effects of cell phone use and automation on driver performance and subjective state in simulated driving. In: Proceedings of the Human Factors and Ergonomics Society Annual Meeting 2012 Sep, vol. 56, no. 1, pp. 1987–1991. Sage Publications, Sage CA, Los Angeles
7. Radlmayr, J., Gold, C., Lorenz, L., Farid, M., Bengler, K.: How traffic situations and non-driving related tasks affect the take-over quality in highly automated driving. In: Proceedings of the Human Factors and Ergonomics Society Annual Meeting 2014 Sep, vol. 58, no. 1, pp. 2063–2067. Sage Publications, Sage CA, Los Angeles (2014)
8. Ho, C., Tan, H.Z., Spence, C.: The differential effect of vibrotactile and auditory cues on visual spatial attention. Ergonomics **49**(7), 724–738 (2006)
9. Politis, I., Brewster, S., Pollick, F.: Language-based multimodal displays for the handover of control in autonomous cars. In: Proceedings of the 7th International Conference on Automotive User Interfaces and Interactive Vehicular Applications 2015 Sep 1, pp. 3–10 (2015)
10. Petermeijer, S., Bazilinskyy, P., Bengler, K., De Winter, J.: Take-over again: investigating multimodal and directional tors to get the driver back into the loop. Appl. Ergon. **1**(62), 204 215 (2017)
11. Eriksson, A., Stanton, N.A.: Takeover time in highly automated vehicles: noncritical transitions to and from manual control. Hum. Factors **59**(4), 689–705 (2017)
12. Gold, C., Berisha, I., Bengler, K.: Utilization of drivetime–performing non-driving related tasks while driving highly automated. In: Proceedings of the Human Factors and Ergonomics Society Annual Meeting 2015 Sep, vol. 59, no. 1, pp. 1666–1670. SAGE Publications, Sage CA, Los Angeles (2015)
13. Wan, J., Wu, C., Zhang, Y.: Effects of lead time of verbal collision warning messages on driving behavior in connected vehicle settings. J. Safety Res. **1**(58), 89–98 (2016)
14. Mok, B., et al.: Emergency, automation off: Unstructured transition timing for distracted drivers of automated vehicles. In: 2015 IEEE 18th International Conference on Intelligent Transportation Systems 2015 Sep 15, pp. 2458–2464. IEEE (2015)
15. Gold, C., Körber, M., Lechner, D., Bengler, K.: Taking over control from highly automated vehicles in complex traffic situations: the role of traffic density. Hum. Factors **58**(4), 642–652 (2016)
16. Barton, B.K., Ulrich, T.A., Lew, R.: Auditory detection and localization of approaching vehicles. Accid. Anal. Prev. **1**(49), 347–353 (2012 )
17. Ho, C., Spence, C.: Assessing the effectiveness of various auditory cues in capturing a driver's visual attention. J. Exp. Psychol. Appl. **11**(3), 157 (2005 )
18. Zhang, Y., Yan, X., Yang, Z.: Discrimination of effects between directional and nondirectional information of auditory warning on driving behavior. Discret. Dyn. Nat. Soc. 1, 2015 (2015 )
19. Liu, Y.C., Jhuang, J.W.: Effects of in-vehicle warning information displays with or without spatial compatibility on driving behaviors and response performance. Appl. Ergon. **43**(4), 679–686 (2012 )
20. Marshall, D.C., Lee, J.D., Austria, P.A.: Alerts for in-vehicle information systems: annoyance, urgency, and appropriateness. Hum. Factors **49**(1), 145–157 (2007 )

21. Burt, J.L., Bartolome, D.S., Burdette, D.W., Comstock, J.R., Jr.: A psychophysiological evaluation of the perceived urgency of auditory warning signals. Ergonomics **38**(11), 2327–2340 (1995)
22. Haas, E.C., Casali, J.G.: Perceived urgency of and response time to multi-tone and frequency-modulated warning signals in broadband noise. Ergonomics **38**(11), 2313–2326 (1995)
23. Garzonis, S., Jones, S., Jay, T., O'Neill, E.: Auditory icon and earcon mobile service notifications: intuitiveness, learnability, memorability and preference. In: Proceedings of the SIGCHI Conference on Human Factors in Computing Systems 2009, 4 April, pp. 1513–1522 (2009)
24. Vengust, M., Kaluža, B., Stojmenova, K., Sodnik, J.: NERVteh compact motion based driving simulator. In: Proceedings of the 9th International Conference on Automotive User Interfaces and Interactive Vehicular Applications Adjunct 2017, 24 September, pp. 242–243 (2017)
25. AV Simulations. SCANeR DT. https://www.avsimulation.fr/
26. SAE: Taxonomy and definitions for terms related to on-road motor vehicle automated driving systems. SAE Standard J3016, USA (2014)
27. Draft, ISO Working. Road vehicles-Ergonomic aspects of transport information and control systems-Specifications for in-vehicle auditory presentation, ISO Standard 15006:2011(E)
28. Kilinç, A.S., Baybura, T.: Determination of minimum horizontal curve radius used in the design of transportation structures, depending on the limit value of comfort criterion lateral jerk. TS06G-Engineering Surveying, Machine Control and Guidance, Rome, Italy (2012)

# Demystifying Interactions Between Driving Behaviors and Styles Through Self-clustering Algorithms

Yu Zhang[1,2], Wangkai Jin[1], Zeyu Xiong[1], Zhihao Li[1], Yuyang Liu[1,3P], and Xiangjun Peng[1,4P(✉)]

[1] User-Centric Computing Group, University of Nottingham, Ningbo, China
xjpeng@cse.cuhk.edu.hk
[2] School of Computer Science, University of Nottingham, Nottingham, UK
[3] Department of Computer Science, University of Machester, Machester, UK
[4] Department of Computer Science and Engineering,
The Chinese University of Hong Kong, Shatin, Hong Kong

**Abstract.** We argue that driving styles demand adaptive classifications, and such mechanisms are essential for adaptive and personalized Human-Vehicle Interaction systems. To this end, we conduct an in-depth study to demystify complicated interactions between driving behaviors and styles. The key idea behind this study is to enable different numbers of clusters on the fly, when classifying driving behaviors. We achieve so by applying Self-Clustering algorithms (i.e. DBSCAN) over a state-of-the-art open-sourced dataset of Human-Vehicle Interactions. Our results derive 8 key findings, which showcases the complicated interactions between driving behaviors and driving styles. Hence, we conjecture that future Human-Vehicle Interactions systems demand similar approaches for the characterizations of drivers, to enable more adaptive and personalized Human-Vehicle Interaction systems. We believe our findings can stimulate and benefit more future research as well.

**Keywords:** Driving behaviors · Driving styles · Adaptive & personalized human-vehicle interactions

## 1 Introduction

Responses from Driver-Vehicle Interactions, whether they satisfy drivers' expectations or not, have significant impacts on users' trusts in terms of Autonomous Driving. To deliver user-expected interactions, detailed insights from Driving statistics are the most critical parts of modern Human-Vehicle Interaction systems. For instance, users' trust in Autonomous Vehicles are highly dependent with such responses, which rely on the detailed insights from driving statistics [16,30,37]. Recent efforts characterize driving behaviors empirically, and further classify them into multiple driving styles in static partitions. However, with the growing popularity of Autonomous Vehicles, computational methods, rather

© Springer Nature Switzerland AG 2021
H. Krömker (Ed.): HCII 2021, LNCS 12791, pp. 335–350, 2021.
https://doi.org/10.1007/978-3-030-78358-7_23

than empirical methods, can potentially fit better within personalized Human-Vehicle Interactions, in practice.

With such a mindset, we argue that **conventional classifications of driving styles are not suitable for adaptive and personalized Human-Vehicle Interaction systems**. We disagree with the conventional approach from the following two aspects. First, static classifications of driving styles are not adaptive during the driving procedures; and second, driving styles, derived from empirical studies, are insufficient to contribute to personalized Human-Vehicle Interaction techniques. We believe that the root causes of the above issues are because complicated interactions between driving styles and behaviors remain under-studied, in terms of both mechanisms and findings.

Our goal is to demystify complicated interactions between driving behaviors and driving styles, to reveal the opportunities for adaptive and personalized Human-Vehicle Interactions. We make the **key observation** that the conventional classifications of driving styles rely on static partitions of driving behaviors, obtained from empirical studies. In other words, the problem is abstracted as clustering techniques, with the pre-determined number of clusters. To this end, our key idea is to apply computational techniques to eliminate the needs for pre-determined number of clusters. Hence, we utilize self-clustering algorithms, Density-Based Spatial Clustering of Applications with Noise (DBSCAN), for adaptive classifications of driving styles and hidden patterns of driving behaviors.

We perform our studies over BROOK, a state-of-the-art and open-sourced dataset for Human-Vehicle Interactions. Our studies have included 34 drivers in 11 dimensions of driving statistics [23]. In total, we make 8 key findings through our studies. Our studies start with rigorous examinations of the impacts from different DBSCAN configurations, representative driver groups, time-series variations, road conditions and etc. Furthermore, we characterize in-depth characteristics of driving styles, by breaking down detailed features and analyzing the overlap across different styles. Based on the above findings, we confirm that the interactions between driving behaviors are more complicated, and our DBSCAN-based approach is more applicable in this context, compared with conventional partitions of driving styles.

We make the following **key contributions** in this paper:

- We address the problem that conventional classifications of driving styles overlook the opportunities for adaptive and personalized Human-Vehicle Interactions, and identify that the static partitions of driving styles are the key limitation in conventional approaches.
- To the best of our knowledge, we are the first to propose and utilize Density-Based Spatial Clustering of Applications with Noise (DBSCAN), for adaptive classifications of driving styles and hidden patterns of driving behaviors.
- We experimentally characterize and examine the effects of our DBSCAN-based approach over BROOK, a state-of-the-art and open-sourced dataset for Human-Vehicle Interactions.

– We retrieve 8 key findings from the above studies, by rigorously changing different configurations. These observations can serve as starting guidelines for adaptive and personalized Human-Vehicle Interactions systems in the future, for both research and industrial communities.

The rest of this paper would be organized as follow. Section 2 introduces the related works in studying driving behaviors and driving styles. Section 3 gives details about the experiment methodology. Results are shown in Sect. 4. Section 5 reports an discussion in order to inspire potential principles. Section 6 presents the conclusions and future work.

## 2    Background and Motivation

Modern Methodology, to evaluate driving style, can be divided into Subjective Evaluation and Objective Evaluation. Subjective Evaluation is carried out through quesionnaires and surveys, to obtain empirical results. For instance, [24] first proposed Driver Behavior Questionnaire (DBQ), to reflect bad driving behavior by self-reporting. Since then, follow-up efforts, based on DBQ, investigates on the impacts of regions, cultures, ages and genders, through the variations of drivers' behaviors. For instance, [13] developed Driver Style Questionnaire (DSQ), to study correlations between Traffic Accidents and Driving Behaviors (e.g. Speed and Distance of Vehicles). Another example is Multidimensional Driving Style Inventory (MDSI) [31] enables the capability to evaluate driving styles from multiple dimensions. More specifically, it defines the structure of driving styles and explicitly classifies them into four categories.

However, these methods are subjective and could be influenced by some external factors. Hence, Subjective Evaluations demand high standards of the effectiveness of the driving questionnaire and experts' experience. To this end, Objective Evaluation is proposed to complement this method. Objective Evaluations analyze driving styles through driving statistics, which are obtained from a driving simulator [4] or in-field vehicles [15]. In the context of driving styles, [5] proposes a classification and recognition model for driving behavior based on sparse representations. More specifically, the vehicle motion tracks, obtained by vision, are used as input in this model, and the sparse representation approach is used to mine the features of the driving behavior decision. Another example is as follow. [33] proposes a pattern recognition method, which utilizes triaxial accelerometers' statistical data to evaluate normal and aggressive driving styles. They further discuss time-domain feature extraction. Also, [20] divides the following behavior according to the difference of patience and puts forward the view, that the following time can measure the driving style. The Mean and Standard Deviations of relevant indicators are often used to differentiate the driver's driving styles, based on the assumption of a normal distribution for these indicators [17,34]. However, these measures, derived directly from sequential observations, are based on static criterion, which can be inconsistent with established parametric distributions [7].

Previous attempts, to characterize driving styles, focus on the differentiated trends of driving statistics (e.g. driving speeds, headway distance, following time, multi-modal information and etc.) [3,5,6,10,12,20,33]. Such studies overlook the effects of time-series and results in coarse-grained decision-making procedures, for the determinations and classifications of driving styles. Hence, we make the key observation that static partitions of driving styles/behaviors may not be suitable for adaptive and personalized Driver-Vehicle Interactions. To this end, we propose to utilize self-clustering algorithms, and compare auto-generated patterns on-the-fly instead of statically partitioned driving styles.

## 3    Study Methodology

### 3.1    Dataset Description

Our study uses a public multi-modal database for Human-Vehicle Interaction – BROOK [23]. BROOK contains 34 drivers' data under four driving scenarios (both manual and automated driving), such as Time, Vehicle Speed, Vehicle Acceleration and Vehicle Coordinates. We utilize representative drivers' driving statistics as input, in terms of time series.

### 3.2    Dataset Pre-processing

Before the bulk of this study, we first normalize all statistics based on the following insights.

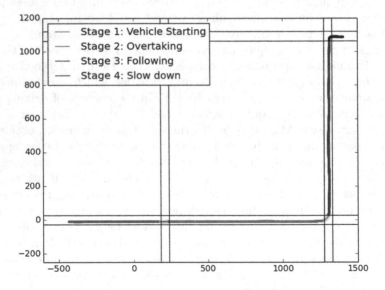

**Fig. 1.** Movement track of vehicle.

**Data Cleanup.** The original database records consist of both only stable stages and star-up stages and parking stages. Since the driving styles are characterized within relatively stable stages, we eliminate the unstable stages to ensure our studies are consistent with others. This representative length of each driving scenario is enough for driving behavior data analysis. The route map is fixed and we show it in Fig. 1.

As shown in Fig. 1, the whole driving route can be divided into four stages[1], which are marked in different colors. The statistics reflect that, there are huge gaps between stable stage and unstable stage, in terms of driving behaviors. For instance, driving speed usually fluctuates within a certain range, where unstable stages drift more randomly. Hence, we consider the statistics, without the fixed range, as a noise source, which is groundless for the characterizing driving styles.

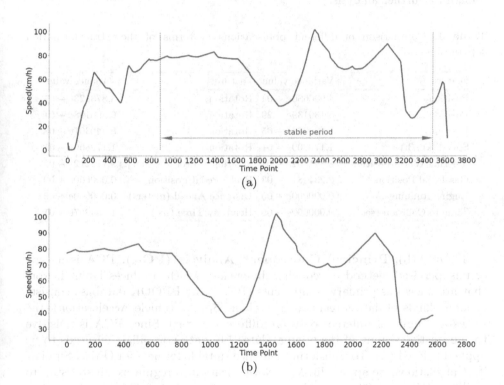

(a)

(b)

**Fig. 2.** Filter driving data

**Feature Selection.** We perform feature selections based on the following insights, where we present backup information in Fig. 2.

Figure 2-(a): **One-way Analysis of Variance (ANOVA).** Feature selection, as a pre-processing stage, aims to select the most discriminative features.

---

[1] The stages don't take traffic lights into account.

From the perspective of clustering, removing irrelevant features won't negatively impact the accuracy of clustering. This is because irrelevant variables may increase noise and mask the underlying pattern or structure in the dataset, as suggested by [36]. Moreover, such cleanup can reduce the required storage and processing time. Our decisions are to utilize Wrapper Approach [2] on the whole dataset, and use these selected features to construct the clusters. The quality of clustering is an indicator of whether the subset of features is satisfactory, where ANOVA method was used. Table 1 presents Variance Values of corresponding features. To this end, those featuress with small variance values (i.e., a variance value $<0.01$) are regarded as meaningless features and then removed [9, 28]. Therefore, the features, such as $Position_x$, $Position_z$, $Rotation_y$, $Rotation_w$, Speed (km/h), Steering wheel position$_1$ and Headaway time (sec), remain for further analysis.

**Table 1.** Comparison of different obfuscations in terms of their transformation capabilities

| Feature | Variance value | Feature | Variance value |
|---|---|---|---|
| $Position_x$ | $4.906680e + 04$ | $Rotation_x$ | $3.873823e - 07$ |
| $Position_y$ | $8.139139e - 29$ | $Rotation_y$ | $6.590198e - 01$ |
| $Position_z$ | $1.024064e + 05$ | $Rotation_z$ | $9.440301e - 08$ |
| Speed (km/h) | $1.172809e - 01$ | $Rotation_w$ | $1.172809e - 01$ |
| Steering Wheel Position$_-$ | $8.841402e - 04$ | Steering Wheel Position$_+$ | $9.772212e - 02$ |
| Gas Pedal Position | $2.234481e - 03$ | Brake Pedal Position | $0.000000e + 00$ |
| Engine Running | $0.000000e + 00$ | Distance Ahead (meters) | $0.000000e + 00$ |
| Time to Collision (sec) | $0.000000e + 00$ | Headway Time (sec) | $1.534207e - 01$ |

Figure 2-(b): **Principal Component Analysis (PCA).** PCA is a type of unsupervised method of reducing dimension, which produces latent factors that are known as primary components (PCs). The BROOK database consists of many kinds of data streams (e.g., Vehicle Speed, Vehicle Acceleration), the scales and units of different data are different as well. Since PCA is sensitive to the relative scaling of the original data [14], data normalization needs to be applied before PCA. To transform them into suitable formats for Object Similarity Calculations, we apply Min-Max Normalization to regularize all statistics to facilitate this need [22]. After that, we project the data onto the maximum feature vector to obtain a one-dimensional feature space to find the principal components representing each sample. With each subsequent component explaining less, the first component explains most of the variance in the data.

### 3.3    Clustering Algorithm

One of the most common clustering strategies is the K-Means Clustering, which requires a pre-determined number of clusters. However, preconditioning the number of clusters are quite challenging since sophisticated knowledge of the domain is required. In our context, we aim to relax such constraints so that we are capable to obtain more insights of the spatial correlations among driving behaviors. To this end, we choose Density-Based Spatial Clustering of Applications with Noise (DBSCAN) [8], for adaptive classifications of driving styles and hidden patterns of driving behaviors. In this way, we aim to demystify the patterns within driving procedures. DBSCAN is a self-clustering algorithm, where the number of clusters is not necessary to be pre-determined. DBSCAN continuously merges two most similar clusters into a new cluster in each iteration until satisfying certain termination criterion (e.g. distance threshold) [19]. Hereby, we elaborate more details of this algorithm and our design choices as follow.

**Distance Measurement.** When performing data clustering, a basic step is to choose an appropriate distance calculation method that quantifies how similar individuals concern measurements provided in the variables. The most commonly used distance measurement is Euclidean Distance, where we take into account at the first place. For completeness of our study, we also utilize Manhattan Distance and Chebyshev Distance to quantify these effects [1] and the results are reported in Table 2. Hereby, we demonstrate these distances to the DBSCAN algorithm mathematically, as shown in the following equations.

$$D_{\text{Euclidean}} = \sqrt{\sum_{i=1}^{N} (x_{1i} - x_{2i})^2} \tag{1}$$

$$D_{\text{Manhattan}} = \sum^{n} |x_{1i} - x_{2i}| \tag{2}$$

$$D_{\text{Chebyshev}}(x, y) = \max_i (|x_i - y_i|) \tag{3}$$

**Parameters Setting.** DBSCAN also needs to take *minPoints* and *epsilon* as input parameters. The *minPoints* refers to the minimum number of data points within the cluster, and *epsilon* refers to the max radius of the cluster. If the *minPoints* value is too small, more core objects will be generated, leading to too many clusters. On the contrary, two adjacent clusters with higher density may be merged into the same cluster, resulting in fewer clusters. The influence of the value selection of *epsilon* also has similar effects. We use the principle *minPoints* = 2·*dim* [26] to select an appropriate range of *minPoints*. After intensive rounds of rigorous testing, we identify NINE different setups of both parameters to serve as the representatives of all possible combinations. This is because our goal is not to provide a recommended, near-optimal setup but to demystify the interactions of driving behaviors and styles in detail.

## 4    Experiment Results

In this section, we present several key results and relevant findings, to showcase the complicated interactions between driving behaviors and driving styles.

## 4.1 Conventional Classification Against Self-clustering

### Finding 1: Heterogeneous Styles can be Generated from DBSCAN.

We select the most important dimension (Principle Component 1) after PCA dimension reduction to represent the original driving behavior feature set. The larger the feature variance, the more original data information can be retained. Although the percentage of Principle Component 1's variance in all feature variances is only 0.65990205, it can also retain most of the required information for information visualization. After using DBSCAN algorithm to assign data points, we showcase the clustering results (as shown in Fig. 3-(a)) and the characteristic examples of the change, in terms of microtubule length versus time, are shown in Fig. 3-(b).

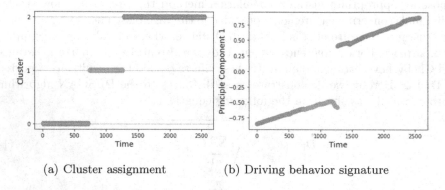

(a) Cluster assignment  (b) Driving behavior signature

**Fig. 3.** Driving-style quantification results of Driver Group 3 under road condition 4.

**Table 2.** Clustering results for different driver groups under the same road conditions.

| Distance measurements | Parameters combination identification | | | | | | | | |
|---|---|---|---|---|---|---|---|---|---|
| | *eps* | 0.125 | 0.25 | 0.5 | 0.125 | 0.25 | 0.5 | 0.125 | 0.25 | 0.5 |
| | *minPoints* | 3 | 3 | 3 | 6 | 6 | 6 | 9 | 9 | 9 |
| Eulidean distance | Group 1 | 6 | 5 | 3 | 6 | 5 | 3 | 7 | 5 | 3 |
| Manhattan distance | | 8 | 5 | 3 | 10 | 5 | 3 | 10 | 5 | 3 |
| Eulidean distance | | 5 | 3 | 3 | 5 | 3 | 3 | 6 | 3 | 3 |
| Eulidean distance | Group 2 | 12 | 3 | 3 | 12 | 3 | 3 | 13 | 3 | 3 |
| Manhattan distance | | 15 | 5 | 3 | 17 | 6 | 3 | 17 | 6 | 3 |
| Eulidean distance | | 10 | 3 | 3 | 10 | 3 | 3 | 10 | 3 | 3 |
| Eulidean distance | Group 3 | 6 | 3 | 3 | 6 | 4 | 3 | 6 | 5 | 3 |
| Manhattan distance | | 6 | 5 | 3 | 6 | 6 | 3 | 6 | 6 | 3 |
| Eulidean distance | | 5 | 3 | 3 | 6 | 4 | 3 | 6 | 5 | 3 |

Different from conventional methods, self-clustering method automatically divides the whole driving stage's data into three categories. We report similar results as presented in Table 2 by adopting the same research method for multiple drivers' data.

**Finding 2: Time-Series Variation is Considered by DBSCAN.** As displayed in Fig. 3, there are certain continuities in behaviors from a single complete time interval. For instance, the clustering part with green color shows that the drivers exhibits the same driving style over this period. However, behaviors for the whole timeline do not exhibit the same characteristics, while they change dramatically across all driving events. This phenomenon coincides with the different driving stages presented in Fig. 1: the driver entered the following stage after overtaking, which further expounds that drivers will give different driving styles in various driving events. What's more, as is evident from Fig. 3-(b), the overtaking stage also consists of different driving styles.

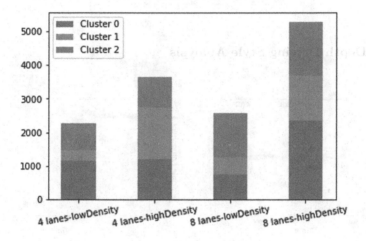

**Fig. 4.** Clustering results under different road conditions.

**Finding 3: Driving Styles can be Customized for Different Roads.** In order to verify whether the clustering results of drivers change under different traffic conditions, we conduct experiments on driving data under different road conditions. We observe that, though the proportion of different driving styles in the whole driving event has changed, the clustering results in a single driving event are still three. Nevertheless, the same driving style has different characterization probability under different road conditions. As shown in Fig. 4, the drivers will show the driving style 1 (cluster 1) with a smaller-time period, under relatively small traffic density.

**Finding 4: Degree of Expression of Different Styles.** We also observe that different styles have different degrees of representation in the whole driving event. Taking the clustering results, obtained from this setting (i.e. eps = 0.5, minPoints = 9, Distance Measures = Eulidean Distance) as an example, the three driver groups lead to three clustering results as driving styles. But the degrees of different styles are different. Table 3 backs up this finding: Drivers 1 is style 1 56% of the whole time, with the time of style 2 and style 3, 30% and 14% respectively. The pattern for drivers 2 and drivers 3 appear to be reversed. Style 3 is the domain style during this period, with 54% for drivers 2 and 50% for drivers 3. It can be seen that the differences between individuals are significant.

**Table 3.** Different levels of style representation.

|         | Style 1 | Style 2 | Style 3 |
|---------|---------|---------|---------|
| Drivers 1 | 56%   | 30%     | 14%     |
| Drivers 2 | 18%   | 28%     | 54%     |
| Drivers 3 | 29%   | 21%     | 50%     |

## 4.2  In-Depth Driving Style Analysis

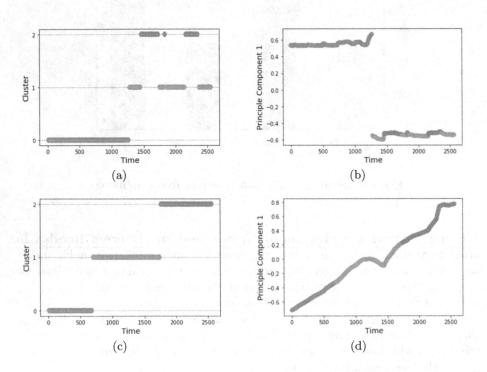

**Fig. 5.** Quantification results of driving styles after breakdown analysis.

**Finding 5: Isolated Features can Greatly Impact the Classifications.**
We perform breakdown analysis by removing key features. Hereby, we isolate

Position-related features or Rotation-related features. Figure 5-(a) and Fig. 5-(b) report the results, where we remove Position-related features; and Fig. 5-(c) and Fig. 5-(d) report the results, where we remove Rotation-related features. We observe that, though the classification results change, the clustering results are still complicated. After comparing Fig. 5-(a)/(c) with Fig. 5-(b)/(d), though the driving styles are still classified into three clusters, removing features have significantly impacted the robustness of classifications, in terms of timeline, due to greatly-impacted features.

**Finding 6: Transient Effects form Mutable Driving Styles.** [21] considers driving styles are transient, and explain that a driver can be aggressive at one period but normal for others. Our experiments also back up that, it's possible for the mutation of driving style, from computational perspectives. As shown Fig. 5-(a), two different driving styles alternately characterize the driver's behavior in the second half part of the whole procedure. This can be explained as the driver's step on the brake and the oil port in congestion, showing a particular different style compared with drivers with stable driving speed on highway. To this end, our mechanism are more adaptive and robust compared with conventional methods.

## 4.3   Driving Styles Overlap

**Finding 7: Clusters That are More Likely to Overlap.** We extend our studies by changing the time window to re-cluster the statistics, and combine the results after using different sizes of time window. We observe that the same driving behavior is possible to be classified into different styles. The part between the two red lines in Fig. 6 reflects this complex clustering situation, some data points are clustered such that they belong to two different clusters. More specifically, Cluster 1 is more likely to overlap with the other two driving styles.

**Finding 8: Scenarios That are More Likely to Occur Overlap.**
In Fig. 6-(c), the overlap occurs when the vehicles encounters a traffic light. This is because the steering of the car is not completely independent of acceleration or braking under the driving event. Thus, wrongly classifying the steering as acceleration or braking leads to this overlap. While in the situations at Fig. 6-(a) and -(b), there are complicated traffic conditions, like congestion leading the driver to have driving behaviors, that are not routine.

So far, we have compared the clustering results based on multiple drivers and a single time series. We find that not only different drivers will show various styles, but also the same driver will show style differences among driving periods.

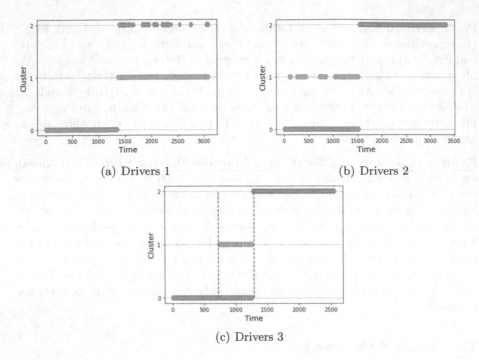

(a) Drivers 1                    (b) Drivers 2

(c) Drivers 3

**Fig. 6.** Overlaps between driving styles.

## 5    Discussion

This work analyze the clustering results of driving behaviors and extensively
retrieve multiple findings from computational perspectives. Unlike conventional
classifications of driving styles, our findings show that our proposed mechanisms
for driving styles reveal more opportunities for adaptive and personalized driving
styles' characterizations. Driving styles will migrate under the influence of traffic
environments, road conditions, and other environmental factors. We can only
conjuncture that drivers are more likely to be in a specific driving style, instead
of classifying them into one specific type. Based on our findings, we further relate
the key points hereby to stimulate new insights and follow-up investigations.

### 5.1    Driving Behavior Combination

Driving behaviors can describe combined events, because different driving behaviors are not always distinguished and independent. For example, acceleration is
often followed by turning at traffic lights, and deceleration behavior is accompanied by stopping in front of traffic lights. After turning to the other direction, the
driver can adjust the speed and lane to better drive experience. This phenomenon
makes it difficult to distinguish steering from acceleration or deceleration behavior under certain conditions.

## 5.2  Traffic Environment

The rising number of vehicles can exacerbate road congestion and render the flow of traffic more complicated. Certain events may be detrimental in some situations. The difference in road conditions will bring difficulties in identifying driving style because driving behavior will change. For instance, the calm type driver will frequently step on the brake and the oil port in the case of congestion, showing a particular aggressive style, while the aggressive type will maintain a relatively stable driving speed on highway. Our studies reveal that, the analysis of driving behaviors in specific road conditions is more critical for style representation.

## 5.3  Driving Behavior Levels

[32] divides the completion of a driving main task into four main levels: strategic level, mode level, operational level and scene awareness level. Driving style can be reflected on any level. There are: (1) Decision preferences at the strategic level, such as selecting short-distance routes [18]; (2) Driving mode preferences at the mode level, such as frequent lane change, near-following, and far-following [27]; and (3) Operating mode preferences at the operational level, such as uniform acceleration, rapid acceleration, and whether to turn on the turn signal in time [25]. At the level of perceptions, there are recognition preferences such as whether to observe the external area adequately before the lane changes and whether the sight line deviates from the path for a long time [35].

We also vision that our study is complementary to other relevant works as well. [11] provides alternative mechanisms to obtain drivers' multi-modal statistics in a more user-friendly manner. [30] examines the influences of user trust in auto-vehicles, by applying BROOK [23] as the dataset. [29] also ignites the opportunities for more practical infrastructure to enhance the dataset. We believe future works are both essential and promising.

# 6  Conclusions

We argue that driving styles demand adaptive classifications, and such mechanisms are essential for adaptive and personalized Human-Vehicle Interaction systems. To this end, we conduct an in-depth study to demystify complicated interactions between driving behaviors and styles. The key idea behind this study is to enable different numbers of clusters on the fly, when classifying driving behaviors. We achieve so by applying Self-Clustering algorithms (i.e. DBSCAN) over a state-of-the-art open-sourced dataset of Human-Vehicle Interactions. Our results derive 8 key findings, which showcases the complicated interactions between driving behaviors and driving styles. Hence, we conjecture that future Human-Vehicle Interactions systems demand similar approaches for the characterizations of drivers, to enable more adaptive and personalized Human-Vehicle Interaction systems. We believe our findings can stimulate and benefit more future research as well.

**Acknowledgements.** We thank for the anonymous reviewers from HCI'21 Regular Paper Track and all members of User-Centric Computing Group for their valuable and insightful feedbacks, especially Mr. Zhentao Huang. This project is a part of the BROOK project from the User-Centric Computing Group in the University of Nottingham Ningbo China [23].

# References

1. Aghabozorgi, S., Shirkhorshidi, A.S., Wah, T.Y.: Time-series clustering-a decade review. Inf. Syst. **53**, 16–38 (2015)
2. Alelyani, S., Tang, J., Liu, H.: Feature selection for clustering: a review. Data Clustering: Algorithms Appl. **29**(1), 230 (2013)
3. Augustynowicz, A.: Preliminary classification of driving style with objective rank method. Int. J. Automot. Technol. **10**(5), 607–610 (2009)
4. Chen, S.W., Fang, C.Y., Tien, C.T.: Driving behaviour modelling system based on graph construction. Transp. Res. Part C: Emerging Technol. **26**, 314–330 (2013)
5. Chen, Z.J., et al.: Vehicle behavior learning via sparse reconstruction with $\ell_2 - \ell_p$ minimization and trajectory similarity. IEEE Trans. Intell. Transp. Syst. **18**(2), 236–247 (2016)
6. Constantinescu, Z., Marinoiu, C., Vladoiu, M.: Driving style analysis using data mining techniques. Int. J. Comput. Commun. Control **5**(5), 654–663 (2010)
7. Driggs-Campbell, K., Govindarajan, V., Bajcsy, R.: Integrating intuitive driver models in autonomous planning for interactive maneuvers. IEEE Trans. Intell. Transp. Syst. **18**(12), 3461–3472 (2017)
8. Ester, M., Kriegel, H.P., Sander, J., Xu, X., et al.: A density-based algorithm for discovering clusters in large spatial databases with noise. KDD **96**, 226–231 (1996)
9. Hua, J., Tembe, W.D., Dougherty, E.R.: Performance of feature-selection methods in the classification of high-dimension data. Pattern Recogn. **42**(3), 409–424 (2009)
10. Huang, J., Lin, W.C., Chin, Y.K.: Adaptive vehicle control system with driving style recognition based on headway distance (Oct 2 2012), uS Patent 8,280,560
11. Huang, Z., et al.: Face2multi-modal: In-vehicle multi-modal predictors via facial expressions. In: Adjunct Proceedings of the 12th International Conference on Automotive User Interfaces and Interactive Vehicular Applications, AutomotiveUI 2020, Virtual Event, Washington, DC, USA, 21–22 September, 2020, pp. 30–33. ACM (2020). https://doi.org/10.1145/3409251.3411716
12. van Huysduynen, H.H., Terken, J., Eggen, B.: The relation between self-reported driving style and driving behaviour: a simulator study. Transp. Res. Part F: Traffic Psychol. Behav. **56**, 245–255 (2018)
13. Ishibashi, M., Okuwa, M., Doi, S., Akamatsu, M.: Indices for characterizing driving style and their relevance to car following behavior. In: SICE Annual Conference 2007 pp. 1132–1137. IEEE (2007)
14. Janžekovič, F., Novak, T.: PCA-a powerful method for analyze ecological niches. Principal component analysis-multidisciplinary applications, pp. 127–142 (2012)
15. Johnson, D.A., Trivedi, M.M.: Driving style recognition using a smartphone as a sensor platform. In: 2011 14th International IEEE Conference on Intelligent Transportation Systems (ITSC), pp. 1609–1615. IEEE (2011)
16. Lee, J.G., Kim, K.J., Lee, S., Shin, D.H.: Can autonomous vehicles be safe and trustworthy? effects of appearance and autonomy of unmanned driving systems. Int. J. Hum.-Comput. Inter. **31**(10), 682–691 (2015)

17. Lefevre, S., Carvalho, A., Borrelli, F.: A learning-based framework for velocity control in autonomous driving. IEEE Trans. Autom. Sci. Eng. **13**(1), 32–42 (2015)
18. Li, G., Li, S.E., Cheng, B.: Field operational test of advanced driver assistance systems in typical Chinese road conditions: the influence of driver gender, age and aggression. Int. J. Automot. Technol. **16**(5), 739–750 (2015)
19. Lin, Q., Wang, W., Zhang, Y., Dolan, J.M.: Measuring similarity of interactive driving behaviors using matrix profile. In: 2020 American Control Conference (ACC), pp. 3965–3970. IEEE (2020)
20. MacAdam, C., Bareket, Z., Fancher, P., Ervin, R.: Using neural networks to identify driving style and headway control behavior of drivers. Veh. Syst. Dyn. **29**(S1), 143–160 (1998)
21. Murphey, Y.L., Milton, R., Kiliaris, L.: Driver's style classification using jerk analysis. In: 2009 IEEE Workshop on Computational Intelligence in Vehicles and Vehicular Systems, pp. 23–28. IEEE (2009)
22. Patro, S., Sahu, K.K.: Normalization: A preprocessing stage. arXiv preprint arXiv:1503.06462 (2015)
23. Peng, X., Huang, Z., Sun, X.: Building BROOK: A Multi-modal and Facial Video Database for Human-Vehicle Interaction Research. the 1st Workshop of Speculative Designs for Emergent Personal Data Trails: Signs, Signals and Signifiers, co-located with the 2020 CHI Conference on Human Factors in Computing Systems, (CHI), Honolulu, HI, USA, 25–30 April 2020 pp. 1–9 (2020), https://arxiv.org/abs/2005.08637
24. Reason, J., Manstead, A., Stradling, S., Baxter, J., Campbell, K.: Errors and violations on the roads: a real distinction? Ergonomics **33**(10–11), 1315–1332 (1990)
25. Sagberg, F., Selpi, Bianchi Piccinini, G.F., Engström, J.: A review of research on driving styles and road safety. Human Factors **57**(7), 1248–1275 (2015)
26. Sander, J., Ester, M., Kriegel, H.P., Xu, X.: Density-based clustering in spatial databases: the algorithm gdbscan and its applications. Data Min. Knowl. Disc. **2**(2), 169–194 (1998)
27. Satzoda, R.K., Gunaratne, P., Trivedi, M.M.: Drive quality analysis of lane change maneuvers for naturalistic driving studies. In: 2015 IEEE Intelligent Vehicles Symposium (IV), pp. 654–659. IEEE (2015)
28. Scholz, M., Gatzek, S., Sterling, A., Fiehn, O., Selbig, J.: Metabolite fingerprinting: detecting biological features by independent component analysis. Bioinformatics **20**(15), 2447–2454 (2004)
29. Song, Z., Wang, S., Kong, W., Peng, X., Sun, X.: First attempt to build realistic driving scenes using video-to-video synthesis in opends framework. In: Janssen, C.P., Donker, S.F., Chuang, L.L., Ju, W. (eds.) Adjunct Proceedings of the 11th International Conference on Automotive User Interfaces and Interactive Vehicular Applications, AutomotiveUI 2019, Utrecht, The Netherlands, 21–25 September 2019, pp. 387–391. ACM (2019). https://doi.org/10.1145/3349263.3351497
30. Sun, X., et al.: Exploring personalised autonomous vehicles to influence user trust. Cogn. Comput. **12**(6), 1170–1186 (2020). https://doi.org/10.1007/s12559-020-09757-x
31. Taubman-Ben-Ari, O., Mikulincer, M., Gillath, O.: The multidimensional driving style inventory–scale construct and validation. Accid. Anal. Prev. **36**(3), 323–332 (2004)
32. Toledo, T., Koutsopoulos, H.N., Ben-Akiva, M.: Integrated driving behavior modeling. Transp. Res. Part C Emerg. Technol. **15**(2), 96–112 (2007)

33. Vaitkus, V., Lengvenis, P., Žylius, G.: Driving style classification using long-term accelerometer information. In: 2014 19th International Conference on Methods and Models in Automation and Robotics (MMAR), pp. 641–644. IEEE (2014)
34. Wang, J., Lu, M., Li, K.: Characterization of longitudinal driving behavior by measurable parameters. Transp. Res. Rec. **2185**(1), 15–23 (2010)
35. Xian, H., Jin, L.: The effects of using in-vehicle computer on driver eye movements and driving performance. Adv. Mech. Eng. **7**(2), (2015)
36. Xie, B., Pan, W., Shen, X.: Penalized model-based clustering with cluster-specific diagonal covariance matrices and grouped variables. Electron. J. Stat. **2**, 168 (2008)
37. Zheng, P., McDonald, M.: Manual vs. adaptive cruise control-can driver's expectation be matched? Transp. Res. Part C: Emerg. Technol. **13**(5–6), 421–431 (2005)

# Interactive Framework of Cooperative Interface for Collaborative Driving

Jun Zhang[1,2], Yujia Liu[1,2], Preben Hansen[3], Jianmin Wang[2], and Fang You[2(✉)]

[1] College of Design and Innovation, Tongji University, Shanghai 201804, China
{alexmaya,liuyujia}@tongji.edu.cn
[2] Car Interaction Design Lab, College of Art and Media, Tongji University, Shanghai, China
{wangjianmin,youfang}@tongji.edu.cn
[3] The Department of Computer and Systems Sciences,
Stockholm University, Stockholm, Sweden
preben@dsv.su.se

**Abstract.** Automobile intelligence improves the perception and decision-making capabilities of cars. This change of technology makes human and machine a joint cognitive and decision-making system, consequently changing the paradigm of human-machine interaction. The machine is no longer a mere tool but a team partner. Both academia and industry are actively investigating human-machine cooperative driving, such as take over, shared control and cooperative driving. However, currently, there is no cooperative driving framework that considers the cognitive characteristics of the interaction between humans and agents. We propose a cooperative interface interaction framework based on human-machine team cognitive information elements that need to be exchanged by both parties in cooperative driving, such as intention, situation awareness, prediction, and their impact on driving tasks. The proposed framework can provide a cognitive dimension for cooperative driving research, which can be used as a reference for the design of interaction in highly automated vehicles.

**Keywords:** Human-machine interaction · Interface · Cooperative interface · Human-machine cooperative · Cooperative driving

## 1 Introduction

The development of automobile intelligence makes advanced driver assistance systems (ADAS) shift towards fully automated vehicles [1]. ADAS improves human perception and control ability. But any form of automation will not simply enhance the ability to perform tasks, because it changes the task itself [2]. For example, the role of the driver has changed from the physical control of the vehicle to the supervision of driving tasks and automation systems [3, 4]. Supervision automation means the task changes from behavioral task to a cognitive task. However, this shift in tasks also brings new problems. What kind of mechanism can make people not out of the loop when the person leaves the control interface, and can simultaneously cooperate with the intelligent system to control the car?

© Springer Nature Switzerland AG 2021
H. Krömker (Ed.): HCII 2021, LNCS 12791, pp. 351–364, 2021.
https://doi.org/10.1007/978-3-030-78358-7_24

According to SAE's Level 0-Level 5 autonomous driving classification, starting from level 3, the guidance and control tasks are performed by the system. however, at this stage, drivers have to supervise the driving tasks of the system and are used as a reserve to take over control at any time [5]. Since the technical advances are not progressed towards full automation (level 5), the human driver has to be available to take over the driving task (level 3) or when the requirements of the system are not satisfied (e.g. not on a highway) (level 3 and 4) [6]. In other words, before real level 5 autonomous driving is realized, human and machine shared control of the vehicle. Shared control means that in certain critical states, driving rights will be switched between man and machine. For this critical state of handover, there are currently two main research directions: (1) The first one is takeover, which is the black and white switching mode. Although takeover is one method, even if there is enough time for the driver to take over the car, it may fail because of the inability to quickly restore situation awareness [7]. Moreover, frequent switching of driving rights will affect driver's trust in the system and damage the user experience. The second one refers to cooperative driving, which is a model of human-machine integration. There are some perspectives for cooperative driving. One is the cooperation of V2V (vehicle-vehicle) or V2I (vehicle-infrastructure) [8], and the other perspective deals with the cooperation between the driver and the intelligent system in the car to form a joint team in order to perform a collaborative driving tasks together. Hoc et al. [9, 10] believe that "machine should support human operators as part of the team, not replace them."

Walch et al. [6] proposed to use the cooperative interface to collaborate in abstract level (plan, decision-making) to overcome the limitations and uncertainties of the system. In addition, they believe that a lot of scenarios that need to be handed over are caused by the system's lack of perception information or the inability to correctly judge the current situation and unable to make the next decision. For example, the car cannot accurately identify whether the person blocking the front or the flying plastic bag. In a scenario like this, only the driver needs to approve the system to execute the next strategy. Subsequently, the system controls the car in accordance with the established strategy. It appears that the cooperative interface does not concentrate on specific car control right, but the cooperation of human and machine cognitive dimensions (the exchange of opinions considered in their minds).

Essentially, the cooperation between human and intelligent system is based on the cooperation in the cognitive dimension of the two worlds (the world felt by the machine and the world felt by the human). In order to build an interactive framework for cooperative interface, we may first highlight two human abilities: (1) *Cognitive ability*, which refers to cognitive tasks (perception, consciousness, reasoning and judgment, etc.) and (2) *Behavioral abilities*, refers to the activities that a person's body functions can complete. When a machine is an extension of the human body's abilities, we can call it "tools". If it is an extension of cognitive ability, we would prefer to call them "partners" because it seems that they have seen what we saw, thought about what we thought, and did what we wanted to do. In such teams, technology can work with humans as team partners, not just as automated tools [11].

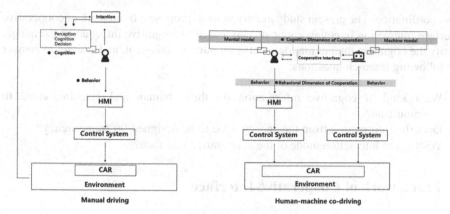

**Fig. 1.** Structure comparison between manual driving and human-machine co-driving

As shown in Fig. 1, in manual driving, inspired by Guy A. boy [12]'s framework of car driving activity as a "Perception-Cognition-Action" dynamic regulation loop, the driver first performs cognitive tasks (perception, cognition, decision-making) based on his own intentions, and then performs behavioral tasks and interacts with the control system through human-machine interface (HMI) and controls the movement of the car.

In human-machine co-driving, both human and intelligent system can perform cognitive tasks and control the movement of the car though the control system. Imagine this is a parallel cognitive and control system. Because even if one of them does nothing (cognition and action), the established driving task may be successfully completed. Currently, the intelligent system is like a "virtual driver" [13], which seems to be capable of performing driving tasks alone.

It can be seen from Fig. 1 that this structure changes the interactive objects that driver face, and the human interactive objects change from a control system to a "virtual driver". Obviously, they need an interface for communication. However, this interface is not for controlling the vehicle, but for two cognitive systems to communicate cognitive information (intention, perception, and decision-making) during the execution of tasks. The medium that carries this kind of communication is regarded as "cooperative interface".

This shift means that people and machines need to be considered together, rather than separate entities linked together through HMI [14]. At this time, the view of human-machine interaction changes from the Human-Machine System (HMS) view to the Joint Cognitive System (JCS) view [15]. JCS considers that cognition is a goal-oriented interaction between people and artifacts in order to generate work in a specific context and at the level of ongoing work [16, 17].

We claim that in human-machine cooperative (HMC) driving, people and intelligent system form a joint cognitive and decision-making team system. Automation and people in the team must be coordinated as a joint cognitive body [18]. The result of cognition directly outputs the strategy of decision-making and the affects the behavior. Therefore, the cooperation of cognitive dimension is an important dimension of team cooperation between human and agents. The cooperative interface will be the carrier of this coordination. The present study mainly aims to propose a framework of cooperative interface interaction in collaborative driving based on cognitive information exchange, clarify the cognitive interaction logic between human and agent, and attempt to answer the following research questions:

1. What kind of cognitive information do they (human and machine) need to communicate?
2. Does the information from the agency need to be designed for transparency?
3. What is the interaction mode of the cooperative interface?

## 2 Framework of Cooperative Interface

Inspired by Flemisch's hierarchical framework of human-machine co-driving cooperation [19], which divided driving tasks into four levels: "navigation level, maneuver level, trajectory level and control level", and Wang's Non-critical Spontaneous Situations (NCSSs) cooperation framework [20], which is intended to provide an overview over the most relevant elements that influence the NCSSs. Combining these two frameworks, we propose an interactive framework of human-computer cooperative driving interface.

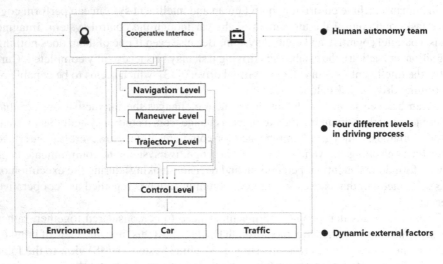

**Fig. 2.** Interaction framework of cooperative driving interface

As shown in Fig. 2, it contains three parts: human autonomy team (HAT), the four level of interaction in driving process, and the dynamic external factors. We believe that the study of the overall framework elements and the influence factors of interface information can give some inspiration to the interaction design. The following contents will explain the elements in the framework in detail.

## 2.1 Human Autonomy Team

### Team

Salas et al. [21] provided a definition of team: "a distinguishable set of two or more people who interact dynamically, interdependently and adaptively toward a common and valued goal, who have each been assigned specific roles or functions to perform and who have a limited life span of membership". Therefore, a team is a goal-oriented combination, where members can be human or non-human. In smart cars, human and smart system can form a Human-Autonomy Team (HAT). Then, Hoc [9] then put forward the minimum requirements for teamwork:

(1) *Each one strives towards goals and can interfere with the other on goals, resources, procedures, etc.*
(2) *Each one tries to manage the interference to facilitate the individual activities and/or the common task when it exists.*

His definition means that the activity has a goal. In order to achieve the goal, the two or more agents interfere with each other and adjust their actions.

This interference can be cognitive or behavioral. Fiebich et al. [22] summarized the cooperation between the cognitive dimension and the behavioral dimension, pointing out that the minimum premise of the cognitive dimension cooperation is: (1) two (or more) agents perform actions to pursue the same goal, and (2) the agents know that they have the same goal. The premises of cooperation in the behavioral dimension are: (1) Two (or more) agents coordinate their behavior in space and time. (2) It can be observable from the outside. (3) Bringing about a change in the environment.

Synthesizing their viewpoints, the "Team" can be regarded as an adaptive dynamic system, and they adjust their state through interaction to achieve a common goal [23]. In the driving task, the human and machine share control of the vehicle, and the team's goal are constantly changing at different levels (Navigation level, Maneuver level, trajectory level and control level) of the driving task based on the corresponding intention. Zimmermann et al. [24] proposed five layers of cooperation elements for human-machine cooperation in highly complex and unpredictable dynamic scenarios during driving including the user and machine intention, the mode of cooperation, the dynamic task and action allocation, the human-machine interface and the contact between human and machine.

To sum up, in the team tasks, both parties of the human-machine team are goal-oriented. Both of them contribute their own abilities, communicate perceptions, plans and decisions with each other, and adjust their own status to achieve the goal of successfully completing the tasks.

### Cooperative Interface

The cooperative interface is the carrier of human-machine cognitive information communication. Team cognition is a cognitive activity at the team level, formed by team members through clear communication activities (such as talking and face-to-face communication.) [25]. In other words, team cognition is the product of interaction between team members. Research shows that effective communication and coordination are positively related to team performance [25]. From the human-machine team, members exchange the cognitive resources of both parties through team cognitive activities with the aim reach a common understanding.

Walch et al. [26] believe that the purpose of the cooperative interface is to keep people in the loop and realize real-time communication between team members. The purpose is to improve situation awareness and the understanding of intentions and actions by both parties. In addition, they further put forward four basic requirements for designing a cooperative interface, respectively, mutual predictability, directability, shared situations representation, and calibrated trust in automation. Wood et al. [27] consider that in order to promote teamwork, the cooperative interface needs to provide a shared status representation, including information about the partner's status, plans, goals, and activities, as well as information about the current task status, environment, and situation. Zimmermann et al. [26] concluded that the human-machine cooperative interface should (1) display the intention of the user and the machine and (2) convey the reconfiguration of the dynamic adaptive system.

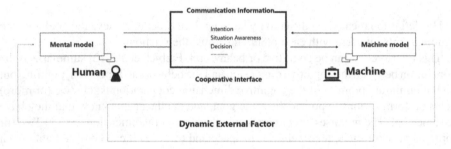

**Fig. 3.** Human-machine cognitive information exchange

Based on the above analysis of the literature, the main cognitive information elements that need to be communicated in the human-machine cooperative interface include: intention, situation awareness, planning, decision-making, task assignment and cooperation mode. In the meanwhile, the transparency of information can increase people's understanding of machines and increase the trust in machine decisions. Finally, what kind of interaction mode is used to facilitate the communication of this information is also an important aspect. Therefore, the following discusses the cooperative interface from the three aspects of cognitive information element composition, transparency, and interaction mode.

### Cognitive Information
**Intention:** The individual's intentions determine the action goals, plans, and cognitive tasks. Intentions are determined by external attributes (environment, self-vehicle dynamics and surrounding traffic) and are affected by internal factors. In terms of drivers, internal factors are their motivation, short-term goals, mood, experience, etc. For the system, internal factors may be preset codes of conduct [21].

Understanding the intentions of others in the team and the intentions that can be characterized as understood are considered key factors for the cooperation. This understanding is two-way. Both parties need to establish common beliefs and joint intentions through interactive activities. Hoc et al. [9] divide cooperative activities into action level, planning level and meta-level by level. Therefore, intentions can occur at any level. Action level is a short-term task. Intentions change rapidly. This intent can be communicated using verbal commands, such as rapid lane changes and overtaking. The planning level is a medium- and long-term task, and the two parties can exchange intentions and coordinate plans, such as navigation and route planning. As a long-term team model, Meta-level bases on training and experience.

However, intention is a mental activity and cannot be directly observed. To recognize the intention of the other party, it is of necessity to use interface (voice, text, etc.) or measurement and reasoning based on behavior indicators. K. Bengler et al. [28] proposed that the machine must infer intentions in the following three ways: (1) The machine must measure and interpret human intentions according to its tasks and goals (2) The machine needs to communicate its intentions in a reliable way, allowing users to obtain Information (3) The machine needs to obtain information, regardless of whether the information has been noticed and processed by the human-computer interaction partner.

In summary, understanding the intentions of cooperating members exerts an important impact on team joint activities. Consequently, the cooperative interface needs to accurately convey and represent intentions in a suitable way (visual, auditory, and tactile).

**Situation Awareness:** In general, situation awareness (SA) is a person's perception, understanding and prediction of the surrounding changing environment. It determines the person's next decision and action. Thus, it is an extremely important psychological activity. Endsley [29] defines three levels of SA. On the first level, the driver perceives each element in the environment. On the second level, the driver has a comprehensive understanding of the current situation. On the third level, the driver makes a prediction for the possible future development of these elements. For non-human actors, the ability to provide mechanisms for predicting future actions based on prior knowledge, perception and prediction is called Autonomy Situation Awareness (ASA) [30].

From the perspective of a team, SA currently does not have a unified definition. The most influential related theory is the SA research based on the expansion of individuals to teams, such as: Endsley's shared SA, Salas et al.'s team SA model [31] as well as Shu and Furuta's mutual awareness model [32]. The other one is based on system SA research, for instance, Stanton's distribution situation awareness theory (DSA) [33].

However, team SA (TSA) is more than just combining SA of individual team members. Team SA consists of collective awareness of the team. Salas et al. [34] believe that the key to team SA lies in the information interaction process of team members. The SA obtained by an individual forms a team SA through the interaction team, and the individual SA can be modified in real time. Consequently, team members must have SA related to their roles and goals, as well as SA associated with other team members. Anticipation of the behavior and information needs of other team members in the team is of much importance for effective TSA [35]. The SA among members is distributed and exchanged in the team through Team Processes (communication, communication, coordination, etc.). Therefore, team SA includes three aspects, respectively, SA of a single member, SA shared between team members, and SA of the entire team combination.

Since the results of human-machine SA may be different, the two parties can cooperate to make up for the other's possible shortcomings. For example, Walch et al. [36] studied how to make the driver's perception cooperate with the automation system through a collaborative interface, aiming to overcome the limitation of automatic perception in the overtaking scene. Chao et al. [37] studied people's prediction of the intentions of other traffic participants, and integrated the prediction judgment into the operation of the automation plan level to make up for the lack of judgment ability of the automation system.

Nevertheless, in HAT, the driver's over-reliance on ADAS will lead to a mental model of over-dependence, which will make the driver spend less time focusing on the road and more time engaged in non-driving tasks. Once the machine has insufficient capacity to correctly understand the current situation, and the driver lacks SA, being unable to quickly take over the vehicle, it will cause a serious accident. Rinta Kridalukmana et al. [38] put forward a supporting SA model (SSA) of multi-agent cooperation to provide support for man-machine team cooperation in driving.

Briefly, SA is an important cognitive activity of information acquisition and analysis. For the elements of the current changing scene, the two parties should fully exchange information and reach a consistent understanding with the aim to take the correct action. The cooperative interface plays a role in full communication, allowing both parties to complement each other and promote the formation of shared SA.

**Decision:** The intelligent system undertakes part of the cognitive work and can make decisions. Adaptive automation can replace the operator's perception as well as assist or replace decision-making processes and actions [39, 40]. However, the effectiveness of this automated decision-making will have an impact on the human mental model, resulting in excessive trust or distrust. In the cooperation between humans and agents, they need to exchange plans and decisions for the following step. Walch et al. [6] investigated the interaction of the cooperative interface to achieve the next step plan when the system remains uncertain or limited. First, the system will alert the driver and inform him of restrictions or uncertainties. Second, it will provide the driver with solutions such as overtaking, route selection, takeover, and other decisions, and finally wait for the driver to approve one of them. If the driver chooses one of the recommendations, the automation will be performed automatically, thus avoiding manual takeover. If the driver does not select any item, the system will automatically downgrade and allow the driver to resume manual driving.

### Information Transparency

As the degree of intelligence increases, agents can make complex decisions, and such decisions and actions may be challenging to understand. Therefore, proper characterization of the reasons for the overthrowing process and behavior behind the decision contributes to people's understanding. One method is to design the machine "transparency module" and thus the behavior, reasoning and expected results of the intelligent robot are transparent to the operator [41, 42]. As a result, people's understanding of machines can be enhanced. Therefore, transparency can provide a mechanism for understanding and monitoring agent behavior, including the predictability of recent actions [43].

Chen and her colleagues [44] took inspiration from the three-level situation aware-ness model and developed an Agent Transparency (SAT) model based on situation aware-ness. It supports human understanding of machines by emphasizing the intent, reasoning, future plans and uncertainties in association with intelligent machines. The SAT frame-work is divided into three levels: Level 1: Intelligent machines provide people with basic information about their current status, goals, intentions, and plans. Besides, they can help people perceive the current actions and plans of intelligent machines. Level 2: Intelligent machines provide people with their reasoning process and the environmental constraints considered when planning their behavior. It benefits people in understanding the current behavior of the machine. Level 3: Intelligent machines provide people with predictions for the future, their consequences, and the possibility of success and failure, also containing some uncertainties. It is helpful for people to make the prediction of future results.

Based on the requirements for the duration of the driving task, the time is often extremely short, and the driver may not have enough time to absorb the transparency information provided by the machine, especially the first level of SAT information. RintaKridalukmana et al. proposed a Time-Constraint-Driven Transparency framework (TCDT), dividing the theme of transparency into four categories, respectively, interpre-tation and performance, planning, decision-making and coordination as well as intention and outcome prediction. According to time constraints, the transparency theme is set priority and visibility [38].

In general, transparency exerts an important role in explaining the behavior and decision-making of agents, which can improve the trust of the driver in the machine and the performance of the team [44]. Therefore, in the cooperative interface, appropriate transparency design for information is of great necessity.

## *Interactive Mode*

The interaction between humans and smart products are widely used in multimodal-ity such as gestures, voice, and expressions. Multi-modal interaction integrates human vision, hearing, touch, and other channels, which can completely simulate the natural way of interaction between people. As said by Norman [45]: "The machine should not allow the operator to leave the loop, but should continue to interact with it in a normal and natural way."

Bengler et al. [34] proposed that multi-modal interaction methods should be estab-lished, including auditory, tactile, and visual channels, instead of reducing the informa-tion between humans and machines to visual interfaces. Numerous models of car are also equipped with gesture recognition technology, which operates the overall display screen located in front of the driver through gesture recognition. Voice is the mode of expressing the user's intention. In human-computer interaction, the user's voice infor-mation is translated into a certain semantic after being accepted by the agent to help the agent understand the user's intention. Flemisch [19] proposed the H-mode mode inspired by the collaboration with the horse through the reins. People can use the tactile interface to convey the driving intention through force in order to change the level of automation. Additionally, vehicle-mounted robots or virtual robots can also be used as an emerging interactive method, which can communicate intentions with the driver in an anthropomorphic way. For example, the positive or negative expressions of virtual characters can improve the driving behavior of the driver [46]. Research by Johannes-MariaKrau et al. [47] proves that the introduction of anthropomorphic design functions may lead to an increase in people's trust in autonomous driving systems.

In addition, from the perspective of the interaction medium between the driver and the car, in future smart car applications, interaction is no longer limited to media that require physical control such as touch screens, knobs, and levers. In the fully automatic driving mode, traditional control equipment may be hidden or disappear and replaced by other methods. Any physical equipment in the car may become an interactive medium. For example, in terms of media materials, the material itself, flexibility, and curvature of the inner surface of the car realize diversified entities and build new forms of interaction.

Overall, the future smart car is not only a means of transportation, but also a mobile place for emotional interaction that can satisfy people's multi-level needs. The design of multi-mode cross-sensory should fully consider the design of the cooperative interface.

## 2.2 Level of Interaction

When in a driving task, long-term tasks and short-term tasks are involved according to the length of time, which usually refer to the three-layer hierarchical control structure of "strategy, tactics, and control" [48, 49]. Flemisch et al. [19] proposed four levels of classification on the three-level model including Navigation level, Maneuver level, trajectory level and control level.

Navigation level refers to the long-term plan of the driver to choose the destination. The interaction frequency is not high and the interaction time is not long. The driver can interact with the intelligent system through multi-modal interaction with the aim to select the destination.

Maneuver level represents a medium-term plan. It refers to the driver's intermittent handling of the detection and response tasks of incidents that occur in road conditions such as turning left and right, overtaking and merging.

Trajectory level is a short-term route planning, which contains space and time dimensions. For the same scene, humans and intelligent systems have the same intentions, but may adopt different path motion strategies. For example, in the overtaking scenario, the system provides a variety of path planning, allowing the driver to select one of them through the touch interface or take over driving the vehicle in accordance with his own intentions [6].

The control level is the level with the most frequent interaction. The driver directly controls the lateral and longitudinal movement of the vehicle through the interface.

The cooperation at these four levels does not remain static, but can include dynamic changes at each level of cooperation, and the four loops are related to each other.

## 2.3 Dynamic External Factors

According to Guy A. boy [12]'s description that car driving activity is defined as an continuous dynamic loop of regulation between (1) inputs, coming from the road environment, and (2) outputs, corresponding to the driver's behaviors implemented into the real world via the car, which generate (3) feedback, in the form of a new inputs, requiring new adaptation (that is, new outputs) from the driver, and so on. Compared with the human-machine team, we can consider the input part as dynamic external factors, environment, vehicle and traffic conditions.

**Environment:** Environment refers to the external conditions of the vehicle, including weather and roads. Among them, weather conditions will affect the perception of humans and intelligent systems, such as rain, snow and heavy fog. The road will affect the control strategies of people and intelligent systems, such as the width, curve, and type of the road.

**Vehicle:** Vehicle is dynamically changing, which is composed of internal factors and external factors. Internal factors include the state of the vehicle and its own performance, and external factors contain changes in position and speed caused by humans as well as intelligent systems controlling it. In addition, these changes also change the traffic environment and affect the intentions and decisions of other traffic participants.

**Traffic Conditions:** Traffic conditions refer to all participants in the traffic system, including various other vehicles and pedestrians on the road. These elements share road resources with the vehicle, and their location changes, speed, and intentions will affect the vehicle's action decisions.

The above dynamic influencing factors will affect the situation awareness of people and intelligent systems, consequently affecting their intentions and control plans.

## 3 Conclusion and Future Work

The present study distinguishes human-computer cooperation from the cognitive dimension and behavioral dimension, and combines it with driving tasks to propose a general cooperation framework, concentrating on the human-computer cognitive information exchange before and during the cooperation process. The human-machine team is goal-oriented, and joint intentions can point to any level of driving task cooperation. We believe that cognitive cooperation occurs earlier than behavioral cooperation. Therefore, we did not consider cooperative driving from the perspective of driving task cooperation level, but from the perspective of cognitive information exchange. Moreover, we considered the information elements needed for this kind of cognitive information exchange and its influential factors. In the future, we will use this framework to conduct research based on corresponding cooperation scenarios.

**Acknowledgements.** This work was supported by The National Key Research and Development Program of China (No. 2018YFB1004903), National Social Science Fund (No. 19FYSB040), Shanghai Automotive Industry Science and Technology Development Foundation (No. 1717). Excellent Experimental Teaching Project of Tongji University and Graduate Education Research and Reform Project of Tongji University (No. 2020JC35).

# References

1. Kyriakidis, M., Weijer, C.V.D., Arem, B.V., Happee, R.: The deployment of advanced driver assistance systems in Europe. In: ITS World Congress 2015 Proceedings. SSRN, Bordeaux, France (2015). https://doi.org/10.2139/ssrn.2559034
2. Norman, D.A.: Cognitive artifacts. In: Carroll, J.M. (ed.) Designing Interaction: Psychology at the Human–Computer Interface, pp. 17–38. Cambridge University Press, Cambridge (1992)
3. Saffarian, M., de Winter, J.C., Happee, R.: Automated driving: human-factors issues and design solutions. In: Proceedings of the Human Factors and Ergonomics Society Annual Meeting, vol. 56, no. 1, pp. 2296–2300. Sage, Los Angeles (2012)
4. Strand, N., Nilsson, J., Karlsson, I.M., Nilsson, L.: Semi-automated versus highly automated driving in critical situations caused by automation failures. Transp. Res. Part F Traffic Psychol. Behav. **27**, 218–228 (2014)
5. SAE: Taxonomy and definitions for terms related to driving automation systems for on-road motor vehicles. Standard J3016_201609. SAE International (2016)
6. Walch, M., Sieber, T., Hock, P., Baumann, M., Weber, M.: Towards cooperative driving: involving the driver in an autonomous vehicle's decision making (2016). https://doi.org/10.1145/3003715.3005458
7. Walch, M., Colley, M., Weber, M.: Driving-task-related human-machine interaction in automated driving: towards a bigger picture. In: The 11th International Conference (2019)
8. Aramrattana, M., Larsson, T., Jansson, J., Englund, C.: Dimensions of cooperative driving, ITS and automation. In: 2015 IEEE Intelligent Vehicles Symposium (IV), pp. 144–149 (2015)
9. Hoc, J.-M.: Towards a cognitive approach to human-machine cooperation in dynamic situations. Int. J. Hum. Comput. Stud. **54**(4), 509–540 (2001). https://doi.org/10.1006/ijhc.2000.0454
10. Hoc, J.-M., Young, M.S., Blosseville, J.-M.: Cooperation between drivers and automation: implications for safety. Theor. Issues Ergon. Sci. **10**(2), 135–160 (2009). https://doi.org/10.1080/14639220802368856
11. Mcneese, N.J., Demir, M., Cooke, N.J., Myers, C.: Teaming with a synthetic teammate: insights into human-autonomy teaming. Hum. Factors **60**(2), 262–273 (2018)
12. Boy, G.A.: The Handbook of Human-Machine Interaction A Human-Centered Design Approach (2011)
13. Verberne, F.M.F., Ham, J., Midden, C.J.H.: Trusting a virtual driver that looks, acts, and thinks like you. Hum. Factors J. Hum. Factors Ergon. Soc. **57**(5), 1–15 (2015)
14. Woods, D.D., Hollnagel, E.: Joint Cognitive Systems: Patterns in Cognitive Systems Engineering. CRC Press, Boca Raton (2006). ISBN 9780849339332
15. Jones, A.T., Romero, D., Wuest, T.: Modeling agents as joint cognitive systems in smart manufacturing systems. Manuf. Lett. (2018). https://doi.org/10.1016/j.mfglet.2018.06.002
16. Hollnagel, E., Woods, D.D.: Cognitive systems engineering: new wine in new bottles. Int. J. Man-Mach. **18**(6), 583–600 (1983)
17. Hutchins, E.: How a cockpit remembers its speeds. Cogn. Sci. **19**(3), 265–288 (1995). https://doi.org/10.1016/0364-0213(95)90020-9
18. Billings, C.E.: Aviation Automation: The Search for a Human-Centered Approach. Erlbaum, Hillsdale (1996)
19. Flemisch, F.O., Bengler, K., Bubb, H., Winner, H., Bruder, R.: Towards cooperative guidance and control of highly automated vehicles: H-Mode and conduct-by-wire. Ergonomics **57**(3), 343–360 (2014)
20. Wang, C.: A framework of the non-critical spontaneous intervention in highly automated driving scenarios. In: 11th International ACM Conference on Automotive User Interfaces and Interactive Vehicular Applications. ACM (2019)

21. Salas, E., Prince, C., Baker, D.P., Shrestha, L.: Situation awareness in team performance: implications for measurement and training. Hum. Factors **37**, 1123–1136 (1995)
22. Fiebich, A., Nguyen, N., Schwarzkopf, S.: Cooperation with robots? A two-dimensional approach. In: Misselhorn, C. (ed.) Collective Agency and Cooperation in Natural and Artificial Systems. PSS, vol. 122, pp. 25–43. Springer, Cham (2015). https://doi.org/10.1007/978-3-319-15515-9_2
23. Salas, E., Dickinson, T.L., Converse, S.A., Tannenbaum, S.I.: Toward an understanding of team performance and training. In: Swezey, R.W., Salas, E. (eds.) Teams: Their Training and Performance, pp. 3–29. Ablex, Norwood (1992)
24. Zimmermann, M., Bengler, K.: A multimodal interaction concept for cooperative driving. In: 2013 IEEE Intelligent Vehicles Symposium (IV), pp. 1285–1290 (2013). https://doi.org/10.1109/IVS.2013.6629643
25. Cooke, N.J., Gorman, J.C., Myers, C.W., Duran, J.L.: Interactive team cognition. Cogn. Sci. **37**(2), 255–285 (2013). https://doi.org/10.1111/cogs.12009
26. Walch, M., Mühl, K., Kraus, J., Stoll, T., Baumann, M., Weber, M.: From car-driver-handovers to cooperative interfaces: visions for driver–vehicle interaction in automated driving. In: Meixner, G., Müller, C. (eds.) Automotive User Interfaces. HIS, pp. 273–294. Springer, Cham (2017). https://doi.org/10.1007/978-3-319-49448-7_10
27. Christoffersen, K., Woods, D.D.: 1. How to make automated systems team players. In: Advances in Human Performance and Cognitive Engineering Research, vol. 2, pp. 1–13 (2002). https://doi.org/10.1016/S1479-3601(02)02003-9
28. Hoc, J.M.: Towards a cognitive approach to human-machine cooperation in dynamic situations. Int. J. Hum. Comput. Stud. **54**, 509–540 (2001)
29. Bengler, K., Zimmermann, M., Bortot, D., Kienle, M., Damböck, D.: Interaction principles for cooperative human-machine systems. IT—Inf. Technol. **54**(4), 157–164 (2012). https://doi.org/10.1524/itit.2012.0680
30. Endsley, M.R., Kiris, E.O.: The out-of-the-loop performance problem and level of control in automation. Hum. Factors J. Hum. Factors Ergon. Soc. **37**(2), 381–394 (1995). https://doi.org/10.1518/001872095779064555
31. McAree, O., Chen, W.H.: Artificial situation awareness for increased autonomy of unmanned aerial systems in the terminal area. J. Intell. Rob. Syst. **70**(1–4), 545–555 (2013). https://doi.org/10.1007/s10846-012-9738-x
32. Shu, Y., Furuta, K.: An inference method of teams situation awareness based on mutual awareness. Cogn. Technol. Work **7**(4), 272–287 (2005)
33. Stanton, N.A.: Distributed situation awareness. Theor. Issues Ergon. Sci. **17**(1), 1–7 (2016)
34. Salas, E., Sims, D.E., Burke, C.S.: Is there a "big five" in teams work? Small Group Res. **36**(5), 555–599 (2005)
35. Endsley, M.R., Robertson, M.M.: Situation awareness in aircraft maintenance teams. Int. J. Ind. Ergon. **26**(2), 301–325 (2000). https://doi.org/10.1016/S0169-8141(99)00073-6
36. Walch, M., Woide, M., Muehl, K., Baumann, M., Weber, M.: Cooperative overtaking: overcoming automated vehicles' obstructed sensor range via driver help, pp. 144–155 (2019). https://doi.org/10.1145/3342197.3344531
37. Wang, C., Krüger, M., Wiebel-Herboth, C.: "Watch out!": prediction-level intervention for automated driving (2020). https://doi.org/10.1145/3409120.3410652
38. Kridalukmana, R., Hai Yan, L., Naderpour, M.: A supportive situation awareness model for human-autonomy teaming in collaborative driving. Theor. Issues Ergon. Sci. (2020). https://doi.org/10.1080/1463922X.2020.1729443
39. Parasuraman, R., Bahri, T., Deaton, J., Morrison, J.G., Barnes, M.: Theory and design of adaptive automation in aviation systems. Naval Air Warfare Center, Warminster, PA, Technical report NAWCADWAR-92 033–60 (1992)

40. Wickens, C.D., Gordon, S.E., Liu, Y.: An Introduction to Human Factors Engineering. Addison-Wesley Educational Publishers Inc, New York (1997)
41. Bitan, Y., Meyer, J.: Self-initiated and respondent actions in a simulated control task. Ergonomics **50**(5), 763–788 (2007). https://doi.org/10.1080/00140130701217149
42. Stanton, N.A., Young, M.S., Walker, G.H.: The psychology of driving automation: a discussion with professor Don Norman. Int. J. Veh. Des. **45**(3), 289–306 (2007). https://doi.org/10.1504/IJVD.2007.014906
43. Endsley, M.R.: From here to autonomy: lessons learned from human-automation research. Hum. Factors J. Hum. Factors Soc. **59**(1), 5–27 (2017)
44. Chen, J.Y.C., Procci, K., Boyce, M., Wright, J., Garcia, A., Barnes, M.: Situation awareness based agent transparency. Aberdeen Proving Ground, MD: U.S. Army Research Laboratory. Report No. ARL-TR-6905 (2014)
45. Norman, D.A.: The 'problem' with automation: inappropriate feedback and interaction, not 'over-automation'. Philos. Trans. R. Soc. B Biol. Sci. **327**(1241), 585–593 (1990). https://doi.org/10.1098/rstb.1990.0101
46. Li, X., Vaezipour, A., Rakotonirainy, A., Demmel, S., Oviedo-Trespalacios, O.: Exploring drivers' mental workload and visual demand while using an in-vehicle hmi for eco-safe driving. Accid. Anal. Prev. **146**, 105756 (2020)
47. Kraus, J.M., Nothdurft, F., Hock, P., Scholz, D., Baumann, M.: Human after all: effects of mere presence and social interaction of a humanoid robot as a co-driver in automated driving. In: Automotive'UI 2016 (2016)
48. Allen, T.M., Lunenfeld, H., Alexander, G.J.: Driver information needs. Highway Res. Rec. **366**, 102–115 (1971)
49. Hollnagel, E., Woods, D.D.: Joint Cognitive Systems: Foundations of Cognitive Systems Engineering. Taylor and Francis, Boca Raton (1995)

# Studies on Intelligent Transportation Systems

# What Humans Might Be Thinking While Driving: Behaviour and Cognitive Models for Navigation

Arun Balakrishna[1]($\boxtimes$) and Tom Gross[2]($\boxtimes$)

[1] Automotive Software Development Services, HERE Technologies, Frankfurt, Germany
arun.balakrishna@here.com
[2] Human Computer Interaction Group, University of Bamberg, Bamberg, Germany
tom.gross@uni-bamberg.de

**Abstract.** In an optimally integrated HMS (Human Machine Systems) human and machine understand each other to provide an optimum integration. This is one of the core principles which is applicable for the research frameworks in vehicle navigation domain for effectively conducting research for creating optimal guidance information for the human driver. Creation and integration of human cognitive models for navigation is necessary to follow this principle effectively. BeaCON: Behaviour-and Context-Based Optimal Navigation is an existing research framework in the car navigation domain, for conducting analysis for the research problem *"Giving the driver adequate navigation information with minimal interruption"*. Currently BeaCON does not use the human cognitive models for navigation for the creation of guidance information and because of that the integration with the human driver is not achieved to an optimum level. In this paper, we present enhancement of BeaCON by integrating behaviour and cognitive models of navigation. Understanding the human thoughts while driving enables BeaCON to have a granular analysis of user cognitive state while creating guidance information, which results further cognitive load reduction for navigation tasks by creating more effective guidance information.

**Keywords:** BeaCON · Navigation system · HMS · Entity of interest · Cognitive load · Cognitive architecture · Machine learning

## 1 Introduction

A car Navigation System (NS) shows the user the current location on the map and gives both audio and visual information for providing guidance for travelling from one location to another [4]. *"Giving the driver adequate navigation information with minimal interruption"* in car navigation domain is one of the identified research gaps in HMS [3]. This research problem can be divided into following three sub problems as described in [1].

1. Given a set of route and map information, what is an optimal guidance information for the user?

© Springer Nature Switzerland AG 2021
H. Krömker (Ed.): HCII 2021, LNCS 12791, pp. 367–381, 2021.
https://doi.org/10.1007/978-3-030-78358-7_25

2. How to find optimal guidance information for the user, provided set of inputs to create the same is given?
3. How to find when, how and what guidance information must be given to the user?

Optimal guidance information eliminates the driver distraction created by NS [5]. To conduct effective research in the above-mentioned problems an understanding of the driver cognitive state for different contexts is necessary. Cognitive models enable the understandability of the user thoughts while driving and enable the extraction of the current and target cognitive state of the user. Identification of past, current and target cognitive state of the user while driving helps to deduce the root causes for the increased cognitive load as well as the creation of the guidance information which addresses the root cause in a more effective way. BeaCON is an existing research framework which enables the analysis of the research problem "*Giving the driver adequate navigation information with minimal interruption*" in car navigation domain [1]. One of the main principles in BeaCON is to understand the human user while creating guidance information which enables creation of optimal guidance information. Currently BeaCON does not use the human cognitive models for navigation for creating guidance information. Lack of cognitive models for navigation as well as its integration with behaviour models limit the effectiveness of the created guidance information. The work presented here addresses this gap by creating and integrating the cognitive models for navigation with the enhanced behavioural models, which can be together visualized as Behaviour and Cognitive Models for Navigation (BCMN) enhancement for BeaCON. Cognitive models in BCMN are created using ACT-R (Adaptive Control of Thought Rational) which is a well-known cognitive architecture for understanding and modelling human cognition process [6]. Currently BeaCON holds the static and dynamic behaviour models created using machine learning algorithms [2]. WEKA machine learning suite [8] is used for the creation of static and dynamic behaviour models. Static behaviour models are created based on statistical survey which includes questions in the following areas as described in [2].

- Driver distraction by the NS
- Extent to which NS understands the user intentions
- Optimal integration between the user and NS

Dynamic behaviour models recreated every time when a new user behaviour is observed [2]. Dynamic behaviour models are created based on the user context and the measured user cognitive load for the context.

Currently BeaCON creates the guidance information for a context, based on the created static and dynamic behaviour models as well the current and upcoming user contexts [1]. This paper also presents the improvements observed in the created guidance information because of creation and integration of BCMN, as well as the visualization and simulation support implemented for the created cognitive models in BCMN.

## 2   Related Work

The work conducted here already considers the research findings in the domain of optimum integration between human and navigation system as well as takes the research further. Balakrishna & Gross [1] provides a novel research framework which enables efficient analysis for creating optimal guidance information. But [1] does not contain the human cognitive models of navigation integrated for improving the efficiency of created guidance information. The analysis about the influence of navigation system behaviour on human behaviour [9] proves that an understanding of granular human behaviour is necessary for designing intelligent navigation system. Enhancement of BeaCON with BCMN enables granular understanding of human cognition while driving where the human cognitive state is created based on current and future navigation contexts. Yoshida, Ohwada et al. [10] also considers the cognitive state for improving the in-vehicle systems. But [10] is not using cognitive models which incorporates the contribution of real-world map entities (Manoeuvres, Roundabout etc.) while driving. The framework introduced for automated driving system testable cases and scenarios [11] provides bench marking strategies where some of them are applicable for human based driving also. The major focus of [11] is the automated driving scenarios, and because of that [11] does not provide human cognition models as a part of the framework. Haring, Ragni et al. [12] creates a cognitive model for human attention during driving, which divides the driving attention process to three sub process which are control, monitoring and decision-making process. The models created in [12] are not integrated to a research framework for enabling further research in the field of optimal guidance information as well as does not consider majority of the driving contexts. Krems & Baumann [13] propose a theory for creation and prediction of mental representation of objects and situation awareness for the driving contexts. But [13] is not concentrating to a granular level on the vehicle navigation tasks, as well as lacks proper experimental foundation. A cognitive model for car drivers which incorporates result from cognitive psychology and human machine interaction is presented in [14]. But [14] is neither considering many of the navigation related contexts which must be modelled very granularly nor conducting research on cognitive load measurement and reduction mechanisms.

## 3   Behaviour and Cognitive Models for Navigation (BCMN)

BCMN enables granular understanding of human cognitive state during navigation. The other components of BeaCON use these cognitive models while creating optimal guidance information. Detailed description of BCMN is given in the following sub sections.

### 3.1   Architecture of Cognitive Model for Navigation in BCMN

ACT-R [7] based cognitive model is created for the navigation process which is implemented using the LISP programming language. Currently declarative, goal and procedural modules provided by ACT-R are used. Some of the major ACT-R features from [7] are given below.

- ACT-R based cognitive architecture is a specification of the structure of the brain at a level of abstraction necessary enough to describe the function of the mind. Different ACT-R modules are associated with corresponding brain regions.
- The declarative module holds and retrieves critical information from the memory and the goal module keeps track of current user intentions.
- Communication among these modules is achieved by using the procedural modules.
- In ACT-R, chunks represent the knowledge a user already has, while solving a problem.
- Chunks also can be visualized as small unit which contains small amount of information.
- Sub symbolic level activation of the chunks and utility-based rule selection of the production rule shall be used for enabling learning for the created cognitive model.

A brief description about the ACT-R based architecture of cognitive model for navigation is given in the following sub sections. The declarative module in the cognitive model for navigation contains mainly the following chunk types.

**Chunk Types**

- State

A chunk type named state which contains a state name and a description is given below

```
(chunk-type state stateName stateDescription)
```

The chunk type *state* is used for representing the existing knowledge of a user about different cognitive states for navigation i.e., "announcement-active", "understanding-announcement". An example of state chunk created is given below

```
(state-1 ISA state stateName "state-1" stateDescription
"announcement-active")
```

- *Transition*

A chunk type named transition which contains current state, trigger, and next state is given below

```
(chunk-type transition currentState trigger nextState)
```

The chunk type transition represents the existing knowledge of the user about the state transition based on a trigger. An example of transition chunk is given below

```
(transition-1 ISA transition currentState "state-1" trig-
ger "trigger-1" nextState "state-2")
```

• *Navigation*

The chunk type named navigation represents the existing knowledge of the user about the navigation state as well as transition based on the trigger. An example of navigation chunk is given below.

```
(chunk-type navigation navigationStart navigationEnd nav-
igationCurrent navigationTrigger)
```

An example of navigation chunk is given below

```
(navigation-input ISA navigation navigationStart "state-1"
navigationEnd "state-16" navigationTrigger "trigger-2")
```

Navigation chunk type is also used to set the contents of the goal buffer which currently act as one of the interfaces to the cognitive model. The cognitive state of the user can be used by the navigation system for deciding the next guidance information as well to decide when to present the next guidance information.

**Production Rules**

ACT-R production rules contains the condition and the corresponding action [7]. Conditions specifies the patterns in the buffer associated with different modules which must be matched for the production to fire [7]. An example of a production which handles different navigation triggers (i.e., Audio announcement) are given below

```
(p navigation
   =goal>
     ISA              navigation
     navigationCurrent    =var1
     navigationTrigger    =var3
   - navigationEnd        =var1
    =retrieval>
     ISA              transition
     currentState     =var1
     nextState        =var2
   ==>
   =goal>
     ISA              navigation
     navigationCurrent    =var2
   +retrieval>
     ISA              transition
     currentState =var2
     trigger      =var3
     !output!         (=var2)
     !output!         (=var3)
)
```

In the production above, based on the navigation trigger the user knowledge about the next transition is retrieved as well as the navigation state is set accordingly. The cognitive state holds the current state of the user with the navigation task.

## 3.2 BCMN Integration with Other Components of BeaCON

The big picture level block diagram of BCMN, integrated into other components of BeaCON is shown in Fig. 1. The purpose of BeaCON is to conduct experiments using Human-in-the-loop systems, collect behavioural data and based on that find user cognitive load points and optimize the machine learning behavioural models, which are the models used by the system to understand the user [1]. Figure 1 also shows the interconnection between different components of the BCMN enhanced BeaCON framework. The BCMN enhanced BeaCON framework consists of 5 main components as described in Table 1.

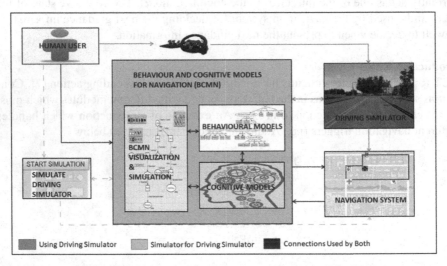

**Fig. 1.** BCMN integration with other components of BeaCON

BeaCON supports different driving simulation environments corresponding to low, medium, and high cognitive load inducing situations [1] while conducting the experiments.

## 3.3 Communication Between BCMN and the Navigation System Application

The integration of the cognitive model for navigation with the navigation system application is given in Fig. 2. The steps shown in Fig. 2 is repeated for every EOI [2] during the navigation process. The navigation system gets the following inputs for creating the next guidance information.

**Table 1.** Components of BCMN enhanced BeaCON

| SN | Component | Description |
| --- | --- | --- |
| 1 | Behaviour and Cognitive Models for Navigation (BCMN) | Purpose of BCMN is to identify the cognitive load and the root cause by using the following subcomponents<br>• Static behaviour models<br>• Dynamic behaviour models<br>• Human cognitive models for navigation |
| 2 | Navigation System (NS) | The navigation system tracks the car position in the route and highlights the path to be taken by the human user. Navigation system also presents the guidance information to the user as per the input from BCMN as well as based on the current driver context [1]. The inputs from NS are used by BCMN for getting the driving context information |
| 3 | Driving Simulator (DS) | CARLA driving simulator based custom implementation is used in BeaCON [14]. CARLA simulator is primarily designed for developing and testing autonomous driving agents, because of that extensive enhancements for existing interfaces are done to support the human driver interface using Logitech G920 driving hardware [1]. The inputs from DS are used by BCMN for getting the driver context information |
| 4 | Simulator for Driving Simulator (SDS) | SDS enables faster research and development for BCMN enhanced BeaCON features [1]. Using SDS, the developers can add and test new features without invoking the DS component |
| 5 | Human User (HU) | BCMN enhanced BeaCON is a human oriented research framework for the effective analysis of the research problem *"Giving the driver adequate navigation information with minimal interruption"* |

- Current cognitive state of the user from BCMN.
- Cognitive load for the next EOI as per the previous learning.
- Characteristics of the next EOI for which the user might need guidance information.

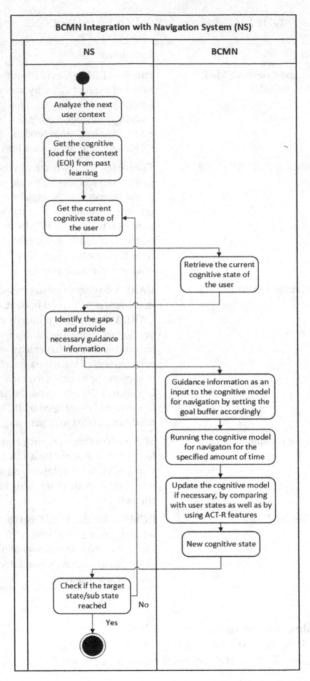

**Fig. 2.** Communication between BCMN and NS for each EOI

## 4    Conducting Human-In-The-Loop Experiments with BCMN Enhancement

The algorithm for conducting Human-in-the-loop experiments are given below, the steps for conducting experiments using the research framework is shown in Fig. 3.

//**Input:** Candidate drives between two selected points on the map

//**Output:** Optimized guidance information is generated using the driver context, behaviour models and human cognitive models

**Step 1:** Two points in the map are selected for conducting human experiments for evaluation of optimum guidance information generation.

**Step 2:** Candidate drives between the selected points.

**Step 3:** Generate and present guidance information for every user context (EOI) [2], based on current behaviour models, user contexts and cognitive state.

**Step 4:** Collect the behavioural data for updating the behaviour models as per latest behaviour observed for the EOI.

**Step 5:** Once user reached the destination, stop collecting behavioural data.

**Step 6:** Give the behavioural data input to the bench marking tool.

**Step 7:** Bench marking tool creates necessary logs for driving behaviour.

**Step 8:** Measure cognitive load at different contexts of driving (EOI) and create report.

**Step 9:** Replay the driving behaviour to re-verify the findings from the report.

**Step 10:** Driver cognition state for EOIs from the cognitive models shall be reverified and shall be enhanced when necessary.

**Step 11:** Cognitive load values at different points are used to recreate the dynamic behavioural models.

## 5    Visualization and Simulation for Cognitive Models for Navigation

Visualization of the created cognitive models in BCMN are very much necessary while conducting research as well as review of the cognitive process associated with navigation tasks. Visualization enables review and discussion of cognitive models in BCMN, with users who are not involved or familiarized with computer programming. But ACT-R based implementation of cognitive models uses LISP programming language and there is no visualization and simulation support are available out of the box which can be

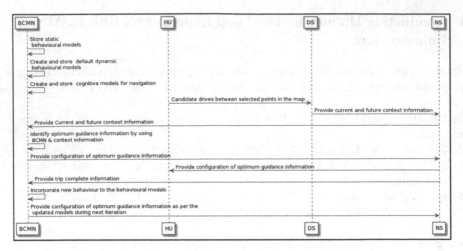

**Fig. 3.** Conducting human-in-the-loop experiments with BCMN enhancement

used by BCMN. Because of that, separate tools are developed for visualization and simulation of the created BCMN models, which are not directly a part of the BeaCON framework. The created tools cannot simulate all the ACT-R based functionalities, but it is extensively useful while visualizing and analyzing part of the cognitive models (i.e., Cognitive state during high speed and bad weather input). The cognitive models for navigation can be visualized and simulated by using the colored Petri net models created using the Petri net formalism, using the Graphical User Interface (GUI) provided by CPN Tools [17], since the design of the created cognitive models is aligned with Petri net based principles. An example of visualization of cognitive model (Part of the main model), which is responsible for creation of guidance information during bad weather and high-speed condition is given in Fig. 4.

One of the drawbacks of GUI based modelling of cognition above is that the modelling is static, which means all the inputs to be visualized and simulated must be decided during the creation of the models. To solve this, a dynamic simulation facility is also created using a JAVA based custom tool. The GUI shown above in CPN Tools creates a Petri Net Mark-up Language (PNML) file which can be given as an input to this custom JAVA based library for parsing and simulating the NS behaviour along with cognitive model behaviour. A subset of Petri net based formalism called BCMNML (BCMN Mark-up Language) is defined which has to be followed while creating the cognitive models in the GUI for enabling subsequent use of JAVA based library for dynamic simulation. The JAVA based library provides more simulation facilities as well as provides API for giving simulation input dynamically, which is not supported in the static GUI based simulation. On the other hand, the GUI based static simulation increases the understandability of the cognitive models which can be used in the initial phases of development, review, discussion, and enhancement. JAVA based library also provides extensive logging facility which can be used to understand the cognitive model's behaviour for different input and different contexts. For example, the logs created for "Reduce Speed" voice guidance during the simulation of the BCMN cognitive model shown in Fig. 3 is given below.

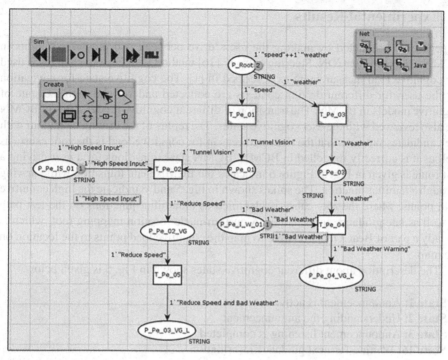

**Fig. 4.** Visualization and simulation of cognitive model for *"bad weather and high-speed condition"* using CPN tools

```
TRANSITION: Calling Arc T_Pe_02-P_Pe_02_VG from Transi-
tion T_Pe_02
ARC: Arc input T_Pe_02-P_Pe_02_VG input data [Reduce
Speed]
ARC: Arc expression checking T_Pe_02-P_Pe_02_VG Expres-
sion [Reduce Speed] And input [Reduce Speed]
ARC: Triggering arc expression for name T_Pe_02-
P_Pe_02_VG with expression [Reduce Speed]
ARC: Enabling T_Pe_02-P_Pe_02_VG
ARC: Setting triggered true for T_Pe_02-P_Pe_02_VG
PLACE: Setting token for P_Pe_02_VG with token [Reduce
Speed]
PLACE: Voice guidance P_Pe_02_VG with token [Reduce
Speed]
```

Once the cognitive models are created and reviewed using the visualization and simulation facilities above, mapping to the LISP programming language-based implementation in BeaCON is done manually because of the complexity involved for the mapping.

## 6  Experimental Results

Experiments conducted by making the users drive between two predefined points in BCMN enhanced BeaCON. Steering entropy [16] method with custom changes is used for cognitive load measurement as described in [1]. The cognitive state for navigation for the user, for different driving contexts are collected and verified with the state of cognitive models in BCMN. The transition of different cognitive model states in BCMN are also reviewed with the user cognitive states. The replay of the driving behaviour with the candidates indicate that the deducted reason for cognitive load is the root cause, as well as it is correctly modelled in BCMN. BCMN state transitions for a user driving a test route is given in Fig. 5. Figure 6 indicates the BCMN state transition while driving through a familiar junction. The spikes shown in Fig. 5 and 6 indicate the high cognitive load points, where the root cause is identified as unfamiliar junctions on the test path as well as late guidance information. A custom-made location mapping tool, which is already a part of BeaCON [1] is used to map the cognitive load points to the location on the map.

The description about different cognitive states shown in Fig. 5 is given below.

- **State 1:** Announcement is active
- **State 2:** Understanding the announcement
- **State 3:** Announcement listening is completed
- **State 16:** Waiting for next guidance information

The description about different cognitive states shown in Fig. 6 is given below.

- **State 7:** Identifying familiarity of the upcoming junction
- **State 11:** Recollecting the distance to the junction
- **State 12:** Junction is visible
- **State 13:** Preparing to take the junction
- **State 14:** Navigating the vehicle through the junction
- **State 15:** Navigating through the junction is completed

Some of the facts deduced through the experiments conducted with BCMN enhanced BeaCON are given below.

- The cognitive load associated with a specific cognitive state also depends on the previous cognitive states in most of the scenarios.
- Cognitive load associated with cognitive state depends highly on the familiarity of the route.
- The time at which the guidance information is provided has an influence on the cognitive load associated with different cognitive states.
- Speed of driving in the test route has an impact on the cognitive load at different cognitive states. Even for familiar route, high speed indicated high cognitive load at different cognitive states.
- Cognitive load associated with the already crossed EOI has an impact on the cognitive load associated with the upcoming EOIs.

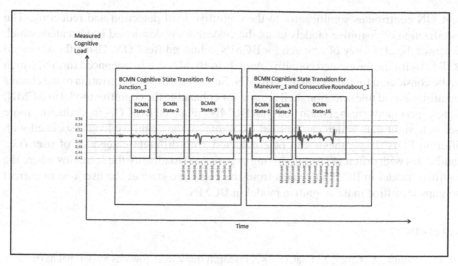

**Fig. 5.** BCMN state transitions while driving a test route with different EOIs

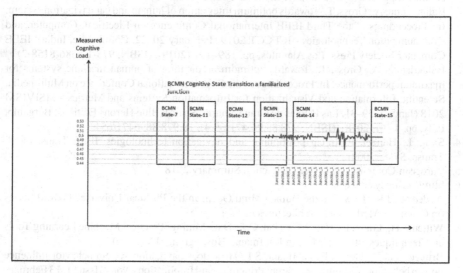

**Fig. 6.** BCMN state transitions while driving a test route with a familiar junction

## 7   Conclusion and Future Work

The BCMN enhanced BeaCON for the analysis of the research problem *"Giving the driver adequate navigation information with minimal interruption"* is presented. The value addition by BCMN for the BeaCON framework for analyzing the research problem is presented as well as its importance is justified. Comparison with the state of the art is done as well as uniqueness of BCMN for analyzing human cognition state for navigation tasks is presented. The experimental results confirm that the integration of BCMN for

BeaCON contributes significantly to the cognitive load detection and reduction. The visualization of cognitive models using the custom tools developed is presented, which increases the efficiency of research for BCMN enhanced BeaCON. Using the advanced ACT-R facilities, the created cognitive models in BCMN can be enhanced further, which can be considered for the roadmap of BCMN. Sub symbolic level activation of the chunks and utility-based rule selection shall be used for enhancing the cognitive model in BCMN. More experiments can be conducted with BCMN enhanced BeaCON for deducting more user behaviour data which can be used for reducing the cognitive load associated with different EOIs. Experiments can be conducted with different categories of user (i.e., Candidates with different level of driving experience) to identity the scenarios where the cognitive model in BCMN deviates from the cognitive state of the user and to correct the gaps identified in the cognitive model in BCMN.

# References

1. Balakrishna, A., Gross, T.: BeaCON – A research framework towards an optimal navigation. In: 22nd International Conference on human computer interaction HCII 2020, Copenhagen, Denmark, pp. 556–574 (2020). (ISBN: 978–3–030–49064–5)
2. Balakrishna, A., Gross, T.: Towards optimum integration of human and car navigation system. In: Proceedings of the Third IEEE International Conference on Electrical, Computer, and Communication Technologies - ICECCT 2019 (February 20–22, Coimbatore, India). IEEE Computer Society Press, Los Alamitos, pp. 189–197 (2019). (ISBN: 978-1-5386-8158-2)
3. Balakrishna, A., Gross, T.: Towards optimum automation of human machine systems for maximum performance. In: Proceedings of the Fifth International Conference on Multimedia, Scientific Information and Visualization for Information Systems and Metrics - MSIVISM 2018 (January 29–31, Las Palmas de Gran Canaria, Spain). Blue Herons Editions, Bergamo, Italy, pp. 1–14 (2018). (ISBN: 978.88.96.471.65.4). 10.978.8896471/654
4. Skog, I., Händel, P.: In-car positioning and navigation technologies. IEEE Trans. Intell. Transp. Syst. (2009)
5. European Commission: Driver Distraction Summary 2018
6. http://act-r.psy.cmu.edu/
7. Anderson, J.R.: How Can the Human Mind Occur in the Physical Universe? Oxford Series on Cognitive Models and Architectures (2007)
8. Witten, I.H., Frank, E., Hall, M.A., Pal, C.J.: Data Mining: Practical Machine Learning Tools and Techniques, 4th edn. Morgan Kaufmann, Burlington, MA (2016)
9. Brügger, A., Richter, K.F., Fabrikant, S.I.: How does navigation system behavior influence human behavior? Cognitive Research: Principles and Implications, vol. 4, Issue 1, 13 February 2019
10. Yoshida, Y., Ohwada, H., Mizoguchi, F., Iwasaki, H.: Classifying cognitive load and driving situation with machine learning. Int. J. Mach. Learn. Comput. 4(3) (2014)
11. National Highway Traffic Safety Administration (NHTSA): A framework for Automated Driving System Testable Cases and Scenarios, September 2018
12. Haring, K.S., Ragni, M., Konieczny, L.: A cognitive model of drivers attention. In: Proceedings of the 11th International Conference on Cognitive Modeling, January 2012
13. Krems, J.F., Baumann, M.R.K.: Driving and Situation awareness: A cognitive model of memory-update processes. In: Kurosu, M. (ed.) HCD 2009. LNCS, vol. 5619, pp. 986–994. Springer, Heidelberg (2009). https://doi.org/10.1007/978-3-642-02806-9_113
14. Daniel, K., Kühne, R., Wagner, P.: A Car Drivers CognitionModel. In: ITS Safety and Security Conference, vol. CD (2004)

15. Ros, G., Codevilla, F., López, A., Koltun, V.: CARLA: An Open Urban Driving Simulator, Alexey Dosovitskiy. CoRL 2017
16. Nakayama, O., Boer, E.R., Nakamura, T., Futami, T.: Development of a steering entropy method for evaluating driver workload. Human Factors in Audio Interior Systems, Driving, and Vehicle Seating (SAE-SP-1426). Warrendale, SAE, pp. 39–48 (1999)
17. http://cpntools.org/

# A Wizard-of-Oz Experiment: How Drivers Feel and React to the Active Interaction of AI Empowered Product in the Vehicle

Qihao Huang[(✉)], Ya Wang, Xuan Wang, Zijing Lin, Jian He, Xiaojun Luo, and Jifang Wang

Baidu Apollo Design Center, Shenzhen 518000, China
huangqihao@baidu.com

**Abstract.** While driving, the drivers have to put their hands on the steering wheel and keep their eyes on the forward. Besides, the distance from the driver to the screen of the central control panel (center stack) of a vehicle is also limited, which lowers the precision of the interaction between drivers and the wireless communication, entertainment, and driver assistance systems in the vehicle. Fortunately, the idea of Active Interaction of AI-Empowered Product, which enables the systems to provide scenario-based recommendations, could be a solution, which is aiming to enhance the user experience of the interaction between the drivers and the systems in the vehicle during driving. So how could it be created to better meet the behavior habits of drivers? Twenty-four drivers were recruited as participants and were asked to give subjective evaluations about their satisfaction and the degree of disturbance of the scenario-based recommendation function. We simulated the AI Empowered Product, which provides a precise recommendation to meet the drivers' needs in the vehicle through a Wizard-of-Oz Experiment, and explored factors which we proposed that may have an influence on the drivers feeling and reaction to the active interaction of AI-Empowered Product in the vehicle, including (a) the real-time traffic conditions outside, (b) the in-vehicle driver distraction, for instance, whether the driver was listening to music or not, (c) the content of the suggested information, and (d) different ways of information transmission. Furthermore, we also observed and analyzed the drivers' reactions to the active interaction. With the ANOVAs of the satisfaction scores and the degree of disturbance evaluated by participants, combined with the analysis of the participants' reactions, the results show that (a) regarding safety, drivers are not willing to accept recommendation during a traffic jam in which the drivers have to focus on the road conditions; (b) they are more reluctant to be bothered while listening to music; (c) they prefer to accept information that helps them boost driving efficiency (d) they like audio message best, because it is the most efficient way for them to acquire and understand the information, and it would be better if with the visual presentation, such as pictures, to improve the efficiency of information acquisition. Based on the experimental results and behavior analysis, we concluded suggestions for the product design process.

**Keywords:** Wizard-of-oz · AI-empowered product · Active interaction · Scenario-based recommendations · In-vehicle · Driving

© Springer Nature Switzerland AG 2021
H. Krömker (Ed.): HCII 2021, LNCS 12791, pp. 382–400, 2021.
https://doi.org/10.1007/978-3-030-78358-7_26

# 1 Introduction

## 1.1 Background

In the traditional "stimulus-feedback" human-machine interaction (HMI) system, the user's input is the very first step, after which the system processes and execute the user's order. While in the in-car environment, the users (drivers) have to put their hands on the steering wheel and keep their eyes on the forward, and the distance from the driver to the screen of the central control panel (center stack) of a vehicle is also limited, which lowers the precision of the interaction between drivers and the HMI systems. Overmuch inputting will bring distraction and safety hazards to the drivers, especially when the driving task has already occupied their hands and vision.

The development of artificial intelligence (AI) technology allows intelligent vehicles to have the abilities of autonomous learning, autonomous decision-making, active interaction, and situational awareness, which enable them to proactively offer scenario-based service/recommendation to meet the driver's potential needs during a journey. The idea of Active Interaction [1] of AI-Empowered Product could be a solution to enhance the user experience of the interaction between the drivers and the systems in the vehicle during driving. A previous study tells that some people do non-driving behaviors when driving, and although knowing that it is illegal, they will still use other devices to find non-driving-related information [2]. To clarify: our purpose of this research is only to make this process safer, not to discuss and argue for such services.

## 1.2 Related Works

How could the AI-Empowered Product be created to be safer and better meet drivers' behavior habits? The vehicle system's active interaction with the driver may be a new opportunity for smarter driving assistants or infotainment systems. Many scholars are exploring how these smart HMI systems can serve users in a proactive way and with a better user experience.

**The Appropriate Time for Active Interaction.** Previous research has revealed that the active interactions in an in-car environment should be presented appropriately because poorly timed interactions could cause stress and even danger to drivers. In the study conducted by Semmens et al. [1], sixty-three drivers, during a 50-min drive, were asked if it was a good point to deliver messages that were not related to driving. Based on the analysis of more than 2,700 responses and synchronized vehicle data, the results show that the best time to actively interact with drivers is when driving straightly, but not when they were in the middle of a driving maneuver or trying to figure out the right direction.

**The Understandability of the System.** Besides an appropriate time, proactive interactions had better be carried out in a way that does not surprise or frustrate users. Abowd et al. [4] indicated that the key challenges in the design of a scene aware system [3] is the fear of people not knowing what some smart system is doing in the "blackbox". Cheverst et al. [5] investigated the tension between human control and proactive services and found more than 90% of the participants in the research sample stated their desire to control the system, and almost 70% of them intended to know the "specific principle" of active interaction. Roland et al. [6] mentioned that if the user does not know why to make the recommendation to him/her, the acceptance of proactive recommendation will not be high, and proposed a method to improve the transparency and understandability of proactive recommendation.

### 1.3 The Current Research

Existing studies mainly focused on when and what to recommend to better meet the users' needs at a proper time and on users' comprehensibility with the proactive recommendation (why to recommend) to enhance the trust between the system and user. But no such research has explored how different ways of information transmission modulate the users' feelings and reactions to the active interaction in the vehicle.

At an early design stage of the AI-Empowered Product, we together with the designers, aimed to figure out how to transfer information conveniently and safely from the HMI system to the driver in a driving scene. We sorted out the process of the smart HMI system's proactive recommendation and divided it into six links (Fig. 1). At the very beginning, the Smart HMI System 1) recognizes the scene and 2) triggers the recommendation. The driver 3) then perceives the recommendation, and 4) the System waits for the driver while understanding it. After 5) the driver responds, 6) the System executes the task according to the driver's feedback. The main purpose of this research was to explore the proper HMI solutions for the proactive recommendation of the AI-Empowered Product, so this experimental research mainly focused on the three links that involve the drivers such as "Trigger," "Perceive", and "Respond". To gain a whole view about the proactive recommendation, we focused on figuring out these three questions concerning each link:

1) What Situations Are Suitable for Triggering Proactive Recommendations?
2) How Do People Perceive Different Ways of Information Transmission?
3) How do people feel and react to proactive recommendations?

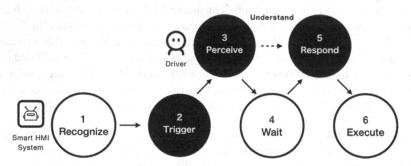

**Fig. 1.** Six links of the process of the smart HMI system's proactive recommendation to the users

In the current study, we collected drivers' subjective evaluations about their satisfaction and the degree of disturbance of the scenario-based recommendation function and observed the drivers' reactions to the active interaction through a Wizard-of-Oz Experiment. By comparing the impact of different interactive prototypes on drivers' subjective evaluations, combined with the analysis of the participants' reactions, we concluded suggestions to optimize the current HMI solutions for the product design process.

## 2   Method

### 2.1   Participant

Twenty-four drivers [8 men and 16 women, mean age = 32.88 ± 3.71 (SD) years], recruited by a third-party agency, participated in the experiment. To guarantee safety, all participants were with 5–10 years of driving experience and were able to drive cars without barriers.

### 2.2   Experimental Design

As mentioned above, this experimental research mainly focused on figuring out these three questions concerning the three links that involve the drivers such as "Trigger," "Perceive," and "Respond." (Fig. 2)

TRIGGER. What situations are suitable for triggering proactive recommendations?

PERCEIVE. How do people perceive different ways of information transmission?

RESPOND. How do people feel and react to proactive recommendations? Based on these three questions, we designed the experiment.

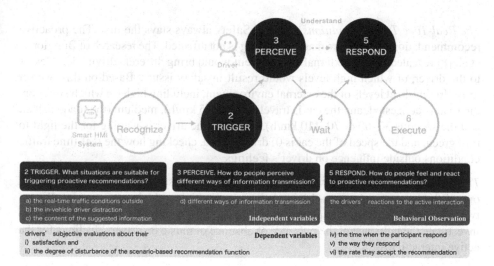

**Fig. 2.** Research topics in the current research

**Independent Variables.** As revealed by the previous studies, when and what to offer the users are the top topics. In this study, combined with the current business goal, we also considered these points. We tried to figure out *"What situations are suitable for triggering proactive recommendations?"* including (a) the real-time traffic conditions outside, (b) the in-vehicle driver distraction, for instance, whether the driver was listening to music or not, and (c) the content of the suggested information. We aimed to answer another question in this study was "*How do people perceive different ways of information transmission?*" We also investigated the influence of (d) different ways of information transmission on drivers' feelings (Fig. 3).

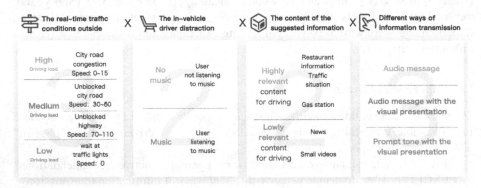

**Fig. 3.** Independent variables of the experiment

*The Real-Time Traffic Conditions Outside.* Safety always stays the first. The proactive recommendation must ensure that safe driving is not affected. The research of Bongiorno et al. [8] revealed that the external environment would bring different driving load levels to the driver, of which high levels would result in safety issues. Based on this, we set three driving-load levels of the external environment, including high (in which situation, the road is congested, and the car is traveling at 0–15 km/h), medium (the road is clear, and the speed is 30–60 or 70–110 km/h), and low (the driver is waiting for the light to turn green, and the speed of the car is 0) driving load, checking how the real-time traffic conditions outside influence on driver's feelings.

*The In-Vehicle Driver Distraction.* By our previous study, listening to music is considered one of the most common non-driving behaviors in the in-vehicle environment. We set two situations, including the presence or absence of music in the car environment, to investigate the influence of the presence or absence of music on drivers' feelings for the active interaction.

*The Content of the Suggested Information.* The proactive recommendation by hand-held smart devices is everywhere and at any time in our daily life. We are familiar with it. However, things change in the in-car environment while driving: the recommendation should be taken carefully due to safety issues. We sorted out the possible recommendation scenarios during the process of travel and proposed that the relevance between recommended content and travel affects drivers' feelings. So, the contents were divided into two parts, the content that is highly related to travel such as recommending online restaurant reservations, better route based on real-time traffic situation, and nearby gas station, etc., and the content that is not relevant to driving, such as recommending news and thumbnail videos.

*Different Ways of Information Transmission.* Combine with the business goal, we choose three ways of information transmission and made the interactive prototypes. In the view of the ways of information transmission, there were three types of the recommended information, including audio message, the audio message with the visual presentation, and prompt tone with the visual presentation.

**Dependent Variables.** After every trial, participants were asked to use the Likert 5-point scale to describe the satisfaction (1 indicated very dissatisfied, five indicated very satisfied) and driving disturbance (1 indicated completely no interference, 5 indicated very interference).

**Behavioral Observation.** *"How do people feel and react to proactive recommendations?"* After each trial of proactive recommendation, the user's response to the proactive recommendation will be observed and recorded, including whether to give feedback to the proactive recommendation, the way of response, and the status of acceptance. After each round of testing, each participant conducted a short post-test interview and completed a subjective assessment of interference and satisfaction under the interviewer's guidance.

**Controlled Variables.** To control the influence of other irrelevant variables on the experiment, participants should pass several trial tests to familiarize themselves with the test process before starting the test. They will also drive the test vehicle for 15–20 min to familiarize themselves with the test vehicle and get started quickly. Moreover, twenty-four participants will be equally divided into four groups to test the outside scenes under four driving loads (the medium driving load included two speed levels). The ratios of gender, age, family status, and driving experience in each group are the same. Overall, the Latin square design was used to exclude material presentation order's influence on experimental results.

### 2.3  Experimental Environment

**External Environment.** According to the three driving-load levels of *the real-time traffic conditions outside,* a total of four routes were selected in the experiment: for the low driving load level, a part of the Beiqing Road, started from NavInfo Building to Beijing Wangfu School, was designated as the Round-1 test road, during which there are 13 traffic lights; as for the medium driving load level, the same test road as the low level one was used as for the participants to drive at the speed of 30–60 km/h (Round 2), while in order to allow the other participants in this group to experience unobstructed high-speed driving (70–110 km/h), G7 Jingxin Expressway, starting from Nanjian Road to Yuanmingyuan, was selected; at last, for the high driving load level, we choose a road starting from Dongbeiwang Nursery to China Overseas International Center, which is often in congestion situation particularly in rush hours.

For the purpose of making the experiment process focused more on the test, the four routes designated by us allowed the participants to drive straightly during most of the test time (as shown in Fig. 4).

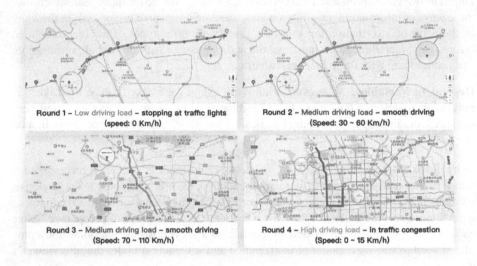

Round 1 – Low driving load – **stopping at traffic lights** (speed: 0 Km/h)

Round 2 – Medium driving load – **smooth driving** (Speed: 30 ~ 60 Km/h)

Round 3 – Medium driving load – **smooth driving** (Speed: 70 ~ 110 Km/h)

Round 4 – High driving load – **in traffic congestion** (Speed: 0 ~ 15 Km/h)

**Fig. 4.** Test routes

**Internal Environment.** Based on the research scenarios and recommended content, this experiment used the keynote to create five proactive recommendation cases (see Fig. 5), such as recommending online restaurant reservations, better routes based on the real-time traffic situation, and nearby gas station, news, and thumbnail videos.

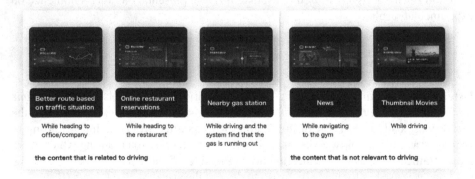

| Better route based on traffic situation | Online restaurant reservations | Nearby gas station | News | Thumbnail Movies |

| While heading to office/company | While heading to the restaurant | While driving and the system find that the gas is running out | While navigating to the gym | While driving |

the content that is related to driving                    the content that is not relevant to driving

**Fig. 5.** Proactive recommendation cases

*The Simulated AI Empowered Product Through the Wizard of Oz Method.* Active interaction takes the smart HMI system as the starting line and actively output execution results or suggestions to users. In order to simulate the active interaction between smart HMI system and driver at the very early design stage of the AI Empowered Product, the entire study introduced the Wizard of Oz method [7]. A modification in this experiment was that we went out of the laboratory and carried out experiments in real scenarios, aiming to directly collect the feelings and observe users' reactions to the active interaction under different real-time traffic conditions outside the vehicle.

We used an Aeolus AX7 as our test car (Fig. 6). Two cameras were installed in the car to record the participants' responses and facial expressions during the test. The proactive recommendation cases were displayed in the central control through an iPad, and the participants were informed that the device is the vehicle's central control panel. The "proactive recommendation" would be triggered at a specific time by experimenter 2 in the rear. Also, experimenter 2 was responsible for monitoring the vehicle speed and the users' response, controlling the play of the music and the keynote (recommendation content) to make the HMI system looks smart.

**Fig. 6.** In-car experimental environment

## 3   Procedure

### 3.1   Experiment Process

Every participant would be informed of the experiment's start time and location and was required to arrive at the test point at the appointed time. The experimenter was planned to lead the participant into the test car that had been set up and familiarize the participant with the driving function of the test car. Before the entire test, the participant was asked to drive for 15–20 min to get familiar with the test car. The experimenter and recorder on the car would also be ready to test materials and debug the equipment. After the participant finished the test procedure and signed an informed consent form, the test officially started.

Each participant ran two rounds of the experiments with or without music. The music playlist was acquired from the participant him/herself one day before the test date. In the two test rounds, participants received proactive recommendations through the tablet above the central control during driving. Each proactive recommendation was triggered only when the driving load outside the vehicle meets the test requirements; for instance, if the car was currently stopping at a traffic light and the speed is 0, experimenter two would artificially trigger the proactive recommendation in meanwhile. Each proactive recommendation content would be triggered in turn according to the preset order. During the whole process, participants were told to maintain their daily driving status. After each recommendation, they respond to or ignore the recommendation information in a natural and authentic state. Experimenter 1 would observe and record the participant's response during the whole procedure. In addition, each time after receiving a recommendation, experimenter one would ask the participants to give subjective evaluations about their satisfaction and the degree of disturbance and use the Likert 5-point scale to describe the participant's satisfaction (1 point for very dissatisfied, 5 points for very satisfied) and the degree of disturbance (1 point for completely no interference, 5 points for very interference). In addition, after each round of testing, experimenter 1 would conduct interviews with the participants about their attitudes and opinions on the recommendation form and content, as well as explore how they feel and why they showed some specific behaviors.

## 3.2 Interview Outline

**The Specific Questions About the Recommendation form are as Follows:** 1) Do you remember that there were several different recommendation forms just now? 2) Which form of recommendation do you prefer? Why? 3) Which form of recommendation is the most impressive one? Why? What is the impact on the driving process? Have you received the content conveyed by this recommendation? Do you think there are any suggestions and ideas that can be optimized? 4) What do you think of the other three forms of recommendation? What is good, what is terrible, and why? How did it influence you at that time?

**The Specific Questions About Recommended Content are as Follows:** 1) Do you remember that there were several different recommendation contents just now? Is the timing of recommendations for these types of recommended content appropriate? Why? 2) Which recommended content is the most impressive one? Why? What is the impact on the driving process? Do you think there are any suggestions and ideas that can be optimized for the recommended content? 3) What do you think of the other three recommendation content? What is good, what is terrible, and why? How did it influence you at that time?

**Overall Evaluation:** 1) In general, what do you think of the idea of proactive recommendation in the car? 2) Do you think the proactive recommendation in the car will affect the driving process? What is the impact? 3) Do you think there are differences in the impact of different recommendation forms? What are the differences? 4) Do you think there are differences in the impact of different recommendation content? What are the differences?

# 4 Result

Based on the core process of proactive recommendation, the analysis of the results will be disassembled into two parts: the first part focused on the trigger and perceive links to analyze the user's Satisfaction and the Degree of Disturbance ratings for different recommended content and different recommended methods in different outside and inside environments; the second part is to analyze the user behavior observation data in the respond link, including the user's respond timing, respond method and acceptance rate for the proactive recommendations.

## 4.1 The Degree of Disturbance and Satisfaction

A 3 (the real-time traffic conditions outside: high, medium, low driving load) × 2 (the in-vehicle driver distraction: music, no music) × 2 (the content of the suggested information: highly, lowly relevance content) × 3 (ways of information transmission: audio message, the audio message with the visual presentation, prompt tone with the visual presentation) analysis of variance (ANOVA) on the subjective ratings for participants' degree of disturbance and satisfaction was conducted, and the results are as follows.

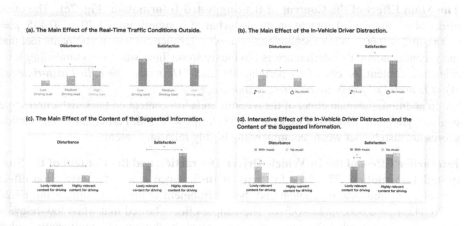

**Fig. 7.** Main effects of (a). the real-time traffic conditions outside, (b). the in-vehicle driver distraction and (c). the content of the suggested information, and (d). the interactive effect of the in-vehicle driver distraction and the content of the suggested information

**The Main Effect of the Real-Time Traffic Conditions Outside (Fig. 7a).** The main effect of the environment outside the vehicle on the disturbance is significant, $F_{2,20} = 5.685$, $p = 0.011$, $\eta p2 = 0.362$, when in high driving load situation, the participants' disturbance score is higher than in low driving load situation, $p = 0.009$. When in a medium driving load situation, the disturbance score is not significantly different from the other two road conditions. The real-time traffic conditions outside bring varying degrees of cognitive load to the driver, and as a result, impact participants' feeling on the proactive recommendation. By observing, the current study found that when congested (high driving load), participants need to focus on handling behaviors such as skipping queues of surrounding vehicles, and there is no chance to think about other things. At this time, the recommendation becomes more disturbing.

The impact of the environment outside the car on satisfaction is not significant, $F_{2,20} = 2.788$, $p = 0.086$, $\eta p2 = 0.218$. However, a similar trend was also shown when the vehicle was parking at a traffic light, and participants were most satisfied with proactive recommendations, followed by an unobstructed driving situation. In contrast, in congested road conditions, users showed the lowest satisfaction with proactive recommendations.

**The Main Effect of the In-Vehicle Driver Distraction (Fig. 7b).** The main effect of the in-vehicle driver distraction on the disturbance was significant, $F_{1,20} = 5.390$, $p = 0.031$, $\eta p2 = 0.212$. When making recommendations while listening to songs, participants felt that the recommendations' interference was significantly higher than when they were not listening.

The main effect of the in-vehicle driver distraction on the satisfaction score was also significant, $F_{1,20} = 10.931$, $p = 0.004$, $\eta p2 = 0.353$. When listening to songs, the satisfaction scores of participants are also lower.

**The Main Effect of the Content of the Suggested Information (Fig. 7c).** The main effect of recommended content on the disturbance is significant, $F_{1,20} = 30.435$, $p = 0.000$, $\eta p2 = 0.604$. When recommending high-relevance content, participants feel that the recommendation's interference is obviously lower than when recommending low-relevance content. As one of the participants said: *"How can it be called disturbance? What is useful to you is not a kind of disturbance."*

In addition, the main effect of the recommendation content on the satisfaction score is also significant, $F_{1,20} = 74.230$, $p = 0.000$, $\eta p2 = 0.788$. Participants' satisfaction scores are also higher when recommending highly relevant content.

**Interactive Effect of the In-Vehicle Driver Distraction and the Content of the Suggested Information (Fig. 7d).** In terms of disturbance score, the interaction effect between recommendation content and the environment in the car is significant, $F_{1,20} = 20.845$, $p = 0.000$, $\eta p2 = 0.510$. The simple effect showed that when low-relevant content is recommended, the interference with music background is significantly higher than without music background, $F_{1,20} = 12.223$, $p = 0.002$, $\eta p2 = 0.379$.

In terms of satisfaction ratings, the interactive effects of recommendation content and the in-car environment are also significant, $F_{1,20} = 8.368$, $p = 0.009$, $\eta p2 = 0.295$. The simple effect showed that when recommending low-relevant content, the satisfaction with music background is significantly lower than without music background, $F_{1,20} = 12.963$, $p = 0.002$, $\eta p2 = 0.393$.

In conclusion, when recommending low-relevance content, if the participant is listening to a song, the recommendation will interfere more with the participant, and the participant's satisfaction will be lower. When recommending high-relevance content, there is no significant difference in interference and satisfaction between participants who listen to songs and not listen to songs. It may be that for participants, high-relevance content has a higher priority than listening to songs.

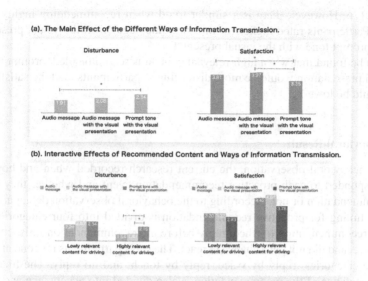

**Fig. 8.** (a). The main effect of the different ways of information transmission, and (b). the interactive effect of the content of the suggested information and different ways of information transmission

**The Main Effect of the Different Ways of Information Transmission (Fig. 8a).** On the disturbance score, the main effect of the information transmission is significant, $F_{2,40} = 5.641, p = 0.007, \eta p2 = 0.220$. The simple effect shows that the prompt tone with the visual presentation's interference is significantly higher than that of the audio message, $p = 0.010, \eta p2 = 0.406$.

On the other hand, on the satisfaction score, the main effect of the information transmission is also significant, $F_{2,40} = 11.494, p = 0.000, \eta p2 = 0.365$. The simple effect shows that participants' satisfaction with the prompt tone with the visual presentation is significantly lower than that of the audio message ($p = 0.002$) and audio message with the visual presentation ($p = 0.002$).

**Interactive Effects of the Content of the suggested Information and Different Ways of Information Transmission (Fig. 8b).** In terms of the disturbance score, the interaction effect between the recommendation content and the ways of information transmission is significant, $F_{2,40} = 3.904, p = 0.028 \eta p2 = 0.163$. The simple effect shows that, when recommending highly relevant content, the prompt tone with the visual presentation's interference degree is significantly higher than that of the audio message ($p = 0.017$) and audio message with the visual presentation ($p = 0.013$). Compared with the other two, the prompt tone with the visual presentation will cause more serious interference to participants. As one of the participants pointed out, the prompt tone is *"too easy to be ignored! You may not see the information. If this is critical information, what should I do if I miss it?"*.

In terms of satisfaction score, the interaction effect between recommendation content and the ways of information transmission is not significant, $F_{2,40} = 2.881, p = 0.068$

$\eta p2 = 0.126$. However, there is a similar trend when recommending highly relevant content. Participants rated higher for the audio message (with the visual presentation) than the prompt tone with the visual presentation.

It can be found that when high-relevant content is recommended, prompt tone with the visual presentation would be more disturbing to participants, and the satisfaction of users would be lower.

## 4.2  Behavioral Result

Through behavioral observation, the current research recorded when and how participants responded to each proactive recommendation event and whether they accepted the recommendation or not. According to the behavioral observation data, participants' response timing for proactive recommendation is divided into four categories: Reply after the recommendation finished, reply before the recommendation finished, deliberately ignore, and did not get a chance to react. The ways participants respond are divided into three categories: reply by voice, reply by touch, and no reply. The final participant's acceptance of the recommendation is also divided into acceptance and rejection (including non-response).

**Response Time (Fig. 9a).** Among all the trials, the proportion of replies after the recommendation is finished is the highest, reaching 78.39%, followed by deliberately ignoring it (10.26%), did not get a chance to react (4.21%), and respond before the recommendation is finished (7.14%).

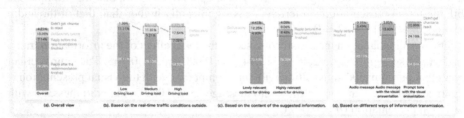

Fig. 9. The results of respond time in different views

*The Influence of the Real-Time Traffic Conditions Outside* (Fig. 9b). As the driving load increases from low to medium, then high, participants will deliberately ignore more recommendations (from 1.39% to 11.81%, then 17.54%). It is increasingly difficult for participants to respond normally (including responses after and before the recommendation is finished). As the participants stated that they chose to focus on driving and deliberately ignore recommendations to ensure driving safety at that point. Also, the possibility of participants' failure to respond raised when driving load increases.

*The Influence of the Content of the Suggested Information* (Fig. 9c). When receiving low-relevance content, participants would be more likely to intentionally ignored it

(12.25%). When receiving highly relevant content (8.48%), participants are more likely to reply before the recommendation is finished than lowly relevant content (4.90%). For messages that require a timely response or useless information, participants will give feedback in advance at the point they understand the content of the recommendation. For instance, once participants hear/see that it is a thumbnail movie, they will reply in advance and say "no need" immediately; News would be more acceptable to participants than thumbnail videos, because it only occupies the auditory channel. Participants will be more likely to choose to listen to it, because it has a less bad influence on driving. For high-relevant content, such as recommending nearby gas stations, participants will respond in advance when hearing it, in case they miss the recommended route change, because they need to respond as soon as possible to ensure that they keep up with changes in road conditions and navigation.

*The Influence of Different Ways of Information Transmission* (Fig. 9d). When the recommendation form is the audio message (with the visual presentation), the user will be more likely to reply in advance (15.93%). Through interviews, it was found that audio message with the visual presentation allows users to obtain information more quickly, so they were able to respond in advance. When the recommended form is the prompt tone with the visual presentation, the situation of getting no chance to reply reaches 10.99%, and the situation of deliberately ignoring reach 24.18%. As participants pointed out in the post-test interview, there was no time to look at the central control panel because of the need to pay more attention to the current road conditions, which resulted in deliberately ignoring the situation.

**The Ways Participant Responds (Fig. 10a).** We observed and recorded how the participants responded: 79.49% of the participants would reply by voice, 14.47% reply by touch, and 6.04% did not reply to the recommendation no matter it is because they deliberately ignore it or they did not get a chance to react.

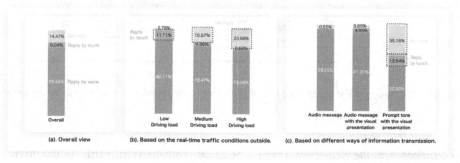

**Fig. 10.** The results of the ways participant respond in different views

*The Influence of the Real-Time Traffic Conditions outside* (Fig. 10b). With the increase of driving load, participants were more likely to not reply (from 2.78% to 16.67%, then 23.68%). While when the driving load is at a low level, the percentage of trials replied by touch reached 11.11%.

*The Influence of Different Ways of Information Transmission* (Fig. 10c). When it is an audio message, more than 90% of the participants will give feedback by voice, as well as audio message with the visual presentation, and only a small part (less than 5.00%) will try to respond to the touch screen. It seems voice is the primary response method. When it is the prompt tone with the visual presentation, participants' willingness to reply is reduced by more than 30%. Moreover, the probability of participants responding by touch has also increased to 12.64%. Nevertheless, interesting, most participants initially think that they can reply by voice without any prompting. When the recommended method is a prompt tone with the visual presentation, some participants will also have doubts- *"Can I reply by voice?"*.

**The Acceptance Rate of Proactive Recommendation Events (Fig. 11a).** Overall, the participant has a 50.55% chance of accepting it, a similar probability of accepting the recommendation and not accepting the recommendation.

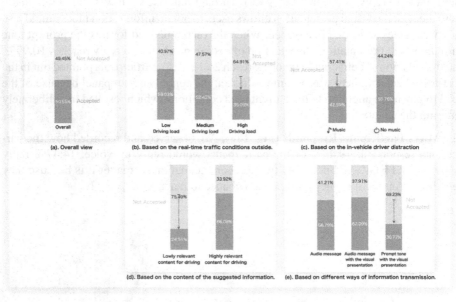

**Fig. 11.** The results of the acceptance rate of proactive recommendation events in different views

*The Influence of the Real-Time Traffic Conditions Outside* (Fig. 11b). As the driving load increases, the acceptance rate gradually decreases, especially when in the high driving load, the acceptance rate is as low as 35.09%. In the low driving load scenario, participants' acceptance rate reached 59.03%, which was higher than the random level.

*The Influence of the In-Vehicle Driver Distraction* (Fig. 11c). When the participants were listening to the song, only 42.59% of the participants accepted the recommendation. When they were not listening to the song, the acceptance rate was 55.76%.

*The Influence of the Content of the Suggested Information* (Fig. 11d). With an acceptance rate of only 24.51%, low-relevance content is more likely to be rejected; as for content that is highly relevant to driving, participants are more willing to accept, of which the rate is reaching 66.08%. Among the low-relevance content, the thumbnail video function has been criticized by users, "recommending small videos while driving is not appropriate at all and will cause safety hazards," it should be strictly closed (the acceptance rate is only 10.87%). Among the high-relevance content, the online restaurant reservations function was found to be a surprise feature for participants. Participants said that they needed such a feature a lot (81.82% acceptance rate).

*The Influence of Different Ways of Information Transmission* (Fig. 11e). The results show that when the information was transferred by the prompt tone with the visual presentation, there would be a lower chance for participants to accept the recommendation (only 30.77%). As mentioned above, prompt tone with the visual presentation requires participants to turn their eyes to the screen and read it actively. As a result, participants tend to ignore the recommended content in such situations during driving and miss some recommendations.

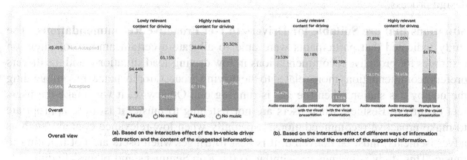

**Fig. 12.** The results of the acceptance rate of proactive recommendation events in different views

*Interactive Effect of the In-Vehicle Driver Distraction and the Content of the Suggested Information* (Fig. 12a). Based on the interactive effect of the recommendation content and the environment in the car on the subjective scores, we further explored the behavioral data and found that for low-relevant content, the acceptance rate was only 5.56% when participants were listening to music, which was lower than 34.85% when they were not listening to music, and it is a 30% difference. For high-relevant content, the acceptance of listening to music and not listening to music is similar, 61.11% and 69.70% respectively. This is consistent with finds in the trigger link. When listening to music, participants feel that they are more disturbed by low-relevance content. It seems that if participants are listening to music, it will be more likely to arouse their disgust and further lead to recommended rejection for the low-relevant content.

*Interactive Effect of Different Ways of Information Transmission and the Content of the Suggested Information* (Fig. 12b). We further found that when recommending highly relevant content, the gap of the acceptance rate between different recommendation methods are enlarged: it is mainly manifested in the acceptance rate of the prompt tone with the visual presentation and that of the audio message (with the visual presentation). Besides the prompt tone is easily missed or ignored by participants, participants shared another reason: when high-relevant content is recommended, participants need to acquire more information, but the prompt tone showed first without any valuable content in it. For example, when recommending a better route based on the traffic situation, the participants think that the prompt tone is only a recommendation to start navigation. However, when it is audio message with or without the visual presentation, they will know that it is a recommendation of a better route to help them avoid congested road conditions, and of course, they are happy to accept it.

## 5 Conclusion

Through experimental design and Wizard of Oz's method, the current research simulated the core process of proactive recommendation and explored participants' attitudes and behaviors toward the AI-Empowered Product and its proactive service. And in this part, we concluded some suggestions to optimize the current HMI solutions for the product design process.

**Situations that are Suitable for Drivers to Get Proactive Recommendations.** The study indicated that participants were able to allocate a certain amount of attention resources for proactive recommendations in low driving-load situations, and the drivers prefer recommendations more related to the driving process, for instance, recommending the nearby gas station when the gas is running out. Otherwise, it will make the users feel disturbed. Besides, when users are not listening to music, it is more appropriate to make proactive recommendations; especially when low-relevance content (such as advertisements) must be delivered, it is best to trigger when users are not listening to music. This study's findings are similar to the Rob Semmens and others' findings [1] that users prefer to receive this proactive recommendation in a less energy-intensive situation.

**The Safe and Efficient Method to offer Recommendations in the Vehicle.** While driving, it seems that audio message is users' favorite way of perceiving the active recommendation because it is the most efficient way for them to acquire and understand the information, and it would be better if with the visual presentation, such as pictures, to improve the efficiency of information acquisition. On the contrary, users showed the least satisfaction and felt the highest level of disturbance to the prompt tone with the visual presentation. Furthermore, when the recommended content is highly related to travel or any other valuable information, the users' dissatisfactory and disturbance level rise even higher. Therefore, the recommended way of prompt tone with the visual presentation can basically be eliminated from the selection scheme. The next step in actively recommending product design can improve the user experience of proactive service when the recommendation is offered through the audio message with visual presentation.

**Evidence from Observing Participants' Behavior.** Based on the participant's behavior observation results, it can be seen that the behavior observation results correspond to the previous experimental data results. On the whole, users will miss or deliberately not reply to the recommendation when a prompt tone with the visual presentation appears, resulting in the recommendation's invalidation (rejected or missed by the user). In contrast, the audio message improves the acceptance rate because it helps users obtain information and reduces the bad impact on driving safety.

Moreover, if you want to increase the participant's acceptance rate to low-relevant content, we suggest recommending it when participants are not listening to music; if you want to ensure the participant's acceptance rate of high-relevant content, you had better not use a prompt tone with the visual presentation to deliver the information.

**Other Suggestions to Optimize the Current HMI Solutions for the Product Design Process.** Based on the above research findings, researchers and designers collaborate in-depth to optimize the proactive recommended experience design. For example, based on behavioral observations, the study found that participants will interrupt the on-board machine broadcast and answer in advance. What we found from the post-test interview is that not answering does not mean the user does not need it, but because they ignore it (no matter on purpose or not). Although this situation does not happen a lot in this study, interruptions frequently occur in daily conversations. The fact that participants lack cognitive resources in the driving scene determines that participants will inevitably miss recommendations. Therefore, the experience under these two behaviors is also very important. Designers and product managers can focus more on these details to improve the overall product experience. Based on the above research findings, researchers and designers collaborate in-depth to optimize the details of the experience that are actively recommended. Based on the above research findings, researchers and designers collaborate in-depth to optimize the details of the experience that are actively recommended.

# 6 Discussion

Proactive services by the smart HMI system in the vehicle have gradually been integrated into car owners' lives. To enhance driving efficiency and experience requires continuous thinking and research. Current research has certain limitations due to equipment and time. The evaluation of driving performance does not include physiological indicators. Instead, we focused on the subjective feelings of users, combined with behavior observation date. Moreover, the test materials in this experiment are relatively preliminary. Thus, the Wizard of Oz method has been used to simulate active recommendations.

On the one hand, we will supplement and improve the evaluation equipment to include more objective indicators for driving task performance inspections to verify each other with subjective feelings. On the other hand, with the advancement of actively recommending products, we will develop real products to understand the user's attitudes and opinions. It is also hoped that based on research capabilities and product realization, the research and exploration of different forms of in-vehicle interaction will be broadened so as to continue to help product optimization and iteration.

# References

1. Semmens, R., Martelaro, N., Kaveti, P., Stent, S., Ju, W.: Is now a good time? An empirical study of vehicle-driver communication timing. In: Proceedings of the 2019 CHI Conference on Human Factors in Computing Systems, pp. 1–12 (2019)
2. Tison, J., Chaudhary, N., Cosgrove, L.: Preusser Research Group.: National phone survey on distracted driving attitudes and behaviors. The United States. National Highway Traffic Safety Administration (2011)
3. Dey, A.K., Abowd, G.D.: The context toolkit: aiding the development of context-aware applications. In: Workshop on Software Engineering for wearable and pervasive computing, pp. 431–441 (2000)
4. Abowd, G.D., Mynatt, E.D.: Charting the past, present, and future: research in ubiquitous computing. ACM Trans. Comput.-Hum. Interact. (TOCHI) 7(1), 29–58 (2000)
5. Cheverst, K., Byun, H.E., Fitton, D., Sas, C., Kray, C., Villar, N.: Exploring issues of user model transparency and proactive behavior in an office environment control system. User Model. User-Adap. Inter. 15(3–4), 235–273 (2005)
6. Bader, R., Karitnig, A., Woerndl, W., Leitner, G.: Explanations in proactive recommender systems in automotive scenarios. In: First International Workshop on Decision Making and Recommendation Acceptance Issues in Recommender Systems (DEMRA 2011), 11 (2011)
7. The Usability Body of Knowledge. https://www.usabilitybok.org/wizard-of-oz. Accessed 9 Feb 2021
8. Bongiorno, N., Bosurgi, G., Pellegrino, O., Sollazzo, G.: How is the driver's workload influenced by the road environment? Proc. Eng. 187, 5–13 (2017)

# Evaluation Driver Mental Load: A Survey Study of Cyclists Who Require to Repair the E-Bike

Fei-Hui Huang[✉]

Oriental Institute of Technology, New Taipei City 22061, Taiwan R.O.C.
Fn009@mail.oit.edu.tw

**Abstract.** As the market share of electric bicycles (e-bikes) continues to rise, the number of related deaths and injuries e-bike also increase. Road safety for e-bike cyclists constitutes an emerging public health challenge. This survey study adopted the Driving Activity Load Index (DALI) measure and used subjective anxiety and arousal perception measures to investigate the psychological factors influencing e-bike cycling crashes. The survey was recruited cyclists who visited repair stations to maintain their e-bikes. A total of 180 individual e-bike owners completed a paper-and-pencil version of the survey. A Python-based Random Forest Regression algorithm from the scikit-learn library was adopted for predicting regression. Results showed that the model constructs of DALI, age, anxiety, and arousal were useful predictors of e-bike crashes in traffic environments. Among these factors, the DALI value was the strongest predictor. It verified that cyclist-perceived mental load while e-bike ride on the road is very important in the context of road safety.

**Keywords:** Electric two-wheelers · Cycling safety · Mental load · Subjective rating scale

## 1 Introduction

Road safety for electric bicycle (e-bike) cyclists constitutes an emerging public health challenge in countries that promote the use of two-wheeled low-carbon vehicles. E-bikes are a relatively new form of transportation, with vehicles containing two wheels that are propelled by human pedaling and supplemented by electrical power from a storage battery. E-bike designs can roughly be divided into three types, including bicycle-style, scooter-like, and motorcycle-like. Among these, bicycle-style e-bikes are popularly sold throughout areas of Asia [1] and Europe [2]. As the market share of e-bikes continues to rise, the number of related deaths and injuries e-bike will also increase. This is because e-bikes can reach higher maximum speeds through motor support when compared to conventional bicycles [3], thus increasing the potential for severe crashes and injuries. The higher speeds may result in riskier behaviors, especially in complex traffic situations that require operators to quickly process information [4]. It is unclear how e-bikes alter the crash risks when compared to conventional cycling and scooter riding. However, e-bikes are already considered dangerous [5].

© Springer Nature Switzerland AG 2021
H. Krömker (Ed.): HCII 2021, LNCS 12791, pp. 401–407, 2021.
https://doi.org/10.1007/978-3-030-78358-7_27

# 2   Literature Review

## 2.1   Road Safety

Road traffic injuries claim more than 1.2 million lives each year globally. Most of these deaths occur in low- and middle-income countries, which lose approximately 3% of total GDP as a result [6]. Most traffic accidents are predictable and preventable, with about 90% being caused by human failure [7] or behavior [8]. Many countries have adopted traffic laws to enforce safety interventions. For instance, there are restrictions and prohibitions on speed, driving after consuming alcohol, helmet usage, seat-belts, and child restraints, each of which are designed to target human risk factors and prevent road traffic injuries and deaths [6]. More than half of road traffic deaths occur among vulnerable road users, especially motorcyclists, cyclists, and pedestrians [9]. More attention must be given to the needs of these users. It is therefore crucial to make walking and cycling safer [6].

Introducing the innovative e-bike products into the two-wheeler market makes make cycling more accessible to people more people in general such as the elderly and those with disabilities, to car-free households in urban environments, and to anyone who might need a bike ride. However, as more riders adopt e-bikes that make up a greater share of the two-wheeler market, new safety concerns have cropped up, i.e. involved in more crashes and cyclist fatalities. For example, in China with the highest sales of e-bikes, e-bike deaths increased almost seven times between 2004 and 2010, moving from 589 to 4,029, respectively (CONEBI). In Netherlands with famously bike-friendly country, e-bike deaths increased nearly doubled between 2016 and 2017. About 75 percent of the victims were men aged 65 and older. Those trends continued into 2019 [10]. E-bikes are much heavier than regular commuter bikes, reach higher top speeds, and make usually manageable corners and obstacles more dangerous. Therefore, it is important for riders interested in going e-bike should have a skill and abilities to handle the e-bike products to ensure road safety of them and road users.

## 2.2   Mental Workload

90% of traffic accidents are caused by human failure that are predictable and preventable (Smiley). Mental workload is a key factor influencing the occurrence of human failure. Both low and high mental workload has been found to disrupt performance in a nonlinear fashion at a given task [11]. Mental workload refers to a set of factors that affect mental information processing, an which thus lead to decision-making and individual reactions within the working environment. Workload depends upon both the individual involved and interactions between the operator and task structure. In mental workload measurement, processing effort, and resource allocation [12], the mobilization of additional resources works as a compensatory process [13]. Most mental workload assessment methods fall into the three categories of performance-based measures, subjective measures, and physiological measures [14]. The existing design for subjectively measuring mental load includes multi- and single-dimensional instruments.

The most outstanding subjective multidimensional measures include the Cooper-Harper Scale [15], the subjective workload assessment technique (SWAT) [16], and

NASA-TLX (Task Load Index) [17]. Driving Activity Load Index (DALI) is a revised version of the TLX. The DALI was specifically created for the driving context [18]. There are six pre-defined DALI factors for evaluating a driver's mental workload, i.e. effort of attention, visual demand, auditory demand, temporal demand, interference, and situational stress.

## 2.3 Machine Learning

Machine learning is a process of instructing machines what to do and make it learn from past experiences [19]. The machine learning algorithms are computed from previous data and are unique because it does not rely on the predetermined equations models. Also, machine learning algorithms have been verified that perform better than predictions from empirical methods. In recent years, the commonly used machine learning algorithms include Artificial Neural Nets, Bayesian Network, Support Vector Machine, and Random Forest (RF). The Artificial Neural Nets have been criticized for the lengthy training process, since the optimal configuration is not known a priori [20]. The Bayesian Network and the Support Vector Machine are suitable for problems with the amount of sample data, but time-consuming when solving complex problems [21, 22]. The RF is a powerful ensemble-learning method that use multiple Decision Trees with the same distribution to set up a forest to train and predict the sample data [23]. Each decision tree will be full growth, it do not need to cut processing, the more tree it has the more accurate the result will be, and it will not over fitting. Moreover, Decision Tree is a non-parametric supervised learning method that can summaries decision rules from a series of data with features and labels, and use the structure of the tree to present these rules to solve classification and regression problem [24]. RF can be applied for classification, regression, and unsupervised learning [25]. Such learning method has been widely used in many fields and exhibited nice performance [26]. When solving classification problems, RF prediction is considered the unweighted majority of class votes [27]. In other words, Random Forest belongs to the combination algorithm and do the overall estimate to have the advantages of automatic balance of error and automatic selection of features [28].

## 3  Methods

This survey study adopted mental workload measure of DALI to investigate user-perceived the riding task difficulty while e-bike ride. An anonymous survey among e-bike riders was conducted in the Taiwanese context. In this study, cyclists who visited e-bike repair stations to repair the e-bikes were recruited. A total of 180 individual e-bike owners (90 females and 90 males) completed a paper-and-pencil version of the survey.

### 3.1  Measurements

The questionnaire contained the following eight sections: (1) personal information, including two items designed to collect sociodemographic data on gender and age; (2) DALI, including six items designed to measure the total workload, effort of attention, visual demand, auditory demand, temporal demand, interference, and situational stress, which were assessed using a 10-point Likert scale; (3) risk estimate, including two items designed to measure the respondent's levels of subjective anxiety and subjective arousal, which were assessed using a 10-point Likert scale ranging from very high to very low.

## 3.2 Data Analysis

A Python-based Random Forest Regression algorithm from the scikit-learn library was adopted for predicting regression. Statisticians usually study random forests as a practical method for non-parametric conditional mean estimation: Given a datagenerating distribution for $(Xi, Yi) \in X \times R$, forests are used to estimate $\mu(x) = E \, Yi; Xi = x$. Several theoretical results are available on the asymptotic behavior of such forest-based estimates $\hat{\mu}(x)$, including consistency, second order asymptotics, and confidence intervals [29].

# 4  Results

In this study, respondents who had crash experience were randomly selected from e-bike require stations. Because survey data of this study with the 180 respondents are small compared to the total study area, in this study, different ratios of respondents who never had crash experience to respondents who had crash experience should be considered to obtain the highest classification accuracy [30]. The 180 survey data were randomly divided into a ratio of 80/20 to build training and validation datasets (test_size = 0.2, random_state = 2). Run the RFA on Train (i) to build RF Model. And then, using the RF Model performs the classification process on the test data. Finally, calculating the accuracy and sorting the feature variables by importance. RF directly performs feature selection while classification rules are built. Feature selection is an important step in applications of machine learning methods. That is because that most data sets are often described with lots variables for model building. However, most of the variables are irrelevant to the classification. Dealing with large feature sets slows down algorithms, takes too many resources, is inconvenient, and exhibits a decrease of accuracy.

In the RF learning model, the results of this study indicated that rider-perceived mental load while riding received from DALI (with a coefficient of 0.306) is the most important factor that may reflect whether he/she has experience in e-bike crashes, followed by age (with a coefficient of 0.270), perceived anxiety (with a coefficient of 0.175), perceived arousal (with a coefficient of 0.173), and gender (with a coefficient of 0.075). The predictive accuracy value is 0.833. In addition, the value of Mean Squared Error (MSE) for predicted scores is 0.008, the Mean Absolute Error (MAE) value is 0.034, and $R^2$ value is 0.850. Based on the results, the gender variable is further deleted for building a new RF leaning model. The results indicated that DALI (with a coefficient of 0.346) is the most important factor that may reflect whether the rider has experience in e-bike crash, followed by age (with a coefficient of 0.320), perceived arousal (with a coefficient of 0.197), and perceived anxiety (with a coefficient of 0.137). The predictive accuracy value is 0.833. In addition, the value of Mean Squared Error (MSE) for predicted scores is 0.010, the Mean Absolute Error (MAE) value is 0.037, and $R^2$ value is 0.813.

## 5   Discussion

In this study, psychological measurements of DALI, anxiety, and arousal were used to obtain cyclist's mental load and riding task difficulty while e-bike ride. A Python-based Random Forest Regression algorithm from the scikit-learn library was adopted for predicting regression. Results revealed that DALI, anxiety, arousal, and age had correlation with cyclist's e-bike crash experiences on the road.

DALI was the strongest predictor, thus suggesting that cyclist-perceived mental load while riding an e-bike was the most important factor affecting the possibility of e-bike crash on the road. Results showed a significant relationship between DALI value and e-bike traffic crashes, with a regression coefficient of 0.346. The impact of mental load on e-bike crash on the road highlights its importance in the context of road safety. The construct of mental load has been defined as the ratio between task demands and user capacity [31]. Since human processing resources are limited, the level of mental load is influenced by difficulty and complexity of the task as well as by the individual's skills, abilities and motivation [32]. This study adopted self-report DALI assessment to reflect cyclist's mental load in subjective aspects. Anxiety and arousal are another two important psychological factors affecting cyclist's e-bike traffic crashes. Results showed a significant relationship between cyclist-perceived arousal and e-bike traffic crashes, with a regression coefficient of 0.197, and a significant relationship between cyclist-perceived anxiety and e-bike traffic crashes, with a regression coefficient of 0.137. An individual's level of arousal has a significant effect on his or her performance in a number of areas. Cyclist's arousal levels are associated with fluctuations in performance on a variety of e-bike riding tasks. The low arousal levels has been found may be produced by sleep deprivation [33] or fatigue to impair performance on a variety of cognitive tasks. However, factors that increase arousal levels, such as noise, can have a negative impact on user's performance [34]. Anxiety is accompanied by a characteristic set of behavioral and physiological responses including avoidance, vigilance and arousal, which evolved to protect the individual from danger. These anxiety-related responses seem to be part of a universal mechanism by which organisms adapt to adverse conditions [35]. The anxiety in driving concerned about personality typologies, disorders, stress, and fear. Drivers with anxiety feelings may have some facilitate or positive effects that are specific to driver behavior and driving skills. The previous studies have verified that drivers with driving anxiety may make more errors in the on-road assessment compared to the control group, e.g., [36].

In addition, results show gender variable with a coefficient of 0.075 that has few influence on predicting cyclist's experience of e-bike crash on the road. It is found that there is no difference in predictive accuracy value between the model with gender variable and the model without gender variable. The gender variable has been verified to be not one of the important factors affecting cyclist's experience of e-bike crash.

## 6   Conclusion

Subjective rating scales of DALI, anxiety, and arousal were used to investigate the factors in the e-bike riding context. Results showed the usefulness of subjective measurements for increasing our understanding of the factors contributing to e-bike traffic crashes. This study provided evidence that the DALI index and psychological feelings (i.e., perceived anxiety levels and arousal levels) were important factors that predicted e-bike crashes in traffic environments. In addition, cyclist's age was one of the important influence factors as well. In order to improve road safety, the design and development of the road user education and training course and road management should be based on the psychological feedback mechanism of road users.

# References

1. Wu, C., Yao, L., Zhang, K.: The red-light running behavior of electric bike riders and cyclists at urban intersections in China: an observational study. Accid. Anal. Prev. **49**, 186–192 (2012)
2. Fyhri, A., Fearnley, N.: Effects of e-bikes on bicycle use and mode share. Transp. Res. Part D: Transp. Environ. **36**, 45–52 (2015)
3. Bai, L., Liu, P., Chen, Y., Zhang, X., Wang, W.: Comparative analysis of the safety effects of electric bikes at signalized intersections. Transp. Res. Part D Transp. Environ. **20**, 48–54 (2013)
4. Vlakveld, W.P., et al.: Speed choice and mental workload of elderly cyclists on e-bikes in simple and complex traffic situations: a field experiment. Accid. Anal. Prev. **74**, 97–106 (2015)
5. Popovich, N., Gordon, E., Shao, Z., Xing, Y., Wang, Y., Handy, S.: Experiences of electric bicycle users in the Sacramento California area. Travel Behav. Soc. **1**(2), 37–44 (2014)
6. World Health Organization: Global status report on road safety. World Health Organization (2015)
7. Lewin, I.: Driver training: a perceptual-motor skill approach. Ergonomics **25**(10), 917–924 (1982)
8. Treat, J.R.: Tri-level study of the causes of traffic accidents: Final Report. Volume I: Causal factor tabulations and assessments (1977)
9. World Health Organization: Global status report on road safety 2018: Summary (No. WHO/NMH/NVI/18.20). World Health Organization (2018)
10. Hurford, M.: How to Ride an E-Bike Safely – electric bikes are seriously fun as long as you don't get seriously injured. Bicycling https://www.bicycling.com/culture/g20085571/ride-electric-bike-safely/. Accessed 24 Dec 2020
11. Gagnon, J.F., Durantin, G., Vachon, F., Causse, M., Tremblay, S., Dehais, F.: Anticipating human error before it happens: towards a psychophysiological model for online prediction of mental workload. Open Archive Toulouse Archive Ouverte (OATAO) (2012)
12. Norman, D.A., Bobrow, D.G.: On data-limited and resource-limited processes. Cogn. Psychol. **7**(1), 44–64 (1975)
13. Aasman, J., Mulder, G., Mulder, L.J.: Operator effort and the measurement of heart-rate variability. Hum. Factors **29**(2), 161–170 (1987)
14. Meshkati, N., Hancock, P.A., Rahimi, M., Dawes, S.M.: Techniques in mental workload assessment (1995)
15. Cooper, G.E., Harper Jr, R.P.: The use of pilot rating in the evaluation of aircraft handling qualities (No. AGARD-567). Advisory group for aerospace research and development Neuilly-Sur-Seine (France) (1969)
16. Reid, G.B., Nygren, T.E.: The subjective workload assessment technique: a scaling procedure for measuring mental workload. Adv. Psychol. (North-Holland) **52**, 185–218 (1988)
17. Hart, S.G., Staveland, L.E.: Development of NASA-TLX (Task Load Index): results of empirical and theoretical research. Adv. Psychol. (North-Holland) **52**, 139–183 (1988)

18. Pauzie, A., Marin-Lamellet, C.: Analysis of aging drivers' behaviors navigating with in-vehicle visual display systems. In: CONFERENCE 1989, VNIS 1989, pp. 61–67. IEEE, September 1989
19. Ching, J., Phoon, K.K.: Constructing site-specific multivariate probability distribution model using Bayesian machine learning. J. Eng. Mech. **145**(1), 04018126 (2019)
20. Zhang, W.G., Goh, A.T.C.: Multivariate adaptive regression splines for analysis of geotechnical engineering systems. Comput. Geotech. **48**, 82–95 (2013)
21. Hu, J.L., Tang, X.W., Qiu, J.N.: A Bayesian network approach for predicting seismic liquefaction based on interpretive structural modeling. Georisk Assess. Manag. Risk Eng. Syst. Geohazards **9**(3), 200–217 (2015)
22. Martens, D., De Backer, M., Haesen, R., Vanthienen, J., Snoeck, M., Baesens, B.: Classification with ant colony optimization. IEEE Trans. Evol. Comput. **11**(5), 651–665 (2007)
23. Breiman, L.: Random forests. Mach. Learn. **45**, 5–32 (2001). https://doi.org/10.1023/A:101 0933404324
24. Zhang, W., Wu, C., Li, Y., Wang, L., Samui, P.: Assessment of pile drivability using random forest regression and multivariate adaptive regression splines. Georisk Assess. Manag. Risk Eng. Syst. Geohazards **15**, 1–14 (2019)
25. Liaw, A., Wiener, M.: Classification and regression by random forest. R News **2**(3), 18–22 (2002)
26. Calderoni, L., Ferrara, M., Franco, A., Maio, D.: Indoor localization in a hospital environment using random forest classifiers. Expert Syst. Appl. **42**(1), 125–134 (2015)
27. Kohestani, V.R., Hassanlourad, M., Ardakani, A.: Evaluation of liquefaction potential based on CPT data using random forest. Nat. Hazards **79**(2), 1079–1089 (2015). https://doi.org/10. 1007/s11069-015-1893-5
28. Lin, W., Wu, Z., Lin, L., Wen, A., Li, J.: An ensemble random forest algorithm for insurance big data analysis. IEEE Access **5**, 16568–16575 (2017)
29. Athey, S., Tibshirani, J., Wager, S.: Generalized random forests. Ann. Stat. **47**(2), 1148–1178 (2019)
30. Bui, D.T., Pradhan, B., Lofman, O., Revhaug, I., Dick, O.B.: Spatial prediction of landslide hazards in Hoa Binh province (Vietnam): a comparative assessment of the efficacy of evidential belief functions and fuzzy logic models. CATENA **96**, 28–40 (2012)
31. Gopher, D., Donchin, E.: Workload: An examination of the concept (1986)
32. Grassmann, M., Vlemincx, E., von Leupoldt, A., Van den Bergh, O.: The role of respiratory measures to assess mental load in pilot selection. Ergonomics **59**(6), 745–753 (2016)
33. Caldwell, J.: Assessing the impact of stressors on performance: observations on levels of analyses. Biol. Psychol. **40**(1–2), 197–208 (1995)
34. Wilkinson, R.T.: Effects of up to go hours' sleep deprivation on different types of work. Ergonomics **7**(2), 175–186 (1964)
35. Gross, C., Hen, R.: The developmental origins of anxiety. Nat. Rev. Neurosci. **5**(7), 545–552 (2004)
36. Taylor, J.E., Deane, F.P., Podd, J.V.: Driving fear and driving skills: comparison between fearful and control samples using standardized on-road assessment. Behav. Res. Ther. **45**(4), 805–818 (2007)

# In-Vehicle Information Design to Enhance the Experience of Passengers in Autonomous Public Buses

Myunglee Kim[1], Jeongyun Heo[1(✉)], and Jiyoon Lee[2]

[1] Kookmin University, 77, Jeongneung-ro, Seongbuk-gu, Seoul 02707, Republic of Korea
{kme5246,yuniheo}@kookmin.ac.kr
[2] Monash University, 900 Dandenong Road, Caulfield East, VIC 3145, Australia
jiyoon.lee@monash.edu

**Abstract.** The purpose of this study is to propose a concept design for an In-Vehicle Information System (IVIS) of an autonomous bus that provides an Autonomous Mobility on Demand (AMoD) service. First, we conducted a literature study to investigate what information is salient to the driving conditions of autonomous buses followed by a user diary method and in-depth interview approach to investigate what information passengers need. The key findings are as follows. First, autonomous buses operate on different operating conditions from existing city buses; thus, new information needs to be delivered to passengers accordingly. Second, the information changes in real-time as autonomous buses operate based on user demand. Third, the information must be delivered to passengers such that they reach their destination. Finally, all of the information passengers need should be communicated in in an easy-to-understand and convenient way. Based on these findings, the IVIS concept design was proposed from a user-centered design perspective. To verify user acceptance of the proposed concept design, we configured five scenarios and conducted one-on-one interviews. The proposed concept design contributes insofar as it investigated what kind of information is needed for passenger experience before the commercialization of autonomous buses. We hope that our findings will help designing of an IVIS for autonomous buses for a better bus travel experience for passengers in the future.

**Keywords:** In-vehicle information · Internal HMI · Autonomous buses · Passenger needs · User-centered design

## 1 Introduction

### 1.1 Background

Autonomous driving technology is recognized as an essential technology for future transportation, and various studies have proliferated on how the technology could enhance our mobility experience [1]. In brief, self-driving public transport is suitable for purposes such as a small number of people sharing [2], a large number of people sharing [3, 4], and rural use [5] and is expected to emerge as a viable technology. Particularly, Autonomous

© Springer Nature Switzerland AG 2021
H. Krömker (Ed.): HCII 2021, LNCS 12791, pp. 408–424, 2021.
https://doi.org/10.1007/978-3-030-78358-7_28

Mobility on Demand (AMoD) has great potential to characterize public transportation for future cities by improving urban mobility by utilizing the existing infrastructure [6].

The AMoD is a service in which autonomous mobility transports passengers from the origin to the user-requested destination according to the user's demand [7]. Unlike traditional public transport, the AMoD stops only at stops where the user demands; consequently, traveling time can be shortened even when driving the same distance. Additionally, although the usage fee is low, it is possible to reach the destination quickly and conveniently, which increases the possibilities for high user inflow [5]. If the influx of users is high, the number of vehicles driving on the road can be reduced, thereby reducing emissions [6]. Moreover, as the demand for parking is reduced, parking lots are freed up to be used as public facilities [8].

## 1.2 Problems and Research Questions

This study focused on autonomous buses providing AMoD services. Unlike traditional public buses, autonomous buses use autonomous driving technology and flexible routes [9]. (Table 1) The flexible route uses a variable schedule operating autonomously driving buses according to a fixed route but stops only at stops requested by passengers [10]. When passengers make a reservation through the app or on the web, it is reflected in the schedule of the autonomous bus [11]. Because of the difference in the operational terms, the In-Vehicle Information System (IVIS) of the autonomous bus must be redesigned for AMoD services. The need for redesign has also increased because the information and delivery methods provided by existing city buses are unsuitable for autonomous buses [12]. Therefore, in this study, the user-centered IVIS is designed such that users of autonomous buses can understand and accept the technology effortlessly. We have structured the following questions to proceed with this study.

- RQ1. What information should be provided to AMoD passengers through an IVIS for a satisfying autonomous bus experience?
- RQ2. What in-vehicle information is to be provided to passengers because of the different operational terms between the AMoD service and traditional public buses?
- RQ3. What information are passengers checking for in each phase of the ride as they use the bus?
- RQ4. What information should passengers on autonomous buses be provided with for each phase of the ride to reach their destination?

**Table 1.** Differences in the operational terms between autonomous buses and traditional public buses

| Operational terms | Traditional public bus | AMoD |
|---|---|---|
| Schedule | Fixed | Flexible |
| Route | Fixed | Fixed |
| Stop | All bus stops | On demand |
| How to use the bus | Wait and board | Reservation type |
| Driving | Non-autonomous driving | Autonomous driving |

### 1.3  Research Overview

The aim of this study is to investigate the information required for passengers on autonomous buses to travel to their destination and process newly created information according to the changed operational conditions of the autonomous bus. The IVIS of traditional public buses delivers information that passengers need but inconveniently and is unsuitable for autonomous buses. Therefore, this study focuses on the IVIS design of an autonomous bus considering the mobility experience of passengers. Section 2 reviews the literature focusing on designing IVIS for autonomous vehicles from a user-centric perspective. Section 3 describes the research methods, user research, concept design, and validation process of the design outcome. Section 4 focuses on the information required by passengers and the delivery method found in user research. Section 5 proposes a concept design based on the insights found in user research and verifies the acceptance of the proposed concept design, and Sect. 6 outlines the conclusions of this study and discusses its limitations and future directions.

## 2  Related Works

### 2.1  Design Purpose and Necessity of IVIS

An IVIS is an information system delivering most of the information generated in the vehicle to passengers in an integrated manner. It communicates the intentions and plans of autonomous vehicles to passengers and allows machines and people to interact [12]. The design purpose of the IVIS for autonomous vehicles is to deliver information that users need and improve reliability, acceptance, and safety. For example, providing passengers with the information necessary to use the service can help in decision-making and reduce the time required to search for information [13, 14]. Moreover, it is possible to reduce the passengers' anxiety and stress by communicating to them information about the intention or status of an autonomous vehicle [15, 16]. In other words, the IVIS is an essential device required to provide passengers with a good mobility experience. However, autonomous buses do not have a visible driver, increasing passengers' dependence on the IVIS [13]. Therefore, the IVIS for autonomous buses should allow passengers and vehicles to interact effectively and communicate information in an easy-to-understand way for passengers [17].

### 2.2  Interactive IVIS Design

Several methods regarding the interaction of autonomous vehicles and passengers have been studied. In Niculescu's study, a robotic assistant allowed passengers to interact with the autonomous vehicles [15]. The robot assistant assisted in setting the route to the destination, guiding the route, and making safety checks through dialog with the passengers in the vehicle. The authors report that the interaction between robotic assistants and passengers can build a high level of trust in autonomous vehicles and reduce safety concerns.

This study used digital assistants to allow passengers and autonomous vehicles to interact [18], studying the impact of the amount and quality of the information provided by the digital assistant and the level of personification on passengers' reliability.

As a result of the experiment, passengers reported that the reliability and their acceptance of autonomous vehicles increased when a highly anthropomorphic digital assistant provided high-quality information.

Robot assistants and digital assistants have the advantage that an anthropomorphic system can communicate directly with passengers. However, interactive interactions can be suitable when dealing with one or a few passengers. As autonomous buses accommodate many people, it should be borne in mind that there are a large number of information recipients. Oliveira et al. studied a Human–Machine Interface (HMI) that allows passengers to control autonomous vehicles and deliver real-time information effectively to passengers in the vehicle. According to their research, when there are many passengers on an autonomous bus, it is possible to deliver information quickly and accurately through a digital display installed inside. Furthermore, it is effective in communicating information required by individual passengers as they directly interact with the vehicle through a mobile app [16]. In other words, information that all passengers need to know, such as the real-time information of an autonomous bus, must interact with several passengers through a digital display. At the same time, individual passengers need to be able to enquire about journey information or requested inquisitive information through a mobile app. (Fig. 1).

Fig. 1. An effective way to interact with autonomous buses and passengers

## 2.3 Easy-to-Understand IVIS Design

Autonomous driving technology is a fusion of Information and Communications Technology (ICT) [1], which is perhaps difficult for the general public to understand. Therefore, it is vital to communicate the intention, plan, and status information of the autonomous vehicle in an easy-to-understand way to the passengers.

**Provides Information on Why It Works.** Passengers find it easier to understand when information about the driving situation and the reason for the operation are delivered together [18, 19]. According to Koo et al.'s study, the "Why" information such as "an obstacle ahead!" contains information about the environment the vehicle is trying to encounter, and it reports that passengers can respond quickly through this. Häuslschmid et al. used an "anthropomorphic 3D driver avatar," "world in miniature," and "left and right indicators" to convey information about driving intentions and evaluate which factors are understood easily [20]. The participants reported that it was easy to understand the intention of the autonomous vehicle because the "world in miniature" factor could tell why the autonomous vehicle was operating. The "world in miniature" is a method of conveying information to passengers through a reduced graphic of the surrounding environment recognized by autonomous vehicles. As autonomous buses use flexible routes, there can arise a situation where the schedule changes in real-time because of external factors and if the IVIS provides feedback on the reason for the change, passengers will respond quickly.

**Familiar Interface.** Mirnig et al.'s study proposed three IVISs with different graphic levels (Traditional List/Abstract Map/Realistic Map) and evaluated whether passengers can press the stop button when they get close to the destination stop [13]. In the experiment, the participants pressed the stop button at the closest distance to the stop when the Traditional List was provided. Conversely, exposure to the Abstract and the Realistic Map led the passengers to press the stop button when they were relatively far from the stop. The authors reported that the IVIS, designed with an interface that passengers are familiar with, impacts user reliability and acceptance of autonomous buses. In other words, an IVIS, designed with a familiar interface, is easy to understand and accept because passengers navigate based on experience.

**Providing the Information Users Need.** Several preceding studies have proven that user observation must be performed for the IVIS design of an autonomous vehicle. User observation must be done to understand who the user is and to provide the information they want. Ayoub et al.'s study attempted to provide information that could improve the parents' confidence with children on autonomous school buses [21]. Through user observation, they identified the anxiety factors of the children and parents, who are the direct and indirect information recipients, respectively, of the autonomous buses. Through interviews and the Wizard of Oz method, they understood the relationship between the two groups of information recipients and designed the IVIS to provide the information the recipients' needed. It is essential to observe how users are using autonomous buses. Also, it is important to understand how they have used traditional public buses and what information they are looking for. According to Hoff and Bashir's research, users claim to build trust by exploring automated systems based on past experience and knowledge [22]. This implies that designers should design an IVIS considering the learning and knowledge gained through the user's experience. Specifically, to provide the necessary information to the user, it is important to observe how the IVIS of the traditional public bus and passengers exchange information and to incorporate the improved information into the autonomous vehicle.

Previous research has confirmed that the reliability and acceptance of autonomous vehicles was deemed high when providing a user-centered IVIS. We intend to expand further on previous studies to propose an IVIS concept design centered on passenger mobility and user experience. It is important to consider reliability and acceptability for early users of autonomous vehicles, but the main purpose of using autonomous buses is to move passengers to their destination. When using public transportation, the required information is different in three stages: before travel, waiting before boarding, and on board [23]. Therefore, we investigated the usage context, needs, pain points, and required information by observing traditional public bus passengers. Then, we proposed an IVIS concept design that reflected the results of user research and delivered necessary information to passengers on autonomous buses promptly.

## 3   Method

**Fig. 2.** Research process

This study was conducted through three processes: user research, concept design, and acceptance evaluation (Fig. 2). First, we carried out user research using user diary and in-depth interviews targeting traditional public bus passengers. Also, we proposed the IVIS concept design of an autonomous bus based on the insights derived from the user research process. Finally, one-on-one interviews were conducted to verify the acceptability of the concept design proposed in this study.

### 3.1   User Research

**User Diary.** A user diary is a method for indirectly observing a user and helps understand the user's experience, behavior, and attitudes [24, 25]. We deployed the user diary to investigate the necessary information at each step of the passenger's use of the city bus. The user diary documents were provided to the participants in texts containing the following questions: 1) where do the participants usually go by public bus, and why did they decide to use the city bus? 2) what information was needed for these four stages of use: before boarding, when boarding, while moving, and when getting off, and 4) how was the information delivered? Participants who used city buses more than once a week were asked to fill out a draft form and after revising, complete the user diary document.

Participants in the experiment were recruited via an online recruitment announcement from a cohort who used city buses at least twice a week. A total of 13 participants were recruited (7 males, 6 females) aged between 24–32 years (average: 26.9 years) and were students or professionals. The recruited participants were instructed to select at least one option from a place they had visited for the first time, 1 or 2 times, and frequently. The reason is that as the journey became more familiar, the amount of information required decreases and the content differs [23]. After selecting the option, the schedule was discussed with the participants to set the date of the experiment. We explained the purpose and instructions of the experiment to the participants by phone or messenger during the morning of the promised date. After the explanation, the user diary document was sent through the messenger.

Participants were asked to complete a user diary as soon as possible after getting off the traditional public bus. However, when writing a user diary, we asked them for a brief memo or photo to be taken when using the bus to remind the user of the experience. The reason was that writing a diary while on the go could be dangerous and the necessary information should not be missed while writing. After completing the diary, the participants delivered it back to the researchers via messenger.

**In-Depth Interview.** From the user diary, it was possible to know what information the participant was checking at each phase of the ride, but it was difficult to confirm in detail how the user needed the information to be delivered. Follow-up in-depth interviews were conducted with user diary participants to obtain more specific responses about the information delivery method, information provider, and information recipient. Therefore, the interviewees consisted of a total of 13 participants from the user diary phase. We contacted the interviewees in advance to explain about the online in-depth interview and to ask for consent to participate and record. Online in-depth interviews were conducted with questions based on user diary results.

## 3.2  Concept Design Suggestion and Evaluation

We proposed the IVIS concept design with insights derived from user research data. One-on-one interviews were conducted to verify the acceptability of the concept design proposed in this study. The questions for acceptance verification were constructed by citing Venkatesh et al.'s research [26]. Three questions were included in all the five prepared scenarios, and one particular question was asked only at the time of leaving the bus. Therefore, the interviewees answered a total of 16 questions about the evaluation of concept design acceptance.

- Questions

  - Is the information provided to be checked at the phase of the ride in the scenario?
  - Did the information provided enable you to complete the each task to be performed during the phase of the ride in the scenario?
  - Could you understand the AMoD operating system at the phase of the ride in the scenario from the information provided?
  - (Question provided only at the drop-off stage) Is there any information you would like to obtain in connection with the drop-off?

Participants in the acceptance evaluation interview were newly recruited through online announcements to prevent data contamination. Recruitment conditions included use of the traditional public bus more than twice a week, and five people were randomly selected from such users (2 men and 3 women). The interview lasted about 40 min and proceeded in the following steps:

1. The researchers explained the definition of the AMoD, its driving system, and usage so as to help participants who have no experience of autonomous vehicles to understand the process.
2. The concept design proposed in this study was explained.
3. The problem and the situation found in user research were explained to the participants at each step of using city buses. Then, we asked the participants to evaluate their understanding of the scenario and need for problem-solving on a Likert 5-point scale.
4. We explained the ideas suggested based on the insights from the problem. Then, the interviewee evaluated the acceptance of the proposed concept design.

# 4  Results

## 4.1  User Diary

We asked the participants their reasons for using the bus to which they responded that the city bus is the fastest way to reach a given destination or it is used when there is no subway station at the boarding or destination stop. It is apparent that the users use public transport but prefer the quickest and most convenient way to their destination.

- Shortest travel distance and time [P2, P4, P9, P10, P11, P13]
- Unable to move by subway [P2, P3, P6, P7]
- Low usage fees [P3, P8]
- Short walking distance [P6]
- No need to transfer to other transport mode [P1]

**User Behavior Based on the Frequency of Visits to the Destination.** The partici-pants were found to have different behaviors and attitudes when using city buses depend-ing on the frequency of visits to their destinations. They constantly checked the IVIS to make sure they were on the right bus when going to the place they were visiting for the first time. Particularly, they had a very tense attitude when getting on or off because they could get off by mistake as well. When going to a place they had visited once or twice, they showed almost the same behavior and attitude as when the former scenario (first time visited place) but the frequency and tension of checking information decreased. On the other hand, when they went to frequently visited places, they rarely checked the IVIS and performed other tasks such as sleeping or reading while on the move.

**Information that the User Confirms.** The participants checked similar information regardless of the frequency of visits to their destinations (Fig. 3). However, higher the frequency of visits, fewer were the information types to be checked. Particularly, when going to frequently visited places, they tend not to check information related to the route. It was confirmed that the user did not check the information well because they had already learned about the route. However, information related to bus operation, such as the bus number, expected arrival time, and the name of an approaching stop, was checked regardless of the frequency of visits. This is because the information is necessary for the user to arrive at the destination by using the city bus.

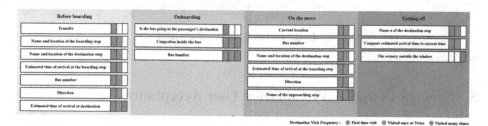

**Fig. 3.** Information required by passengers

## 4.2 In-Depth Interview

We confirmed how passengers receive the information they need through in-depth interviews (Fig. 4). Passengers primarily checked the information on the direction, route, their current location, and distance to the destination required to travel to the destination. They were able to find out the direction, current location, and remaining distance to the destination by comparing various types of operational information, such as information on the currently approaching stop and on the destination stop and the route. Consequently, the participants were mainly using them in the order of "the map application," "IVIS," and "talk with bus drivers." The maps application was preferred as that information could be assimilated visually. The participants preferred to check the information visually because the background noise or they were wearing earphones. Additionally, IVIS provides information only when it is close to a stop; hence, when passengers miss information, they use the map application a lot. Through the in-depth interview, we identified the following insights:

1. We need to provide information that can confirm intuitively that passengers are on the right bus.
2. The real-time operation information of the bus should be delivered using graphics for easy understanding.
3. Visual and auditory information must be communicated together because passengers may otherwise miss the information they need.
4. Real-time information should be visible at all times so that passengers can check it at any time.

| Phase of ride | Required information | Provider | Way to information delivered |
|---|---|---|---|
| Onboarding | Estimated time of arrival at the boarding stop | Map app | Check the estimated arrival time of the bus to be boarded through the map app [P1, P3, P6, P8, P9, P11, P12] |
| | Information that allows passengers to verify that they are on the correct bus | Bus driver | Talk to the bus driver to see if they are driving to their destination [P11] |
| | | IVIS | Check the route map and bus number inside the bus [P1, P2, P3, P7, P8] |
| | | Map app | Turn on the map app to check the direction of operation of the bus and the movement of the bus using GPS [P1, P2, P4, P8, P12, P13] |
| On the move | Remaining distance to destination stop | IVIS | By looking at the route map and the information display device inside the bus, count the number from the current stop to the drop-off stop [P7, P12] |
| | | Map app | Visually check the remaining distance from the current location to the destination on the route [P2, P3, P4, P5, P6, P7, P9, P10, P13] |
| | | | Use the map app to check your current location with GPS [P1, P2, P3, P4, P6, P8, P10, P11, P13] |
| | | | Using the map app, search for the name of the stop to drop off and check the location [P1, P2, P9, P11] |
| | | | Use the map app to find out the estimated time required after searching for a destination [P2, P3, P4, P8, P10, P11] |
| | | | Check the bus stop through the voice announcement on the bus and find out the current location through the route map [P5, P7, P9, P12] |
| | Transfer | Map app | Search in advance for walking or transfer information from the drop off stop to the destination[P1, P2, P3, P8, P13] |
| Drop-off | Information that allows passengers to verify that they are getting off the correct destination stop | IVIS | Check the name of the stop to drop off from the bus internal information display device or voice announcement [P2, P3, P5, P7, P9, P10, P11] |
| | | Map app | Checked the current location icon is close to the destination stop [P1, P4, P6, P8, P12, P13] |
| | | | Compare the estimated time of arrival with the current time.[P2, P11] |

**Fig. 4.** Results of in-depth interviews

## 5 Concept Design and Evaluation User Acceptance

### 5.1 Concept Design Suggestion

In this study, we propose an IVIS concept design that provides necessary information for autonomous bus passengers promptly. The concept design proposed in this study is assumed to be an IVIS provided by an Autonomous Midi bus that provides the AMoD service. The information to be provided to all passengers is transmitted through the HMI inside the vehicle, and passengers can interact directly with the autonomous bus through the mobile app. There are two situations in which information is provided to passengers (Fig. 5). First, information that passengers need to know is provided in advance whenever a variable occurs while using the bus. We defined variables into five categories and organized the information to be provided to passengers (Table 2). Second, if passengers have any questions, information is provided through direct interaction with the IVIS. This information is not provided in advance but detailed information is provided at the request of the user.

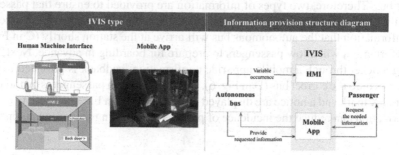

**Fig. 5.** Assuming HMI installation location

**Table 2.** Composition of the proposed IVIS

| Phase of ride | Variable | Provided information |
|---|---|---|
| Onboarding | Bus dispatch | User's journey information, expected arrival information |
| | Arriving at the boarding stop | Notification of arrival at bus stops, feedback on correct boarding |
| On the move | Moving along route | Real-time information, driving status information |
| | Boarding of new passengers | Changed bus schedule information, Reason for variable occurrence |
| Getting off | Arriving at the destination stop | Notification of arrival at the bus stop, feedback on correct drop-off |

**Autonomous Bus Dispatch Information and Real-Time Location Information.** According to user research, passengers wait at the bus stop or use a smartphone after checking the estimated arrival time of the bus marked on the digital display at the station. However, the estimated arrival time is not accurate and, thus, the passengers rush to board the bus. Therefore, we additionally provided visual information about the current location of the autonomous bus, allowing passengers to predict when the bus will arrive at the stop. In Fig. 6, A and B are the screens provided when an autonomous bus is dispatched after a passenger requests a ride. The information on the dispatched autonomous bus is divided into upper and lower sections, and real-time location information about the autonomous bus is provided at the top (B in Fig. 6). The lower section displays the boarding pass Quick response code (QR code) and reserved itinerary information. This innovation allows passengers to perform other tasks while waiting for the autonomous bus and use their time efficiently.

**Two Types of Information to Confirm the Correct Ride.** As a result of user research, passengers report experiencing the inconvenience of constantly checking the direction of travel, bus number, and where the next stop is to confirm that they have boarded the correct bus. Therefore, two types of information are provided to ensure that passengers are on the correct bus. First, the push notification of the mobile app provides passengers with information that the autonomous bus will arrive at the station shortly (C in Fig. 6). This information will allow passengers to prepare for boarding in advance. Next, when tagging a ticket through a mobile app, ride confirmation feedback is provided to guide passengers onto the correct bus (D in Fig. 6). However, when a passenger is on the wrong bus, a red X mark and a notice are displayed on the ticket (E in Fig. 6). These two pieces of information can reduce the incidence of passengers boarding the wrong autonomous bus.

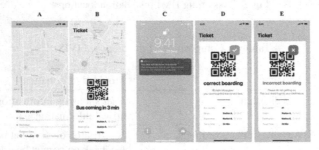

**Fig. 6.** IVIS provided when boarding (Color figure online)

**Provides Real-Time Information on Autonomous Buses.** Through a HMI inside the vehicle, real-time information can be delivered such that passengers can easily understand the driving information and plan of autonomous buses (A in Fig. 7). Real-time information was produced by utilizing existing route map graphics that passengers are familiar with. The real-time information was divided into a left section comprising the overall route and operation status and a right section comprising the driving status. The reason is that the information on driving situations can encapsulate the overall flow of the autonomous driving bus, and the driving situation is information that can encapsulate in cognitive terms the driving state and intention of the autonomous driving bus. In the left section, the following information is displayed. First, the serial number of the autonomous bus is displayed such that passengers can check which bus they had taken. Second, it displays the current time such that passengers can compare the expected arrival time and the current time. Third, it marks the current location of the autonomous bus such that passengers can see where they are. Fourth, the route covered and the route to be taken are displayed such that the passengers can check the driving situation. Fifth, stop icons are represented by filled circles and unstopped stop icons by empty circles, allowing passengers to check at which station the autonomous bus stops. Sixth, the estimated arrival time is displayed at the bottom of the stop such that passengers can check when to get off. Finally, a public transport transfer icon has been added to the stops such that passengers can check the public transport available after getting off. The right section displays the following information. First, the name of the autonomous bus approaching and the remaining travel time and travel distance are displayed. Through this, the passengers who need to get off at the next stop can prepare, and the autonomous bus can recognize the operating schedule and inform the passengers that it is driving as planned. Second, the actual driving speed and the speedometer graphic allows the passengers to check the driving speed. Using the speedometer graphic, passengers without driving experience can intuitively experience the speed.

**Information on Schedule Changes.** As the AMoD operates according to a variable schedule, it is essential to quickly deliver information that has changed due to new ride requests or cancellations to passengers. In situations where the operation schedule was changed, voice announcements, interactions, and colors were used to attract passengers' attention (B in Fig. 7). Feedback was provided to passengers regarding the reason for the change in information, enabling them to comprehend the situation and respond quickly. A blinking alert was provided at the stops stopping because of a change in the schedule to check the changed information.

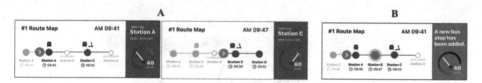

**Fig. 7.** IVIS provided when on the go

**Two Types of Information to Confirm Getting Off At the Correct Stop.** Some participants have experience of getting off in a hurry or passing the bus stop because of performing other tasks such as sleeping and reading on the move. As a result of user research, the participants responded that they had an experience of suddenly getting off or passing their destination while sleeping or reading. Therefore, two types of information are delivered through mobile apps such that passengers can get off at the correct stop even though they are not focused on the IVIS (Fig. 8). Push notifications are provided when the bus is close to the passenger's destination stop that allows passengers to recognize that the time has come to get off and prepare (A in Fig. 8). If the passenger has disembarked at the correct stop, the transparency of the boarding pass will be reduced, indicating that it is a completed boarding pass, and a green check box will be displayed

(B in Fig. 8). However, when a passenger gets off at the wrong bus stop, a red X mark and a notice are displayed on the ticket (C in Fig. 8). These information will help reduce the number of passengers getting off in a hurry or at the wrong stop.

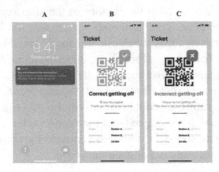

**Fig. 8.** IVIS provided when alight (Color figure online)

## 5.2 Evaluation of User Acceptance

Five scenarios were selected to evaluate the acceptability of our proposed IVIS. The selected scenarios were constructed based on the occurrence of variables specified in 5.1, and when passengers found the information through user research, they confirmed it (Fig. 9).

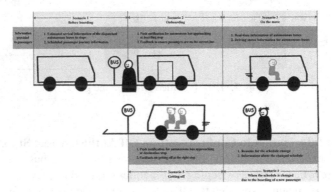

**Fig. 9.** Scenario selection criteria

**Results.** The participants in the acceptance evaluation understood the situation of the problem specified in this study. The participants positively evaluated that the proposed concept design can deliver the necessary information on time such that they can move to their destination conveniently. One of the results that should be noted during the acceptance evaluation is delivering two types of information that recognize getting on and off in scenarios 2 and 5. Although there were a needs for participants to clearly check the boarding and getting off on traditional public buses, it was evaluated that it must be provided on autonomous buses. The reason being that passengers do not continue to focus on IVIS, hence, the need for information to induce correct getting on and off.

For Scenario 1, the participants reported that they could wait while performing other tasks before boarding because they could check the expected arrival time and real-time location of the bus to be boarded. One participant reported that digitally vulnerable groups such as the elderly would find it difficult to use. Two participants thought it was similar to the interface of the service they are currently using. This shows the possibility that users might find our proposed IVIS design a little more familiar. However, it is necessary to consider the differentiation of the information delivery method applying the characteristics of the AMoD service.

In Scenario 3, the participants reported that they could easily understand the driving situation by being provided with real-time information visually. They also reported that could understand the AMoD's flexible route because the authorized and unauthorized stops were separated. One participant reported that it would be easier to check if each passenger was color coded. It is apparent that the passengers have expectations of personalization but designers will have to provide it in considering scenarios when the number of passengers increase.

In Scenario 4, the participants reported that notification and reason information is required when a variable situation occurs. However, one participant reported that the blinking interaction at the additional stop seemed to only mean getting off. Therefore, it is effective to provide interaction when a change in the schedule occurs, but it should be designed so as not to confuse the passengers. Finally, in response to whether the participants would like to receive any additional information after getting off the autonomous bus, the replies included "Route to Transfer Point," "Transfer Schedule," and "Save Frequently Used Routes." The participants need to receive the necessary information through one IVIS by the end of the journey.

## 6  Conclusion

This study investigated the design of an IVIS that considers mobility by further extending passenger trust in and acceptance of autonomous buses. We found the need to design an IVIS suitable for autonomous buses through literature studies and user research. Autonomous buses and traditional public buses have different operational conditions. Further, passengers were experiencing inconvenience in the information transmission method found in existing city buses. Autonomous buses run based on user demand and use flexible routes. This is an entirely different operational condition from traditional public buses, and new information must be delivered to passengers. Moreover, in autonomous buses with no visible driver, passengers have a strong dependence on an IVIS [13]. If the IVIS delivers the necessary information to passengers on time, they can understand and trust how to use autonomous buses. Therefore, we conducted user research on existing city bus passengers to investigate what kind of information is required before travel, when boarding, while moving, and when getting off. This information is also provided on traditional public buses, but the needs and pain points of passengers must be identified, improved upon, and delivered. The passengers' needs and pain points were identified as being that most of the necessary information was conveyed inconveniently. Therefore, we propose some recommendations when designing the IVIS for autonomous buses. First, it is necessary to provide the estimated arrival time and current location information together such that it is possible to predict clearly when the bus to be boarded will arrive at the stop. Second, information confirming that passengers are boarding and traveling on the correct bus should be communicated. Third, the real-time information about the bus must be transmitted using graphics to be understood easily. Fourth, real-time information on the operation should be exposed at all times, such that passengers can check it at any time. Lastly, when the schedule is changed because of external factors during driving, information must be delivered quickly. Additionally, the reason for changing the schedule must be communicated for the user to respond quickly.

The proposed IVIS concept design could enhance the ease of using autonomous buses and increase the reliability and acceptance of autonomous buses. It could be used for reference when designing the IVIS for autonomous buses in the future. Although the research results constitute a contribution to the design of IVIS for autonomous buses, limitations still remain, namely, that the participants in this study's user research were in their 20s and 30s. In further, we plan to expand user research targets to investigate various age groups and traffic weaknesses and propose solutions. Despite these limitations, this study has made progress in that it investigates information and delivery methods for passengers before the introduction of autonomous buses. It is expected that the information and delivery method proposed in this study can contribute to the future design of IVIS for autonomous buses for users.

# References

1. Yaqoob, I., Khan, L.U., Kazmi, S.M.A., Imran, M., Guizani, N., Hong, C.S.: Autonomous driving cars in smart cities: recent advances, requirements, and challenges. IEEE Netw. **34**, 174–181 (2020). https://doi.org/10.1109/MNET.2019.1900120
2. WAYMO. https://waymo.com/tech/. Accessed 24 Mar 2021
3. NAVYA. https://navya.tech/en/technology/software/. Accessed 24 Mar 2021
4. Mercedes-Benz Future Bus: Revolutionary design. https://www.mercedes-benz.com/en/des ign/vehicles/future-bus-revolutionary-design/. Accessed 24 Mar 2021
5. Hinderer, H., Stegmüller, J., Schmidt, J., Sommer, J., Lucke, J.: Acceptance of autonomous vehicles in suburban public transport. In: 2018 IEEE International Conference on Engineering, Technology and Innovation (ICE/ITMC), pp. 1–8 (2018). https://doi.org/10.1109/ICE.2018.8436261
6. Salazar, M., Rossi, F., Schiffer, M., Onder, C.H., Pavone, M.: On the interaction between autonomous mobility-on-demand and public transportation systems. In: 2018 21st International Conference on Intelligent Transportation Systems (ITSC), pp. 2262–2269 (2018). https://doi.org/10.1109/ITSC.2018.8569381
7. Pavone, M.: Autonomous mobility-on-demand systems for future urban mobility. In: Maurer, M., Gerdes, J.C., Lenz, B., Winner, H. (eds.) Autonomes Fahren, pp. 399–416. Springer, Heidelberg (2015). https://doi.org/10.1007/978-3-662-45854-9_19

8.  Oh, S., Seshadri, R., Azevedo, C.L., Kumar, N., Basak, K., Ben-Akiva, M.: Assessing the impacts of automated mobility-on-demand through agent-based simulation: a study of Singapore. Transp. Res. Part A Policy Pract. **138**, 367–388 (2020). https://doi.org/10.1016/j.tra.2020.06.004

9.  Kim, M., Schonfeld, P.: Conventional, flexible, and variable-type bus services. J. Transp. Eng. **138**, 263–273 (2012). https://doi.org/10.1061/(ASCE)TE.1943-5436.0000326

10. Koffman, D., Transit Cooperative Research Program: Operational Experiences with Flexible Transit Services. Transportation Research Board (2004)

11. Hensher, D.A.: Future bus transport contracts under a mobility as a service (MaaS) regime in the digital age: are they likely to change? Transp. Res. Part A Policy Pract. **98**, 86–96 (2017). https://doi.org/10.1016/j.tra.2017.02.006

12. Harvey, C., Stanton, N.A., Pickering, C.A., McDonald, M., Zheng, P.: In-vehicle information systems to meet the needs of drivers. Int. J. Hum.-Comput. Interact. **27**, 505–522 (2011). https://doi.org/10.1080/10447318.2011.555296

13. Mirnig, A.G., Gärtner, M., Wallner, V., Trösterer, S., Meschtscherjakov, A., Tscheligi, M.: Where does it go? A study on visual on-screen designs for exit management in an automated shuttle bus. In: Proceedings of the 11th International Conference on Automotive User Interfaces and Interactive Vehicular Applications, pp. 233–243. Association for Computing Machinery, New York (2019). https://doi.org/10.1145/3342197.3344541

14. Fonzone, A., Schmöcker, J.-D.: Effects of transit real-time information usage strategies. Transp. Res. Rec. **2417**, 121–129 (2014). https://doi.org/10.3141/2417-13

15. Niculescu, A.I., Dix, A., Yeo, K.H.: Are you ready for a drive? User perspectives on autonomous vehicles. In: Proceedings of the 2017 CHI Conference Extended Abstracts on Human Factors in Computing Systems, pp. 2810–2817. Association for Computing Machinery, New York (2017). https://doi.org/10.1145/3027063.3053182

16. Oliveira, L., et al.: Evaluating how interfaces influence the user interaction with fully autonomous vehicles. In: Proceedings of the 10th International Conference on Automotive User Interfaces and Interactive Vehicular Applications, pp. 320–331. Association for Computing Machinery, New York (2018). https://doi.org/10.1145/3239060.3239065

17. Carsten, O., Martens, M.H.: How can humans understand their automated cars? HMI principles, problems and solutions. Cogn. Technol. Work **21**(1), 3–20 (2018). https://doi.org/10.1007/s10111-018-0484-0

18. Alpers, S., et al.: Capturing passenger experience in a ride-sharing autonomous vehicle: the role of digital assistants in user interface design. In: 12th International Conference on Automotive User Interfaces and Interactive Vehicular Applications, pp. 83–93. Association for Computing Machinery, New York (2020). https://doi.org/10.1145/3409120.3410639.

19. Koo, J., Kwac, J., Ju, W., Steinert, M., Leifer, L., Nass, C.: Why did my car just do that? Explaining semi-autonomous driving actions to improve driver understanding, trust, and performance. Int. J. Interact. Des. Manuf. (IJIDeM) **9**(4), 269–275 (2014). https://doi.org/10.1007/s12008-014-0227-2

20. Häuslschmid, R., von Bülow, M., Pfleging, B., Butz, A.: SupportingTrust in autonomous driving. In: Proceedings of the 22nd International Conference on Intelligent User Interfaces, pp. 319–329. Association for Computing Machinery, New York (2017). https://doi.org/10.1145/3025171.3025198.

21. Ayoub, J., Mason, B., Morse, K., Kirchner, A., Tumanyan, N., Zhou, F.: Otto: an autonomous school bus system for parents and children. In: Extended Abstracts of the 2020 CHI Conference on Human Factors in Computing Systems, pp. 1–7. Association for Computing Machinery, New York (2020). https://doi.org/10.1145/3334480.3382926

22. Hoff, K.A., Bashir, M.: Trust in automation: Integrating empirical evidence on factors that influence trust. Hum. Factors **57**(3), 407–434 (2015)

23. Grotenhuis, J.-W., Wiegmans, B.W., Rietveld, P.: The desired quality of integrated multimodal travel information in public transport: customer needs for time and effort savings. Transp. Policy **14**, 27–38 (2007). https://doi.org/10.1016/j.tranpol.2006.07.001

24. Czerwinski, M., Horvitz, E., Wilhite, S.: A diary study of task switching and interruptions. In: Proceedings of the SIGCHI Conference on Human Factors in Computing Systems, pp. 175–182. Association for Computing Machinery, New York (2004). https://doi.org/10.1145/985692.985715

25. Sohn, T., Li, K.A., Griswold, W.G., Hollan, J.D.: A diary study of mobile information needs. In: Proceedings of the SIGCHI Conference on Human Factors in Computing Systems, pp. 433–442. Association for Computing Machinery, New York (2008). https://doi.org/10.1145/1357054.1357125

26. Venkatesh, V., Morris, M.G., Davis, G.B., Davis, F.D.: User acceptance of information technology: toward a unified view. MIS Q. **27**, 425–478 (2003)

# Do German (Non)Users of E-Scooters Know the Rules (and Do They Agree with Them)?

Tibor Petzoldt[1]([⊠]), Madlen Ringhand[1], Juliane Anke[1], and Nina Schekatz[2]

[1] Technische Universität Dresden, 01062 Dresden, Germany
tibor.petzoldt@tu-dresden.de
[2] Technische Universität Chemnitz, 09111 Chemnitz, Germany

**Abstract.** Despite being a comparatively recent phenomenon, e-scooters enjoy an immense popularity in the cities in which they are available. As a result, their potential impact on road safety has been questioned, as injury crashes and violations of road rules are reported with increasing frequency. It can be suspected that at least to some degree, a lack of awareness with regard to the existing rules, coupled with a certain lack of agreement with these rules, might be the cause. Aim of the study presented in this paper was to quantify this issue with the help of an online survey. With a usable sample of 337 participants (105 users of e-scooters, 232 non-users), we looked into rule knowledge, agreement with the rules (users only) and past behaviour, including violations, while riding (users only). The results indicate that there might be reason for concern. While users seemed to outperform non-users on the knowledge questions, their rates of correct responses were far from perfect. Among the users, agreement with the rules was generally high, yet there were also clearly visible minorities who considered several rules as being too strict. It seems that in order to improve compliance, there is a need for a structured instruction of the users that addresses issues of rule knowledge, but also instils the motivation to follow these rules.

**Keywords:** Micromobility · Traffic safety · Violations

## 1 Introduction

Although e-scooters have been available only for a few years, they have seen a massive growth in usage since their introduction. For the US, NACTO (2019) reports that in 2019, 109 cities had at least one dockless e-scooter program. People took about 86 million trips on shared e-scooters, which is an increase of over 100% from 2018. In Germany, e-scooters were available in 39 cities in summer 2020, according to the apps of eight sharing providers. While comprehensive statistics are hard to come by, it is known that in Berlin alone, about 11 000 shared e-scooters of three providers (Lime, Tier, Voi) were active in September 2019 (Tack et al. 2019).

As with any new mode of transport, questions about its safety arise immediately. However, due to the novelty of e-scooters, they are not yet covered in most official crash statistics. Instead, we have to rely on hospital data and information on exposure to get

© Springer Nature Switzerland AG 2021
H. Krömker (Ed.): HCII 2021, LNCS 12791, pp. 425–435, 2021.
https://doi.org/10.1007/978-3-030-78358-7_29

a general idea about e-scooters' safety impact. For the city of Austin, Texas, a rate of 20 injured individuals per 100 000 trips has been estimated based on injury reports and usage data (APH 2019). Of the 190 injured riders, one third crashed during their first ride. About half of the hospitalisations were the result of single vehicle crashes. In an analysis of Californian data, about 80% of the 228 injuries recorded were found to be the result of single vehicle events (Trivedi et al. 2019). Similar proportions have been reported from German analyses, with single vehicle crashes making up about 75% of all injury crashes in Berlin (Uluk et al. 2020) and more than 90% in Frankfurt, again with one third of the injured being first time users (Störmann et al. 2020).

Most of the investigations of injury data imply that e-scooter crashes might, to a substantial degree, be caused by or at least go along with illegal behaviour on behalf of the users, such as riding while intoxicated (Kobayashi et al. 2019; Uluk et al. 2020) or on the pavement (Bloom et al. 2020). This impression is supported by police reports, which typically find that the overwhelming proportion of reported e-scooter crashes is caused by the riders themselves, and that intoxication is highly prevalent (Polizei Berlin 2019; Polizei Dortmund 2019; Polizei Hamburg 2020). While this is certainly not an occurrence that is exclusive to this specific mode of transport, it raises the question of whether riders are actually aware of the existing regulations, and whether or not they are accepting these rules.

In Germany, the use of e-scooters is legal, but tightly regulated, since June 2019. The so-called "Elektrokleinstfahrzeuge-Verordnung" (eKFV, loosely translated as "Personal Light Electric Vehicles Regulations") stipulates that e-scooters might be used at a speed of max. 20 km/h, by only one person, preferably on cycling infrastructure. If no cycling infrastructure is available, e-scooters are supposed to be used on-road, whereas riding on the pavement or in pedestrian areas is illegal. The minimum age to use an e-scooter is 14 years. Neither a driving license of any sorts, nor a specific test on rule knowledge or similar is required. A helmet is not required, either. However, e-scooters need to be insured, like other small motorized vehicles. The legal limit for blood alcohol content is 0.5‰ (0.0‰ for riders below the age of 21 or during the probation period for a driving license). Police reports indicate that many of these rules are violated on a regular basis, by a substantial number of users (Polizei Berlin 2019; Polizei Dortmund 2019; Polizei Hamburg 2020).

Information on how well users of e-scooters as well as the general population in Germany is aware of these rules is limited. Siebert et al. (2021) surveyed only e-scooter users, and found considerable gaps in rule knowledge for a sample of very young age. From Austria (which differs, to some degree, in how e-scooters are regulated), it has been reported that users of e-scooters lack knowledge especially with regard to the legal age of use and existing speed limits (Mayer et al. 2020; Mayer et al. 2019). In general, for any given rule, the rate of incorrect responses was at least about 20%. Moreover, non-users seemed to be as knowledgeable as users, with practically no differences between these two populations. In the US, a survey of users in Portland, Oregon found that about 14% of local users, and close to 39% of visitors who used an e-scooter, did not "know what the e-scooter laws are in Portland or Oregon" (Portland Bureau of Transportation 2020). Not surprisingly, participants in focus groups expressed the need for more information on existing regulations (Portland Bureau of Transportation 2019).

While, to our knowledge, there are no studies on the connection between e-scooter users' rule knowledge and road safety available yet, we know that for cyclists, lower levels of knowledge are linked to risky behaviours and crash involvement (Useche et al. 2019). Similarly, it has been reported that car drivers' scores in a knowledge test correlate with on-road test performance (Wolming and Wiberg 2004).

It should be noted, however, that rule knowledge might be necessary, but not sufficient for actual compliance. Even road users that are aware of the rules might disregard and deliberately violate them. While knowledge on e-scooter users' actual acceptance of the rules is limited, a look into cycling might provide some clues as to what to expect. A survey study by Huemer et al. (2019) indicates that German cyclists might have a positive attitude towards certain infringements. Similarly, a considerable portion of Norwegian cyclists seems to have pragmatic attitudes towards rule violations, and also expresses dissatisfaction with the rules as they currently are (Kummeneje and Rundmo 2020).

Overall, the studies so far indicate that riders of e-scooters do not have a sufficient level of knowledge on the regulatory framework surrounding the use of these vehicles. It also can be suspected that a certain lack of agreement with or acceptance of these rules might exacerbate the problem. Investigations of crashes and crash related injuries of e-scooter users suggest that disregard for the rules, either deliberately or due to lack of knowledge, might be a road safety issue. In the study presented in this paper, we tried to shed some light on the issue of rule knowledge and rule acceptance of e-scooter users in Germany. Our goal was to get a general idea of how much they know about the existing regulations, and whether or not they actually know more than non-users. In addition, we wanted to find out if users considered these regulations appropriate, a judgment which might have an impact on the users' willingness to follow these rules.

## 2   Method

### 2.1   Survey

To address our research questions, we decided for an online survey that was implemented in LimeSurvey. After a short introduction on the welcome page, which included terms of data protection and processing instructions, a filter question was applied to separate non-users and users of e-scooters. For the users, this was followed by several questions regarding their frequency of use, characteristics of their last trip (helmet use, used traffic infrastructure) and specific rule violations (riding an e-scooter in pairs, alcohol use; see Table 3 for the wording of questions). Both users and non-users then had to answer questions on their knowledge of e-scooter related rules and regulations. These questions included items on the maximum speed permitted, the legality of riding on certain types of road infrastructure, the minimum age required to ride, helmet use, the number of users allowed per vehicle and the legal limit for blood alcohol content (see Table 4 in appendix for a complete list of questions and answer categories). Users were then asked to provide their opinion on these rules, i.e. whether they considered them appropriate, too strict, or not strict enough (Table 5 in appendix). Both groups then finished the survey by providing basic socio-demographic information.

## 2.2  Sample

The survey was online from mid-June to early July 2020. It was distributed through social networks such as Facebook and Instagram. A link to the survey was sent to a lobby group for users of e-scooters. In addition, we used an internal mailing list through which we regularly recruit cyclists for our studies. The survey took 5 min for e-scooter non-users and 10 min for e-scooter users. No incentives were given. After the removal of unusable datasets (e.g., incomplete surveys, obvious non-sense responses such as a three-digit age), the usable sample contained 337 participants (102 female, 211 male, 24 NA) with a mean age of 43.7 years (SD = 13.8). Out of all respondents, 105 (31.2%) reported to have used an e-scooter at least once, and hence constitute our "user" group, with the remaining 232 being the "non-user" group. E-scooter users had a mean age of 39.4 years (SD = 14.3), whereas the non-users were 45.5 years (SD = 13.3) old on average.

## 3  Results

### 3.1  Rule Knowledge

Table 1 shows the respondents' knowledge of e-scooter related regulations. As can be seen, the proportion of correct responses differs considerably between the different rules. Whereas nearly all participants were aware that only one rider is allowed on a single e-scooter, less than half knew about the legal limit for blood alcohol content or speed. Even fewer knew the legal age for use. Also interesting to note is that for nearly all rules, users showed a higher level of awareness of the rules than non-users. However, not all of these differences were statistically significant when comparing the proportions of correct and incorrect responses (maximum speed: $\chi^2$ (1) = 32.03, $p < .001$/legal age: $\chi^2$ (1) = 10.52, $p = .001$/helmet use: $\chi^2$ (1) = 3.41, $p = .065$/users per vehicle: $\chi^2$ (1) = 0.19, $p = .664$/blood alcohol limit: $\chi^2$ (1) = 2.55, $p = .110$). Still, it must be highlighted that the users' knowledge on most issues was far from perfect. In light of its practical implications, the fact that about 10% overestimated the legal limit for blood alcohol content, and nearly a quarter of respondents reported to simply not know the limit, is especially worrisome.

When asked about what types of road infrastructure would be (il)legal for use, only six of our e-scooter users (5.7%) responded that they didn't know, whereas more than one fifth of the non-users (49/21.1%) selected this option, indicating a clear difference in knowledge between these two groups ($\chi^2$ (1) = 12.56, $p < .001$). Among the remaining participants, knowledge about what infrastructure to (not) use was high throughout, both for users and non-users (see Table 2), with only the shared pedestrian / bike path producing a proportion of correct responses below 90%.

**Table 1.** Rule knowledge for maximum speed, allowed age, compulsory helmet use, persons per vehicle and blood alcohol limit. *Bold highlighted differences between users and non-users are significant on $\alpha = 0.05$ level –for Chi-squared test: correct response vs. incorrect response.*

| Category | Response | Users (N = 105) | | Non-users (N = 232) | |
|---|---|---|---|---|---|
| | | N | % | N | % |
| Maximum speed (20 km/h) | Correct response | **69** | **65.7** | **76** | **32.8** |
| | Less than 20 km/h | 7 | 6.7 | 19 | 8.2 |
| | More than 20 km/h | 11 | 10.5 | 45 | 19.4 |
| | Don't know | 18 | 17.1 | 92 | 39.7 |
| Minimum legal age (14 years) | Correct response | **33** | **31.4** | **37** | **15.9** |
| | Less than 14 years | 9 | 8.6 | 15 | 6.5 |
| | More than 14 years | 33 | 31.4 | 89 | 38.4 |
| | Don't know | 30 | 28.6 | 91 | 39.2 |
| Compulsory helmet use (none) | Correct response | 89 | 84.8 | 176 | 75.9 |
| | Incorrect response | 4 | 3.8 | 22 | 9.4 |
| | Don't know | 12 | 11.4 | 34 | 14.7 |
| Maximum number of users per vehicle (one) | Correct response | 97 | 92.4 | 211 | 90.9 |
| | More than one person | 4 | 3.8 | 2 | 0.9 |
| | Don't know | 4 | 3.8 | 19 | 8.2 |
| Blood alcohol limit[a](0.5‰) | Correct response | 56 | 53.3 | 102 | 44.0 |
| | Less than 0.5‰ | 13 | 12.4 | 28 | 12.1 |
| | More than 0.5‰ | 11 | 10.5 | 19 | 8.2 |
| | Don't know | 25 | 23.8 | 83 | 35.8 |

[a] Above age of 21 years and outside of probationary period.

**Table 2.** Absolute and relative numbers of correct responses on rule knowledge questions regarding use of traffic infrastructure (participants that stated that they "did not know" not included). No significant differences between users and non-users on $\alpha = 0.05$ level.

| Traffic infrastructure | Users' correct responses (N = 99) | | Non-users' correct responses (N = 183) | |
|---|---|---|---|---|
| | n | % | n | % |
| Pedestrian area (not permitted) | 91 | 91.9 | 171 | 93.4 |
| Pavement (not permitted) | 90 | 90.9 | 174 | 95.1 |
| Shared pedestrian / bike path (permitted) | 78 | 78.8 | 160 | 87.4 |
| Separated pedestrian / bike path (permitted) | 91 | 91.9 | 170 | 92.9 |
| Bike path (permitted) | 97 | 98.0 | 172 | 94.0 |

## 3.2 Rule Assessment

When asked about whether or not they considered the existing rules and restrictions appropriate, the majority of our users responded approvingly. However, the degree of approval varied between rules, and the direction of disapproval differed as well (see Fig. 1). Agreement was highest for the restriction to one user per e-scooter. Still, about one

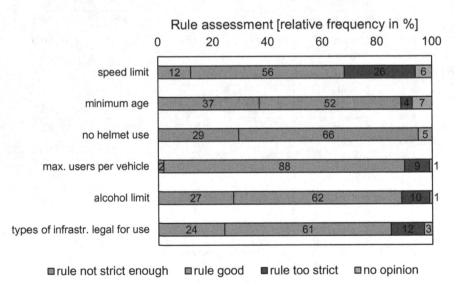

**Fig. 1.** Rule assessment in relative frequency of e-scooter users for the following rules: speed limit and max. users per vehicle, types of infrastructure legal for use (N = 100), no mandatory helmet use and alcohol limit (N = 99), minimum legal age (N = 98).

in ten respondents considered this rule too strict. Similarly, about one in ten respondents felt that legal limit for blood alcohol content was set too low (although it should be noted that more than twice as many would like to see an even lower limit). More than one quarter considered the speed limit overly restrictive. In other instances, considerable minorities of the users indicated that they actually would like to see stricter rules. About one third expressed the desire for compulsory helmet use. Only about half of the users agreed that the legal age of use, currently at 14 years, was appropriate, while about four in ten users would like to see that number increased.

### 3.3 Reported Behaviour

When asked about aspects of their rider behaviour during their last ride, about 18% (n = 19) of the users reported to have primarily used the e-scooter on types of infrastructure (the pavement or a pedestrian area) where riding is illegal. Fourteen riders stated that they had used a helmet (see Table 3). About one sixth (n = 17) admitted to have been riding the e-scooter together with another person at least once, with twelve users having done so during their last 1–3 rides. Similar proportions were found for riding under the influence of alcohol, with 14 participants stating that they have used an e-scooter after consuming alcohol at least once, and six to have done so during the last 1–3 rides. It should be noted, however, that this does not necessarily mean that the level of blood alcohol content was above the legal threshold, i.e., it is unclear if these rides actually constitute incidents of "riding under the influence" in the legal sense.

**Table 3.** Relative frequencies of reported behaviour of e-scooter users.

| Reported behaviour items | Responses of e-scooter users | | |
|---|---|---|---|
| | N | Yes | No |
| Did you wear a helmet during your last ride with an e-scooter? | 105 | 13.3% | 86.7% |
| Have you ever used an e-scooter with another person? | 105 | 16.2% | 83.8% |
| Have you used an e-scooter with another person during your last 1–3 e-scooter trips? | 17 | 70.6% | 29.4% |
| Have you ever used an e-scooter under the influence of alcohol? | 105 | 13.3% | 86.7% |
| Have you used an e-scooter under the influence of alcohol during your last 1–3 e-scooter trips? | 14 | 42.9% | 57.1% |

# 4  Discussion and Conclusion

The goal of this study was to get a better understanding of how much users and non-users of e-scooters know about e-scooter related rules and regulations, and whether or not they agree with them. The results show that knowledge varied considerably from rule to rule. More than 90% of both groups were aware that only one user is allowed per e-scooter, but even among the users, less than one third was aware of the minimum legal age for using an e-scooter. To some degree, this is not surprising, as not all rules are relevant for every individual to the same degree. E.g., a 35-year old simply has not much reason to care about the minimum legal age for usage, so, given the mean age of the sample, a rather low rate of correct responses might have been expected. Nevertheless, the fact that knowledge was limited even for some obviously safety relevant rules, such as the legal limit for blood alcohol content, is worrisome.

Still, it is encouraging to see that users showed a higher (though not always significantly higher) awareness of the rules than non-users. Likewise, most riders also either approved the existing rules, or even would like to see them tightened further, which should result in a rather high rate of compliance. However, the small but clearly visible portions of the sample that feel that rules are too strict with regard to speed, infrastructure use or blood alcohol content must be considered problematic. It must be assumed that, as with any other vehicle, violations of the existing rules are often conscious decisions resulting from a lack of acceptance, rather than a lack of knowledge.

At the same time, it must be highlighted that, as road infrastructure and the traffic environment has not been designed with e-scooters in mind, following the rules is not always easy. Investigations of violations among cyclists point to the fact that at least in some scenarios, violating a rule might be considered the safest option (Chaloux and El-Geneidy 2019). When it comes to e-scooters, this might be especially the case in situations in which there is no cycling infrastructure, and users are forced to ride along other motorised vehicles in heavy traffic, while being limited to a maximum speed of 20 km/h. While there seems to be no obvious regulatory solution for this issue, it underscores the need for infrastructure that accommodates modes that are moving considerably faster than pedestrians, but still much slower than cars.

It must be acknowledged that our approach to participant recruitment might have resulted in a somewhat biased sample. Our sample's mean age of about 39 years is much higher than other researchers report for Germany (Siebert et al. 2020) (23 years), which might explain our comparatively high rate of agreement with the rules, as well as the comparatively low rate of reported lifetime violations. In addition, the fact that we recruited among regular cyclists might have resulted in some participants taking the perspective of another road user (e.g., a cyclist) rather than an e-scooter user when assessing e-scooter related rules. This might also explain the slight discrepancy between reported agreement with the rules and actual behaviour (e.g., about 16% reporting that they have used an e-scooter with another person, but about 90% agreeing with the rule that makes this illegal). Overall, we must assume that our rates of agreement with the rules are overestimations with respect to the overall population of e-scooter users, while the reported violation rates are most likely underestimations.

Still, overall, it appears that there is a clear need for some form of structured instruction regarding the rules and regulations surrounding e-scooter use. This instruction must, however, not only cover rule knowledge, but also address the users' motivation to follow these rules. The questions of how to design such an instruction and when to give it need further scientific assessment. Examples for instructional approaches exist, such as the campaign "Roll ohne Risiko" ("roll without risk") from the German Road Safety Council which designed small flyers presenting central rules that were placed on the handlebars of rental e-scooters (Deutscher Verkehrssicherheitsrat e.V. 2020). Rental companies have also integrated relevant information in their apps, but it can be quickly clicked away. At the same time, authorities must start to find better ways to consider the needs of slower modes such as e-scooters when planning roads and the road environment. We cannot just rely on road users doing the right thing under any circumstances. Instead, we must help them in any way we can to make the right thing to do also the easiest and most obvious thing to do. Only then can we expect violations to be reduced, and road safety to be preserved.

# Appendix

**Table 4.** Items of questionnaire regarding rule knowledge.

| Questions/items | Response categories |
|---|---|
| *Rule knowledge* | |
| What is the maximum speed an e-scooter is allowed to be used at? | 15 km/h |
| | 20 km/h |
| | 25 km/h |
| | 30 km/h |
| | Don't know |
| Which areas is an e-scooter permitted to be used on? | Pedestrian area |
| | Sidewalk |
| | Common foot and cyclepath |
| | Divided foot and cyclepath |
| | Cyclepath |
| | Don't know |
| What is the minimum age for riding an e-scooter? | There is no minimum age |
| | 12 years |
| | 14 years |
| | 16 years |
| | 18 years |
| | Don't know |

(*continued*)

**Table 4.**  (*continued*)

| Questions/items | Response categories |
|---|---|
| Is it a legal requirement to wear a helmet when riding an e-scooter? | Yes |
| | No |
| How many people are allowed on one e-scooter? | One |
| | Two |
| | Three |
| | There is no limit |
| | Don't know |
| What is the alcohol limit for riding an e-scooter? (for people who are at least 21 years old)? | 0.0‰ |
| | 0.5‰ |
| | 1.1‰ |
| | 1.6‰ |
| | Don't know |

**Table 5.**  Items of questionnaire regarding rule assessment.

| Questions/items | Response categories |
|---|---|
| ***Rule assessment*** | |
| E-scooters are permitted to drive at a maximum speed of 20 km/h | For all items: |
| E-scooters can be ridden on the following areas: (pictures of the traffic signs were shown): common foot and cyclepath, divided foot and cyclepath, cyclepath | - This rule not strict enough<br>- This rule is good<br>- This rule is too strict<br>- I have no opinion on this |
| The minimum age for riding an e-scooter is 14 years | |
| Users don't have to wear a helmet when riding an e-scooter | |
| Only one person is allowed per e-scooter | |
| An alcohol limit of 0.5‰ applies when using an e-scooter | |

# References

APH. Dockless electric scooter-related injuries study, Austin, Texas, September–November 2018 (2019). https://www.austintexas.gov/sites/default/files/files/Health/Epidemiology/APH_Dockless_Electric_Scooter_Study_5-2-19.pdf

Bloom, M.B., et al.: Standing electric scooter injuries: impact on a community. Am. J. Surg. (2020). https://doi.org/10.1016/j.amjsurg.2020.07.020

Chaloux, N., El-Geneidy, A.: Rules of the road: compliance and defiance among the different types of cyclists. Transp. Res. Rec. J. Transp. Res. Board **2673**(9), 34–43 (2019). https://doi.org/10.1177/0361198119844965

Deutscher Verkehrssicherheitsrat e.V. Roll ohne Risiko (2020). https://www.dvr.de/praevention/kampagnen/roll-ohne-risiko

Huemer, A.K., Gercek, S., Vollrath, M.: Secondary task engagement in German cyclists – an observational study. Saf. Sci. **120**, 290–298 (2019). https://doi.org/10.1016/j.ssci.2019.07.016

Kobayashi, L.M., et al.: The e-merging e-pidemic of e-scooters. Trauma Surg. Acute Care Open **4**(1), e000337 (2019). https://doi.org/10.1136/tsaco-2019-000337

Kummeneje, A.-M., Rundmo, T.: Attitudes, risk perception and risk-taking behaviour among regular cyclists in Norway. Transp. Res. F Traffic Psychol. Behav. **69**, 135–150 (2020). https://doi.org/10.1016/j.trf.2020.01.007

Mayer, E., Breuss, J., Robatsch, K., Salamon, B., Soteropoulos, A.: E-Scooter: Was bedeutet das neue Fortbewegungsmittel für die Verkehrssicherheit. Zeitschrift Für Verkehrssicherheit **66**(3), 153–164 (2020)

Mayer, E., Breuss, J., Robatsch, K., Zuser, V., Kaltenegger, A.: E-Scooter: Auswirkungen des Trends auf die Verkehrssicherheit. Zeitschrift Für Verkehrsrecht (12), 417–424 (2019). https://www.kfv.at/download/zvr-12-2019-e-scooter-auswirkungen-des-trends-auf-die-verkehrssicherheit/?wpdmdl=6352&refresh=5eeb57faf180f1592481786

NACTO. Shared micromobility in the U.S.: 2019 (2019). https://nacto.org/wp-content/uploads/2020/08/2020bikesharesnapshot.pdf

Polizei Berlin. Drei Monate E-Scooter in Berlin – Polizeiliche Zwischenbilanz (2019). https://www.berlin.de/polizei/polizeimeldungen/pressemitteilung.847172.php

Polizei Dortmund. Verkehrsbericht 2019 (2019). https://dortmund.polizei.nrw/sites/default/files/2020-02/Verkehrsbericht%202019_0.pdf

Polizei Hamburg. Sicherheit im Straßenverkehr 2019. Bilanz und Ausblick (2020). https://www.polizei.hamburg/contentblob/13594608/e3071bda8b79a47cacc82ac812be0acd/data/vks-2019-ppt-preko-do.pdf

Portland Bureau of Transportation. 2018 - E-scooter pilot: User survey results (2019). https://www.portlandoregon.gov/transportation/article/700916

Portland Bureau of Transportation. 2019 E-Scooter Report and Next Steps (2020). https://www.portland.gov/transportation/escooterpdx/2019-e-scooter-report-and-next-steps

Siebert, F.W., Ringhand, M., Englert, F., Hoffknecht, M., Edwards, T., Rötting, M.: Braking bad - ergonomic design and implications for the safe use of shared e-scooters. Saf. Sci. **140**, 105294 (2021)

Siebert, F.W., Ringhand, M., Englert, F., Hoffknecht, M., Edwards, T., Rötting, M.: Einführung von E-Tretrollern in Deutschland - Herausforderungen für die Verkehrssicherheit. In: Trimpop, R., Fischbach, A., Selinger, I., Lynnek, A., Kleineidam, N., Große-Jäger, A. (eds.) 21. Workshop: Psychologie der Arbeitssicherheit und Gesundheit: Gewalt in der Arbeit verhüten und die Zukunft gesundheitsförderlich gestalten, pp. 207–210 (2020). Asanger Verlag, Heidelberg

Störmann, P., et al.: Characteristics and injury patterns in electric-scooter related accidents - a prospective two-center report from Germany. J. Clin. Med. **9**(5) (2020). https://doi.org/10.3390/jcm9051569

Tack, A., Klein, A., Bock, B.: E-Scooter in Deutschland: Ein datenbasierter Debattenbeitrag (2019). http://scooters.civity.de

Trivedi, T.K., et al.: Injuries associated with standing electric scooter use. JAMA Netw. Open **2**(1), e187381 (2019). https://doi.org/10.1001/jamanetworkopen.2018.7381

Uluk, D., et al.: E-Scooter: erste Erkenntnisse über Unfallursachen und Verletzungsmuster. Notfall + Rettungsmedizin **23**(4), 293–298 (2020). https://doi.org/10.1007/s10049-019-00678-3

Useche, S.A., Alonso, F., Montoro, L., Esteban, C.: Explaining self-reported traffic crashes of cyclists: an empirical study based on age and road risky behaviors. Saf. Sci. **113**(4), 105–114 (2019). https://doi.org/10.1016/j.ssci.2018.11.021

Wolming, S., Wiberg, M.: The Swedish driver licensure examination: exploration of a two-stage model. J. Saf. Res. **35**(5), 491–495 (2004). https://doi.org/10.1016/j.jsr.2004.08.003

# Are E-Scooter Riders More Oblivious to Traffic Than Cyclists? A Real World Study Investigating the Execution of Shoulder Glances

Maximilian Pils[1], Nicolas Walther[2], Mathias Trefzger[1(✉)], and Thomas Schlegel[1]

[1] Institute of Ubiquitous Mobility Systems, Karlsruhe University of Applied Sciences, Moltkestr. 30, 76133 Karlsruhe, Germany
{maximilian.pils,mathias.trefzger,thomas.schlegel}@h-ka.de
[2] Karlsruhe University of Applied Sciences, Moltkestr. 30, 76133 Karlsruhe, Germany
wani1024@h-ka.de

**Abstract.** E-scooters are a comparatively new means of transportation. At the same time, it is considered more dangerous than established means of transport such as bicycles. To find out more about the traffic behavior of e-scooter riders, we investigate the performance of shoulder glances. To do this, we first had to define when a turning over the shoulder was counted as complete shoulder glance. In a study with 21 subjects, we used mobile eye-tracking glasses to investigate whether the gaze behavior of e-scooter riders and cyclists differs. In the situations we investigated where shoulder glances are necessary to protect oneself from other road users, we found little difference between the cyclists and e-scooter drivers. On average, the test persons performed more than half of all possible shoulder glances: 57.93% for the e-scooters and 60.71% for the bicycles. In addition, the difference in the average shoulder glance duration is negligible. We also observed that car drivers seem to be more likely to stop at crosswalks for cyclists than e-scooter riders.

**Keywords:** Gaze behavior · Eye-tracking · Cyclists · Shoulder glance · E-scooter

## 1 Introduction

One of the current topics in the transport sector is the approval of e-scooters. The opinions are polarizing. Experts and societies debate whether e-scooters are a useful addition to the existing means of transport or whether they pose an increased safety risk. Since June 15th 2019, e-scooters have been allowed to operate on German roads. With the increasing number of sharing service providers e-scooters are now part of the traffic environment in many cities. The question now arises whether this introduction will change the ubiquitous road safety.

In this paper we contribute an eye-tracking study, evaluating a key indicator for traffic safety: the shoulder glance. In order to get an indication of the traffic behavior of e-scooter riders we compare them with cyclists. Both means of transport share similarities in speed and used infrastructure.

© Springer Nature Switzerland AG 2021
H. Krömker (Ed.): HCII 2021, LNCS 12791, pp. 436–445, 2021.
https://doi.org/10.1007/978-3-030-78358-7_30

Besides the collected studies for the doctoral thesis of Vansteenkiste et al., [6] there is still a small number of papers that investigate the gaze behavior of cyclists [1–4]. An eye-tracking study with e-scooter drivers was published by Trefzger et al. [5].

Cyclists are significantly affected especially at turning accidents [8]. The focus of these accidents is the missing shoulder glance. Especially when changing lanes, turning and starting from the edge of the road, shoulder glances are essential to see the following traffic in the blind spot. In order to record this shoulder glance accordingly, the test persons of this study were equipped with eye-tracking glasses.

## 2 Study Design

### 2.1 Participants

21 participants (22–59 years of age) took part in the study. The test persons can be divided into two age groups: 17 participants were between 22–29 years old and the remaining four were between 52–59 years old. Out of the 21 participating test persons, 20 possessed a driver's license and 15 owned their driver's license for less than ten years. The remaining test persons owned their driver's license for more than ten years. With the possession of the driver's license, we assume that the test persons are familiar with the generally existing road traffic regulations. The test persons take part in road traffic several times a week, and most of them are mobile on a daily basis.

The participants mainly use their own vehicle and their bicycle. Closely followed by walking. The other means of transport, such as e-scooters, skateboards and public transport are scarcely used. Four test persons had already driven an e-scooter at least once prior to the study. Every third respondent stated that the traffic rules for the e-scooters are the same as for riding a bicycle. Two-thirds of the test persons stated that the general German road traffic regulations [7] apply to e-scooters. Based on the statements of the participants we conclude that many test persons were not aware of the exact traffic rules for using the e-scooters. About 40% of the test persons knew that there are differences in the rules between the use of a bicycle and the use of an e-scooter. With the exception of two participants, they were able to name at least one of the most significant differences between the two means of transport. The following aspects were mentioned:

- As long as there is a bicycle infrastructure, an e-scooter is obliged to use it.
- The speed of e-scooter is limited to 20 km/h.
- The use of an e-scooter is permitted from the age of 14 years.
- There is an insurance and registration requirement.

### 2.2 Protocol

At the beginning of the study, the consent of the subjects was obtained to use the collected data for research purposes. The test persons were equipped with the Tobii Glasses 2 Eye-Tracker. These were adapted to the respective visual impairment. To get the subjects used to the glasses, they were asked to put the glasses on before answering the first questionnaire. Subsequently, the first questionnaire was handed out to the test persons to collect personal metadata.

Afterwards the functionality of the e-scooter was explained to the participants: braking, ringing and acceleration were discussed. Then, the subjects were provided up to five minutes to test the functionalities and become accustomed to the e-scooter. This was done to create a certain sense of security. Before the study was conducted, the test persons had to sign a user agreement for the e-scooter. This included all relevant traffic regulations for the use of e-scooters. The content of the agreement also included safety aspects, including the risk of wet surfaces or unevenness and the insurance obligation. Following the user agreement, the test tracks were explained to the test persons with the help of an OSM map extract (see Fig. 1).

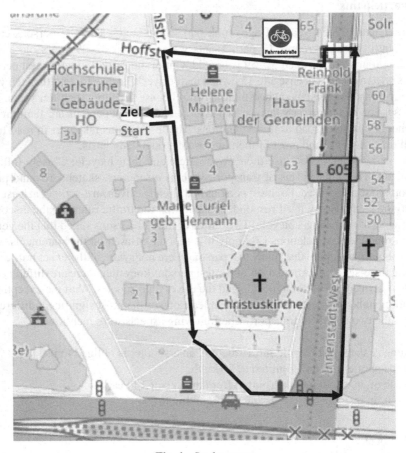

**Fig. 1.** Study course

After verbal confirmation of the test persons, the eye-tracking glasses were calibrated and then the recording was started. Each participant completed the test track by bicycle and e-scooter. The use of the means of transport changed after a complete run by one test person in order to minimize the learning effect during the study. At the end of the study we collected subjective behavioral data of the test person handing out a final questionnaire.

### 2.3  Definition of a Shoulder Glance

In order to accurately record shoulder glances via eye-tracking, the question arises: What is a shoulder glance? A shoulder glance is a complete look over the shoulder. However, since there is no further definition as to when exactly a shoulder gaze is complete, a threshold value was determined by means of a self-experiment. The experiment was based on the generally known visual conditions of the eye for peripheral vision.

Eye-tracking glasses were put on participants and calibrated accordingly. Furthermore, different points of view were placed at eye level at a previously determined distance. The distance from the test persons to the viewing point was 2.50 m. This corresponds approximately to the distance of an e-scooter driver or a cyclist to a motor vehicle at a junction in real road traffic. Here, different viewpoints were marked on a wall, which the test persons had to focus on. These viewpoints were placed in 0°, i.e. perpendicular to the test persons, 45°, i.e. a slight turn to the left and finally 90°, a complete turn to the left. Now, according to reality, objects were placed parallel to the participants, in order to simulate parallel traffic. Afterwards, the test persons had to fix the individual viewpoints and describe which objects they were taking in their peripheral vision. The results of the experiment were clear.

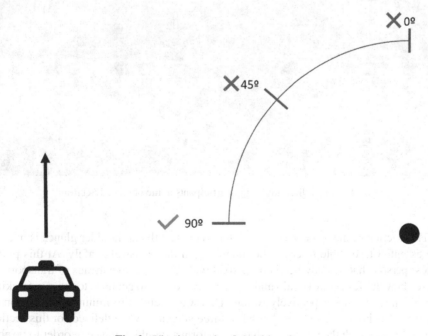

**Fig. 2.** Viewing angle of shoulder glances

As can be seen in Fig. 2, the test persons could only clearly perceive the "simulated" parallel traffic from a rotation angle of at least 90°. Based on this result, the shoulder glances were further evaluated in the study as follows. Only when the head of the participants turn at least 70° in the direction of vision, this shoulder gaze is considered complete. All other smaller rotations or jerky shoulder glances are treated as "half" or "not full" shoulder glances.

## 2.4  Examined Situation

A previously defined section of the test track was evaluated during the recording of the shoulder glances. Figure 3 shows the selected section of the test track including junction and route.

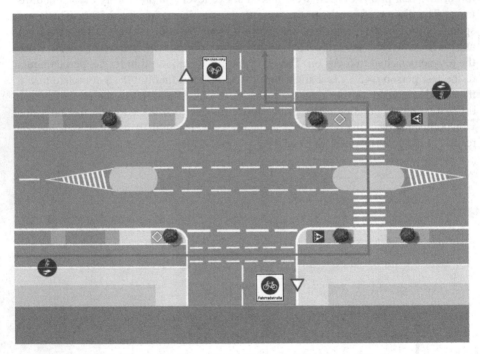

**Fig. 3.** Travel direction of the participants at the observed Section

This section of the route was chosen due to the fact that a shoulder glance is in each case essential to be able to cross the junction and the crosswalk safely. At this point, the test persons had already been on the road with the respective means of transport (e-scooter/bicycle) for about 10 min and were therefore able to get used to the eye-tracking glasses while driving respectively riding. This was intended to minimize distortion of the results by the response bias. A total of three situations were defined for this section of the route in which the test person had to perform a shoulder glance in order to be able to detect the vehicles in the blind spot (see Fig. 4).

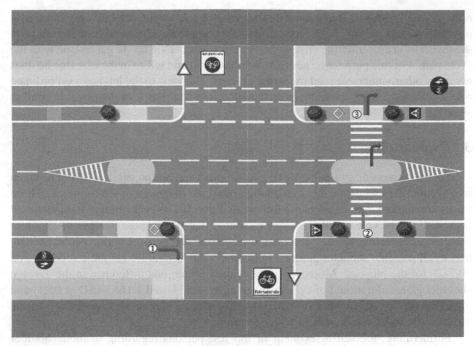

**Fig. 4.** Important safety-related shoulder glance positions

These are numbered from 1 to 3 in Fig. 4. A full shoulder glance at a crossing should include at least a left and right shoulder glance. In the figure above this is indicated by an orange (for left) and a blue (for right) arrow. Situation 1 could be solved by two different shoulder glance sequences. Either right-left-right or left-right. In the following situations, a test person was able to perform multiple shoulder glances, but the focus of the study was on the recording of the minimum necessary number of shoulder glances. Therefore only these relevant shoulder glances were evaluated.

### 2.5 Description of the Selected Situation

In situation 1, the test person rides on a bicycle path and has right of way when crossing the junction. Vehicles coming from the direction of the junction must respect the right of way of the other road users, whether it is the cyclist on the cycle path or a vehicle on the priority road. Despite the right of way regulation, cyclists should perform a shoulder glance when crossing the junction in order to detect vehicles which are about to turn into the junction as well as the vehicles driving out of the junction. The next situation starts at the crosswalk. Here the test persons have to stop and establish a visual relation to the vehicles to their left. The moment the test person crosses the crosswalk by foot, he is counted as a pedestrian. Thereupon the right of way according to §26 paragraph 1 StVO applies, i.e. the drivers must allow the test person to pass the crosswalk and if necessary stop in front of it with a sufficient distance. If a safe crossing of the lane is possible, the test person can drive to the center island. There he has to take another shoulder glance in order to establish a visual relation with the vehicles to his right. If a safe crossing is also possible, the test person can cross the crosswalk completely. The third situation starts on the installation area after the crosswalk. Here the test person has to get in lane with the cycle path. For this purpose he has to look to the right and left to make sure that he does not cross other vulnerable road users like cyclists. The view to the left has a subordinate role, because the test person drives in this direction and thus sees the possible conflict situation in time.

## 3  Evaluation

Four of the 42 recordings (including three e-scooters and one bicycle) could not be taken into account for evaluation due to tracking errors and disregard of the test track. Those tracks are marked in brackets in the following Table 1. The evaluation of the individual trips with the bicycle and the e-scooter showed that the average number of shoulder glances is approximately the same. On average, the test persons performed more than half of all possible shoulder glances. In numbers this is 57.93% for the e-scooters and 60.71% for the bicycles. Moreover, there is only a marginal difference in the average shoulder glance duration. For cyclists, a shoulder gaze lasted 1.168 s (SD = 0,21) and for e-scooter drivers 1.064s (SD = 0,34). This is not a significant discrepancy, as the difference is in the hundredths of a second range.

Furthermore, the self-assessment of the test persons regarding to their shoulder glances was evaluated using the second questionnaire. The self-assessment options were categorized as follows. At least seven shoulder glances had to be performed in the selected scenario. This leads to the following grouping:

- $\geq 6 \triangleq$ very good
- $5 \triangleq$ good
- $4 \triangleq$ satisfactory
- $\leq 3 \triangleq$ sufficient

The subjective self-assessment corresponds approximately to the in reality performed shoulder glances. In general, the test persons rated themselves slightly better. In total, only four test persons rated themselves very well. However, two of these test persons show a significant discrepancy between the objective and subjective perception. Based on the shoulder glances carried out, the two test persons can be classified as satisfactory or sufficient.

More than 80% of all test persons (17 in total) preferred to use the bicycle because of its simple handling and the existing routine. Likewise, the subjective feeling when using the bicycle turned out to be more positive, which can be concluded based on many years of experience.

Heat maps were created to visualize the shoulder views. These contain all shoulder glances performed at the respective position per means of transport. The comparison of the heat maps (compare Fig. 5) shows that the test persons with e-scooters in situation 1 had previously performed their shoulder glances. This can be recognized by the slight rash in situation 1. In situation 3, the direction of the shoulder glance differs. This is shown by the center of the cluster.

**Table 1.** Comparison of self-estimated shoulder glances vs. executed shoulder glances

| Participant | Self-estimation | Number shoulder glances e-scooter | Number shoulder glances bicycle |
|---|---|---|---|
| 1 | Good | 6 | 4 |
| 2 | Good | (0) | (0) |
| 3 | Good | 4 | 2 |
| 4 | Good | (0) | 6 |
| 5 | Satisfactory | 3 | 2 |
| 6 | Good | 4 | 4 |
| 7 | Very good | 5 | 6 |
| 8 | Good | 4 | 3 |
| 9 | Satisfactory | 4 | 3 |
| 10 | Very good | (0) | 4 |
| 11 | Good | 4 | 4 |
| 12 | Very good | 3 | 2 |
| 13 | Good | 4 | 6 |
| 14 | Good | 4 | 5 |
| 15 | Very good | 6 | 6 |
| 16 | Good | 4 | 4 |
| 17 | Satisfactory | 5 | 2 |
| 18 | Good | 3 | 2 |
| 19 | Good | 4 | 4 |
| 20 | Good | 2 | 2 |
| 21 | Good | 6 | 4 |

## 3.1 Further Observations

During the evaluation of the video footage, we got the impression that the vehicles arriving at the crosswalk stopped more often for bicyclists in compliance with the rules than for e-scooter riders. Therefore, we reviewed this observation. We went back through all videos and counted how many vehicles stopped for the cyclists and e-scooter riders. Vehicles were included in the count if the bicyclists and e-scooter riders had dismounted at the crosswalk and the cars were still far enough ahead to brake in time. Since 50 km/h may be driven on the road, cars were included that were more than 40 m (15 m reaction time + 25 m braking distance) away from the crosswalk.

In total, there were 37 arriving motor vehicles in the evaluation. For the cyclist, out of 15 arriving motor vehicles, 11 (73.3%) stopped for the subjects and four motor vehicles (26.7%) passed through. For the e-scooter drivers, a total of 22 arriving motor vehicles were counted, with 12 motor vehicles (54.5%) waiting in accordance with the rules and

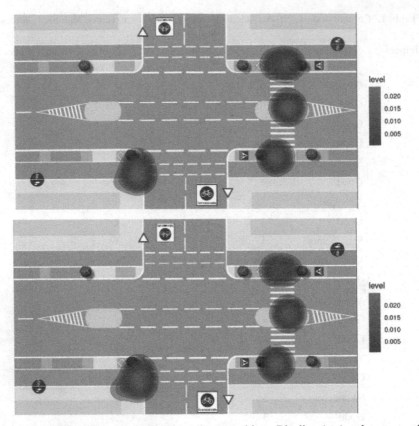

**Fig. 5.** Heatmaps of the performed shoulder glance positions. Bicylists (top) and e-scooter riders (bottom)

10 motor vehicles (45.5%) driving through and thus disregarding the right of way of the participants. In our sample, car drivers show a significantly lower willingness to stop for e-scooter riders. This raises the question of what this might be due to. Perhaps the acceptance of the newer means of transport is not yet there, or perhaps the negative image of e-scooter riders as comparatively inconsiderate road users may have an influence on car drivers. It was also noticeable that when a car crossed the crosswalk, other cars usually followed immediately behind. Since the study was conducted in daylight and the crossing is well equipped with cantilevers, signs, additional lighting, etc., it cannot be assumed that following vehicles would not have seen the crossing.

## 4   Conclusion

Our results show that the gaze behavior in terms of shoulder checks does not differ between cyclists and e-scooter riders. Thus, we rule out the possibility that a potentially greater hazard to e-scooter riders is due to greater inattention to traffic events. The number of shoulder glances as well as the duration of shoulder glances do not or only slightly differ. However, the study shows that the minimum number of shoulder glances is to be considered critically, as it is only 60% - but for both bicyles and e-scooters. Furthermore, we found that the test persons generally assessed their traffic behavior better than it was in reality. A further observation of the evaluation of the eye-tracking recordings was that the car drivers represent the greatest potential danger for vulnerable road users. Especially when crossing the crosswalk, significant rule breaches were detected. Almost every second trip with the e-scooter, the car driver ignored the right-of-way regulation at the crosswalk. For bicyclists, this was only the case for every seventh trip. This crossing has a high potential of danger, therefore a center island was already integrated to enable vulnerable road users to cross the road safely.

## 5   Future Work

In our study, we investigated the performance of shoulder glances in different situations. Thus, we have first examined one of many aspects that can have an impact on traffic safety. Therefore future studies are needed including evaluating Attention distribution, Sensor data (speed, breaking) and accident statistics.

## References

1. Igari, D., Shimizu, M., Fukada, R.: Eye movements of elderly people while riding bicycles. In: Proceedings of the 6th International Conference of the International Society for Gerontechnology, pp. 1–4 (2008)
2. Mantuano, A., Bernardi, S., Rupi, F.: Cyclist gaze behavior in urban space: an eye-tracking experiment on the bicycle network of Bologna. Case Stud. Transp. Policy 5(2), 408–416 (2016)
3. Schmidt, S., Von Stülpnagel, R.: Risk perception and gaze behavior during urban cycling – a field study. In: Eye Tracking for Spatial Research, Proceedings of the 3rd International Workshop, pp. 34–39. ETH Zürich (2018)
4. Trefzger, M., Blascheck, T., Raschke, M., Hausmann, S., Schlegel, T.: A visual comparison of gaze behavior from pedestrians and cyclists. In: Proceedings of the Symposium on Eyetracking Research & Applications. ACM (2018)
5. Trefzger, M., Titov, W., Sgodzaj, K., Schlegel, T.: Analysis and comparison of the gaze behavior of e-scooter drivers and cyclists. Lecture Notes in Informatics (LNI). Gesellschaft für Informatik, Bonn (2020, in print)
6. Vansteenkiste, P.: The role of visual information in the steering behavior of young and adult bicyclists. Doctoral thesis, Ghent University (2015)
7. Bundesministerium der Justiz und für Verbraucherschutz & Bundesamt für Justiz (6. März 2013) -Straßenverkehrs-Ordnung (StVO) - Straßenverkehrs-Ordnung vom 6. März 2013 (BGBl. 1 S. 367), die zuletzt durch den Artikel 4a der Verordnung vom 6. Juni 2019 (BGBl. 1 S. 756) geändert worden ist -www.gesetz-im-internet.de
8. Statisches Bundesamt (Destatis) - Verkehrsunfälle – Kraftrad und Fahrradunfälle im Straßenverkehr 2018 – Artikelnummer: 5462408-18700-4, 19 August 2018

# Safety Related Behaviors and Law Adherence of Shared E-Scooter Riders in Germany

Felix Wilhelm Siebert[1]([✉]) [iD], Michael Hoffknecht[2] [iD], Felix Englert[2] [iD], Timothy Edwards[2] [iD], Sergio A. Useche[3] [iD], and Matthias Rötting[2] [iD]

[1] Department of Psychology, Friedrich Schiller University Jena, Fürstengraben 1, 07743 Jena, Germany
felix.siebert@uni-jena.de
[2] Department of Psychology and Ergonomics, Technische Universität Berlin, Berlin, Germany
[3] Faculty of Psychology, University of Valencia, Valencia, Spain

**Abstract.** Shared e-scooters, whose supply and coverage keeps increasing in many cities around the globe, are rapidly changing mobility in urban road environments. As rising injury rates have been observed alongside this new form of mobility, researchers are investigating potential factors that relate to safe/unsafe e-scooter use. In Germany, e-scooter sharing platforms were only recently permitted in the middle of 2019, and their number has increased steadily since then. The aim of this study was to assess key factors that relate to their safe use, through a direct observation of e-scooters conducted at three observation sites around Berlin. Helmet use, dual use, type of infrastructure use, and travel direction correctness were registered for 777 shared e-scooters during 12.5 h of observation. Results reveal a high level of rule infractions, with more than one quarter of observed shared e-scooter riders using incorrect infrastructure, and one in ten e-scooter users riding against the direction of traffic. Dual use (i.e., two riders per e-scooter), was observed for 5.1% of shared e-scooters. Moreover, none of the riders observed in this study used a helmet on their shared e-scooter. These results point to a need for better communication and enforcement of existing traffic rules regarding infrastructure use and dual use. Further, they indicate a lack of efficacy of safety-related advice of shared e-scooter providers, who promote helmet use in their smartphone application and directly on their e-scooters.

**Keywords:** E-scooters · Helmet use · Law adherence · Observational study

## 1 Introduction

Electric scooters, or *e-scooters*, are part of a larger micro-mobility wave that has primarily hit urban regions around the world in recent years (Gössling 2020; Tuncer and Brown 2020). The scooter rental market has grown rapidly, and today e-scooters from sharing providers can be easily found across many cities of Asia, Australia, Europe, and North and South America. The use of e-scooters by the general public has been facilitated through a number of sharing providers, who have supplied e-scooters in urban environments. These are for-profit companies that, based on convenience strategies (i.e.,

© Springer Nature Switzerland AG 2021
H. Krömker (Ed.): HCII 2021, LNCS 12791, pp. 446–456, 2021.
https://doi.org/10.1007/978-3-030-78358-7_31

relatively cheap prices, quick accessibility and reduced travel times) rent out shared e-scooters (or *rental scooters*), which they distribute station less around cities, and which can be activated simply through a smartphone application. After each ride, e-scooters can be parked directly at the destination, where they are then ready for the next customer. In Germany, shared e-scooters have been allowed after the *Elektrokleinstfahrzeuge-Verordnung* (eKFV, engl. ordinance on small electric vehicles) was enacted on June 15, 2019. Only two months later, five sharing providers were active in more than 20 German cities (Agora Verkehrswende 2019). At that time, the total number of e-scooters in Germany was already around 25,000 (Civity 2020).

Despite their popularity among urban users, shared e-scooters have received a fair share of criticism in the media. One reason for this is the uncompromising strategy of the sharing providers which are profit driven and have at times introduced e-scooters in cities without first consulting city administrations (Fearnley 2020). Furthermore, recent studies have found rising numbers of e-scooter related injuries in hospital-based studies, raising further questions about the safety of shared e-scooter use in urban environments (Bekhit et al. 2020; Mayhew and Bergin 2019; Moftakhar et al. 2020; Uluk et al. 2020).

While both national and local authorities have been strict on the regulation of users' behaviors, such as the correct use of infrastructure, they count on the voluntary compliance of shared e-scooter riders for other safety-related behaviors, such as using helmets. In light of the increasing number of shared e-scooter related injuries, and existing differences in terms of regulations, the aim of this study is to assess the safety related behavior of shared e-scooter riders in real traffic situations.

## 2 Background

Since e-scooters are an emerging form of mobility in Europe, specific laws and regulations concerning their use have just recently been implemented (most of them with little empirical support). Responsibility for the regulation of e-scooter in the European Union is placed with individual member states according to the EU's Type Approval Regulation of January 2016 (Bierbach et al. 2018). Consequently, on June 15, 2019, the eKFV was enacted by the German legislator, which regulates the use of *small electric vehicles* on German roads, which includes e-scooters. For legal participation in road traffic, small electric vehicles and accordingly also e-scooters need a general operating permit, which is issued by the Federal Motor Transport Authority if the technical requirements specified in the eKFV are fulfilled (eKFV §2(1)).

According to the eKFV, no driver's license is required to drive an e-scooter, but the law stipulates a minimum age of 14 years (eKFV §3). The transport of passengers on the scooter, so called *dual use*, is prohibited (eKFV §8). In urban environments, e-scooter riders must use the bicycle infrastructure. If no bicycle infrastructure is available, e-scooter riders are allowed to use the road (eKFV §10(1&2)). Unlike cyclists, e-scooter riders are not allowed to choose freely between a bicycle lane and the road (StVO §(4)). In addition, e-scooters can be assigned additional traffic areas by the road traffic authorities, which are marked with the sign "Elektrokleinstfahrzeuge frei" (engl. "small electric vehicles free").

There is no requirement to equip e-scooters with a physical turn indicator (eKFV §8 and StVZO §67). Nonetheless, e-scooter riders are required to indicate their turns by using their hands, according to eKFV §11(3). Since small electric vehicles are considered motor vehicles, the general alcohol limit of 0.5 per mill applies, analog to driving a car (StVG §24a). The *0.0 per mill* limit also applies to persons under 21 or during the probationary period after the driving test (StVG §24c). The eKFV applies to vehicles with a maximum speed of 20 km/h, hence faster e-scooters are outside of the regulatory focus of the eKFV (eKFV 63 §1(1)).

In addition to the eKFV, some cities have entered into voluntary agreements with e-scooter sharing providers. For example, the agreements of the cities of Hamburg, Munich, and Stuttgart -which share general concepts and wording- define (dynamic) fleet limits, requirements for the installation of e-scooters and an obligation for providers to pick up defective or incorrectly parked e-scooters. Local authorities also design maps with (e.g.) no-parking, no renting, and no driving zones, which are to be enforced by the providers by means of geofencing or visual inspections, among other things. An example of such no-parking zones is shown in Fig. 1 for a temporary change of prohibited e-scooter parking and driving zone during the 2019 Oktoberfest in Munich (Abendzeitung München 2019). In addition, the providers commit themselves to inform their customers about how the E-Scooters work and to educate them about the main traffic rules. This is usually done via the providers' smartphone applications, as shown in Fig. 2. Finally, the agreements state that a long-term evaluation of the integration of e-scooters in urban traffic will be carried out in cooperation with the providers (Freie und Hansestadt Hamburg 2019; Landeshauptstadt München 2019; Landeshauptstadt Stuttgart 2019).

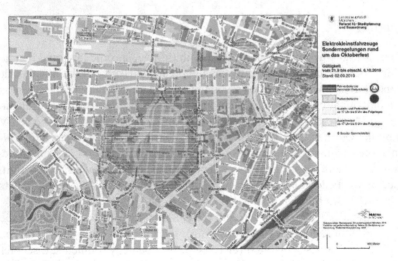

**Fig. 1.** Visualization of temporary changes to permitted shared e-scooter operating area during the 2019 Oktoberfest in Munich (red areas: no driving allowed; dark blue: no parking; light blue: no renting of e-scooters between 5 pm and 6 am; yellow: no renting and parking of e-scooters between 5 pm and 6 am). (Color figure online)

## 2.1 Helmet Use

A highly relevant passive safety-related behavior of e-scooter riders is helmet use, as helmets can decrease injury severity of riders in case of a crash. Studies have found a high frequency of head injuries among hospitalized e-scooter riders, highlighting the need of riders to use helmets when riding e-scooters (Aizpuru et al. 2019; Trivedi et al. 2019). One of the first studies on general e-scooter helmet use was conducted in Brisbane, Australia, in early 2019 (Haworth and Schramm 2019). At the time of the study, rental scooters had been available in Brisbane for three months and helmet use was mandatory for private and shared e-scooters. About 800 private and shared e-scooters were observed, registering helmet use among their riders. Helmet use for shared e-scooters was found to be 61%, that is significantly lower if compared to the rate of 95% observed for private e-scooters. The authors concluded that helmet use is related to the type of e-scooter, i.e., if a private of rental/shared e-scooter is used (Haworth and Schramm 2019). Other studies have found even lower helmet use for shared e-scooters, e.g., between 2% and 10.9% in California, U.S.A. (Arellano and Fang 2019; Todd et al. 2019) and 0.4% in Berlin, Germany (Siebert et al. 2020). While helmet use is not mandatory in Germany, e-scooter providers advise for helmet use in their apps and directly with pictograms on e-scooters (Fig. 2 and Fig. 3).

**Fig. 2.** Screenshots of safety related information found in the Lime smartphone app during the time of this study. Left: "Drive careful! We suggest using a helmet"; Center left: "Use bike lanes, not the sidewalk"; Center right: "Tandem riding is forbidden! Just one person per scooter is allowed"; Right: "Rules and regulation. Please agree to the following rules before you start your Lime ride. Use a helmet. I won't ride on sidewalks or in pedestrian zones. Just one person per scooter. I am sober. [...]". Button text: "I agree".

## 2.2 Dual Use

Dual use, i.e., the simultaneous use of an e-scooter by two people, has been an early focus of observational studies on e-scooters use, as it can obstruct access to the foot brake on the rear-wheel of some e-scooter models, and decrease stability and maneuverability of the scooter. In Germany, eKFV §8 explicitly prohibits transporting other passengers.

In Brisbane, Australia, Haworth and Schramm (2019) registered dual use in 2% of all shared e-scooters observed. In California, USA, Todd et al. (2019) registered dual use in 1.8% of observations, while in Berlin, Germany, Siebert et al. (2020) registered 3.1% of dual use. The explicit ban on dual use is generally mentioned in shared e-scooter apps (Fig. 2), as well as directly on the scooters themselves (Fig. 3).

**Fig. 3.** Driving instructions on a Jump e-scooter. "Minimum age 18+"; "Wearing a helmet is safer"; "1 person per scooter"; "Don't ride on the sidewalk"; "Follow all traffic rules"; "Only park in areas designated in the smartphone app".

## 2.3  Infrastructure Use

While most countries have different rules on permitted infrastructure use for e-scooters, and differ in regulation on allowed directions of traffic, illegal infrastructure use has been observed among e-scooter riders of various countries. In Brisbane, Australia, 6.9% of e-scooters were observed to use prohibited infrastructure (i.e., driving on the road instead of the footpath; Haworth and Schramm 2019). In California, USA, 6.7% of riders were observed to drive in the opposite direction of traffic, a finding similar to a video-based observation study in Berlin, Germany, where 5.5% of e-scooters were observed to drive opposite the direction of traffic (Siebert et al. 2021). In Germany, e-scooters riders generally need to adhere to right-hand side traffic unless a specific exemption is made. As stated before, riders also must use the bicycle infrastructure, and can only use the road if no bicycle infrastructure is available. Also, information on the regulations on infrastructure use is presented in the e-scooter smartphone app (Fig. 2), as well as directly on the e-scooter (Fig. 3).

## 2.4  Study Aim

In light of existing regulation and multiple advisory instructions in e-scooter smartphone apps and directly on shared e-scooters, the goal of this study was to assess actual behavior of shared e-scooter riders in urban environments concerning four safety-related behaviors: helmet use, dual use, type of infrastructure use, and direction of travel.

## 3  Methods

In order to register e-scooter riders' behavior in traffic, a direct observation was conducted at three observation sites in Berlin, Germany between September and October 2019. In line with earlier studies and existing regulation for shared e-scooter use, five parameters were observed: (1) e-scooter provider, (2) infrastructure used [bicycle lane/sidewalk/road], (3) direction of travel [correct/incorrect], (4) helmet use [yes/no], and (5) dual use [yes/no].

### 3.1  Selection of Observation Sites

Observation sites were selected with two variables in mind, frequency of shared e-scooter use at individual sites, and distance between different observation spots. At the time of this study, six providers offered shared e-scooters in Berlin: Bird, Circ, Jump, Lime, Tier, and Voi. Through geofencing, all providers limited the operational area of e-scooters to the (large) urban center of Berlin, only differing slightly in operational area boundaries. Within providers' operational areas, main points of e-scooter use were identified around transportation hubs of subway/urban railway stations, tourist attractions, and main shopping streets.

Three sites with high inter-site distance were selected, in order to facilitate a broad data collection of Berlin e-scooter riders, and to ensure that different e-scooters riders would be observed at the different sites (Fig. 4). The first observation site (Observation Site 1 in Fig. 4) was selected in the west of Berlin at the *Kurfürstendamm*, a main shopping street adjacent to a main transportation hub for interurban and interregional trains (*Bahnhof Zoologischer Garten*). Available infrastructure at Observation Site 1 consisted of a sidewalk and the street, according to the eKFV, e-scooter riders must use the street at this site.

The second observation site (Observation Site 2 in Fig. 4) was selected in central Berlin (*Potsdamer Platz*), at a main public square with multiple public transport connections. Available infrastructure at Observation Site 2 consisted of a sidewalk, a bicycle path, and a multi-lane road. According to the eKFV, e-scooter riders must use the available bicycle path at this site.

The third observation site (Observation Site 3 in Fig. 4) was selected in the eastern part of Berlin, it is a smaller transport hub with adjacent shopping streets. Available infrastructure at Observation Site 3 consisted of a sidewalk and a multi-lane road, hence e-scooter users must use the road at this location according to the eKFV. At all three observation sites, shared e-scooter traffic was observed from the roadside, between September 21, 2019 and October 23, 2019. Traffic was observed on ten afternoons for a total of 12.5 h. The total observation time at the individual observation locations was 5 h and 35 min (four observations) at Observation Site 1, 4 h and 55 min (five observations) at Observation Site 2, and 2 h (one observation) at Observation Site 3. The exact positions of the observation sites are shown in Fig. 4. Observation variables were collected on notepads and transferred to an Excel table after each observation. Only *shared* e-scooters were registered; therefore, data on private e-scooters was not collected.

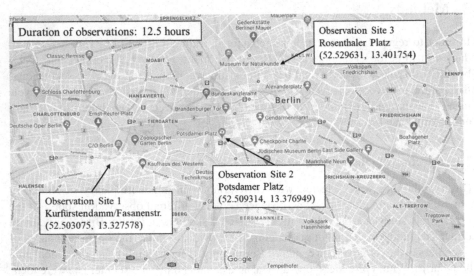

**Fig. 4.** Distribution of observation sites within Berlin (street names and latitudinal and longitudinal coordinates).

## 4 Results

A total of 777 shared e-scooters were observed during this study. Of these, $n = 20$ were *Bird* scooters, $n = 25$ were *Circ* scooters, $n = 51$ were *Jump* scooters, $n = 446$ were *Lime* scooters, $n = 150$ were *Tier* scooters, and $n = 82$ were *Voi* scooters. For $n = 3$ scooters, the provider was not clearly recognized. Hence, the large majority of shared e-scooters at the three observation sites consisted of *Lime* e-scooters (57.6% of identified scooters). Data for helmet use, type of infrastructure, direction of travel, as well as dual use is presented in Table 1.

**Table 1.** Observed e-scooter rider behavior at the three observation sites.

| Variable | | Observation site 1 | Observation site 2 | Observation site 3 |
|---|---|---|---|---|
| # E-scooters observed | | 270 | 405 | 102 |
| Helmet use | Yes | 0% (n = 0) | 0% (n = 0) | 0% (n = 0) |
| | No | 100% (n = 270) | 100% (n = 404) | 100% (n = 102) |
| Dual use | Yes | 5.2% (n = 14) | 5.2% (n = 21) | 4.9% (n = 5) |
| | No | 94.8% (n = 256) | 94.8% (n = 384) | 95.1% (n = 97) |
| Infrastructure use | Correct | 68.1% (n = 184) | 78.0% (n = 316) | 55.9% (n = 57) |
| | Incorrect | 31.9% (n = 86) | 22.0% (n = 89) | 44.1% (n = 45) |
| Direction of travel | Correct | 90.7% (n = 245) | 82.7% (n = 335) | 93.1% (n = 95) |
| | Incorrect | 9.3% (n = 25) | 17.3% (n = 70) | 6.9% (n = 7) |

Overall helmet use of shared e-scooter riders was observed to be 0%, i.e., not a single observed rider at any of the three observation sites used a helmet. Overall dual use was observed to be 5.1% ($n = 40$) varying only slightly between observation sites (4.9–5.2%). Overall, shared e-scooter riders used prohibited infrastructure in 28.3% ($n = 220$) of all observations, with relatively large variations between different observation sites, ranging from 22.0% to 44.1%. The observation of direction of travel revealed opposite direction (wrong-way) driving for 13.1% ($n = 102$) of riders, with relatively large variation between sites (6.9–17.3%).

## 5  Discussion

With this study we have conducted one of the first in-traffic observational data collections on shared e-scooter use in Germany. While the study was exploratory in nature, a number of important results related to rule adherence and safety-related behavior of shared e-scooter riders were found. These results mainly indicate a lack of rule compliance observed among a relatively large proportion of riders, underlining the inefficacy of shared e-scooter providers' safety-related information approaches commonly facilitated through their smartphone applications and on-board driving hints available on their e-scooters.

For adherence to the prevailing legal regulation for e-scooters in Germany, we found that a relatively large percentage of shared e-scooters riders violate existing road rules for infrastructure use that, apart from not being exclusive for e-scooters, are of wide knowledge among different road users (Johnson et al. 2010 and 2014). Close to one third of riders use a shared e-scooter on prohibited infrastructure. Similarly, more than one in ten shared e-scooters was observed to be ridden in the wrong direction. Both of these illegal behaviors were found to vary between observation sites, indicating a need to further investigate factors that relate to this kind of illegal infrastructure use by shared e-scooter riders. The share of incorrect road infrastructure use and wrong direction of travel found in this study is considerably higher than what other studies had found on illegal infrastructure use (Haworth and Schramm 2019; Siebert et al. 2021). Illegal dual use was observed for 5.1% of riders, a higher share compared to other observational studies on shared e-scooter dual use (Haworth and Schramm 2019; Siebert et al. 2020; Todd et al. 2019).

As for helmet use, we did not observe a single shared e-scooter user wearing a helmet during the 12.5 h of observation in Berlin. This finding of 0% helmet use, shared with other previous studies addressing e-scooter riders' safety in Germany (Störmann et al. 2020) is critically smaller than findings from other countries (Arellano and Fang 2019; Haworth et al. 2021; Todd et al. 2019). The non-use of helmets by shared e-scooter users in Germany is especially alarming in light of findings of frequent head injuries of hospitalized e-scooter riders (Aizpuru et al. 2019; Trivedi et al. 2019).

For all observed behaviors that impede riders safety, i.e. lack of helmet use, dual use of e-scooters, and incorrect infrastructure use, e-scooter providers include advisory warnings in their smartphone applications (Fig. 2), as well as directly on the e-scooters (Fig. 3). For advisory warnings, shared e-scooter user must explicitly agree that they will adhere to the stipulated rules regarding regulations and road safety. Nonetheless, our observational data indicates that a considerable share of riders disregards these advisory warnings and uses shared e-scooters against prevailing road traffic laws. Regulators as well as shared e-scooter providers are tasked with developing effective strategies and measures to counter these rider behaviors which are detrimental to riders' wellbeing.

This study has a number of limitations. While the sample size for observed shared e-scooter riders is sufficient for a first exploratory analysis of safety-related behavior, data collection was limited to one city that, although registering the highest number of daily e-scooter trips in Europe (Civity 2020), may differ from other German cities' dynamics and behavioral trends. Hence, future studies should collect data in multiple cities, to be able to detect potential regional effects on safety-related behavior of shared e-scooter riders. Our results of high variations in incorrect infrastructure use between different observation sites point to a need to better understand factors in the road environment which potentially relate to safety critical behavior of e-scooter riders.

In conclusion, this observational study found a critically high share of illegal behavior of shared e-scooter riders in Berlin, which can be detrimental to riders' safety. These findings can be used to develop targeted strategies and measures to increase riders' law-abiding behavior.

# References

Abendzeitung München: Oktoberfest 2019: Das bringt die Wiesn heuer (2019). https://www.abe ndzeitung-muenchen.de/muenchen/oktoberfest/oktoberfest-2019-das-bringt-die-wiesn-heuer-art-562897. Accessed 1 Dec 2020

Aizpuru, M., Farley, K.X., Rojas, J.C., Crawford, R.S., Moore, T.J., Jr., Wagner, E.R.: Motorized scooter injuries in the era of scooter-shares: a review of the national electronic surveillance system. Am. J. Emerg. Med. **37**(6), 1133–1138 (2019)

Agora Verkehrswende: E-Tretroller im Stadtverkehr. Handlungsempfehlungen für deutsche Städte und Gemeinden zum Umgang mit stationslosen Verleihsystemen (2019). https://static.agora-verkehrswende.de/fileadmin/Projekte/2019/E-Tretroller_im_Stadtverkehr/Agora-Verkehrswende_e-Tretroller_im_Stadtverkehr_WEB.pdf. Accessed 1 Dec 2020

Arellano, J.F., Fang, K.: Sunday drivers, or too fast and too furious? Transp. Find. (2019)

Bekhit, M.N.Z., Le Fevre, J., Bergin, C.J.: Regional healthcare costs and burden of injury associated with electric scooters. Injury **51**(2), 271–277 (2020)

Bierbach, M., et al.: Untersuchung zu Elektrokleinstfahrzeugen (Berichte der Bundesanstalt für Straßenwesen, Fahrzeugtechnik No. F 125). Bergisch Gladbach (2018). Bundesanstalt für Straßenwesen website: https://www.bast.de/BASt_2017/DE/Publikationen/Berichte/unterreihe-f/2019-2018/f125.html

Civity: E-scooter in Deutschland (2020). http://scooters.civity.de/. Accessed 1 Dec 2020

Fearnley, N.: Micromobility–regulatory challenges and opportunities. In: Shaping Smart Mobility Futures: Governance and Policy Instruments in times of Sustainability Transitions. Emerald Publishing Limited (2020)

Freie und Hansestadt Hamburg: Mikromobilität: Elektro-Tretroller in Hamburg (2019). https://www.hamburg.de/verkehr/12732854/e-tretroller. Accessed 1 Dec 2020

Gössling, S.: Integrating e-scooters in urban transportation: problems, policies, and the prospect of system change. Transp. Res. Part D Transp. Environ. **79**, 102230 (2020)

Haworth, N.L., Schramm, A.: Illegal and risky riding of electric scooters in Brisbane. Med. J. Aust. **211**(9), 412–413 (2019)

Haworth, N., Schramm, A., Twisk, D.: Comparing the risky behaviours of shared and private e-scooter and bicycle riders in downtown Brisbane, Australia. Accid. Anal. Prev. **152**, 105981 (2021)

Johnson, M., Charlton, J., Newstead, S., Oxley, J.: Painting a designated space: cyclist and driver compliance at cycling infrastructure at intersections. J. Australas. Coll. Road Saf. **21**(3), 67–72 (2010)

Johnson, M., Oxley, J., Newstead, S., Charlton, J.: Safety in numbers? Investigating Australian driver behaviour, knowledge and attitudes towards cyclists. Accid. Anal. Prev. **70**, 148–154 (2014)

Landeshauptstadt München: Selbstverpflichtungserklärung der E-Scooter-Verleiher (2019). https://www.muenchen.de/rathaus/Stadtverwaltung/Kreisverwaltungsreferat/Wir-ueber-uns/Pressemitteilungen/06-2019/E-Scooter.html. Accessed 1 Dec 2020

Landeshauptstadt Stuttgart (Hrsg.): Elektromobilität: E-Scooter ausleihen (2019). https://www.stuttgart.de/leben/mobilitaet/elektromobilitaet/e-scooter/. Accessed 1 Dec 2020

Mayhew, L.J., Bergin, C.: Impact of e-scooter injuries on emergency department imaging. J. Med. Imaging Radiat. Oncol. **63**(4), 461–466 (2019)

Moftakhar, T., et al.: Incidence and severity of electric scooter related injuries after introduction of an urban rental programme in Vienna: a retrospective multicentre study. Arch. Orthop. Trauma Surg. 1–7 (2020). https://doi.org/10.1007/s00402-020-03589-y

Siebert, F.W., Ringhand, M., Englert, F., Hoffknecht, M., Edwards, T., Rötting, M.: Braking bad – how ergonomic design is related to the (un)safe use of e-scooters. Saf. Sci. **140**, 105294 (2021)

Siebert, F.W., Ringhand, M., Englert, F., Hoffknecht, M., Edwards, T., Rötting, M.: Einführung von E-Tretrollern in Deutschland – Herausforderungen für die Verkehrssicherheit. In: Trimpop, R., Fischbach, A., Selinger, I., Lynnyk, A., Kleineidam, N., Große-Jäger, A. (Hrsg.) 21. Workshop Psychologie der Arbeitssicherheit und Gesundheit: Gewalt in der Arbeit verhüten und die Zukunft gesundheitsförderlich gestalten, pp. 207–210. Asanger Verlag, Heidelberg (2020)

Störmann, P., et al.: Characteristics and injury patterns in electric-scooter related accidents-a prospective two-center report from Germany. J. Clin. Med. **9**(5), 1569 (2020)

Todd, J., Krauss, D., Zimmermann, J., Dunning, A.: Behavior of electric scooter operators in naturalistic environments (No. 2019-01-1007). SAE Technical Paper (2019)

Trivedi, T.K., et al.: Injuries associated with standing electric scooter use. JAMA Netw. Open **2**(1), e187381–e187381 (2019)

Tuncer, S., Brown, B.: E-scooters on the ground: lessons for redesigning urban micro-mobility. In: Proceedings of the 2020 CHI Conference on Human Factors in Computing Systems, pp. 1–14, April 2020

Uluk, D., et al.: E-Scooter: erste Erkenntnisse über Unfallursachen und Verletzungsmuster. Notfall+ Rettungsmedizin **23**(4), 293–298 (2020). https://doi.org/10.1007/s10049-019-006 78-3

# Qualitative Examination of Technology Acceptance in the Vehicle: Factors Hindering Usage of Assistance and Infotainment Systems

Dina Stiegemeier[1](✉), Sabrina Bringeland[1], and Martin Baumann[2]

[1] Robert Bosch GmbH, 70049 Stuttgart, Germany
dina.stiegemeier@de.bosch.com
[2] Ulm University, 89081 Ulm, Germany

**Abstract.** More and more assisting and entertaining systems find their way into the cockpit [1]. But the proposed benefits of increased safety, efficiency, and comfort can only come into effect if drivers decide to use the systems. Therefore, it is essential to understand what determines drivers' acceptance of technology in the vehicle. A lot of research addresses technology acceptance applying quantitative methods [2–4]. This work gives an outline on the Technology Acceptance Model (TAM) [5] and driving-related adaptations as well as the potential of qualitative research in this field. Further, we conducted a qualitative online study ($N = 600$) on factors influencing technology usage. We examined the reasons why drivers do not use a system although their car is equipped with it. The qualitative statements were analyzed according to Mayring [6]. The category scheme was developed inductively and compared with the TAM 3 [7]. The analyses show that 56.87% of the reported statements address usefulness and 12.57% ease of use. Seven additional categories emerged accounting for 27.85% of the statements. The results reveal what is subjectively important for drivers and enhance our understanding of barriers for technology usage in the car. The work outlines the potential of qualitative insights adding to the existing body of research.

**Keywords:** Driver information and assistance systems · Technology acceptance · TAM · User survey · Qualitative methods

## 1 Introduction

The user experience of driving a car is changing as more and more assisting and entertaining systems find their way into the cockpit [1]. Thereby, advanced driver assistance systems (ADAS) such as the adaptive cruise control or parking assistants increasingly support the driver in the driving task. On the other hand, in-vehicle information systems (IVIS) like the navigation system, the radio, or mobile phone interfaces provide information and entertainment on the go. While the systems are intended to increase the safety, comfort, and efficacy of driving, the decision to buy and to use the technology lies in the consumer's hands. If drivers do not accept the provided systems, the expected benefits cannot come into effect.

© Springer Nature Switzerland AG 2021
H. Krömker (Ed.): HCII 2021, LNCS 12791, pp. 457–466, 2021.
https://doi.org/10.1007/978-3-030-78358-7_32

Technology acceptance is a highly researched field and many efforts aim to identify the factors influencing acceptance. The acceptance models are mainly derived from studies applying general psychological theories to the usage of information technology [4]. Further, driving-specific adaptations were proposed to account for the characteristics of technology use in the car. Thereby, technology acceptance research mostly relies on a quantitative assessment of the proposed constructs. Only a few studies have adopted qualitative methods to investigate acceptance [2, 3]. Especially for driver-vehicle interaction, qualitative methods can add to the current body of research as they allow for an identification of themes that were not already known before.

This work seeks to give an outline on the present quantitative approaches explaining technology acceptance in general and in the vehicle especially focusing on the Technology Acceptance Model (TAM). Further, we outline the potential of qualitative methods adding to the current research. To enhance our understanding of influencing factors, we conducted a qualitative online study with $N = 600$ drivers examining why drivers do not use available technology in their car.

## 1.1 Technology Acceptance

A driver's technology acceptance can be defined as "the degree to which an individual intends to use a system and, when available, incorporates the system in his/her driving" (p. 477) [8]. Several factors supporting or hindering the acceptance of technology have been identified in the literature. The most prominent approach, the TAM [5], was derived from the Theory of Reasoned Action (TRA) [9]. The TAM is based on quantitative studies concerning information technology usage in the workplace. It proposes that the behavioral intention to use a system is predicted by the Perceived Usefulness (PU) and the Perceived Ease of Use (PEoU). PU thereby reflects the belief that a system enhances the user's performance. The PEoU includes the belief that using the system is free of effort [5].

In later adaptations, the TAM was expanded by factors influencing the PU and PEoU. The TAM 2 found that subjective norm, image, job relevance, and output quality influence the PU [10]. The TAM 3 further investigated the impact of computer self-efficacy, perception of external control, computer anxiety, computer playfulness, perceived enjoyment, and objective usability on the PEoU [7].

Another very prominent model of acceptance is the UTAUT [11], whereby the intention to use a system is influenced by the performance expectancy (PE), effort expectancy (EE), social influences (SI), and facilitating conditions (FC). These factors are moderated by gender, age, experience, and voluntariness of use [11].

The basic TAM and UTAUT have been broadly applied and been shown to replicate across different contexts and technologies. Different research groups investigated the models in many different contexts (e.g. web-based training [12], online banking [13], recommender systems [14]). Also in the driving context, the models were applied to automated driving [15], autonomous shuttles [16], and electric vehicles [17].

Especially the popularity of the TAM stems from its parsimony and simplicity while at the same time, the parsimony has been a subject of discussion [2, 18]. The model does not explicitly link system characteristics with acceptance leaving a gap between existing frameworks and design implications. Also, context-dependent characteristics are not taken into account.

## 1.2 Technology Acceptance in Driver-Vehicle Interaction

When investigating drivers' technology acceptance, the characteristics of the driving task need to be considered. To address the characteristics of the driving task, existing models like the TAM are often extended with additional constructs specifically relevant for technology usage in the vehicle. A subset of recent approaches applying the TAM to driver-vehicle interaction is summarized in Table 1.

**Table 1.** A subset of recent work applying the TAM to in-vehicle technology divided into a) automated vehicles and b) other technology. Non-significant findings are in parentheses.

| Model | Technology | $N$ | Source |
|---|---|---|---|
| **a) Automated vehicles (AV)** | | | |
| TAM: PU, PEoU + Trust + (Perceived risk, Sensation Seeking) | AV | 552 | [19] |
| TAM: PU, PEoU + Trust, SI | AV | 483 | [20] |
| TAM: PU, PEoU + Trust, Safety | AV | 300 | [21] |
| TAM: PEoU, (PU) + Trust TPB: Attitude, Social Norm, Perceived Behavioral Control | AV | 74 | [15] |
| **b) Other technology** | | | |
| TAM: PU, PEoU + Trust | Monitoring system | 34 | [22] |
| TAM: Attitude, (PU), PEoU, + Enjoyment | In-vehicle GPS and automated vehicles | 251 | [23] |

What the approaches have in common is the importance of PU and PEoU. Further, trust has been repeatedly shown to have an impact on the acceptance of automated technology [15, 19–22]. Also, anxiety [24–26] is discussed to influence acceptance. As Table 1 shows, many different approaches have emerged from applying the TAM to the context of autonomous driving and single ADAS or IVIS.

## 1.3 Qualitative Examinations of Technology Acceptance

Overall, the described models are mainly derived from studies transferring general psychological models to technology usage. Applying mostly quantitative methods, standardized questionnaires are used to assess the set of proposed constructs [4]. Literature reviews suggest that less than 5% of the acceptance research is qualitative [2, 3]. More recently, some researchers argue for qualitative [2, 27] or mixed methods combining quantitative and qualitative data [3, 4]. Especially, if the existing research is inconclusive or fragmented, mixed methods can provide a more holistic understanding of the subject [3]. While qualitative data can be deductively analyzed, we can also add categories or themes inductively offering deeper insights into acceptance and influencing factors beyond the existing theories.

To examine the acceptance of automated driving, Zmud and colleagues conducted phone interviews with 205 participants on the reasons to use an AV. Identified reasons were the safety of AV compared to human drivers, relieved stress, the comparison with public transport experiences, trust in the testing of the technology, an affinity for new technology, the enabling of seniors, and the ability to be productive in the car [28]. The authors also asked for concerns or anxieties of drivers related to the technology. Here, mentioned themes included malfunctions, hacking of the car, the co-existence of human drivers and AVs, and the handling of unforeseen events [28].

Another example is provided by qualitative semi-structured interviews with 32 older drivers about different ADAS [29]. Here, the lack of perceived usefulness is mentioned most frequently (32%), implying that the subjects did not perceive a personal benefit from using the system. Further barriers to ADAS usage are functional limitations, costs, a lack of trust, a lack of knowledge, problems in the availability, distraction, loss of control, as well as inappropriate system design.

The described studies illustrate that emerging themes specific to the technology can hinder or support technology acceptance. Thereby, existing theories are supported while at the same time, technology- or context-dependent factors can be added and investigated in more detail.

### 1.4 Objective of this Study

Although a lot of research focuses on determining factors of technology acceptance, the investigations often include only single systems or systems still not on the market such as automated vehicles. Therefore, we conducted an online survey to examine the usage of assistance and infotainment systems that are already available in vehicles today. Thereby, we concentrated on the experiences each individual already has with technology in their own car.

Besides, the majority of the studies adopt a quantitative approach [2, 27]. To enhance our understanding of technology acceptance and especially hindering factors, our study qualitatively examines the participants' reasons not to use technology already available in their vehicle. Also, the few qualitative studies on acceptance often include smaller sample sizes. In contrast, we pursued an online approach with 600 drivers to be able to reach a high diversity in individually available systems and thus, identify different factors influencing the technology acceptance.

## 2   Method

In the following, we describe our research design as depicted in Fig. 1.

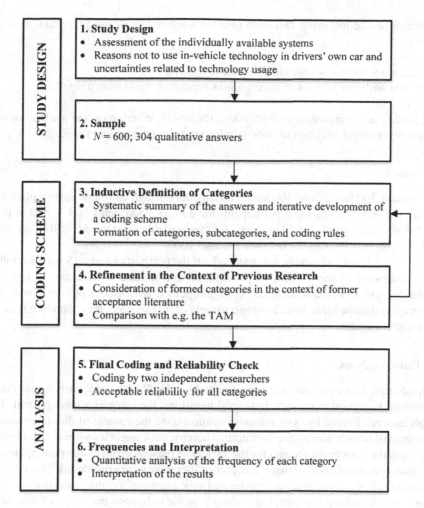

**Fig. 1.** Process and analysis of this research

## 2.1 Study Design

The online study is based on actual experiences drivers make interacting with technology on the market today. First, participants indicated which systems they have in their most used vehicle from a list of in-vehicle systems. The list was based on driver assistance systems or infotainment functions currently on the market including systems like adaptive cruise control, the navigation system, and voice control. For the second question, only the individually available systems were displayed in a list. For these, participants were asked to indicate which of them are used on a regular basis.

Lastly, the assistance and infotainment systems not used regularly were shown as indicated by the participants. The reasons not to use the technology regularly were assessed qualitatively.

Therefore, the following two open questions leaving space for individual answers were shown:

1. What are your reasons for not using these systems?
2. Do you have any worries or uncertainties regarding the system usage?

Here, participants were asked to shortly describe first their reasons not to use the systems and second possible anxieties or uncertainties regarding the system usage.

## 2.2 Sample

We recruited 300 female and 300 male participants through an online survey platform. Overall, the survey reveals that participants have $M = 3.5$ systems ($SD = 3.2$) in their vehicles with a range of 0 to 14 systems. Of the available systems, $M = 70\%$ are used regularly ($SD = 0.30$) varying between 0% and 100%.

For the analyses, the sample was reduced. Of the 600 participants, 15 were excluded due to confusing answers. Then, 93 persons did not indicate to have any system and 188 persons use all available systems. Thus, altogether, 281 persons (48,03%) were not forwarded to the open questions leaving a sample of 304 subjects for examining reasons of not using a system.

## 2.3 Data Analyses

The qualitative data were analyzed according to Mayring [6]. Thereby, each answer could be assigned to several categories. In several iterations, the categories were refined. The categories were formed by systematically summarizing the content of the statements in the context of former acceptance literature. Thereby, the categories were derived inductively from the answers to account for the variety of reasons given by the participants. In the refinement, parallels to existing acceptance literature were considered. This resulted in a coding scheme comprising categories and subcategories, a coding rule, and example statements. Describing the complete coding scheme is beyond the scope of this paper but the emerging categories are presented in the results section.

Two researchers discussed and adapted the coding scheme and finally coded the statements independently. This led to acceptable reliability for all categories. By comparing the findings with the TAM, the categories were clustered among PU, PEoU, and Other. Finally, the frequency of each category was counted.

## 3 Results

### 3.1 Emerging Categories

In total, thirteen categories were identified iteratively. Further, these categories were explicitly compared with the TAM literature.

Thereby, the categories were clustered within PU if they addressed usefulness [5] or one of the antecedents as defined in the TAM 2 [10]. This resulted in the following categories for PU:

1. Need
2. Context and Task
3. Reliability

Categories were subsumed under PEoU if they addressed ease of use or related factors from the TAM 3 [7]. The following three categories reflect aspects of ease of use:

1. Increased Effort
2. Aversion
3. Anxiety

Apart from the categories containing aspects of usefulness or ease of use, the following categories emerged which are not directly associated with the TAM:

1. Preference for Own Action
2. Distrust
3. Safety
4. Knowledge
5. Habit
6. Use Other System
7. No Reason

The first six categories reflect reasons not to use technology as indicated by participants. The last one, no reason, simply reflects the lack of an answer and is thus regarded separately.

## 3.2 Frequencies

Finally, we coded all statements according to the developed category scheme. Overall, 517 statements were assigned to the 13 different categories. The frequencies for PU, PEoU, Other, and No Reason are depicted in Table 2.

Table 2. Frequencies of statements addressing PU, PEoU, other, and no reason

|          | PU     | PEoU   | Other  | No reason |
|----------|--------|--------|--------|-----------|
| Frequency | 56.87% | 12.57% | 27.85% | 2.71%     |

The results show that usefulness accounts for the greatest amount of answers. This is followed by constructs not related to the TAM and the least amount addresses ease of use. Further, 2.71% were not able to describe the reason not to use the technology.

# 4 Discussion

The results emphasize the importance of PU because usefulness and the related constructs are addressed in more than 50% of the statements. Further, ease of use and the associated factors are relevant for technology usage although participants did not mention PEoU as often as PU. This corresponds with the findings of general technology acceptance research [5, 11].

Additionally, both usefulness and ease of use each comprise three different categories in our study. For example, the category "Context and Task" signifies that the task or context in which the system would be required is not encountered often. Thereby, drivers stated for example the speed limiter is not regularly used as drivers do not perceive it as beneficial on urban or rural roads or the navigation is only used in unfamiliar areas. Thus, findings are in line with the TAM [5, 7, 10] while the single categories give deeper insights into potential barriers for technology usage specifically in the vehicle.

Besides, not all statements can be summarized by PU, PEoU, and the related factors. In fact, seven additional categories had to be included to account for the variety of answers. Other categories included the preference for own action, distrust, safety, knowledge, and habit. For example, trust was not part of classic acceptance models but has been shown to impact the acceptance of automated technology repeatedly [15, 19–22]. Our study shows that further constructs such as trust are subjectively relevant for technology acceptance.

We adopted a qualitative instead of a quantitative study design to enrich the prior research. Our study shows that drivers' acceptance is influenced by more than the well-known acceptance constructs. The observed categories were only partly reflected in the TAM and its extensions [5, 7, 10]. Thereby, new categories and subcategories were formed providing a deeper understanding of actual users' problems. From this holistic understanding, practical implications for technology development can be derived. For example, for the system to be beneficial, we need to consider how often a task is executed at all, which contexts are perceived as suitable for a technology, and whether the context is encountered often enough. While outlining all identified categories is not the scope of this work, we wanted to highlight the potential of qualitative studies in both the identification of influencing factors of acceptance and deriving practicable implications.

Further, the presented study enables an investigation of actual experiences that drivers make with their own systems. As outlined before, current studies mainly focus on single systems or systems not yet available. Thereby, participants typically encounter the system in an experimental setting only or read a description of the technology. In contrast to that, our study was based on the experiences drivers already made in their car with a variety of systems. Thus, this may have included the adoption process as well as long-term usage.

Qualitative studies often rely on smaller sample sizes. In this study, an online survey approach was taken in order to approach a greater range of participants. This might be a limitation if the participants lack knowledge about the systems they have. If a driver does not adopt a system, this driver might also not be aware of having it. Also, the qualitative assessment of reasons and worries is limited as in the online approach, no detailing questions could be added like in an interview.

# 5  Conclusion

To conclude, this work presents insights into the quantitative and qualitative examinations of technology acceptance in the vehicle. While most research is quantitative in nature, we outline how qualitative data can add to our current understanding of technology acceptance. Our study with its comparatively big sample size together with the qualitative approach reveals what is subjectively important for drivers and provides a view on barriers for technology usage in the car. Future research might further investigate the identified constructs and their impact on acceptance to deepen our understanding of technology acceptance. Also as has been suggested [3], we argue for the application of mixed methods combining quantitative and qualitative methods to exploit the advantages of both approaches.

# References

1. Bengler, K., Dietmayer, K., Farber, B., Maurer, M., Stiller, C., Winner, H.: Three decades of driver assistance systems: Review and future perspectives. IEEE Intell. Transp. Syst. Mag. (2014). https://doi.org/10.1109/MITS.2014.2336271
2. Lee, Y., Kozar, K.A., Larsen, K.R.T.: The technology acceptance model: Past, present, and future. Commun. Assoc. Inf. Syst. (2003). https://doi.org/10.17705/1CAIS.01250
3. Venkatesh, V., Brown, S.A., Bala, H.: Bridging the qualitative-quantitative divide. guidelines for conducting mixed methods research in information systems. MIS Q. (2013). https://doi.org/10.25300/MISQ/2013/37.1.02
4. Wu, P.F.: A mixed methods approach to technology acceptance research. J. Assoc. Inf. Syst. **13**, 172–187 (2012)
5. Davis, F.D.: A technology acceptance model for empirically testing new end-user information systems: Theory and results. Dissertation, MIT Sloan School of Management (1985)
6. Mayring, P.: Qualitative Inhaltsanalyse [28 Absätze]. Forum Qual. Soc. Res. **1** (2000). http://nbn-resolving.de/urn:nbn:de:0114-fqs0002204
7. Venkatesh, V., Bala, H.: Technology Acceptance Model 3 and a research agenda on interventions. Decis. Sci. **39**, 273–315 (2008)
8. Adell, E.: Acceptance of driver support systems. In: Proceedings of the European Conference on Human Centered Design for Intelligent Transport Systems, pp. 475–486 (2010)
9. Fishbein, M., Ajzen, I.: Belief, Attitude, Intention and Behavior: An Introduction to Theory and Research. Addison-Wesley, Reading (1975)
10. Venkatesh, V.: Determinants of ease of use: Integrating control, intrinsic motivation, and emotion into the technology acceptance model. Inf. Syst. Res. **11**, 342–365 (2000)
11. Venkatesh, V., Morris, M.G., Davis, G.B., Davis, F.D.: User acceptance of information technology: Toward a unified view. MIS Q. **27**, 425–478 (2003)
12. Park, Y., Son, H., Kim, C.: Investigating the determinants of construction professionals' acceptance of web-based training: An extension of the technology acceptance model. Autom. Constr. (2012). https://doi.org/10.1016/j.autcon.2011.09.016
13. Zhou, T., Lu, Y., Wang, B.: Integrating TTF and UTAUT to explain mobile banking user adoption. Comput. Hum. Behav. (2010). https://doi.org/10.1016/j.chb.2010.01.013
14. Hu, R., Pu, P.: Acceptance issues of personality-based recommender systems. In: Proceedings on 3rd ACM Conference on Recommender Systems, pp. 221–224. ACM, New York (2009)
15. Buckley, L., Kaye, S.-A., Pradhan, A.K.: Psychosocial factors associated with intended use of automated vehicles: A simulated driving study. Accid. Anal. Prev. (2018). https://doi.org/10.1016/j.aap.2018.03.021

16. Nordhoff, S., de Winter, J., Madigan, R., Merat, N., van Arem, B., Happee, R.: User acceptance of automated shuttles in Berlin-Schöneberg: A questionnaire study. Transp. Res. F Traffic Psychol. Behav. (2018). https://doi.org/10.1016/j.trf.2018.06.024

17. Schmalfuß, F.: Acceptance of electric mobility system components and the role of real-life experience. Dissertation, Technische Universität Chemnitz (2017)

18. Bagozzi, R.P.: The legacy of the technology acceptance model and a proposal for a paradigm shift. J. Assoc. Inf. Syst. **8**, 244–254 (2007)

19. Choi, J.K., Ji, Y.G.: Investigating the importance of trust on adopting an autonomous vehicle. Int. J. Hum.-Comput. Interact. (2015). https://doi.org/10.1080/10447318.2015.1070549

20. Panagiotopoulos, I., Dimitrakopoulos, G.: An empirical investigation on consumers' intentions towards autonomous driving. Transp. Res. Part C Emerg. Technol. (2018). https://doi.org/10.1016/j.trc.2018.08.013

21. Xu, Z., Zhang, K., Min, H., Wang, Z., Zhao, X., Liu, P.: What drives people to accept automated vehicles? Findings from a field experiment. Transp. Res. Part C Emerg. Technol. (2018). https://doi.org/10.1016/j.trc.2018.07.024

22. Ghazizadeh, M., Peng, Y., Lee, J.D., Boyle, L.N.: Augmenting the Technology Acceptance Model with trust: Commercial drivers attitudes towards monitoring and feedback. Proc. Hum. Factors Ergon. Soc. 56th Annu. Meeting **56**, 2286–2290 (2012)

23. Chen, C.-F., Chen, P.-C.: Applying the TAM to travelers' usage intentions of GPS devices. Expert Syst. Appl. (2011). https://doi.org/10.1016/j.eswa.2010.11.047

24. Osswald, S., Wurhofer, D., Trösterer, S., Beck, E., Tscheligi, M.: Predicting information technology usage in the car. In: Proceedings of the 4th International Conference on Automotive User Interfaces and Interactive Vehicular Applications - AutomotiveUI 2012. The 4th International Conference, Portsmouth, New Hampshire, 17–19 October 2012, pp. 51–59. ACM Press, New York (2012). https://doi.org/10.1145/2390256.2390264

25. Hewitt, C., Politis, I., Amanatidis, T., Sarkar, A.: Assessing public perception of self-driving cars. the autonomous vehicle acceptance model. In: 24th International Conference on Intelligent User Interfaces (IUI 2019), pp. 518–527 (2019). https://doi.org/10.1145/3301275.3302268

26. Venkatesh, V., Thong, J.Y.L., Xu, X.: Consumer acceptance and use of information technology: Extending the unified theory of acceptance and use of technology. MIS Q. **36**, 157–178 (2012)

27. Vogelsang, K., Steinhueser, M., Hoppe, U.: A qualitative approach to examine technology acceptance. In: Proceedings of the 34th International Conference on Information Systems, Milan, Italy (2013)

28. Zmud, J., Sener, I.N., Wagner, J.: Self-driving vehicles: determinants of adoption and conditions of usage. Transp. Res. Rec. (2016). https://doi.org/10.3141/2565-07

29. Trübswetter, N., Bengler, K.: Why should I use ADAS? Advanced driver assistance systems and the elderly: knowledge, experience and usage barriers. In: Proceedings of the 7th International Driving Symposium on Human Factors in Driver Assessment, Training, and Vehicle Design. Driving Assessment Conference, Bolton Landing, New York, USA, 17–20 June 2013, pp. 495–501. University of Iowa, Iowa City (2013). https://doi.org/10.17077/drivingassessment.1532

# User Diversity and Mobility

# Gender, Smart Mobility and COVID-19

Angela Carboni[1]([✉]) [iD], Mariana Costa[2], Sofia Kalakou[3] [iD], and Miriam Pirra[1] [iD]

[1] Department of Environment, Land and Infrastructure Engineering, Politecnico di Torino,
Corso Duca degli Abruzzi 24, 10129 Torino, Italy
{angela.carboni,miriam.pirra}@polito.it

[2] VTM, Ed. Central Plaza - Av. 25 de Abril de 1974, 23 - 2°A, 2795-197 Linda-a-Velha, Portugal
mariana.costa@vtm-global.com

[3] ISCTE Business School - Business Research Unit (BRU-IUL), Avenida das Forças Armadas,
Edifício II, Gabinete D402, 1649-026 Lisboa, Portugal
Sofia.Kalakou@iscte-iul.pt

**Abstract.** The COVID-19 pandemic has strongly impacted people's main routine, which certainly includes their mobility habits. This paper aims to assess the pandemic's mobility impacts and whether these may have increased the already existing inequality between men and women. In particular, the variation of mode choice in a pre-COVID and post-COVID scenario is investigated, focusing on the use of transport mode defined as Smart Mobility. The analysis is performed on data collected in thirteen European countries between July and September 2020 through a survey designed using an intersectional approach. Responses are analyzed to highlight correlations between different factors affecting mobility changes: some interest is reserved to the modes used according to the journey scope (work, errand, shopping). Overall, results reveal more people walking for their daily journeys, while a significant decrease in the use of public transport is observed. Although these changes affect women more, the main reason behind this is the need for more safety in terms of low risk of contagion, irrespective of gender. A specific focus on using modes commonly associated with a Smart Mobility offer (such as shared modes, public transport, walking, and biking) reveals differences originating when comparing men and women responses and various age ranges.

**Keywords:** Gender balance · Smart mobility · COVID-19 · Mobility survey

## 1 Introduction

The impact of the COVID-19 pandemic on the world of transport and mobility has been significant, following the measures that have required many businesses' closure, forcing people to stay at home to reduce the risk of contagion. These measures have been demonstrated to be necessary for limiting contacts and have influenced travel behaviors, mainly in public transport use, due to the lower level of health security perceived associated with specific means of transport and crowded places [1].

Transport systems in most cities were able to cope with the pandemic's immediate challenges, but some adverse effects have been observed. Unequal access to transit, which

© Springer Nature Switzerland AG 2021
H. Krömker (Ed.): HCII 2021, LNCS 12791, pp. 469–486, 2021.
https://doi.org/10.1007/978-3-030-78358-7_33

is commonly affecting users such as people with disabilities, has been exacerbated in the emergency period [2]. Many cities are widely characterized by limited spaces for pedestrians and cyclists in their urban planning. However, cycling and walking can be seen as a great way to stay healthy facing, for example, the closure of gyms in many cities. These two modes are also an effective way to support physical distancing and relieve public transport burden during a pandemic [3].

Also, it is worth investigating how this crisis affected shared modes (as car sharing or bikesharing) that have started to populate the cities in the last decades and how these services are linked with characteristics such as gender and age. The use of these means, which, for the same definition of being shared, would imply treating specific cleaning and sterilizing approaches to avoid infection risk, seems to be perceived as a safer alternative for traveling.

The current work fits into the European TInnGO (Transport Innovation Gender Observatory) project framework, which deals with gender inequalities in smart mobility opportunities and transport employment. The project's primary goal is to create a mechanism for a sustainable game change in European transport through the transformative strategy of gender and diversity-sensitive Smart Mobility (SM) [4]. TInnGO also acts by creating a network of 10 national Hubs in 13 different European countries: each Hub addresses issues of local importance in gender and diversity sensitive Smart Mobility. This latter has become a buzzword of the 21st century; it involves four main contents, such as "vehicle technology, Intelligent Transport Systems, data, and new mobility services" [5].

Smart Mobility, although just one component of a smart city, is seen as a means of delivering key benefits such as reducing air and noise pollution, traffic congestion, transfer costs, increasing transport safety and improving transfer speed [6]. Moreover, it is commonly associated with the offer of various mobility options and services such as on-demand ride services, real-time ridesharing services, car- and bike-sharing programs, multimodal trip-planning apps, smart traffic control, up to the self-driving vehicles concept [7]. Many of these themes have been catapulted to the discussion about urban mobility during (and in the future after) the pandemic. Transport systems have been asked to become "smart", as citizens are changing their commuting preferences to adapt to a new way of life where main changes include people distancing and crowds reduction.

The present paper aims to shed light on how the COVID-19 emergency influences mobility choices (also the "smart" ones) of women and men in Europe, highlighting inequalities. This is done by intersecting users' profiles (gender, age, work status before and after COVID-19) with their mobility patterns and mode choice in the pre- and post-COVID scenario. Additionally, this study will analyze the stated reasons behind these changes and the connection between mode choice, gender, and caregiving roles. Understanding how mobility habits changed during the pandemic and the reasons behind, and analyzing the differences in men and women's behavior is crucial for planning the reboot of cities' transport services consistent with future needs.

The paper is structured as follows: the next section will try to correlate known women's mobility habits with the restrictions due to the COVID-19 emergency, with particular attention to the Smart Mobility concepts. The methodology and the data collection procedure will be described in Sect. 3. Then, the results obtained through analyses correlating various respondents' characteristics are presented in Sect. 4. Finally, conclusions will discuss and elaborate on the impact and implications of this work.

## 2   Women's Mobility Habits and COVID-19 Emergency

Previous studies have shown that different user characteristics, including social roles rooted in society, lead to different activity patterns and a gender mobility gap. Women are known to use Public Transport (PT) more than men, who in a traditional society make commuting trips by car and get the first right to car usage in a household [8]. However, during the pandemic, governments and local authorities implemented restriction measures to reduce the use of PT, as their overcrowding represents a high-level risk of contagion [9]. Thus, women's mobility might be more affected by COVID-19 because they are more frequent public transport users than men [10].

While many employers can rely on smart working, reducing their journeys only to necessary food shopping, other essential service personnel, such as healthcare (hospitals and pharmacies), have continued their trips to the workplaces. In such a context, it is relevant to highlight the differences between genders in the presence in the labor market's different sectors to characterize, for example, the PT users. As stated in [11], women make up almost 70% of the healthcare workforce, so they have to reach hospitals and other care facilities during the emergency. Thus, they are asked to arrive at their workplace, possibly trying to reduce contagion risk. In this context, it is interesting to see if they are willing to rely on the same means they were used to as in the pre-COVID situation.

In general, it is observed that, in need of moving during the emergency, people favored to travel with modes that guaranteed adequate physical distancing, such as cycling, walking, or private cars. At the same time, there was a drop in transit and shared transport use [9]. However, these changes in the mode choices need to be investigated more in detail. Research conducted before the COVID pandemic revealed that gender seems to affect the interest towards various transport means differently. For example, women are commonly not seen as frequent cyclists, mainly due to significant safety concerns and family burden issues [8]. These elements seem to influence bike-sharing services, too; a considerable number of male users commonly characterizes them due to the design of the bikes themselves and the absence of baby saddles that would prevent women from relying on this service [12–14]. However, the changes observed in the city affected by restrictions, as the creation of pop-up bike lanes, with the potential to become permanent cycling infrastructure [3], and a reduced number of cars traveling around, could draw more women to this means of transport. Moreover, the need to travel just for 'mandatory' trips (to reach workplaces or for caring issues), namely without dependents as in the pre-COVID situation, could influence these mobility choices.

As mentioned in the previous section, the ultimate goal of introducing the Smart Mobility concept in the cities is reducing traffic and pollution. So, introducing proper infrastructures, such as new bike lanes, could be seen as a relevant measure to reach a low level of congestion and make citizens flow more fluid. The sharing of vehicles and modes, on-demand services, and micro-mobility are other potential answers of Smart Mobility to environmental and personal sustainability problems. Car sharing systems, a real presence in the SM framework, are commonly characterized by a lower frequency of female users than males. The main features of people joining this service are somewhat similar, despite the gender [15].

Moreover, the costs of car sharing compared to car ownership could play a more significant role for women than for men [16], mainly because the former commonly drive a household's second car and the use of this service could be less expensive than buying a second car [17]. The shared modes seem to have been able to face the change in mobility choices associated with the pandemic better than other means. For example, a report shows that after the first lockdown in Italy in March–April 2020, car sharing had a relatively low recovery in May with only 30% points more compared to the previous months. However, shared bikes and e-scooters showed a more consistent increase, respectively, 60 and 70% more, reaching almost the pre-Covid-19 levels of use [18].

While the female interest in bike-sharing systems has already been cited, the spread of e-scooters and their shared offer is a relatively new trend in urban mobility. This micro-mobility mode has seen a significant increase in association with the pandemic travel restriction, as they are sustainable individual means of transport. A reduced number of studies have investigated the differences in users' gender so far, but some preliminary results are available. For example, research conducted in a big German city showed that three-quarters of the e-scooter sharing customers are male [19]. Moreover, the distribution of ride lengths proves that gender hardly has any influence on usage patterns.

Ride-hailing and on-demand services are other modes in the SM domain that have gained a lot of success in pre-COVID cities worldwide. Despite the fear of traveling with unknown people, women seemed interested in the service. A reduced number of studies has focused on the impacts of ride-sourcing. Still, there have been reports of some considerable safety setbacks as cases of drivers sexually assaulting female passengers have emerged from across the globe [7]. The bad experiences reported have led to increasing demand for safe transport services. Many women-only ride-hailing services (exclusively for females drivers and passengers) have been launched in various countries [7]. Besides, SM could provide a useful contribution, as technology can increase accountability for crimes committed. GPS tracking, trip arrangement through apps, driver and rider feedback, and law enforcement outreach are examples of actions that could be of interest for potential female customers [20]. Compared to other modes in the SM offer, ride-hailing seems to be the service that could be more affected by the fear of contagion: this puts a strain on the sense of trust towards peers commonly required by this type of platform.

The provided overview on mobility habits gendered differences is mainly based on the pre-COVID scenario. A low number of studies are currently focusing on this topic in the post-COVID scenario, whereas some investigations on the general trends on the influence of the emergency on travel behavior are already available. A fundamental point that must be taken into account while conducting these investigations is considering the employment dynamics changes due to the COVID-19 pandemic. As already cited, people have switched to smart working as much as possible following the measures to reduce contacts. However, differences between the genders are found in this context, which would inevitably reflect their mobility. As highlighted in [10], women "will likely experience a significant burden on their time given their multiple care responsibilities as school closures and confinement measures are adopted, possibly leading to reductions in working time and permanent exit from the labor market".

A study conducted in Nederland has already highlighted these differences [21], looking at the distribution between men and women across different professions and estimating the effects of restriction measures. The authors estimated that more than 50% of the men were probably still traveling for their commuting, while this percentage reduced 10 points in women's cases. They also tried to evaluate the distribution of people who stayed at home or were temporarily out of work due to the COVID-19 crisis, saying that it can assess to 21% in the female population, reducing to only 10% among men. The disproportionate impact due to lockdown measures on women, which reduced their mobility more than men, has been verified in this study [22], based on the analysis of mobility indicators derived from anonymized and aggregated data provided by a mobile phone operator. In this case, the authors recognized that the travel patterns' differences are mainly because of the uneven burden in caring for children when schools are closed. They raised the attention to the need for targeted policy intervention to support women during the pandemic, for example, by offering parental leave to both men and women to encourage equal dependents' care, joining with more campaigning and supportive activities connected to the workplaces.

Despite these latter aspects being far from the focus and the current paper's scope, we think it is necessary to consider when analyzing mobility changes due to an exceptional situation like a pandemic affecting peoples worldwide. Thus, in the following, we will present an analysis of a dataset collected at the European level to relate how this pandemic affected women's mobility patterns also in connection with job changes.

## 3 Methodology

The spread of mobility surveys has gained more importance in a period with cities affected by a worldwide pandemic that it is expected to produce considerable changes in daily life and travel patterns. One of the TInnGO project aims is to investigate possible barriers and potential improvements that could help operators create a better and more gender-equal transport system. This knowledge was planned to be gained through a data collection procedure based on a survey investigating which aspects can be modeled to explain and predict mobility choices. Focus group analyses and a detailed literature review were conducted as preliminary activities to the questionnaire design [8]. These allowed to collect information on particular users' mobility needs, gain new knowledge of possible barriers and potential improvements to help transport operators improve their services.

According to the project-planned activities, the data collection campaign was planned to be started in the TInnGO Hubs cities in spring 2020. However, the COVID emergency forced a delay in the data collection and pushed us to reconsider the original plans. Indeed, expecting that the pandemic would produce changes in people's mobility habits, the typical journey characteristics have been investigated in a pre-COVID scenario and in a post-COVID one. This could provide insights into how the emergency changed the mobility scenario, how it could affect different kinds of users, and highlight potential improvements to help transport operators manage city mobility in the future. It is relevant to observe that the reference to what we call 'a post-COVID scenario' does not aim at investigating future intentions but is associated with the condition of life after the worldwide spread of the virus. Thus, the objective is to understand how the respondents have changed their mobility habits due to the cohabiting with the COVID-19.

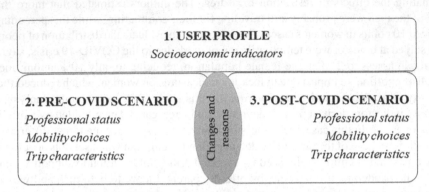

**Fig. 1.** The survey structure.

Before the data collection procedure, the survey structure has been reviewed by the TInnGO Project coordinator (namely Coventry University) to check its compliance with the University's ethical framework. It has been approved through a formal Research Ethics procedure confirming that it is GDPR-compliant too. The respondents are informed about all these points before participating in the survey. They are told that their answers are anonymous and treated confidentially and the information they provide is kept anonymous in any research outputs and publications.

The survey structure can be depicted in Fig. 1. As it can be observed, the first part of the questionnaire aims to outline the respondent profile, investigating her/his socioeconomics characteristics, such as gender, age, social level, education, ethnic origin, family composition, sexual orientation, accessibility to car use. The focus on specific questions, such as *"Do you have children living in your household?"*, *"In your family unit, do you live with any dependent person?"* and *"Which, if any, of the following disabilities do you have?"*, have the objective of better understanding specific user groups' mobility characteristics, according to the literature review and the focus group activities operated in [8].

The focus then goes on to the analysis of typical travel patterns and how they have been affected by the COVID emergency. The respondents have to state which mode they used the most (e.g., motorized and non-motorized, owned, and shared) for different activities (e.g., job, commuting, shopping) in a pre- and post-COVID scenario. Obviously, the possibility of not conducting certain kinds of journeys is proposed as a possible answer. For these cases, a specific question tries to understand the motivation behind this choice: *"Which are the reasons behind your changes in the modes used?"*, including safety, sustainability, and planning issues.

Regarding the trip characteristics, the questionnaire aims to collect a better characterization of the transport chain (activity patterns) to assess possible differences between women and men as reported in the literature. This is done through asking explicitly: "*Do you use to travel with a dependent person on this trip? (children, elderly, caring for disabled people)*", "*Do you use to stop regularly along this journey?*", "*What are the reasons for these stops?*". On the whole, some details about the accessibility to all the possible means of transport are required.

As highlighted in the previous sections, it is necessary to relate how this pandemic affected women's jobs in their mobility patterns. As seen in Fig. 1, changes in the professional status and the organization of respondents' work are investigated. For example, have they chosen some lay-off or part-time? Was the taking care of children and household chores equally distributed in the household?

The data collection activity has been opened from July to September 2020, intending to collect the first wave of a limited number of results in the 10 TInnGO Hubs' working-age population. This allows having insights into the effects of the pandemic on mobility patterns in cities with different restrictions during a so-called $1^{st}$ *lockdown* in Europe. The investigation of the responses collected in 13 EU countries, each of them facing the COVID-19 emergency at various moments and, sometimes, in not uniform ways, can provide a broader view of the pandemic's impact on users' mobility patterns. At the same time, it is well known that the situation influenced employment, producing, for example, an increase in the commonly known inequality.

## 4 Results and Discussions

### 4.1 Sample Characterization

The data collected refer to 208 European respondents, 57% of which were women. The results analysed in this paper are related to the preliminary phases of a more extensive study that will broaden the survey to a much larger sample (around 4000 answers expected). Table 1 provides an overview of the sample's socioeconomic characteristics (Sect. 1 in Fig. 1). The respondents are from different European countries: Portugal, Spain, France, Italy, United Kingdom, Germany, Sweden, Denmark, Romania, Greece, Lithuania, Latvia, Estonia. The TInnGO partners conducted the survey campaign through their personal contacts, professional contacts and social media, thus resulting in a good majority of respondents in the age ranges 25–34 (37%) and 35–44 (26%), mainly with a high-level education (Degree, Master or Doctorate). The investigation on the residential area type revealed a great majority of respondents living in an urban environment (74% of answers), with only 6% of them declaring a rural one. Table 2 shows the sample's breakdown according to age and gender; the respondents in the age ranges 25–34 are mainly women, while the balance is greater for the other age ranges.

### 4.2 Pre-COVID Scenario

As could also be expected based on the respondents' age, most of them stated that they worked in an office in the pre-COVID scenario, around 70% of answers (Table 3). It also emerges that the proportion of women working remotely, self-employed or unemployed, is higher than that of men, although this difference is relatively marginal. The overview

**Table 1.** Descriptive statistics on socioeconomics characteristics of the respondents.

| Feature | Choices | % |
|---|---|---|
| Age bracket | 18–24 years | 14 |
| | 25–34 years | 37 |
| | 35–44 years | 26 |
| | 45–54 years | 11 |
| | 55–64 years | 7 |
| | 65–74 years | 3 |
| | >75 years | 2 |
| Children in the HH | No | 64 |
| | 1 | 17 |
| | 2 | 16 |
| | 3 | 3 |
| Dependent person | No | 71 |
| | Preschool age children (under 5 years) | 13 |
| | School age children (5–10 years) | 10 |
| | School age children (10–16 years) | 12 |
| | Elderly relative | 1 |
| Residence area type | Urban | 74 |
| | Suburban | 20 |
| | Rural | 6 |
| Level of education | Secondary | 13 |
| | Degree | 15 |
| | Master | 46 |
| | Doctorate | 23 |
| | Other | 3 |

of professional status distribution is essential since, as pointed out above, employment status can influence mobility choices, especially in the pandemic scenario.

The focus goes on users' mobility habits, distinguishing the answers according to gender to highlight peculiarities and possible differences. In the pre-COVID scenario, the most used mode of transport for each activity is investigated and the results are reported in Table 4 separately for women and men. The only respondent that preferred not to declare her/his gender was excluded from this analysis.

In general, the private vehicle is the most frequently chosen means of transport, especially for visiting a shopping center, while the second choice is walking. This being said, it emerges, however, in line with what is found in the literature, that men are more

**Table 2.** Age and gender characteristics of the respondents

| Age/gender | Female | Male | Prefer not to answer | |
|---|---|---|---|---|
| 18–24 years | 13 | 17 | | 30 |
| 25–34 years | 47 | 29 | 1 | 77 |
| 35–44 years | 27 | 28 | | 55 |
| 45–54 years | 17 | 6 | | 23 |
| 55–64 years | 9 | 6 | | 15 |
| 65–74 years | 3 | 4 | | 7 |
| >75 years | 1 | | | 1 |
| | 117 | 90 | 1 | 208 |

**Table 3.** Professional status in pre-COVID scenario

| Professional status pre-COVID | Female | Male |
|---|---|---|
| Paid employment - working in an office/plant | 70% | 72% |
| Paid employment - working remotely | 5% | 4% |
| Paid employment - parental leave | 1% | 1% |
| Paid employment - lay off | 0% | 0% |
| Self-employed | 7% | 4% |
| Non-paid work | 2% | 0% |
| Student | 11% | 14% |
| Homemaker | 1% | 0% |
| Retired | 2% | 1% |
| Unemployed (health reason) | 0% | 0% |
| Unemployed (other reason) | 2% | 1% |
| Prefer not to answer | 0% | 1% |
| DK/NA | 0% | 0% |
| Other | 0% | 0% |

likely both to use a private vehicle and to walk. On the other hand, women seem more inclined to use local public transport except to run an errand or go shopping. Sharing solutions seem to be of little attraction to both women and men at less than 3%. More than 70% of the respondents stated that they work in an office (Table 3), so it is interesting to focus on the mobility habits to get to the workplace, which is most probably the main journey. Preferences are quite similar between men and women: the first choice is the private car with, respectively, 40 and 41% of answers, followed by public transport with

**Table 4.** Mobility habits according to gender and scope (pre-COVID scenario)

| Mobility habits pre-COVID [%] | | Private car/moto | Shared bikes | Owned bikes | PT | Shared modes | Walking | Other |
|---|---|---|---|---|---|---|---|---|
| Job/university/school | F | 41 | 1 | 15 | 30 | 2 | 7 | 5 |
| | M | 40 | 3 | 15 | 25 | 0 | 17 | 0 |
| Visiting people another town | F | 55 | 0 | 3 | 33 | 3 | 4 | 2 |
| | M | 55 | 0 | 6 | 32 | 0 | 4 | 4 |
| Running an errand | F | 39 | 1 | 14 | 10 | 3 | 32 | 1 |
| | M | 41 | 1 | 14 | 17 | 0 | 26 | 1 |
| Out dinner | F | 32 | 2 | 10 | 21 | 3 | 29 | 3 |
| | M | 32 | 0 | 11 | 19 | 3 | 33 | 1 |
| Tourism | F | 21 | 3 | 17 | 24 | 2 | 32 | 1 |
| | M | 23 | 2 | 13 | 21 | 0 | 39 | 1 |
| Tourism other city | F | 38 | 2 | 0 | 32 | 3 | 23 | 2 |
| | M | 49 | 0 | 0 | 28 | 0 | 20 | 3 |
| Groceries | F | 41 | 3 | 11 | 3 | 2 | 39 | 1 |
| | M | 34 | 2 | 8 | 7 | 0 | 48 | 1 |
| Shopping center | F | 56 | 0 | 10 | 21 | 1 | 11 | 2 |
| | M | 59 | 2 | 6 | 20 | 0 | 13 | 0 |
| Weekend activities | F | 47 | 2 | 15 | 20 | 2 | 14 | 2 |
| | M | 50 | 1 | 16 | 21 | 1 | 10 | 1 |

25 and 30%. The following options are different, with 17% of men saying they walk and 15% using their bicycles; women, on the other hand, prefer to cycle with 15% and walk with 7%.

### 4.3 Post-COVID Scenario

Excluding the 28% of respondents who do not know or cannot answer, 38% of them work remotely, whereas 35% declare going to the office in the post-COVID scenario. As might be expected, these two percentages are very different from the pre-pandemic scenario in which 71% of respondents went to the office, and only 5% worked smart. Interestingly, in the pre-pandemic, no respondents had selected the DK/NA option; probably this is linked to the still uncertain situation in Europe in the survey period (July–September 2020). Table 5 reports the detailed percentage distribution of replies according to gender and the variation from the declared answers for the pre-COVID scenario.

The mobility habits in post-COVID are expected to change according to different governmental choices to limit non-essential travel. Countries across Europe have significantly limited public life in order to stop the spread of the COVID-19 outbreak. In Italy, France, Belgium, and Spain, for example, people are only permitted to leave the

**Table 5.** Professional status in post-COVID scenario and variation from pre-COVID one.

| Professional status post-COVID [%] | Female | Variation | Male | Variation |
|---|---|---|---|---|
| Paid employment - working in an office/plant | 28 | −42 | 21 | −51 |
| Paid employment - working remotely | 26 | 21 | 29 | 24 |
| Paid employment - parental leave | 0 | −1 | 1 | 0 |
| Paid employment - lay-off | 0 | 0 | 0 | 0 |
| Self-employed | 5 | −2 | 0 | −4 |
| Non-paid work | 2 | 0 | 0 | 0 |
| Student | 4 | −7 | 11 | −3 |
| Homemaker | 0 | −1 | 0 | 0 |
| Retired | 1 | −1 | 1 | 0 |
| Unemployed (health reason) | 0 | 0 | 0 | 0 |
| Unemployed (other reason) | 2 | 0 | 2 | 1 |
| Prefer not to answer | 0 | 0 | 0 | −1 |
| DK/NA | 30 | 30 | 24 | 24 |
| Other | 0 | 0 | 0 | 0 |

house for health reasons or for going grocery shopping. Instead, Germany and Portugal have opted for less restrictive measures, advising to limit unnecessary travel. Sweden, on the other hand, has not imposed severe lockdowns, for example, but adviced people to work from home and avoid crowded public transport.

A possible additional answer, "*I stopped making this trip*", has been introduced to consider that a specific type of journey is no longer carried out. As shown in Table 6, women and men stopped moving mainly for tourism to other cities (24% and 26%), to go to the shopping center (21% and 19%), and to go out to dinner (16% and 20%).

### 4.4  COVID Impacts on Mobility Patterns

Having analyzed the main elements in the pre- and post-COVID scenario about mobility habits disaggregated by gender, in this section the particular aspects of the declared changes are reported and discussed. As shown in Table 7, 62% of the women interviewed stated that they had not changed their professional status but their mobility habits. A similar situation occurs among male respondents, of whom 63% have not changed their job, but 64% have changed their travel choices.

In particular, filtering out the positive responses to the job change, it emerges that the COVID-19 emergency brought changes in the professional situation mainly due to the necessity of starting to work remotely (71% of respondents), but this affected women and men differently, with this percentage coming to 68 and 75, respectively.

Most people, as expected, changed their travel choices, and the graph in Fig. 2 shows the percentage changes between the pre- and post-COVID scenarios net of those who said they would no longer make that journey.

**Table 6.** Mobility habits according to gender and scope (post-COVID scenario)

| Mobility habits post-COVID [%] | | Private veh | Shared bikes | Owned bikes | PT | Shared modes | Walking | Other | I stopped making this trip |
|---|---|---|---|---|---|---|---|---|---|
| Job/university/school | F | 38 | 3 | 15 | 12 | 1 | 9 | 9 | 14 |
| | M | 41 | 1 | 10 | 12 | 0 | 18 | 4 | 13 |
| Visiting people another town | F | 55 | 1 | 11 | 11 | 2 | 8 | 3 | 9 |
| | M | 59 | 2 | 4 | 12 | 1 | 7 | 3 | 11 |
| Running an errand | F | 37 | 1 | 15 | 6 | 0 | 37 | 3 | 3 |
| | M | 44 | 1 | 18 | 6 | 1 | 27 | 1 | 2 |
| Out dinner | F | 28 | 1 | 9 | 12 | 1 | 28 | 4 | 16 |
| | M | 32 | 0 | 9 | 9 | 1 | 27 | 2 | 20 |
| Tourism | F | 17 | 3 | 13 | 11 | 0 | 37 | 3 | 16 |
| | M | 21 | 4 | 14 | 6 | 0 | 38 | 3 | 13 |
| Tourism other city | F | 39 | 2 | 4 | 13 | 0 | 14 | 4 | 24 |
| | M | 38 | 1 | 1 | 10 | 1 | 19 | 4 | 26 |
| Groceries | F | 36 | 0 | 11 | 1 | 1 | 44 | 3 | 4 |
| | M | 40 | 0 | 10 | 4 | 1 | 43 | 1 | 0 |
| Shopping centre | F | 44 | 0 | 7 | 11 | 0 | 12 | 4 | 21 |
| | M | 50 | 0 | 8 | 12 | 1 | 8 | 2 | 19 |
| Weekend activities | F | 45 | 1 | 16 | 10 | 0 | 18 | 3 | 6 |
| | M | 49 | 1 | 18 | 7 | 0 | 18 | 3 | 4 |

It clearly emerges that public transport (yellow bar in Fig. 2) is the mode that has been most affected by the changes in mobility due to the spread of the pandemic, probably because it is considered unsafe in terms of infection. Women were found to be the primary users of this transport solution (Table 4), and indeed they are the ones who report the main changes, especially for work-related trips. Sharing modes of transport, which are already little used, decreased further regardless of the gender. On the other hand, private transport (cars and bicycles) or walking have increased, but not homogeneously by gender or purpose. For example, women started to walk for many of their activities, more than men, with an increase of 7% for shopping for groceries or 11% for tourism in their city. The opposite behavior is observed for the male component, which has stopped going to groceries and shopping centers on foot but seems to have opted for the own bicycle (orange bar in Fig. 2). This alternative is now used more by women to travel mainly to other cities, for tourism or to visit other people.

In general, walking, owned bikes, and private cars have been declared as transport modes used more than in the pre-COVID scenario, as shown in Fig. 3. These plots collect the responses to the question "*Which of the following travel mode have you*

**Table 7.** Change in professional status and mobility habits based on gender and age

| | Change professional status | | | Change mobility habits | | |
| | Female | Male | Female | Male |
| --- | --- | --- | --- | --- |
| No | 62% | 63% | 38% | 36% |
| *Of which* | | | | |
| 18–24 years | 15% | 23% | 2% | 26% |
| 25–34 years | 36% | 26% | 43% | 23% |
| 35–44 years | 22% | 28% | 27% | 32% |
| 45–54 years | 16% | 9% | 18% | 10% |
| 55–64 years | 8% | 9% | 7% | 6% |
| 65–74 years | 1% | 5% | 2% | 3% |
| >75 years | 1% | 0% | 0% | 0% |
| Yes | 38% | 37% | 62% | 64% |
| *Of which* | | | | |
| 18–24 years | 5% | 9% | 15% | 16% |
| 25–34 years | 48% | 32% | 38% | 36% |
| 35–44 years | 25% | 27% | 21% | 30% |
| 45–54 years | 11% | 2% | 13% | 5% |
| 55–64 years | 7% | 2% | 8% | 7% |
| 65–74 years | 5% | 2% | 3% | 5% |
| >75 years | 0% | 0% | 1% | 0% |

*started using more than previously?*", where multiple choices are allowed for the female and male samples. The option "PT" combines various means, namely bus, private bus, train, trolleybus, tram, and metro. The group of users who stated that they use new means of transport more than previously are 25–34 years' women (with 42% of the answers), whereas, for male respondents, the change is more homogeneous in the three minor age groups (18–24 years; 25–34 years; 35–44 years). Moreover, it could be observed that in the female sample, a higher number of respondents have started being passengers, both on private cars and taxis. In the former case, significant differences among age ranges could be observed while comparing the genders. Indeed, a considerable number of people older than 35 years are found in the *"Private car as a passenger"* column in Fig. 3(a), while they are absent in the corresponding column of Fig. 3(b).

A specific focus is proposed on the investigation of modes associated with Smart Mobility, according to the definition provided in [4], namely shared modes, public transport, walking, and biking. Results are proposed in Fig. 3, where SM's means are found before the blue line. The plot clearly shows that female respondents' sample seems less prone to approach some SM modes than men, mostly the shared ones. Among these

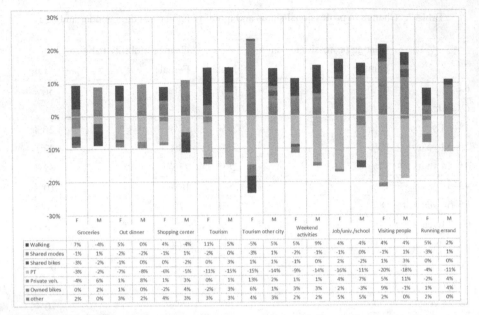

| | F | M | F | M | F | M | F | M | F | M | F | M | F | M | F | M | F | M |
|---|---|---|---|---|---|---|---|---|---|---|---|---|---|---|---|---|---|---|
| | Groceries | | Out dinner | | Shopping center | | Tourism | | Tourism other city | | Weekend activities | | Job/univ./school | | Visiting people | | Running errand | |
| ■ Walking | 7% | -4% | 5% | 0% | 4% | -4% | 11% | 5% | -5% | 5% | 5% | 9% | 4% | 4% | 4% | 4% | 5% | 2% |
| ■ Shared modes | -1% | 1% | -2% | -2% | -1% | 1% | -2% | 0% | -3% | 1% | -2% | -1% | -1% | 0% | -1% | 1% | -3% | 1% |
| ■ Shared bikes | -3% | -2% | -1% | 0% | 0% | -2% | 0% | 3% | 1% | 1% | -1% | 0% | 2% | -2% | 1% | 3% | 0% | 0% |
| ■ PT | -3% | -2% | -7% | -8% | -6% | -5% | -11% | -15% | -15% | -14% | -9% | -14% | -16% | -11% | -20% | -18% | -4% | -11% |
| ■ Private veh. | -4% | 6% | 1% | 8% | 1% | 3% | 0% | 1% | 13% | 2% | 1% | 1% | 4% | 7% | 5% | 11% | -2% | 4% |
| ■ Owned bikes | 0% | 2% | 1% | 0% | -2% | 4% | -2% | 3% | 6% | 1% | 3% | 3% | 2% | -3% | 9% | -1% | 1% | 4% |
| ■ other | 2% | 0% | 3% | 2% | 4% | 3% | 3% | 3% | 4% | 3% | 2% | 2% | 5% | 5% | 2% | 0% | 2% | 0% |

**Fig. 2.** Mobility habits variation based on gender and scope (Color figure online)

modes, those gaining more interest among the women are the shared bikes and services such as Uber and similar. Other means that have started spreading in the cities and are expected to reach a good success after the COVID-19 emergency are shared e-scooters. However, these means seem to attract men more than women, confirming what is assessed in Sect. 2.

After it has become clear that the COVID-19 pandemic has changed mobility habits, it is interesting to understand the motivations behind these changes, and this can be done by assessing the answers to the question: *"Which are the reasons behind your changes in the modes used?"*, with a multiple-choice option. In general, the most selected answer was the motivation related to health issues to prevent infection. However, Fig. 4 shows how all the possible choices separate between gender. For example, most women are found among those respondents declaring a change in the modes due to a longer travel duration while using the previous mode. A similar partition between gender is found for the option *"Increase of conditions in the use of previous modes (more bike lanes, more bus lanes, …)"*. Women seem to be more affected by the reduction of reliability in PT services, obviously connected to the fact that they are known to be more frequent users of this means. In contrast, economic savings move more the choices done by men.

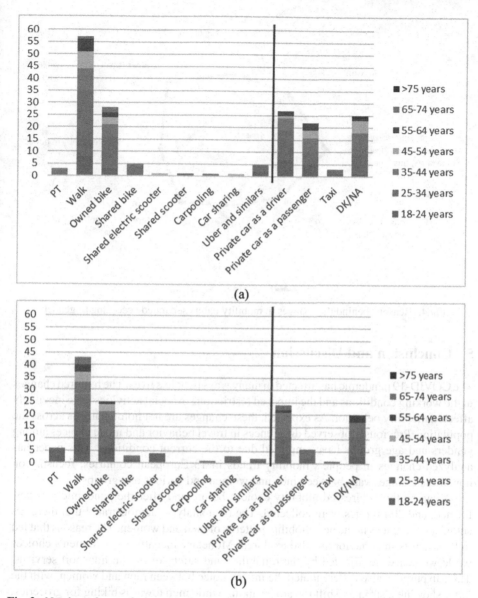

**Fig. 3.** Number of users for means of transport used more than previously based on age for (a) female and (b) male; SM before the blue line (Color figure online)

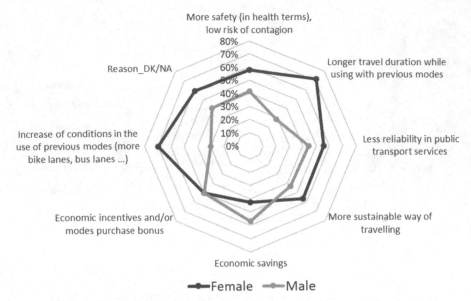

**Fig.4.** Reasons behind the changes in mobility habits separated according to gender.

## 5   Conclusion and Discussion

The COVID-19 pandemic has affected in many ways people's lives. The required changes in the working conditions and the imposed restrictions in social activities have inevitably affected mobility behavior. As expected, these changes are not homogeneous across the population. Previously observed differences in travel behavior and mode choices among genders and age groups have amplified due to the current conditions. This paper has analyzed changes in people's mobility trends in 13 European countries, focusing on how these changes vary between men and women and various age groups.

A survey that addressed mobility choices before and after the pandemic was performed, and 208 replies were collected. The data analysis demonstrated that there are actually differences in the new mobility patterns of men and women. The reasons that led to these shifts in behavior are also explored. Monetary incentives drove men's choices while women were affected by the reliability and safety of public transport services. The trip purpose also differentiated the mode choice between men and women, with the latter showing a stronger shift towards walking while men towards biking for groceries. For shopping and tourism, men revealed a preference for biking.

In general, walking increased while a significant decrease in the use of PT is observed. These results have been expected due to the limitation of the trips performed per day due to the mobility restrictions imposed in many cities and the increase of remote work. These changes seem somehow correlated with gender, considering, for example, that women are commonly more likely to use PT than men [8]. However, the main reason behind the changes in mobility habits is the need for more safety in terms of low risk of contagion, irrespective of gender. A shift towards private car use is evident in the age groups of people older than 35. Given the preventive measures of social distancing and frequency of cleaning commonly used facilities, shared modes use has dropped during the pandemic. However, shared bikes, specifically, have seen an increased interest from male users.

The current exploratory analysis has shown that mobility behavior has been affected by the pandemic's advent and its impact on daily activities and working conditions. Different groups of people, clustered by gender and age, have been experiencing these changes in various ways as far as their mobility is concerned. In general terms, it is seen that walking activities have in general increased, followed by biking while in motorized transport, the private car usage is dominant followed by ride-hailing options. It is still unknown how these changes will evolve in the future and how mobility behavior is expected to be formed under the conditions of the pandemic and the post-pandemic era. Future research aims to analyze the post-COVID19 mobility behavior of people and analyze if the shifts towards active modes will remain or will be absorbed, again, by the use of motorized transport and if the current preference towards private cars could change towards public transport modes.

**Acknowledgements.** The current paper is part of the activities of the H2020 European project TInnGO - Transport Innovation Gender Observatory, funded under grant agreement no. 824349. The Authors are thankful to the TInnGO Hubs and partners for their contribution to the data collection activities.

# References

1. Abdullah, M., Dias, C., Muley, D., Shahin, M.: Exploring the impacts of COVID-19 on travel behavior and mode preferences. Transp. Res. Interdiscip. Perspect. **8**, 100255 (2020). https://doi.org/10.1016/j.trip.2020.100255
2. Cochran, A.L.: Impacts of COVID-19 on access to transportation for people with disabilities. Transp. Res. Interdiscip. Perspect. **8**, 100263 (2020). https://doi.org/10.1016/j.trip.2020.100263
3. Transformative Urban Mobility Initiative (TUMI): COVID-19 and sustainable mobility (2020)
4. Woodcock, A., Christensen, H.R., Levin, L.: TInnGO: challenging gender inequality in smart mobility. Put i saobraćaj **66**(2), 1–5 (2020). https://doi.org/10.31075/PIS.66.02.01
5. Pirra, M., Carboni, A., Diana, M.: Assessing gender gaps in educational provision, research and employment opportunities in the transport sector at the European level. Educ. Sci. **10**(5), 123 (2020). https://doi.org/10.3390/educsci10050123
6. Bencardino, M., Greco, I.: Smart communities. Social innovation at the service of the smart cities. TeMA J. Land Use Mob. Environ. (2014). https://doi.org/10.6092/1970-9870/2533
7. Singh, Y.J.: Is smart mobility also gender-smart? J. Gend. Stud. **29**(7), 832–846 (2019). https://doi.org/10.1080/09589236.2019.1650728
8. Pirra, M., Kalakou, S., Carboni, A., Costa, M., Diana, M., Lynce, A.R.: A preliminary analysis on gender aspects in transport systems and mobility services: presentation of a survey design. Sustainability **13**, 1–20 (2021). https://doi.org/10.3390/su13052676
9. Lozzi, G., Marcucci, E., Gatta, V., Pacelli, A., Rodrigues, M., Teoh, T.: Research for TRAN committee – COVID-19 and urban mobility: impacts and perspectives, Brussels (2020)
10. de Paz, C., Muller, M., Boudet, A.M.M., Gaddis, I.: Gender Dimensions of the COVID-19 Pandemic. World Bank, Washington, DC (2020). https://doi.org/10.1596/33622

11. OECD: Forthcoming OECD policy brief: women at the core of the fight against COVID-19 (2020)
12. Wang, K., Akar, G.: Gender gap generators for bike share ridership: evidence from Citi Bike system in New York City. J. Transp. Geogr. **79**, 1–6 (2019). https://doi.org/10.1016/j.jtrangeo.2019.02.003
13. Ma, X., Yuan, Y., Van Oort, N., Hoogendoorn, S.: Bike-sharing systems' impact on modal shift: a case study in Delft, the Netherlands. J. Clean. Prod. **259**, 120846 (2020). https://doi.org/10.1016/j.jclepro.2020.120846
14. Zhang, L., Zhang, J., Duan, Z., Bryde, D.: Sustainable bike-sharing systems: characteristics and commonalities across cases in urban China. J. Clean. Prod. **97**, 124–133 (2015). https://doi.org/10.1016/j.jclepro.2014.04.006
15. Chicco, A., Pirra, M., Carboni, A.: preliminary investigation of women car sharing perceptions through a machine learning approach. In: Stephanidis, C., Antona, M. (eds.) HCII 2020. CCIS, vol. 1224, pp. 622–630. Springer, Cham (2020). https://doi.org/10.1007/978-3-030-50726-8_81
16. Kawgan-Kagan, I., Popp, M.: Sustainability and gender: a mixed-method analysis of urban women's mode choice with particular consideration of e-carsharing. Transp. Res. Procedia. **31**, 146–159 (2018). https://doi.org/10.1016/j.trpro.2018.09.052
17. del Mar Alonso-Almeida, M.: Carsharing: another gender issue? Drivers of carsharing usage among women and relationship to perceived value. Travel Behav. Soc. **17**, 36–45 (2019).https://doi.org/10.1016/j.tbs.2019.06.003
18. Osservatorio nazionale sharing mobility: quarto rapporto nazionale sulla sharing mobility (2020)
19. Degele, J., et al.: Identifying e-scooter sharing customer segments using clustering. In: Proceedings of the 2018 IEEE International Conference on Engineering, Technology and Innovation, ICE/ITMC 2018 (2018). https://doi.org/10.1109/ICE.2018.8436288
20. International transport forum: women's safety and security: a public transport Priority. https://www.itf-oecd.org/womens-safety-security. Accessed 05 Jan 2021
21. Kloof, A.V.D., Kensmil, J.: Effects of COVID-19 measures on mobility of men and women. https://mobycon.com/wp-content/uploads/2020/06/Effect-of-covid-19-measures.pdf
22. Caselli, F., Grigoli, F., Sandri, Damiano; Spilimbergo, A.: Mobility under the COVID-19 Pandemic: Asymmetric Effects across Genderand Age (2021). https://www.imf.org/en/Publications/WP/Issues/2020/12/11/Mobility-under-the-COVID-19-Pandemic-Asymmetric-Effects-across-Gender-and-Age-49918

# Smart is (Not) Always Digital!

## Expanding the Concept of Assistive Technology: The Roller as an Age-Based, Gendered and Social Innovation

Hilda Rømer Christensen[✉]

University of Copenhagen, Copenhagen, Denmark
hrc@soc.ku.dk

**Abstract.** The value of rollers for the elderly and other groups who need walking assistance has been underestimated both in terms of practice and in knowledge production. This paper aims at scrutinising the roller as an age-based and gendered innovation. Using the theoretical notion of scripts, it demonstrates how rollers and their users are intertwined in everyday practices and how these relationships intersect with notions of age and social welfare provision. Based on contrasting images of rollers and their users, as well as semi-structured interviews, this paper examines the puzzle of how the take-up of new (technological) devices comes about. It argues that the roller can be seen as a simple, disruptive innovation emerging from the bottom up with contradictory scripts of gender and age. The paper concludes with a perspectivation of how the rollers could be made both smarter and digital!

**Keywords:** Assistive technology · Walker · Gendered scripts · Social innovation · Ageism · Stereotyping

## 1 Introduction

Rollers, or four-wheeled rollers, do not exactly constitute a sexy or cutting edge research topic – not even when it comes to studies of the elderly and mobility or those conducted in the fields of design and assistive technologies. In other words, the value of rollers for the elderly and other groups who need walking assistance has been underestimated both in terms of practice and in knowledge production. This paper aims at scrutinising the roller as an age-based and gendered innovation. Using the theoretical notion of scripts, it demonstrates how rollers and their users are intertwined in everyday practices and how these relationships intersect with notions of age and social welfare provision. Based on contrasting images of rollers and their users, as well as semi-structured interviews, this paper examines the puzzle of how the take-up of new (technological) devices comes about. It argues that the roller can be seen as a simple, disruptive innovation emerging from the bottom up with contradictory scripts of gender and age. The paper concludes with a perspectivation of how the rollers could be made both smarter and digital!

© Springer Nature Switzerland AG 2021
H. Krömker (Ed.): HCII 2021, LNCS 12791, pp. 487–499, 2021.
https://doi.org/10.1007/978-3-030-78358-7_34

## 2  The Roller in the Context of Ageism and Stereotyping

The roller, and the numerous stereotypes and interpretations attached to it, can be seen as being aligned with an overall discourse of oppressive ideas about disability in general and old age in particular. Ageism is the theoretical use of "age" as a degrading category, which, in practice, can involve stereotyping and prejudice against older persons. According to the American psychologist Todd D. Nelson, age is used as a category in line with gender and race, but of the three categories, age is the only one in which the members of the in-group – meaning the younger will eventually join the out-group – the older when and if they reach old age [1]. The elderly, he argues, are stigmatised and marginalised, which can have serious effects on their health and well-being. Todd notes that ageism is found cross-culturally and has become associated with depression, fear and anxiety. He also points out that while racism and sexism have drawn a lot of attention in research and in the public debate, old age has lacked attention both in knowledge production and in the popular media. This situation is due to the fact that prejudice is still socially acceptable and widespread. Ageism, in turn, spills over to common attitudes about "rollers" and their users.[1]

In general, the roller is positioned as being residual, both in mobility and age based gerontechnological research. In mobility studies and modelling, for instance, there is a lack of recognition of simple technologies and walking as a significant modality – meaning that the roller as a mobility device, as well as walking itself, is positioned outside of most existing ranks of mobility. Walking, including the walking with a roller, has consequently seldom been included in regular transport surveys at national or European levels; nor does walking often figure in transport budgets and resource allocations[2] [2].

The same pattern is seen in the recent wave of age-based innovation and research, and of gerontechnological research. Although the roller has been around for several decades now, this simple technology has mainly been scrutinised in scattered medical/ergonomic studies [3–5]. Moreover, the device has only caught little attention in ethnographic and cultural studies of old age and assistive designs [6–8].

The following analysis of rollers and their users is inspired by scripts theory and various ideas about innovation, which I will present in the first section. In the following section I will refine the notion of scripts and use them as guidelines for a qualitative analysis of interviews with a sample of roller users in Denmark. The paper will conclude with an assessment of how rollers and their users can be included in the future age including and gender smart mobility city.

## 3  Gendered and Contradictory Scripts

Science and technology studies (STS studies) have made us aware of the co-production of techno scientific objects and social order. STS-informed studies have made a significant

---

[1] While the term "rollator" is the most commonly used term in Scandinavian languages, several terms can be used in English, such as roller, walker, rolling walker or four-wheeled walker. "Roller" is used as the key term for this device throughout this paper.

[2] One exception is Norway, where a national walking strategy has been presented, see Nasjonal gåstrategi (2012) also: https://www.tiltak.no/tag/gange/

impact by opening many "black boxes" in science and technology and revealing the intersecting nature of science, technology and social processes [9–11]. The notion of scripts plays a key role in this relationship and has been applied to empirical analyses of partial predetermined aspects of technology and human action. Scripts are implied in processes where "particular technologies incorporate intentional or intentional 'scripts' that prefigure users' identities and experience, while at the same time being shaped by them". 6:3, [12, 13].

Feminist studies operate with a similar idea of Gender Scripts which address gendered aspects that are glossed over in the general idea of a script. Gender Scripts refer to explicit and implicit gendered bias in technological objects, such as cars and bikes, which are commonly perceived as "neutral" objects. In order to approach the gendered dimension of new technologies, one might understand them in terms of social relations that interact with the technology during its design, development, production and use [14–16]. The social shaping of technologies such as cars, bikes and walking devices includes the material object, the practice of using it and the identity of the (supposed) user. Gender scripts are powerful and used both by the producers and in marketing and can be applied to the study of the gendered character of the roller and walking as a transport modality.

However, the roller and the ways in which it is presented can also illustrate the shifting nature of a script and how technical devices can contain gendered contradictions and paradoxes. The highly gendered and contrasting depictions of the roller in public media is an excellent example of these contradictions, and this is also echoed in the interviews with the users of the rollers presented in this paper.

One high profile example appeared in the NYC court case against the film producer Harvey Weinstein, who was convicted of violent sexual assault and rape of women. The depiction of Weinstein using a roller on his way to the court clearly bolstered the notion of a fallen and deprived man. The New York Times labelled it as a deliberate and degrading depiction of physical weakness and dependence [18]. The roller was seen as part of a theatrical show that could influence the jury to issue a milder sentence. But this particular use of the roller could also reinforce harmful stereotypes of people with disabilities. The negative media attention being paid to the roller in this widely publicised event therefore galvanised negative stereotypes attached to the roller by sending a clear message of vulnerability and de-masculinisation.

The second depiction of the roller, by Danish queen mother Ingrid, presents a contrasting image. When the queen mother, a beloved public figure in Denmark, appeared with a roller at a royal wedding in 1998, it was welcomed and highly acclaimed by the Danish press and public [17, 19, 20][3]. "Queen Ingrid of Denmark uses a loaned roller for first time in public, becoming "a powerful image that encouraged others not to be ashamed of their rollators" (17:71). The photo of the queen mother and her roller became iconic and the occasion is still referred to among roller users. In contrast to Weinstein, she became a figure of dignity and a role model for elderly people in Denmark. A side story was that the event also spurred a debate on whether universal welfare provisions

---

[3] The style-conscious queen mother, even made the colours of her elegant outfit match the green nuance of the standard, off-the-shelf roller. The roller and the outfit later became part of a historical exhibition at the Royal residence in Copenhagen.

also should include the royal family. The queen mother had in line with any other citizen, acquired her roller as a free provision from the municipality [20–22].

The third example depicts a fictional and mundane roller user who is far from the limelight of celebrities and scandals. The depiction is derived from the trailer of a recent Danish film, "Uncle", with the roller and the uncle, as one of the protagonists in the film, playing a crucial role [23–26]. The film is set on stunning, flat marshlands well away from the urban centre, and at first sight the roller aptly conveys the message of the disabled farmer's character as being on the fringe of modern urban culture. Yet eventually the roller enables him to contribute to the running of the farm. From a conceptual angle, the film stages the roller in a hybrid position, acting as a kind of cyborg, where physical abilities and material devices are interwoven and interdependent. I will elaborate on this dimension in the qualitative analysis later in this paper.

Whereas the roller, among welfare professionals, is regarded and distributed as a neutral assistive technology, in the public eye and on the "stage" the roller is a highly gendered device that is full of contradictions. This also complicates the notion of scripts and makes them open to change and interpretation. The analysis of the qualitative data also revealed contradictions along with a more balanced interpretation of the usefulness of the roller in daily life. I will return to these interpretations after presenting an introduction of the roller as a simple technology in the context of disruptive and gendered innovation.

## 4    The Roller as Disruptive Innovation

From a conceptual point of view, I will argue that the manufacturing and actual use of the roller accentuate specific notions of innovation, such as disruptive age-based and gendered innovation. The idea of disruptive innovation was coined by the American scholar Clayton Christensen, who in 1997 described disruptive innovation as providing a "cheaper, easier to use alternative to existing products or services often produced by non-traditional players that target previously ignored systems and/or use in novel contexts and combination" [27]. On this horizon, the notion of disruptive innovation suggests a comprehensive location of the innovation processes in contrast to the weakness of the dominant ideas of innovation, separated from social contexts. Christensen's idea of disruptive innovation also challenged the prevailing definition of innovation as being linked exclusively to technology and business outcome. There was an imperative in this logic for innovation to include broader mobilisations and involvements of people and politics. Besides this, it was urged that innovation should initiate profound changes in everyday life as well as in socioeconomic systems [28, 29]. Disruptive innovation, from this particular angle, was understood as the invention of alternatives to high-tech solutions; with examples such as refrigerators and air conditioning systems using simple materials like water or sand to fuel appliances.

The roller is, to some extent, aligned with this definition of disruptive innovation, evidenced in a new and simple technology which, in an almost Biblical sense, has allowed people who were previously bound to their chairs to walk again. From a technological perspective, the roller is located between the wheelchair, which binds the body to the device and prevents users from walking (if they are able to), and the various forms of "sticks" used to support and balance an individual. The story about the creation of the roller serves to galvanise the notion of a disruptive innovation. The roller was invented by Aina Wifalk in 1978, a Swedish nurse who was neither an insider in the field of high-tech gerontechnology nor a business person [30]. Wifalk suffered from polio and experienced walking difficulties, and she had a strong desire to continue to walk despite her impairment. Wifalk presented her first invention to the public in 1965. This prototype was called a "Manu-ped" and was developed as a training device for people with physical disabilities. Sufferers could use a Manu-ped to train their arms and legs, as well as their coordination. On the basis of the Manu-ped prototype, other training devices for physically disabled people were developed. Some of these devices are still being used today in the health service as well as in special sport schools. When Aina Wifalk's condition worsened and she was no longer able to walk with sticks, she presented the first model of a "rollator" in 1978, which was, at the time, called a walking frame. The roller was gradually put on a business track in cooperation with a state development fund, and a partnership was established with a Swedish company to produce a prototype roller. Mass production of the "roller" began shortly after, and Swedish producers in particular took the lead in producing and distributing the roller over the first decades.

I would argue that the roller presents a good and robust example of disruptive innovation because, as a material device, it provided a simple solution to a walking impairment and was invented by an outsider of the established gerontechnical circles. Second, the roller spurred a social change, a revolution in the lives of the elderly and walking impaired people as evidenced in the qualitative section of this paper article A third aspect relates to politics and social services where in Denmark, for instance, the roller has become part of mainstream provisions and a kind of imperative for elderly walking or physically impaired persons. The Danish Service Law requires municipalities to provide rollers free of charge to the elderly and physically impaired citizens. The aim is to compensate for their reduced mobility and to support their ability to manage activities in the home and to participate in society[4] [31, 33].

Europe in the 21st century, take the lead in using rollers with the Nordic countries and Germany as the leading nations [34]. This is partly due to the merge of individual and recent professional notions of "successful ageing" as well as welfare states' provisions of social and material welfare devices. Sweden was a first mover, both as the country of the invention and the place where rollers were adopted first, and distributed rollers in significant numbers earlier than other countries. The United States remains a laggard in the use of rollers. Despite having a population of over 300 million citizens, roller sales were estimated at 20,000–40,000 annually at the turn of the century (17: 78). An explanation could be both cultural and related to social services. Medical insurance coverage for rollers remains limited in many countries, putting rollers at a major disadvantage compared to wheelchairs, which often have better coverage. This may even be the case in parts of Europe where the take-up of rollers has been low so far.

---

[4] The Danish Social Service law in 2013/2015 (http://english.sm.dk/media/14900/consolidation-act-on-social-services.pdf) - Paragraph 112 it states that the municipality shall grant support for technical aids to persons with permanent impairments to physical or mental function. The provision is meant to remedy the permanent effects of the functional impairments significantly, or facilitate daily life in the home significantly.

## 5  The Roller as Gendered Innovation

Gendered innovation is another recent notion that taps into the creation, interpretation and use of the roller. Gendered innovation is a flexible term that implies gendered user proximity in both methodological and material practices. Gendered innovation was coined by the ground-breaking *Gendered innovations* project that was carried out by a partnership between the EU and the US-based Stanford University. The project addressed the unhappy consequences of gender blind research and provided a rich database of hands-on examples and methodologies for proper and improper gendered innovations [35]. It claimed that lack of gender perspective and, more profoundly, gendered dimensions of technological development and design, were regarded as reasons behind the failure of many start-ups and why they and their products were not as foreseen "changing the world" [36].

The roller also provides an example case here. As mentioned, the roller was invented by a woman, an accomplishment in itself, in the very male dominated field of innovation and entrepreneurship. Still, in the 2010s, patents taken by women equated to under 10% of patents in many western countries [37][5]. What qualifies the roller as a gendered innovation is the interactive and user-driven innovation process as well as the social impact of this simple assistive device.

In line with the residual staging of the roller, hardly any statistics of roller users exist – and demographic information about users, such as gender, age and geography, is both limited and scattered. Figures from a 2006 Danish survey based on reports from 33 municipalities showed that the roller was by far the most frequently provided assistive walking technology. The results of the survey indicated that 70,800 rollers were provided by the municipalities, whereas other assistive technologies provided included manual wheelchairs (53,500), e-scooters (17,800) and walking sticks (9400) [38]. Ishøj municipality, a low income and multi-ethnic suburban municipality located on the outskirts of Copenhagen, provided an updated chart of walkers and e-scooters related to gender for this study. The chart shows that provisions of assistive mobility technologies and devices are increasing, with the roller being the most used device from 2011–2020. Figures also show that women make up a considerable majority of roller users as seen in the chart to the left. Whereas the provision of e-scooters is lower and more gender equal as seen in the chart to the right. However, men are more frequent users of this semi-motorised mobility device. (Blue graph: women, red graph: men, green graph: total).

A Swedish report from 2010 gave a more detailed overall picture. It revealed that users of mobility assistive technology increased according to age, with gender being another significant category, and that 70% of Swedish women and 55% of Swedish men over the age of 85 made use of a mobility assistive technology. 95% of all roller users were over 65 and more than two-thirds were over 80. Around 50% used the roller both outdoors and indoors [39].

The roller seems to meet the needs of elderly women with walking impairments in particular. This is likely to be because many elderly women with impairments do not have other alternatives to mobility inside and outside of their homes. In Denmark and in

---

[5] The female as inventor does not as such qualify as gendered innovation. Women might also invent technology which is useless.

the Nordic countries elderly women belong to a generation that is not used to motorised devices, as car drivers or car owners, which is obvious in the general transport analysis and something that tends to leave them behind in terms of affordable and convenient mobility in old age [40, 41].

Demographic and health factors can also explain why women make up a majority of users, as women typically live longer and are more likely to suffer from osteoporosis than men. Rollers are also part of the welfare states' and municipalities' budgets because they can prevent or reduce the prevalence of falls and hip operations, which are costly [42]. In general, it is interesting that the roller has been transposed from being a universal device for an ageless impaired group into an assistive technology that is mainly used by elderly people today – and mainly by women. More research is needed to find explanations and solutions that address a broader range of needs, lifestyles and preferences.

## 6   Roller Users – Scripts, Boundary-Making and Intra-actions

In the following section I will analyze a couple of cross-cutting themes located in the qualitative interviews (see appendix). I will do so within the conceptual framework of scripts, and supplement these with notions of how to understand intersections of humans and materiality from well-known feminist scholars. Both Donna Haraway and Karen Barad focus on the role of human and non-human materiality and on intra-actions as determinants of social identity [43]. Haraway talks about the body as a boundary project which comes into being when people draw boundaries. According to her, "'objects' do not pre-exist as such; instead, they materialise in social interactions. Thus the boundaries should always be considered as dynamic and complicated" (43:114).

The roller users interviewed for this study handle their relationship with the roller in various ways. In their descriptions and interpretations, it is clear that the roller becomes an agent – and acts both as foe and a friend. For some, the roller acts as a close companion in daily life, where it is characterised in human terms as a good friend and as part of an intra-active material agent for independency and mobility: "Me and my roller are best friends" (female, 90 years, interview 5). "The roller is very social – one can walk and sit down, and people become very helpful and friendly" (female, 98 years, interview 7). In contrast, others see the roller as a threat to their self-esteem and bodily integrity: "The hospital offered me a roller after my operation. But I refused – out of pride and for vanity reasons. I chose sticks instead and now I use the stick to walk to and from the car" (male, 83 years, interview 3). "The roller has its advantages, but the roller adds 10 years to your age. I use a stick and backpack for shopping instead" (female, 77 years, interview 5). "My husband is too proud to use a roller; he uses a stick instead" (female, 70 years, interview 12).

These assessments also indicate a hidden hierarchy related to assistive technologies among the elderly; the use of the stick or sticks among several, was regarded as less stigmatising and intrusive compared with the threat of becoming a roller user. Added to this were a couple of accounts of reluctant roller users, who only took up the roller when all other options were exhausted. "I used my bike for shopping, but at the end I was not able to get from the bike to the shopping basket, which I could lean on. Then I called the municipality and asked for a roller (female, 89 years, interview 2). Another respondent repeatedly said: "I have a stick, which I would rather use, but now I have overcome the shame – and now I use the roller" (female, 86 years, interview 6).

As for daily use of the roller, the interviewees also revealed varying and contrasting experiences. Some of the respondents described the roller as a foreign intruder in their homes, and as a device that should be kept outside – or to the side, out of the way. They were not willing to redesign their home environments according to recommendations from the professional advisers. "My roller is only for outside use – there is not space inside. When I move around inside my home I depend on the furniture" (female, 89 years, interview 2). "I leave the roller in the kitchen – it is dirty and the apartment is not appropriate for a roller" (female, 86 years, interview 9). Others reported a more functionalist and pragmatic approach: "When my husband got his roller, we redesigned our living room and made it more simple, so that he could move around without difficulty" (female, 75 years, interview 1). Yet, again, for several of the respondents the roller had become a dear and a daily companion – and their mobility and wellbeing was totally dependent on this device. "I use the roller for everything; it offers me safety and balance" (female, 84 years, interview 14). "I use the roller every day and I am very fond of it. I use it outdoors and indoors. I use it for sitting and for socialising… I feel like a normal housewife when I go shopping with the roller" (female, 70 years, interview 12).

As for the role of the welfare state and the prevalent mantras of ageing better, there also seem to be variations in experiences and interpretations of the social dogma and imperatives [32]. At one end, there were respondents who saw the municipality and its provisions as an ally and the roller as an opportunity for improvement and optimisation of life and wellbeing. At the other end, there were respondents who believed that the welfare state and care workers should be kept at a distance. "I tell the annual visitor not to come." (female, 86 years, interview 9). Some had bought their own roller and declined the municipality's provisions. Or they felt they were being pushed towards physical activity, rather than being allowed to be old and tired: "At the age of 95 one should be allowed to be tired" (female, 95 years, interview 8).

These examples show that the roller and their users still present a range of paradoxes – and boundary settings. The intra-acting roller user signals both stigma and outsiderness – as well as being the embodiment of inclusion, participation and normality. Such contrasting accounts echo both the general negative notion of old age and public depictions that alternate between images of the disabled and fragile and the dignified and still mobile old body.

## 7 Perspectivation: Smart is (Not) Always Digital!

Initially the aim of this paper was to fill in a gap in the study of walking as a transport modality – and to argue that a simple, non-digital assistive technology might also be disruptive and smart! These ideas matured and were challenged while I was teaching a mixed group of Chinese and Danish/international students in the discipline of social innovation at the Sino Danish Centre in Beijing in the spring of 2019. Here the students brought in ideas of both digital and gendered innovations of the roller. They claimed that rollers currently carry too much of a "one size fits all" connotation and at the same time the roller seems feminised in colour, equipment and use. There is a need for a more masculine aesthetic and features – such as pockets for a beer/or a fishing rod – in order to attract men as well. Another suggestion the students made were that shared rollers could be offered and made available in public spaces, in line with shared bikes and cars, by means of smart technologies – a convincing argument that smart rollers could also be digital.

The TiNNGO project has developed the concept of gender smart mobility to meet such needs[6] [16]. The notion of Gender smart mobility is defined according to 5 indicators, towards which to measure any (smart) transport modality and devise. Meaning that gender smart mobility should meet the following requirements of being 1. Inclusive, 2. Affordable, 3. Effective. 4. Attractive, and 5 Sustainable. The concept sustains a focus on gender and diversity sensitive assessments, which are often lost in existing approaches to smart mobility and cities. Based on the present analysis the roller can be assessed as follows:

| The Roller as gender smart mobility? | |
| --- | --- |
| Inclusive | Yes and no. While the roller provides the user with enhanced mobility, the roller is still regarded as stigmatizing and associated with old age, and lack of strength |
| Affordable | Yes, the roller is a cheap devise (under 300 euro's for a quality roller) compared to e.g. rolling chairs or e-scooters |
| Effective | Yes and no. – Many roller-users manage to get around but are restricted by uneven surfaces, stairs etc. There is a need for smart solutions for roller users in public spaces and transit (accessibility, even surfaces, roller lanes etc.) |
| Attractive | Yes and no. Smarter and more differentiated designs are needed – including digital solutions/ access to shared rollers in cities etc. |
| Sustainable | Yes. Rollers enhance mobility of walking impaired citizens, create better quality of life reduces public/health costs for hip operations e.g. |

All in all this study of the roller, which cuts across innovation, production and consumption accentuates new ideas of the smart city - This implies a more creative and sensitive approach to both gender and age, and the imperative to transcend existing silos of knowledge and practices in social welfare, city planning and transport policy. Including new ways of doing elderly and gender smart cities and spaces.

**Acknowledgements.** This paper forms part of the TiNNGO project, which has received funding from the European Union's Horizon 2020 research and innovation programme under grant agreement no. 824349.

---

[6] This (normative) concept of gender smart mobility as broad, inclusive and dynamic is inspired by a merge of various perspectives found in [16, 44–46].

## Appendix: The Interview Material

During February 2021, a total of 14 respondents in an age range of 70 to 98 were interviewed. They all live in their own homes – including self-owned houses, rented or self-owned apartments, and a few live in municipal elderly apartments. The sample was identified and selected according to the snowball method and with emphasis given to including city, suburban and small town residents, as well as to ensuring social diversity. In addition, three professional care/ergonomic workers from the Copenhagen area were interviewed to provide accounts of how laws were interpreted and their more general views on the roller users and their problems. The design of the interview chart and the professional interviews has been assisted by Stud. Scient. Soc. Stine Petersen. Most of the interviews took place over the phone due to Covid-19 restrictions and followed a questionnaire with open questions. Parts of the interviews were recorded, transcribed or reproduced according to careful note taking done during the interviews and then inserted into a standardised chart. A few of the respondents had access to a computer and provided written answers after a phone introduction. The Covid-19 restrictions also prevented the inclusion of broader constituencies in terms of gender and ethnicity, planned to be approached in public places streets or shopping malls. The aims of the interviews were: 1. Collect personal accounts from roller users. What motivated them to acquire the roller and who was the provider – private or public – and how were they instructed to use it? 2. Description of daily use – both inside and outside the home. 3. Experiences – good and bad – and, where possible, a brief photo elicitation where the respondents commented on a sample of pictures with roller users doing various activities. Basic information provided in chart.

| Gender | Age | Profession | Location | Civil status | Roller user |
|--------|-----|------------|----------|--------------|-------------|
| 1. Female | 75 | Public service/middle manager | Provincial town | Widowed for 7 years | Potential roller user, husband was user |
| 2. Female | 89 | Office worker/hospital aid, domestic work | Small town | Single | 1 year – walking impairment, roller on own initiative |
| 3. Male | 83 | Bank accountant | Cop. Suburbia/middle class/high income area | Divorced | Walking impairment Does not want a roller |
| 4. Female | 71 | Consultant/analyst | Cop. Suburbia/low – middle income | Married | Active, no need for a roller, but is ok with it if needed |
| 5. Female | 90 | School teacher | Provincial town | Single | 5–6 years – roller allocated after fall |

(continued)

*(continued)*

| Gender | Age | Profession | Location | Civil status | Roller user |
|---|---|---|---|---|---|
| 6. Female | 86 | School teacher | Cop. suburbia/middle-high income | Single | 2–3 years – roller allocated after fall/ hospital operation |
| 7. Female | 98 | Housewife/assistant for husband/dentist Clinique | Cop. Suburbia/middle-high income | Widow | Occasional roller user/husband full use after fall |
| 8. Female | 95 (passed away shortly after communication) | Housewife/assistant for husband/clergy | Small town | Widow | Intense roller user – high roller/ low roller 7- 8 years |
| 9. Female | 86 | Social and health assistant/elderly people | Cop. Suburbia/ low-middle income area | Single | 2–3 years - roller allocated after a fall |
| 10. Female | 77 | Flight attendant/stewardess | Cop. Suburbia/middle/high income | Divorced | Potential roller user/reluctantly uses stick currently |
| 11. Female | 89 | Social worker | Cop. Suburbia/middle-high income | Widowed | 1 year – roller user after a fall. Has 2 rollers, indoor/outdoor |
| 12. Female | 70 | Shop assistant, executive secretary, marketing assistant, receptionist | Big provincial town | Married, lives apart – in the same complex | Roller use due to Parkinson's disease – roller user for 8 years. Has 2 rollers, indoor/outdoor |
| 13. Female | 76 | Business education – marketing chief | Big provincial town | Divorced, no kids | Roller provided based on advice after back operation Uses roller together with stick + e-scooter + cars |
| 14. Female | 84 | Textile worker, shop assistant + own shop | Big provincial town | Single – 2 marriages, 3 kids, 0 grandchildren | Roller user – recommended, bought it herself Roller supplemented by a scooter |

# References

1. Nelson, T.D. (ed.): Ageism. A Bradford Book, 2nd edn. MIT Press, Cambridge (2017)
2. Uteng, T., Christensen, H.R., Levin, L. (eds.): Gendering Smart Mobilities. Routledge (2020). https://www.taylorfrancis.com/books/e/9780429466601
3. Salminen, A.L., Brandt, Å., Samuelsson, K., Töytäri, O., Malmivaara, A.: Mobility devices to promote activity and participation: a systematic review. J. Rehabil. Med. **41**, 697–706 (2009)
4. Löfqvist, C., Nygren, C., Brandt, Å., Iwarsson, S.: Very old Swedish women's experiences of mobility devices in everyday occupation: a longitudinal case study. Scand. J. Occup. Ther. **16**, 181–192 (2009)
5. Brandt, Å., Iwarsson, S., Ståhl, A.: Satisfaction with rollators among community-living users: a follow-up study. Disabil. Rehabil. **25**(1), 19–34 (2003)
6. Peine, A., Faulkner, A., Jæger, B., Mors, E.: Science, technology and the 'grand challenge' of ageing – understanding the socio-material constitution of later life science, technology and the grand challenge of ageing. Thematic vol. Technol. Forecast. Soc. Chang. **93**, 1–9 (2015)
7. Woodcock, A., Moody, L., McDonagh, D., Jain, A., Jain, L.C. (eds.): Design of Assistive Technology for Ageing Populations. Springer Nature, Switzerland AG (2020)
8. Loe, M.: Doing it my way: old women, technology and wellbeing. Sociol. Health Illn. **32**(2), 319–334 (2010). https://doi.org/10.1111/j.1467-9566.2009.01220.x
9. Bijker, W.E.: Of Bicycles, Baekelite and Bulbs: Towards a theory of Socio Technical Change. MIT Press, Cambridge (1995)
10. Latour. B.: Re-assembling the Social – An Introduction to Actor Network Theory. Oxford University Press (2005)
11. MacKenzie, D., Wajcman, J. (eds.): The Social Shaping of Technology. Open University Press, Buckingham (1999)
12. Akrich, M.: The de-scription of technological objects. In: Bijker, W.E., Law, J. (eds.) Shaping Technology/Building Society-Studies. Sociotechnical Change, pp. 205–224. The MIT Press. Cambridge (1992)
13. Oudshoorn, N.: Telecare, Technologies and the Transformation of Healthcare. Palgrave Macmillan, Basingstoke (2011)
14. Manderscheid, K.: From the auto-mobile to the driven subject? Transfers **8**(1), 24–43 (2018). https://doi.org/10.3167/TRANS.2018.080104
15. Christensen, H.R., Breengaard, M.H.: TInnGO Road map. Conceptual coordination: translating theory into practice. Gender Smart Mobility (2019)
16. Christensen, H.R., Breengaard, M.H: Gender Smart mobility. Concepts, methods, practices. In: Routledge Transport and Mobility Series (2021, forthcoming)
17. Levsen.N.: Lead Markets in Age-Based Innovations. Demographic Change and Internationally Successful Innovations. Springer, Heidelberg (2015). https://doi.org/10.1007/978-3-658-08815-6
18. New York Times Opinion. The Truth about Harvey Weinsteins walker, 30 January 2020 (2020). Depiction at https://twitter.com/emmavigeland/status/1237771736600514567
19. Lucy Lions: Experiences of Ageing. Exhibition at Medical Museion, 27 April–21 October 2012. https://www.museion.ku.dk/en/whats-on/exhibitions/experiences-of-ageing/
20. https://twitter.com/RoyalArjan/status/1004352099768061952/photo/2
21. https://www.museion.ku.dk/aktuelt/udstillinger/oplevelser-af-aldring/rollatorerr
22. https://www.dr.dk/nyheder/indland/video-se-dronning-ingrids-rollator-paa-amalienborg/
23. https://ordet.net/onkel-soenderjysk-hverdagsmelankoli/https://www.imdb.com/title/tt7523 172/mediaviewer/rm2225180417/?context=default
24. https://www.dr.dk/nyheder/kultur/film/en-aabenbaring-usaedvanlig-dansk-film-vaelter-anm elder-helt-omkuld

25. https://www.screendaily.com/reviews/uncle-tokyo-review/5144470.article
26. https://www.dfi.dk/en/viden-om-film/filmdatabasen/film/onkel
27. Christensen, C.M.: The Innovator's Dilemma. Harvard Business School Press, Cambridge, Massachusetts (1997)
28. Willis, R., Webb, M., Wilddon, J.: The Disrupters: Lessons for Low Carbon Innovation form the New Wave of Environmental Pioneers. NESTA, London (2007)
29. Tyfield, T., Jin, J., Rooker, T.: Game changing China: lessons from China about Disruptive Low Carbon Innovation. Research Report. NESTA, UK (2010)
30. Wifalk, A.: Svenskt upfinnara museum (2016). http://svensktuppfinnaremuseum.se/aina-wif alk/
31. Danish Act on Social Service. http://english.sm.dk/media/14900/consolidation-act-on-social-services.pdf
32. Neven, L.: By any means? Questioning the link between gerontechnological innovation and older people's wish to live at home. Technolo. Forecast. Soc. Chang. **93**, 32–43 (2015)
33. Hansen. E.M.: Mobilitetshjælpemidler til voksne. Hjælpemidler, der virker. Aktuel forskningsbaseret viden til udvikling og planlægning af den kommunale indsats. Socialstyrelsen, Odense (2014)
34. Rollator Walker Market 2020. https://www.wboc.com/story/42954409/rollator-walker-market-2020-report-update-on-top-countries-data-industry-trends-growth-size-segmentation-fut ure-demands-latest-innovation-sales
35. Schiebinger, L., Klinge, I.: Gendered Innovations: How Gender Analysis Contributes to Research. Publications Office of the European Union, Luxembourg (2013)
36. Schiebinger, L. (ed.): Gendered Innovations in Science and Engineering Stanford. Stanford University Press, Redwood (2008)
37. Poutanen, S., Kovalainen, A.: Gender and Innovation in the New Economy. Women, Identity and Creative Work. Palgrave Macmillan (2017)
38. Brandt, Å.: Odense rapporten: Borgeres mobilitet og deltagelse efter tildeling af rollator: Samarbejdsprojekt mellem Odense kommune og hjælpemiddelinstituttet
39. Århus (2008)
40. Hjälpmedelsinstitutet: Äldrestatistik 2010. Äldres hälsa funktionsnedsättning, boende och hjälpmedel. Vällingby: Hjälpmedelsinstitutet (2010)
41. Transgen, Gender Mainstreaming European Transport Research and Policies Building the Knowledge Base and Mapping Good Practices. (EU report by Hilda Christensen, H.R, Oldrup, H.H., Poulsen, H.) Koordinationen for Kønsforskning, Københavns Universitet (2007). https://cordis.europa.eu/docs/results/36/36774/123865391-6_en.pdf
42. Danish National Travel Survey. https://www.cta.man.dtu.dk/english/national-travel-survey
43. Johnsson, L.: The importance of the four-wheeled walker for elderly women living in their home environment – a three year study. Centre for Providing Technical Aids for Elderly and Disabled People. Vällingby Sweden (2007). English edition
44. Suopajarvi, T.: Past experiences, current practices and future design. Technol. Forecast. Soc. Chang. **93**, 112–123 (2015). https://ubicomp.oulu.fi/files/tfsc15.pdf
45. Marsden, G., Reardon, L.: Questions of governance: rethinking the study of transportation policy. Transp. Res. Part A Policy Pract. **101**, 238–251 (2017)
46. Lyons, G.: Getting smart about urban mobility – aligning the paradigms of smart and sustainable. Transp. Res. Part A Policy Pract. **115**, 4–14 (2018)
47. Gendered innovation project. https://genderedinnovations.stanford.edu/

# Discussion of Intelligent Electric Wheelchairs for Caregivers and Care Recipients

Satoshi Hashizume, Ippei Suzuki(✉), Kazuki Takazawa, and Yoichi Ochiai

University of Tsukuba, Tsukuba, Japan
wizard@slis.tsukuba.ac.jp, 1heisuzuki@digitalnature.slis.tsukuba.ac.jp

**Abstract.** In order to reduce the burden on caregivers, we developed an intelligent electric wheelchair. We held workshops with caregivers, asked them regarding the problems in caregiving, and developed problem-solving methods. In the workshop, caregivers' physical fitness and psychology of the older adults were found to be problems and a solution was proposed. We implemented a cooperative operation function for multiple electric wheelchairs based on the workshop and demonstrated it at a nursing home. By listening to older adults, we obtained feedback on the automatic driving electric wheelchair. From the results of this study, we discovered the issues and solutions to be applied to the intelligent electric wheelchair.

**Keywords:** Aging society · Accessibility · Wheelchair · Self-driving

## 1 Introduction

Recently, research and development work has examined practical applications of automotive driving technology. Automotive driving requires recognition of the environment, judgment of the situation, and operations planning by a machine instead of a human driver. Therefore, advanced information processing technologies such as artificial intelligence (AI) and sensor fusion are required. Technology in automatic driving can also be applied to machines other than cars, such as factory machines and personal mobility. However, with current technology, it is still difficult to fully automate the drive function for personal mobility machines; hence, they have not been put to practical use on public roads.

Wheelchairs are important modes of mobility for older adults and for physically handicapped persons with limited movement. Electric wheelchairs began appearing on the market in the 1950s. Since then, people who have difficulty driving non-electric wheelchairs, due to lack of physical strength can easily drive electric wheelchairs. In nursing care facilities, it is necessary for the caregiver to remain beside the electric wheelchair at all times. The global population aged 60 years or over numbered 962 million in 2017. The number of older persons is expected to double again by 2050, when it is projected to reach nearly 2.1

H. Krömker (Ed.): HCII 2021, LNCS 12791, pp. 500–516, 2021.
https://doi.org/10.1007/978-3-030-78358-7_35

**Table 1.** Overview of our experiments

| | Number of participants | Target | Purpose |
|---|---|---|---|
| St1-Problem finding | 6 | Caregivers | To find the problems and solutions in care |
| St1-Problem solving | 2 | Caregivers | To evaluate methods found during the problem finding phase |
| St2 | 2 | Care recipients | Verbal survey of older adult people who experience the intelligent wheelchair |

billion [8]. In these aging societies, there are increasingly urgent demands for wheelchairs and caregivers. In addition, because of a shortage of caregivers, the burden on them is also increasing. How to reduce the burden on caregivers is a challenging problem.

To reduce this burden, a fully-automated electric wheelchair is ideal for caregivers who use wheelchairs in their work. With a fully-automatic wheelchair, a caregiver does not always have to accompany the care recipient. Furthermore, the care recipient does not have to request assistance; thus, the automatic wheelchair reduces their psychological burden. Currently, however, automatic driving techniques are not widely adopted in nursing care fields.

Therefore, we developed an intelligent electric wheelchair called the Telewheelchair. In previous research, we conducted an operability experiment evaluation, where the usefulness of the remote control was demonstrated [13]. Because the Telewheelchair was developed with general-purpose equipment and software, it is possible to add additional functions. In the experiment, We also discussed what new functions should be added to solve problems, specifically in the nursing care environment.

To identify the functions to be installed in an intelligent electric wheelchair, we explored the types of problems encountered in a practical nursing care environment. We conducted two-part studies (Table 1) at a nursing home using the intelligent wheelchair. The first study aimed to find the problems and their solutions in the nursing care facility. Based on the results of that study, we implemented the cooperative operation function with the intelligent wheelchair and demonstrated it at the nursing home. The second study was a verbal survey of older adults who experienced the intelligent wheelchair.

Contributions of this research are as follows.

- We held a workshop at a nursing home, and from the feedback of the experiments, we found that there was a problem of the workload of caregivers in nursing care.
- We proposed a cooperative operation. Based on the implementation and feedback from the experiments, we concluded it was a useful method to solve the problems in nursing care.
- We demonstrated the remote control for the older adults and found that they have positive opinions of the intelligent wheelchair.

## 2   Related Work

### 2.1   Technical Approach for an Aging Society

In an aging society, psychological and physical burdens are placed on older adults and caregivers. Many technological implementations are underway to reduce the burden on these groups. The portal monitor [4] is a monitoring system that addresses the privacy of older adults. Safety and consideration of the psychological problems of older adults have been studied. Ichinotani et al. [17] developed a deformable wheelchair and a nursing care bed. The systems are used to solve the problem of older adults getting hurt when transferring from a bed to a wheelchair. Huang et al. [16] conducted an approach to solve the loneliness and anxiety of the older adult through talking with CG animation agents. This method showed that active listening could be done at the same level as human beings. Conte et al. [6] proposed a system that provided tactile assistance for a tablet according to the learning preferences of older adults.

In addition, research has been conducted to investigate the relationship between technology and the aging society. Senior Care for Aging in Place [4] investigated how older adults live with both self-care and collateral care at the same time. They found that many older adults need a monitoring system that does not violate their privacy. Technological Caregiving [23] conducted an interview survey on the caregiver side and the nursing care side about online support for older adults. They showed that online nursing care activities are generally stressful to caregivers. Buccoliero et al. [3] aimed to clarify the factors that affect the adoption of technology concerning the health of older adults. Kostoska et al. [19] aimed to replace manual data collection with automatic data collection using a distributed system based on mobile phones and biosensors. Researchers, acting as proxies for informal caregivers, [7] examined the specifications of PictureFrame, a new social technology to support the provision of medical examinations for older adults residing at home. Virtual Carer [22], a web service model, generated personalized recommendations for informal caregivers according to the need of older adults. Arif et al. [1] developed a mobile telepresence robot, "Telemedical Assistant", that minimized medical mistakes. Chen et al. [5] conducted an interview study on caregivers. Caregivers maintained the balance of their personal lives with work, family, and their caregiver roles with the concept of giving-impact and visibility-invisibility. Holbo et al. [14] examined factors to allow people with dementia to walk safely from workshops and interview.

Many technologies for the aging society were implemented as described above, and the relationship between older adults, caregivers, and technical approaches has been investigated from various perspectives. However, most of these studies had been limited to finding problems in care from workshops and did not implement functions. We implemented one of the solutions from the proposed problem and held a workshop again. We can solve the problem quickly by turning the loop of problem definition, implementation, and workshop (Fig. 1). Moreover, there is a little practical discussion of how technologies like automatic wheelchairs are

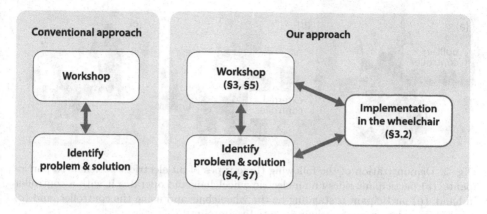

**Fig. 1.** Position of our study. Our approach used turning loop of problem definition, implementation, and workshop.

perceived by older adult participants in nursing homes. We tried the approach of simulating the automatic driving concept in practical experiments to discover methods to reduce the burden on older adult people and caregivers.

## 2.2   Intelligent Electric Wheelchair

Since the 1990s, many researchers have been studying the automatic operation of electric wheelchairs. Gundersen et al. [11] presented an electric wheelchair equipped with a remote-control system via a head-mounted display (HMD) and an obstacle detection system with an ultrasonic sensor. Pires et al. [2] explored the usability of wheelchairs by conducting experiments on their operation using voice, joystick, and on obstacle detection and collision detection. The NavChair [20] enhanced navigational functionality by enabling guided door passage and wall following in addition to obstacle avoidance. Mazo [21] examined automatic driving by using various methods for environmental recognition and user motion detection. Diao et al. [9] developed the Intelligent Wheelchair Bed, which could detect and avoid obstacles using eleven ultrasonic sensors. Hua et al. [15] developed a wheelchair that recognized the environment by analyzing laser range finder and camera data with a neural network. Faria et al. [10] developed a new wheelchair control method for Cerebral Palsy. Telewheelchair [13], an electric wheelchair equipped with a remote control function, a computer-operation support function, and controlled by an HMD, was developed to reduce the burden on the caregivers.

Some research focuses on designing a general-purpose electric wheelchair as well as automatic operation of specific motions. The DECoReS system [12] can be driven using orders such as "go straight, fast" and "make a wide curve to the right", turning corners according to the user's preference. Kobayashi et al. [18] uses laser range sensors to move the electric wheelchair while keeping a certain distance from the accompanying caregiver. Passengers can communicate with their companions.

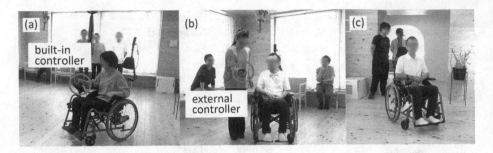

**Fig. 2.** Demonstration of the following three types of an electric wheelchair to participants: (a) participant rides on an electric wheelchair and operates it with a controller in hand, (b) participant is standing by the wheelchair and using the controller, and (c) participant rides on a wheelchair operated remotely.

## 3   Study 1: Problem Finding and Solving

The purpose of this research was to discuss how the intelligent electric wheelchair can be utilized in the nursing care field. First, there was a need to know what problems are present in the nursing care field. Therefore, to identify these problems, we held a workshop at a nursing home. Next, we considered methods that could solve the problems. Lastly, we conducted experiments with an intelligent electric wheelchair. We proposed to discuss the intelligent electric wheelchair more practically by conducting a series of steps from problem finding to practical implementation.

### 3.1   Problems Finding Phase

**Design:** The purpose of this research was to reduce the burden on caregivers using an intelligent wheelchair. We needed to find out what kind of problems there are in the work of nursing care facilities. In the problem finding phase, we aimed to find the problems with wheelchair care at nursing care facilities. Participants of the workshop were staff working at nursing care facilities. Caregivers have a lot of experience and knowledge about nursing care, so it was possible to describe the correct problems in nursing care sites. At the same time, subjects considered problem solving methods based on their experience at nursing care facilities. In problem solving phase, we implemented function into a wheelchair based on problem solving method.

Caregivers participating in the workshop had little knowledge about automatic driving and artificial intelligence. Therefore, we also conducted a technical lecture on these topics during the workshop. Technical training included explanations on artificial intelligence and wheelchair operation, as well as an intelligent wheelchair demonstration. In the intelligent wheelchair demonstration, the caregivers participated in three types of experiments: (i) An occupant manipulated a joystick mounted on an electric wheelchair (Fig. 2(a)), (ii) A caregiver stood at a distance and controlled the wheelchair using the remote controller (Fig. 2(b)), and (iii) An operator simulated automatic driving (Fig. 2(c)). Experiments (i) and (ii) were based on the implementation of previous research [13]. Experiment (iii) was a mode in which a wheelchair automatically moved when the pilot is not nearby. It is noted that instead of implementing a self-driving electric wheelchair, we actually conducted an experiment that simulated the self-driving environment. An operator controlled the wheelchair from a remote location. However, the participants were not told that the control operation was manually controlled from a remote location; therefore, participants believed that the electric wheelchair was being driven automatically. The aim of the experiment (iii) was to encourage participants to experience operations like automatic driving and to have a deeper understanding of automatic driving. The target comprehension level of participants after receiving technical training was an understanding of the meaning and purpose of artificial intelligence and the outline of the intelligent wheelchair. Participants were asked to consider solutions to problems before and after technical training. Following their technical training, we asked them to respond using the artificial intelligence information they learned in the course and their knowledge of automatic driving.

**Procedure:** Each participant was briefly informed of the purpose of the study and was informed that they could stop the study and take a break at any time. Further, they were provided with a consent form to sign and a demographics questionnaire to complete. At the beginning and end of the workshop, the participants were asked to answer their expectation level for AI and robot care using a five-level Likert scale.

Initially, the participants were asked to write out their career experiences and problems they have encountered. The response time was 15 min. Next, the participants were asked to write a method capable of solving a nursing care problem. We asked them to write three problems that they wanted to solve, and their proposed solutions. The problems they wanted to solve were expected to be based on the problem that they wrote. The response time was 30 min. Next, we presented a technical training demonstration using the electric wheelchair. The duration of the technical course was 15 min and the demonstration was 45 min. After the three types of demonstrations were completed, participants were asked to provide their impressions of each demonstration. Lastly, we asked the participants to write a method capable of solving a nursing care problem. The second problem solving method was to be based on the material learned in the technical course. It included up to three problems that participants wanted to solve. The response time was 30 min. All response times were sufficient to answer the questions.

**Participants:** Six participants (three females and three males) with ages between 23 and 63 years ($M = 39$, $SD = 14.2$) participated in the experiment (Table 2). All participants were caregivers. All participants were employed in nursing care facilities that conducted workshops. The average years of employment were 11.5 years ($SD = 7.2$). Four participants did not know about

**Table 2.** Summary of workshop participants on an intelligent wheelchair for caregivers. *"Chief" is responsible for supervising caregivers in nursing homes

|        | Sex | Age | Position         | Length of service (year) |
|--------|-----|-----|------------------|--------------------------|
| $P_11$ | F   | 29  | Caregiver        | 7                        |
| $P_12$ | F   | 23  | Former caregiver | 3                        |
| $P_13$ | M   | 33  | Chief*           | 4                        |
| $P_14$ | M   | 42  | Chief            | 20                       |
| $P_15$ | F   | 63  | Chief            | 20                       |
| $P_16$ | M   | 44  | Chief            | 12                       |

Telewheelchair in advance and two participants knew a little about it. All participants spoke Japanese.

## 3.2   Problems Solving Phase

**Design:** In this phase, the aim was to implement and evaluate methods found during the problem finding phase. Caregivers who participated in the problem finding phase raised a common problem: having fewer hours to care for older adults. For example, the duties of a caregiver consisted of a large number of tasks, such as working alone to assist three older adults with their diets and assisting with the bathing of more than 30 people a day. With this workload, the caregiver may not have time to communicate well with the older adults. Hence, to increase the time spent on meals, nursing care, and bathing assistance, we considered shortening the traveling time of the wheelchair.

To shorten the travel time of the wheelchairs, we implemented an automatic tracking travel system in the intelligent wheelchair. The caregiver could simply operate the first wheelchair, and a second rear wheelchair automatically followed; hence, all the wheelchairs could be operated simultaneously. In a nursing home care facility, many wheelchairs have to be moved at the same time when moving the care recipients from a room to a dining room or a bathroom. At such times, controlling a plurality of wheelchairs simultaneously could shorten travel time.

**Implementation:** We used two electric wheelchairs. We attached the tablet that displayed the artificial reality (AR) marker on the back of the base unit wheelchair. The slave unit wheelchair recognized the AR marker with its attached web camera. The slave unit followed the base unit wheelchair based on the position of the AR marker. The base unit could be remotely operated using HMD or by a controller standing beside a wheelchair.

We acquired three-dimensional (3D) position coordinates and roll, pitch, and yaw with AR markers using a processor running on a Windows laptop. The speed and direction of rotation of the slave unit wheelchair were calculated using the position of the AR marker. The speed of the slave unit was controlled according

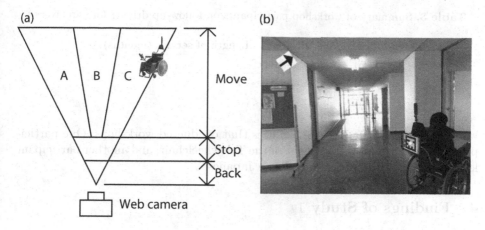

**Fig. 3.** System of automatic tracking travel: (a) rear wheelchair utilizes the AR marker to acquire the position of the front wheelchair. By controlling the rear wheelchair according to the position of the AR marker, we succeeded in following using only the camera, and (b) image of the camera attached to the rear wheelchair.

to the distance between the AR marker and the slave wheelchair. As shown in Fig. 3(a), when approaching the base unit, the slave wheelchair stopped or backed up. The rotation direction of the slave unit was calculated according to the horizontal position of the base unit. The image taken with the camera is divided into three areas as shown in Fig. 3(b). The rotation direction was determined by the area where the base unit wheelchair was located. The slave unit wheelchair turned right if the base unit wheelchair was in the right area, went straight if in the middle area, and turned left if in the left area. The rotation speed for turning right and left increased as the base unit wheelchair moved away from the center of the screen. In this system, we divided the image in the ratio 7:6:7, as left turn:straight ahead:left turn. When the slave unit rotated according to the rotation of the base unit, the slave unit started to turn as soon as the base unit started to turn. By using the proposed divide-by-three method, it was not necessary to record the locus of the base unit, and the slave unit could pass through the path that the base unit passed.

**Procedure:** We informed the participant about the study and obtained their consent as described above. First, participants were asked to introduce themselves. Next, the experimenter introduced the functions of the intelligent wheelchair to the participants and explained the demonstration to be conducted later. After the explanation was completed, we demonstrated the automatic tracking travel system. The demonstration time was 45 min. Finally, we asked the participants to provide their impressions of the demonstration.

**Participant:** Two participants (two males), average age 22 years, participated in the experiment (Table 3). Both participants were caregivers. Both participants

**Table 3.** Summary of workshop participants on follow-up driving for caregivers

|        | Sex | Age | Position  | Length of service (months) |
|--------|-----|-----|-----------|----------------------------|
| $P_2 1$ | M   | 22  | Caregiver | 3                          |
| $P_2 2$ | M   | 21  | Caregiver | 21                         |

were employed in nursing care facilities that conducted workshops. One participant did not have prior knowledge of the Telewheelchair, and another participant knew a little. Both participants spoke Japanese.

## 4    Findings of Study 1

### 4.1    Problem Finding Phase

**Problems of Care:** In this section, we present the problems at the nursing care facility that caregivers identified during the workshop. First, the participants described the privacy of the older adults. $P_1 2$ said, *"We need to give care in the toilet, but from the privacy point of view it is difficult to stay in the toilet."* $P_1 4$ answered that taking care of meals is an example, *"Consideration must be given so that the people do not feel like they are being watched while I am observing the situation."*

Next, participants described the time available for each older adult. $P_1 2$ and $P_1 5$ responded about the amount of work done by one caregiver, *"It is necessary for one person to change more than five diapers."* ($P_1 2$), *"I do three person meals assistance at the same time."* ($P_1 2$), and *"There was no time to communicate with the older adults because only two people assisted the bathing of 30 people a day."* ($P_1 5$). $P_1 4$ said about various assistance, *"the time spent per capita is short because doing a lot of work within the day seems like non-stop work."*

$P_1 2$ and $P_1 3$ felt that caregivers were always considering how to ensure older adults live safely. $P_1 2$ felt, *"Caregivers talk to the older adults sitting on a wheelchair from behind them"*. $P_1 3$ said, *"Caregivers are always working hard because they care about multiple older adult people; but, it is necessary to have the older adults with the caregiver feel safe, so they should always respond slowly and clearly, with a smile."* $P_1 2$ answered, *"Caregivers should always accompany the older adults for their safety."*

**Problem Solving Methods:** We first describe the problem solving methods given by the participants before the technical learning seminar. $P_1 1$, $P_1 2$ and $P_1 6$ cited the problem of sitting when the older adults are in a wheelchair. Because $P_1 1$ wants to have the older adults sit comfortably in the wheelchair with little effort, he proposed a wheelchair whose seat surface could be adjusted. $P_1 6$ replied, *"We will make the seat height changeable and make the cushioning material adjustable."* $P_1 4$ and $P_1 6$ proposed a solution for transferring. $P_1 4$ replied, *"In order to enable transfer without relying on the caregiver's skills, we will make* *wheelchairs with sizes and functions suited to the older adults.*" $P_16$ said, "*I want the wheelchair's footrest and foot support to be able to close together automatically so as to solve the problem of moving the wheelchair from the bed.*" $P_11$ and $P_15$ mentioned the time and labor of movement. $P_11$ said that the trouble with moving the wheelchair can be solved if the wheelchair can be operated close to the older adults automatically. $P_15$ replied, "*I hope all the assistance can be operated by a voice control system.*" $P_13$ answered, "*I want to walk in parallel rather than pushing the wheelchair from the back; I want the wheelchair to be operated remotely.*"

After the technical learning seminar, the problem-solving methods were discussed. $P_11$ and $P_12$ said that the older adults should be able to transfer alone and go to their destination. They said that it would be significantly effective to have automatic driving, voice recognition, wheelchair tracking, and calling for help functions on the wheelchair itself. $P_15$ and $P_12$ added a comment about dementia. $P_15$ said, "*Older adults with dementia will forget their location and wheelchair operation procedures. For this kind of person to maneuver the wheelchair, it is necessary to control it by speech recognition. Furthermore, functions of remembering the route and driving automatically are desirable.*"

**Intelligent Wheelchair Demonstration:** We first performed the demonstration with the control operation of using a joystick mounted on an electric wheelchair. $P_12$ felt, "*I could drive with my own will.*" $P_15$ answered, "*I can operate the driving just with my fingers, so people without paralysis in hand may also control it themselves.*" $P_12$, $P_16$ replied, "*It is peace of mind.*"

Next, we performed the demonstration with the external wheelchair controller operated by an experimenter standing next to the participants. $P_12$ and $P_15$ answered, "*When seated in the wheelchair, I was relieved to see the face of the other person and to talk with him.*" $P_13$ said, "*The operation became easy when I became familiar with the controller.*" $P_15$ answered, "*It is difficult to get used to the operations when I am using the controller.*" $P_23$ and $P_24$ answered, "*The operation of the controller is difficult so sometimes I cannot control the wheelchair well.*"

Finally, the participants experienced a demonstration simulating automatic driving. $P_12$ did not feel like a human. $P_13$ replied, "*It is more comfortable than being operated using the controller.*" $P_15$ answered, "*It will be fun to move in this way because I can do other things while riding.*" On the other hand, there were opinions such as "*I feel an anxiety that there is no trust relationship with someone.*" ($P_14$)

## 4.2   Problem Solving Phase

Participants experienced a wheelchair equipped with cooperative operation functions. $P_21$ felt that "*It could be effective if used in nursing care facilities. In the cafeteria, I felt that this way could lead all of us to the dining room at almost the same time, without deciding the priority among the older adults. It may also*

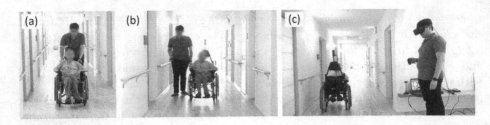

Fig. 4. We carried out the following three pattern experiments with the participants on a wheelchair: (a) push the wheelchair from behind, (b) operate with the controller while walking or standing by the wheelchair, and (c) operate away from the wheelchair using an HMD.

*change the living situation of people who cannot push wheelchairs by themselves. Those who live in nursing homes are often in their rooms, but I think that if they have this wheelchair they will be able to go shopping and go out at any time." $P_2$2 mentioned that "It is natural to go straight, but I hope there will be a function that prevents it from hitting obstacles on the side or getting stuck even if it hits something. For those with a paralyzed body, there is a possibility of falling, in that case, I would like the wheelchair to stop urgently."*

## 5    Study 2: Interview with Older Adults

### 5.1    Goal

The purpose of the experiment in this section was to investigate how older adult people accept intelligent electric wheelchairs. Artificial intelligence and automatic driving are relatively familiar to young people. However, they are not familiar with older adults; hence, it was unknown how the older adults feel about the new wheelchair. By experiencing the new electric wheelchair we developed, we can learn how older adults feel about technologies such as automatic driving and artificial intelligence. In addition, because previous experiments [13] were conducted on the same wheelchair from the viewpoint of the pilot, we can compare those results with the current experiment from the passenger's point of view.

### 5.2    Design

In the previous study [13], an experiment on the operability of the electric wheelchair was conducted. A remotely controllable electric wheelchair was developed. Four operation methods and experiments on operability were conducted. We conducted experiments using three manipulation methods out of the four methods conducted in the previous research [13]: (i) Normal mode. This mode is a general operation method operated with the handle of the wheelchair (Fig. 4(a)); (ii) Stand by mode. Participants operate using the controller while standing next

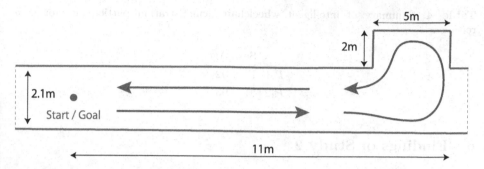

**Fig. 5.** The course we used for the experiment at the nursing home. In the experiment, we traveled back and forth 11 m in a corridor with a width of 2.1 m. The turning point is in the hall and the width of the corridor is widened.

to the wheelchair (Fig. 4(b)); and (iii) HMD mode. Participants wear an HMD and operate the external controller. Participants can operate while freely watching the surroundings (Fig. 4(c)). The older adults rode in the wheelchairs while the experimenter controlled it. The wheelchair traveled straight through at corridor inside the nursing home for 11 m from the start point, performed a U-turn, and returned straight 11 m to the goal point (Fig. 5). The width of the corridor was 2.1 m.

## 5.3  Procedure

We informed the participant about the study and obtained their consent as described above. Participants experienced wheelchair operation in the order of Normal mode, Standby mode, and HMD mode. After experiencing each maneuvering mode, the participants were asked to answer the question about whether the ride was good or fearful, or whether they would want to use the wheelchair again. The answer to each question is based on a five-level Likert scale.

## 5.4  Participants

Two participants (one female, one male), with an average age of 92.5 years, participated in the experiment (Table 4). Both of the participants lived in the nursing facility where the workshop was conducted. Participants required wheelchairs and could maneuver wheelchairs on their own if they were indoors. Participants could conduct commonplace conversations and write without problems. The wheelchair usually used is Next Core Puchi[1]. All participants spoke Japanese.

---

[1] http://www.matsunaga-w.co.jp/search/detail_64.html, (last accessed February 12, 2021. In Japanese).

**Table 4.** Summary of intelligent wheelchair demonstration participants for care receivers

|        | Sex | Age |
|--------|-----|-----|
| $P_3 1$ | F   | 92  |
| $P_3 2$ | M   | 93  |

## 6    Findings of Study 2

First, the participants experienced the wheelchair that the experimenter pushed from behind. $P_3 1$ answered, "*It's the easiest to get on the wheelchair I have ever had.*" $P_3 2$ did not feel any change. Next, the participants got on a wheelchair operated by the external controller. $P_3 1$ said, "*It was exhilarating, and I felt safe because there was the person next to me.*" $P_3 2$ answered, "*The wheelchair was slightly shaky. There was no fear, but I think it is not much relative to the presence of a person next to me.*" Finally, the older adults got on a remotely controlled wheelchair. $P_3 1$ said, "*I think I'm a little afraid when there is no one next to me now, but if I get used to it I think that it is okay. I want to go outside but I was still too scared to do that.*" $P_3 2$ said, "*I was afraid of what will happen because the wheelchair is not under my control. But as the performance of the wheelchair rises, I think that fear may disappear.*" $P_3 1$ answered that the method of using the controller was the best among the three riding methods.

## 7    Discussion

### 7.1    Busyness of Caregivers

In this section, we discuss the problems found during the practical workshop to find out how to make use of the intelligent electric wheelchair. In the opinion of the caregivers, the time to devote to nursing care for each older adults is less because the actual nursing care work includes many aspects such as excretion aid, meal assistance, and bath assistance. Furthermore, in nursing care facilities, the number of older adults who need care from each caregiver is large because of a caregiver shortage. As $P_1 2$ and $P_1 5$ stated, caregivers have shorter contact times with each older adults. Also mentioned from $P_1 5$, it is difficult for a caregiver to communicate sufficiently with older adults. Moreover, if communication becomes impossible and it becomes like routine work, it may lead to a psychological burden on the older adult. In order to increase the time to be in contact with the older adults, it is necessary to reduce the time taken for nursing care.

Thus, we provided possible solutions for using an electric wheelchair to reduce person movement time. In nursing care facilities, there are many opportunities to move multiple wheelchairs at the same time from, for example, the living room to the dining room and the bathroom. If the caregiver can shorten the time it takes to move the wheelchairs, he or she can increase the time for other nursing care work. In order to shorten the time it takes to move, we installed a cooperative-operation function on an electric wheelchair. The function enables the system to move multiple wheelchairs at the same time, so the time it takes the caregiver to make the moves decreases. We conducted a demonstration and interview with caregivers using an electric wheelchair equipped with the cooperative operation function. As a result of the interview, the participants recognized the usefulness of the cooperative-operation function. However, the caregivers believed it to be difficult for a single caregiver to monitor multiple wheelchairs simultaneously. Thus, a future improvement will be a monitoring system to prevent a passenger from falling down and accident.

## 7.2    Operation Method

The participants of the workshop provided many opinions on the operation method. As $P_1 5$ and $P_1 2$ described, there are older adults with dementia living in nursing homes. Such older adults cannot remember how to operate devices, so it is difficult to use an electric wheelchair. In order to enable such older adults to use the electric wheelchair, it is necessary to be able to control the electric wheelchair without remembering the operation method themselves. A voice manipulation function was thought to be an effective solution because even older adults with dementia can control it through voice commands. In addition, older adults with dementia cannot remember the location of their room, so it is difficult for them to act alone. An electric wheelchair that could automatically advance to the destination, by voice control and a navigation system, can reach the destination even for an older adult who has forgotten the location of the room.

From the workshop, changes to the operation method were suggested so that the older adults can be relieved of their fears. As mentioned from $P_1 2$ and $P_1 3$, it is important to make it possible for older adult people to feel secure. Also mentioned from $P_1 2$, when a caregiver is assisting a wheelchair and pushing it from the back, the caregiver often talks to the older adult when behind their back. From behind, the older adults cannot see the caregiver's facial expression, which makes it difficult to communicate. Therefore, this is a possibility of a situation that makes the older adults become uneasy. Hence, we implemented a function that permits the caregiver to stand next to the wheelchair and steer it. Standing next to the wheelchair makes it easier to communicate with older adults. We implemented this control function using an external controller and demonstrated it to the older adults. After this demonstration, the older adults expressed that they felt safe because they could see the face of the operator.

## 7.3    Psychology of the Older Adults

In this part of the study, we conducted experiments with the older adults in the nursing home using the intellectualized electric wheelchair we developed previously. Overall, the older adults expressed a positive opinion. When the experimenter assisted by standing next to the wheelchair, there was a sense of safety, due to the presence of a person. In the case of remote control, the older adults felt fear. However, $P_3$1 thought that the fear may disappear if one became accustomed to the remote operation, and $P_3$2 said that the fear would go away if operational safety is higher. Therefore, we inferred that older adults might be more accepting of the remote control if safety could be ensured. It seemed that older adults positively understand the evolution of nursing care devices by technologies such as remote control. We need to experiment with the electric wheelchair equipped with automatic driving and artificial intelligence, to see if the older adults will accept it. However, we must always be conscious of the necessity of enhancing the safety of intelligent electric wheelchairs.

### 7.4   Limitations of Our Study

In this section, we discuss the limitations of our experiments. First, we only held a one-day workshop and we have not conducted long-term experiments. Whether the intelligent electric wheelchair proposed by us can be used for nursing care work cannot be determined unless caregivers use it for a longer term. We have to conduct more experiments at a nursing facility for a few months.

The participants did not have a wide range of symptoms. The older adults who participated in the experiments were able to talk and able to maneuver wheelchairs on their own. Furthermore, the number of subjects in some experiments was as few as two. There were some older adults living in the facility who had severe dementia and others who could not get up by themselves. For older adults with such severe disabilities, we need to experiment to see if intelligent electric wheelchairs are useful. For that reason, we need to design experimental methods for those who cannot talk well.

## 8   Conclusion

We conducted a workshop for caregivers and tried to identify some problems of nursing care and some solutions. There were some major problems described by participants, however, in the experiments, we focused on the problem of a caregiver's workload. In order to solve this problem, we implemented a cooperative-operation function to the wheelchair and demonstrated it to caregivers. The usefulness of our proposed wheelchair was shown. In the interviews with older adults, we found that the older adults positively accepted the intelligent electric wheelchair.

**Acknowledgements.** This work was supported by JST CREST Grant Number JPMJCR1781, Japan. We would like to thank SILVER WOOD-Corporation. for its cooperation on user studies.

# References

1. Arif, D., Ahmad, A., Bakar, M.A., Ihtisham, M.H., Winberg, S.: Cost effective solution for minimization of medical errors and acquisition of vitals by using autonomous nursing robot. In: Proceedings of the 2017 International Conference on Information System and Data Mining, ICISDM 2017, pp. 134–138. ACM, New York (2017). https://doi.org/10.1145/3077584.3077598
2. Bento, L.C., Pires, G., Nunes, U.: A behavior based fuzzy control architecture for path tracking and obstacle avoidance. In: Proceedings of the 5th Portuguese Conference on Automatic Control (Controlo 2002), Aveiro, p. 341Å346. Citeseer (2002)
3. Buccoliero, L., Bellio, E.: The adoption of "silver" e-health technologies: first hints on technology acceptance factors for elderly in Italy. In: Proceedings of the 8th International Conference on Theory and Practice of Electronic Governance, ICEGOV 2014, pp. 304–307. ACM, New York (2014). https://doi.org/10.1145/2691195.2691303
4. Caldeira, C., Bietz, M., Vidauri, M., Chen, Y.: Senior care for aging in place: balancing assistance and independence. In: Proceedings of the 2017 ACM Conference on Computer Supported Cooperative Work and Social Computing, CSCW 2017, pp. 1605–1617. ACM, New York (2017). https://doi.org/10.1145/2998181.2998206
5. Chen, Y., Ngo, V., Park, S.Y.: Caring for caregivers: designing for integrality. In: Proceedings of the 2013 Conference on Computer Supported Cooperative Work, CSCW 2013, pp. 91–102. ACM, New York (2013). https://doi.org/10.1145/2441776.2441789
6. Conte, S., Munteanu, C.: An interactive tactile aid for older adults learning to use tablet devices. In: Extended Abstracts of the 2018 CHI Conference on Human Factors in Computing Systems, CHI EA 2018, pp. D317:1–D317:4. ACM, New York (2018). https://doi.org/10.1145/3170427.3186548
7. Davis, H., Pedell, S., Lorca, A.L., Miller, T., Sterling, L.: Researchers as proxies for informal carers: photo sharing with older adults to mediate wellbeing. In: Proceedings of the 26th Australian Computer-Human Interaction Conference on Designing Futures: The Future of Design, OzCHI 2014, pp. 270–279. ACM, New York (2014). https://doi.org/10.1145/2686612.2686652
8. Department of Economic and Social Affairs: World population ageing 2017 (2017). http://www.un.org/en/development/desa/population/publications/pdf/ageing/WPA2017_Highlights.pdf
9. Diao, C., et al.: Design and realization of a novel obstacle avoidance algorithm for intelligent wheelchair bed using ultrasonic sensors. In: 2017 Chinese Automation Congress (CAC), pp. 4153–4158 (2017). https://doi.org/10.1109/CAC.2017.8243508
10. Faria, B.M., Reis, L.P., Lau, N.: Adapted control methods for cerebral palsy users of an intelligent wheelchair. J. Intell. Rob. Syst. 77(2), 299–312 (2015). https://doi.org/10.1007/s10846-013-0010-9
11. Gundersen, R., Smith, S.J., Abbott, B.A.: Applications of virtual reality technology to wheelchair remote steering systems. In: Proceedings of 1st Euro Conference of Disability, Virtual Reality & Assoc. Technology, pp. 47–56 (1996)
12. Hasegawa, K., Furuya, S., Kanai, Y., Imai, M.: Decores: Degree expressional command reproducing system for autonomous wheelchairs. In: Proceedings of the 3rd International Conference on Human-Agent Interaction, HAI 2015, pp. 149–156. ACM, New York (2015). https://doi.org/10.1145/2814940.2814942

13. Hashizume, S., Suzuki, I., Takazawa, K., Sasaki, R., Ochiai, Y.: Telewheelchair: the remote controllable electric wheelchair system combined human and machine intelligence. In: Proceedings of the 9th Augmented Human International Conference, AH 2018, pp. 7:1–7:9. ACM, New York (2018). https://doi.org/10.1145/3174910.3174914

14. Holbø, K., Bøthun, S., Dahl, Y.: Safe walking technology for people with dementia: What do they want? In: Proceedings of the 15th International ACM SIGACCESS Conference on Computers and Accessibility, ASSETS 2013, pp. 21:1–21:8. ACM, New York (2013). https://doi.org/10.1145/2513383.2513434

15. Hua, B., Hossain, D., Capi, G., Jindai, M., Yoshida, I.: Human-like artificial intelligent wheelchair robot navigated by multi-sensor models in indoor environments and error analysis. Procedia Comput. Sci. **105**, 14–19 (2017). https://doi.org/10.1016/j.procs.2017.01.181, http://www.sciencedirect.com/science/article/pii/S1877050917301990

16. Huang, H.H., Konishi, N., Shibusawa, S., Kawagoe, K.: Can a virtual listener replace a human listener in active listening conversation? In: Proceedings of the International Workshop on Emotion Representations and Modelling for Companion Technologies, ERM4CT 2015 pp. 33–39. ACM, New York (2015). https://doi.org/10.1145/2829966.2829971

17. Ichinotani, H., Ikeda, N., Ioi, K.: Maneuvering chair transformable to nursing care bed. In: Proceedings of the 2018 10th International Conference on Computer and Automation Engineering, ICCAE 2018, pp. 216–220. ACM, New York (2018). https://doi.org/10.1145/3192975.3192983

18. Kobayashi, Y., et al.: Robotic wheelchair easy to move and communicate with companions. In: CHI 2013 Extended Abstracts on Human Factors in Computing Systems, pp. 3079–3082. ACM (2013). https://doi.org/10.1145/2468356.2479615

19. Kostoska, M., Simjanoska, M., Koteska, B., Bogdanova, A.M.: Real-time smart advisory health system. In: Proceedings of the 8th International Conference on Web Intelligence, Mining and Semantics, WIMS 2018 pp. 47:1–47:4. ACM, New York (2018). https://doi.org/10.1145/3227609.3227686

20. Levine, S.P., Bell, D.A., Jaros, L.A., Simpson, R.C., Koren, Y., Borenstein, J.: The navchair assistive wheelchair navigation system. IEEE Trans. Rehab. Eng. **7**(4), 443–451 (1999)

21. Mazo, M.: An integral system for assisted mobility [automated wheelchair]. IEEE Rob. Autom. Mag. **8**(1), 46–56 (2001). https://doi.org/10.1109/100.924361

22. Moreno, P.A., Chernbumroong, S., Langensiepen, C., Lotfi, A., Gómez, E.J.: Virtual carer: personalized support for informal caregivers of elderly. In: Proceedings of the 9th ACM International Conference on PErvasive Technologies Related to Assistive Environments, PETRA 2016, pp. 38:1–38:4. ACM, New York (2016). https://doi.org/10.1145/2910674.2935855

23. Piper, A.M., Cornejo, R., Hurwitz, L., Unumb, C.: Technological caregiving: supporting online activity for adults with cognitive impairments. In: Proceedings of the 2016 CHI Conference on Human Factors in Computing Systems, CHI 2016, pp. 5311–5323. ACM, New York (2016). https://doi.org/10.1145/2858036.2858260

# Different Types, Different Speeds – The Effect of Interaction Partners and Encountering Speeds at Intersections on Drivers' Gap Acceptance as an Implicit Communication Signal in Automated Driving

Ann-Christin Hensch[(✉)], Matthias Beggiato, Maike X. Schömann, and Josef F. Krems

Department of Psychology, Cognitive and Engineering Psychology,
Chemnitz University of Technology, Wilhelm-Raabe-Strasse 43, 09107 Chemnitz, Germany
Ann-Christin.Hensch@psychologie.tu-chemnitz.de

**Abstract.** To exploit the benefits of automated vehicles (AVs), the systems' functions need to be accepted by the driver and other traffic participants. Thus, the human-machine interaction should be considered as a key issue. Manual driving is often coordinated by implicit communication cues. Therefore, specific parameters such as drivers' gap acceptance (GA) should be investigated to be prospectively implemented in AVs. The current study aimed at identifying the effects of different interaction partners, encountering speeds and participants' age and gender on GA. The video material displayed a left-turn scenario from the drivers' perspective including encountering interaction partners from the left. The study investigated four different interaction partners (passenger car, motorcycle, e-scooter, bicycle), all approaching at four different speeds (10/15/20/25 km/h). In sum, 121 participants contributed to the online study. The results revealed main effects for interaction partner, encountering speeds and participants' age on GA. Participants selected the smallest comfortable time gaps for the bicycle and the largest gaps for the passenger car, indicating that participants anticipated the potential threat of the interaction partner when selecting comfortable gaps. In accordance with previous studies, smaller gaps were accepted at higher speeds resulting in riskier decisions. Younger participants accepted smaller gaps than older participants did. Hence, AVs should consider the types and the speeds of the encountering interaction partners when selecting a comfortable gap as a form of implicit communication. Moreover, drivers' characteristics should also be considered when implemented driving styles in AVs, e.g., by selectable driving style profiles.

**Keywords:** Implicit communication · Gap acceptance · Automated driving

# 1 Introduction

Automated driving offers the potential of increased road safety, traffic efficiency and enhanced driving comfort [1]. However, users need to apply the system's automation

© Springer Nature Switzerland AG 2021
H. Krömker (Ed.): HCII 2021, LNCS 12791, pp. 517–528, 2021.
https://doi.org/10.1007/978-3-030-78358-7_36

to benefit from the advantages of increased automated driving functions. Since users' acceptance of a system depicted as an important precondition for its usage, the human-machine interaction should be considered as a key issue in automated driving [2].

Currently, the interaction between automated vehicles (AVs) and other traffic participants follows predominantly the rational principle of accident avoidance and lacks established communication capabilities of human drivers. For instance, the time-to-collision between different interaction partners could be applied by AVs to initiate an objectively safe driving maneuver; while human drivers would potentially consider additional communication cues during an interaction to decide whether a maneuver should be execute or not. Thus, the vehicles' movement could be perceived as unexpected and artificial and might result in drivers' discomfort. These potentially unpredictable movements of AVs could in turn lead to an impairment of the users' acceptance of AVs [3]. The human-machine interaction needs to be respected as a key issue in automated driving to support the users' acceptance and thus the usage of vehicles' automation [2]. Therefore, elaborated communication capabilities of manual driving are also required in AVs to provide smooth and cooperative interactions with other traffic participants that exceed the mere principle of accidence avoidance and foster drivers' comfort [3].

## 1.1 Communication in Manual Driving

The established forms of communication in manual driving could provide a basis for the communication between AVs and surrounding traffic participants. The identified communication cues could serve in AVs to detect and anticipate prospective movements from other traffic participants. Moreover, the signals could be applied by AVs itself to announce upcoming maneuvers to surrounding traffic participants in a transparent manner [3]. Thus, essential communication parameters need to be implemented in AVs to provide a safe, cooperative and proactive driving style that is familiar and thus accepted by the driver and other traffic participants [4].

Besides explicit signals (e.g., turn indicator), manual drivers mainly use implicit driving cues (e.g., vehicles' trajectory) to communicate with surrounding traffic participants. Implicit communication signals are highly context-dependent and therefore difficult to interpret. However, implicit cues are highly relevant in negotiating priority and anticipating prospective movements of other traffic participants [5]. Hence, these implicit communication parameters are also required to be implemented in AVs.

For instance, gap acceptance (GA) often serves as an implicit signal to initiate a communication process [6]. GA could be defined as the accepted time gap to other traffic participants during a maneuver (i.e., a time gap that is perceived as safe and comfortable by drivers) [7]. Thus, GA could be considered as one parameter of drivers' individual varying safety margins according to the Safety Margin Model by [8]. Corresponding to the model, operating within the individual safety margins would result in drivers' comfort; whereas, operating beyond these margins would imply drivers' discomfort. Therefore, an individually comfortable driving style (i.e., operating within the drivers' individual safety margins) should be provided by AVs to support the drivers' acceptance and thus the usage of the automated driving system [8].

## 1.2 Gap Acceptance as an Implicit Communication Signal

Previous research investigated different influencing factors on drivers' GA. In particular, a) different types of interaction partners (e.g., [9, 10]), b) various speed levels of the encountering interaction partner (e.g., [10]) and c) the effect of different drivers' characteristics (e.g., [6]) on GA were often explored.

With regard to the type of the interaction partner during the encounters, [11] reported a higher number of turn rejections during a left-turn maneuver when interacting with a truck compared to a passenger car and a motorcycle. Moreover, an effect on GA parameters could be shown. In particular, larger time gaps were selected during the interaction with a passenger car in contrast to a motorcycle [9]. [12] hypothesized the anticipated threat of the respective interaction partner in case of an accident to influence GA. Due to the hypothesis, more threatening interaction partners would be given larger safety margins resulting in larger accepted time gaps to execute a maneuver in front of these interaction partners compared to less threatening interaction partners [12]. This effect was also revealed when controlling for vehicles' size by comparing exclusively two-wheelers with the same size of the visual angle [10]. The authors reported larger gaps in front of the investigated motorcycle as a potentially more threatening interaction partner in contrast to the bicycle as a less threatening interaction partner.

In addition, the impact of different speed levels of the encountering interaction partners on GA was addressed in different studies. A stable effect of smaller accepted time gaps at higher speed levels, resulting in riskier decisions, was revealed for different types of interaction partners (e.g., [10]).

Considering personal characteristics, drivers' age seems to influence the selection of time gaps during the interaction with other traffic participants. In particular, different studies reported the general tendency that older participants tended to select larger time gaps than younger participants did, which resulted in increasingly conservative decisions (for a pedestrians' perspective: [6], for a drivers' perspective with a focus on safety issues: [10]). These consistent results could be explained by strategies to compensate for motoric, visual and cognitive declines or a general less risky behavior of older people. In addition, drivers' gender seems to effect the accepted time gaps as well. Previous studies reported smaller accepted time gaps (i.e., riskier decisions) for male participants compared to female participants [7].

## 1.3 Objectives of the Current Study

The current study applies the paradigm of gap acceptance to identify accepted *comfortable time gaps* during the encounter with different traffic participants from a drivers' perspective. Therefore, the study contributes and extents the body of research on implicit communication cues in traffic. The identified parameters could prospectively be implemented in AVs to enhance a safe, efficient and comfortable interaction with surrounding traffic participants, which in turns are assumed to support the drivers' acceptance of AVs. The study included four different interaction partners (i.e., passenger car, motorcycle, e-scooter and bicycle) comprising various characteristics (i.e., different potential threats; vulnerable road users (VRUs) vs. a non-VRU). VRUs comprise traffic participants such as bicycles and motorcycles that are particularly at risks in case of a collision due to a missing protective cover [13]. Moreover, the effect of speed levels during the encounter and drivers' characteristics, i.e., participants' age and gender, on GA was investigated in the present online study. Based on previous findings the following hypotheses (H) were investigated within the current study:

- (H1) Larger time gaps are chosen when interacting with increasingly threatening interaction partners (i.e., passenger car > motorcycle > bicycle).
- (H2) Smaller time gaps are accepted during higher speed levels of the respective encountering interaction partners.
- (H3) Larger time gaps are chosen by older participants.
- (H4) Larger time gaps are chosen by female participants.

In addition to the previously mentioned interaction partners (i.e., passenger car, motorcycle and bicycle), an e-scooter as a rather new and currently little investigated traffic participant was included in the current study. Since the size of this interaction partner is rather similar to the bicycle and the motorcycle but participants potentially lack experience in interacting with this type of vehicle, the following research question (RQ) was explored within the study:

- (RQ1) How does the e-scooter as an interaction partner influence participants' GA with regard to the other, familiar, investigated VRUs of a similar size (i.e., bicycle and motorcycle)?

## 2   Method

### 2.1   Research Design

A 4 × 4 × 2 mixed design was applied in the present online study. The interaction partners (passenger car, motorcycle, e-scooter and bicycle) and the speed levels of the encountering interaction partners served as within-subject factors. To implement detailed and realistic speed levels, particularly for the bicycle and the e-scooter, speed levels of 10/15/20 and 25 km/h were investigated. Each condition was presented twice to the participants in a randomized order, resulting in a total of 32 trials. The participants' gender served as a between-subject factor in the study. Moreover, participants' age was applied as a covariate in further analysis. The participants were instructed to indicate the last accepted time gaps they would still perceive as comfortable to initiate a cooperative left-turn maneuver in front of the approaching interaction partners. Hence, the indicated gap acceptance served as the dependent variable in the current study.

### 2.2   Material

**Video Material.** Pre-recorded video material from a real-world driving scene was used as the study's material (Fig. 1). The material was recorded by a GARMIN VIRB Ultra 30 (1920 × 1080 pixels, 100 fps). The videos displayed a left-turn scenario from a drivers' perspective. The respective interaction partners approached from the left of the ego-vehicles' perspective, which resulted in an overlap of the ego-vehicles' and the interaction partners' trajectory (Fig. 2). All recorded encounters of the investigated interaction partners were driven by the same researcher to control for influencing factors (e.g., driving style) with a constant speed of about 15 km/h. Synchronized protocol video cameras were placed at fixed distances to determine speed precisely. Since the video material exclusively displayed the interaction partners as a moving object, the speed of the encountering vehicle was modified afterwards by manipulating the playback rate of the video material (i.e., reducing or accelerating the video material). The recordings were edited with Adobe Premiere Pro and were presented online in a resolution of 1280 × 720 pixels (30 fps) to the participants.

**Simulation Software and Apparatus.** An online questionnaire was applied to collect sociodemographic data from the participants. In addition, a simulation environment was programmed in jspsych 6.1.0 to present the video material in a randomized order and collected participants' GA precisely in an online format.

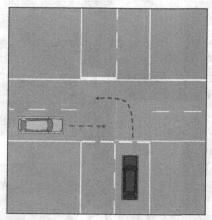

**Fig. 1.** Recorded scenario of the current study. The dark red vehicle displays the participants' perspective. The blue vehicle displays the different encountering interaction partners. (Color figure online)

### 2.3 Instruction and Procedure

At the beginning, participants were informed about the scope of the online study and informed consent was obtained. Holding a valid drivers' license and a minimum screen resolution of 1280 × 720 pixel was required to participate to the online study. In the first part of the study, the participants filled in a sociodemographic questionnaire. In the second part of the study, participants' accepted gaps during the encounter with different interaction partners approaching in various speeds were collected. Participants were instructed to indicate a respective time gap they would still perceive as comfortable to initiate a cooperative and proactive left-turn maneuver in front of the encountering

interaction partner by pressing the enter key. Four test trials were presented to the participants including all four investigated interaction partners to become familiarized with the
task. The participants had the possibility to repeat the test trials if required. Afterwards
the data collection of 32 trials was conducted. In total, the study lasted about 25 min.
Participants did not receive monetary compensation for participating.

## 2.4 Participants

In sum, $N = 127$ participants contributed to the study. Due to extreme outlier in boxplot-
checks, $n = 6$ participants had to be excluded, resulting in a final sample of $n = 121$
(79 women, 42 men) with a mean age of $M = 36$ years ($SD = 19.48$). The participants
reported a mean annual mileage of $M = 8751.07$ km ($SD = 9572.31$).

**Fig. 2.** Examples of the recorded video material displaying all four investigated interaction
partners: a) the passenger car, b) the motorcycle, c) the e-scooter and d) the bicycle.

## 3 Results

The data was analyzed by repeated measures ANOVAs and ANCOVAs with the following independent variables: interaction partner and encountering speed (within-subject
variables), participants' age (covariate) and participants' gender (between-subject variable). The participants' gap acceptance in seconds served as the dependent variable.
Mauchly's test indicated violations for the assumption of sphericity for the factors interaction partner and encountering speed. Thus, Greenhouse-Geisser corrected degrees of
freedom are reported for the results of the ANOVAs and ANCOVAs. An overview of
the results can be found in Table 1.

### 3.1 The Effect of Interaction Partner and Encountering Speed Levels on GA

Figure 3 displays the accepted comfortable time gaps in dependence of interaction partner and their encountering speed levels. Participants indicated the largest comfortable time gaps during the encounter with the passenger car ($M = 9.01$, $SD = 2.95$) followed by the motorcycle ($M = 8.28$, $SD = 2.59$) and the e-scooter ($M = 7.16$, $SD = 2.11$). The smallest gaps were chosen for the bicycle ($M = 6.84$, $SD = 2.30$) as an interaction partner. The ANOVA revealed a significant main effect for the investigated within-subject factor interaction partner (Table 1; H1). Bonferroni-corrected pairwise comparisons depicted significant differences between all investigated interaction partners ($p <$ .001, respectively).

With regard to the encountering speed levels, participants accepted smaller time gaps with higher speed levels of the encountering interaction partner ($M_{10km/h} = 10.60$, $SD_{10km/h} = 3.30$; $M_{15km/h} = 8.04$, $SD_{15km/h} = 2.33$; $M_{20km/h} = 6.64$, $SD_{20km/h} = 1.92$; $M_{25km/h} = 5.60$, $SD_{25km/h} = 1.54$). A significant main effect was also shown by the ANOVA for the within-subject factor encountering speed (Table 1; H2). Again, post-hoc comparisons (Bonferroni-corrected) revealed significant differences between all speed levels ($p < .001$, respectively).

Moreover, a significant interaction of the encountering speed levels and the interaction partner was found (Table 1). In detail, the selected time gaps for the passenger car ($M_{Passenger\ car-10km/h} = 11.71$, $SD_{Passenger\ car-10km/h} = 4.12$) and the motorcycle ($M_{Motorcycle-10km/h} = 11.39$, $SD_{Motorcycle-10km/h} = 3.69$) were quite similar during the lowest speed level of 10 km/h. During higher speed levels the differences in accepted comfortable time gaps between the two interaction partners increased as it is also displayed in Fig. 3. In contrast to the these increased differences in time gaps between the passenger car and the motorcycle during higher speed levels, the differences between the scooter and the bicycle tended to decreased at higher speed levels ($M_{Scooter-15km/h} = 7.31$, $SD_{Scooter-15km/h} = 2.09$; $M_{Bicycle-15km/h} = 6.96$, $SD_{Bicycle-15km/h} = 2.05$; $M_{Scooter-20km/h} = 6.00$, $SD_{Scooter-20km/h} = 1.69$; $M_{Bicycle-20km/h} = 5.67$, $SD_{Bicycle-20km/h} = 1.67$; $M_{Scooter-25km/h} = 4.97$, $SD_{Scooter-25km/h} = 1.37$; $M_{Bicycle-25km/h} = 4.85$, $SD_{Bicycle-25km/h} = 1.33$; RQ1).

### 3.2 The Effect of Participants' Age and Gender on GA

The data revealed a significant main effect for participants' age on the accepted comfortable time gaps (Table 1). Generally, older participants indicated significantly larger time gaps than younger participants did (H3). In addition, a significant interaction between the participants' age and the interaction partner was found (Table 1) that is also displayed in Fig. 4. In particular, the absolute values of the accepted comfortable time gaps during the interaction with the e-scooter and the bicycle depicted a sharper decrease for older participants in contrast to the values of the younger participants. Moreover, a significant interaction between the participants' age and the encountering speed of the interaction partners was uncovered (Table 1). Figure 5 presents the interaction between the participants' age and the encountering speed level of the vehicles. A decreasing difference in accepted comfortable time gaps between the age groups during higher speeds was visible within the data. The analysis did not reveal an effect for participants' gender on GA (Table 1; H4).

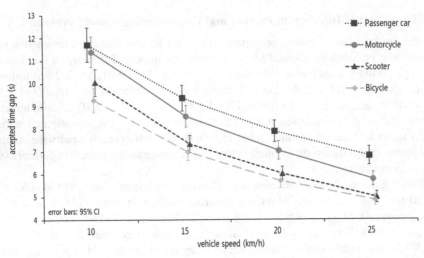

**Fig. 3.** Accepted comfortable time gaps in seconds dependent of interaction partner and encountering speed.

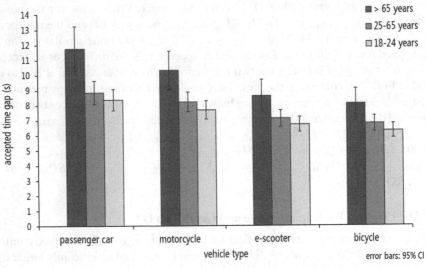

**Fig. 4.** Accepted comfortable time gaps in seconds for different age groups dependent of interaction partner. Group 1 (18–24 years): $N = 58$ ($M = 21$ years, $SD = 1.82$); group 2 (25–65 years): $N = 50$ ($M = 44$ years, $SD = 15.85$); group 3 (>65 years): $N = 13$ ($M = 71$ years, $SD = 6.02$).

## 4 Discussion

The current online study assessed the effects of different interaction partners and their encountering speed in a left-turn maneuver on comfortable accepted time gaps as an implicit communication signal to initiate a cooperative interaction. Since the indicated

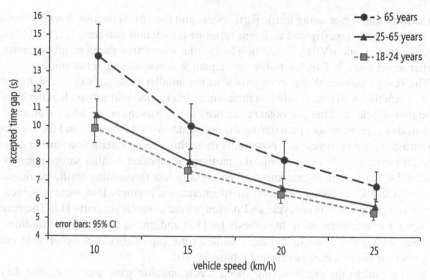

**Fig. 5.** Accepted comfortable time gaps in seconds dependent of participants' age groups and the encountering speed of the interaction partners. Group 1 (18–24 years): $N = 58$ ($M = 21$ years, $SD = 1.82$); group 2 (25–65 years): $N = 50$ ($M = 44$ years, $SD = 15.85$); group 3 (>65 years): $N = 13$ ($M = 71$ years, $SD = 6.02$).

**Table 1.** ANOVA and ANCOVA results showing the main effects and interactions of the investigated factors interaction partners, encountering speed, participants' age and gender on GA.

| Main and interaction effects | ANOVA | | |
|---|---|---|---|
| | $F$ | $p$ | $\eta^2_p$ |
| **Interaction partner[a]** | **$F(2.03, 238.92) = 203.57$** | **<.001** | **.63** |
| **Speed[a]** | **$F(1.27, 149.79) = 619.61$** | **<.001** | **.84** |
| **Age** | **$F(1, 117) = 11.84$** | **.001** | **.09** |
| Gender | $F(1, 117) = 2.12$ | .148 | .02 |
| **Interaction partner x speed[a]** | **$F(6.15, 768.77) = 7.56$** | **<.001** | **.06** |
| **Interaction partner x age[a]** | **$F(2.10, 245.61) = 9.28$** | **<.001** | **.07** |
| **Speed x age[a]** | **$F(1.31, 152.97) = 15.26$** | **<.001** | **.12** |
| Interaction partner x gender[a] | $F(2.03, 237.34) = 0.66$ | .519 | .01 |
| Speed x gender[a] | $F(1.27, 148.46) = 1.07$ | .319 | .01 |

*Note.* Statistically significant results are highlighted in bold. [a]Greenhouse-Geisser-corrected degrees of freedom are reported. $N = 121$.

time gaps were selected with a focus on participants' comfort, the results might prospectively be implemented in AVs to provide cooperative and predictable but also efficient interactions with surrounding traffic participants and therefore enhance driving comfort. The present study investigated different types of interaction partners, i.e., VRUs vs. a passenger car as a non-VRU. In addition to familiar interaction partners, an e-scooter as a rather novel and less familiar traffic participant was considered in the study.

The results showed that participants selected smaller time gaps (i.e., riskier gaps) when interacting with potentially less threatening interaction partners such as the bicycle or the motorcycle (and the e-scooter) compared to the passenger car. Moreover, smaller comfortable gaps were accepted during the encounter with the bicycle and the e-scooter in contrast to the motorcycle as a potentially more threatening interaction partner despite a similar object size. In other words, the participants applied smaller safety margins as comfortable gaps when interacting with potentially less threatening traffic participants (i.e., bicycle and e-scooter) in contrast to interaction partners that were assumed as more threatening (i.e., motorcycle and passenger car), which supports H1. The results correspond to the threatening hypothesis by [12] and are in line with the findings of previous studies that investigated the accepted time gaps between a motorcycle and a passenger car [9] or a motorcycle and a bicycle [10].

With regard to the encountering speed levels, smaller gaps were accepted during higher speeds of the encountering interaction partner, hence resulting in riskier decisions. The results are in line with the assumptions (H2) and previous studies that addressed participants' GA with a focus on traffic safety (e.g., [10]). When considering the results in closer detail, it should be acknowledged that the differences in accepted time gaps between the passenger car and the motorcycle increased within higher speed levels of the encountering traffic participants. Whereas, differences in comfortable time gaps among the VRUs as interaction partners (i.e., motorcycle, e-scooter and bicycle) decreased with higher speed. Moreover, it should be highlighted that significantly larger time gaps were selected at a lower speed level for the e-scooter as a rather unfamiliar interaction partner compared to the bicycle, which depicts a familiar interaction partner. In contrast to the motorcycle, significantly smaller time gaps (i.e., riskier gaps) were accepted for the e-scooter (RQ1). The results imply that, both, the anticipated threat and the speed of the encountering interaction partner should be taken into account in AVs when a cooperative left-turn maneuver is initiated. The application of particular time gaps could provide an increasingly familiar driving style that is accepted by the human driver. However, before implementing specific parameters to AVs, prospective studies should also investigate the accepted time gaps from additional perspectives (e.g., from the perspective of the interaction partner in the current study) to meet the requirements of all involved traffic participants and, for instance, identify gaps that are perceived as uncomfortable from different perspectives.

Beyond that, the results of the current study revealed, as expected, that older participants generally indicated larger time gaps (i.e., more conservative decisions) in front of encountering traffic participants as comfortable gaps than younger participants did (H3). The results correspond to the stable trend in previous studies that investigated the gap acceptance of different age groups from a pedestrians' perspective [6] or with a focus on traffic safety from a drivers' perspective [10]. The results of the current study depicted particular higher accepted time gaps for older participants during lower speed levels and during the encounter with the passenger car as the potentially most threatening interaction partner in this study. This result might be explained by compensational strategies due to age-related declines and a more conservative behavior of older people in general. However, for encounters with the e-scooter and the bicycle, the differences in accepted time gaps decreased between the age groups. These reduced differences in GA might be either explained by the perceived lower threat of the interaction partners and thus resulting in more similar results between the age groups. On the other hand, the reduced GA could be explained by longer identification times due to their smaller object sizes in contrast to a passenger car, which is leading to smaller remaining time gaps in front of these vehicles. With regard to individual safety margins and their implementation in AVs, drivers' age needs to be considered to provide a comfortable driving style (e.g., by maintaining sufficiently larger gaps for older participants to other traffic participants) that supports the acceptance and thus the usage of the system. Therefore, the GA parameters of the different age groups could serve as a basis for different driving style profiles in AVs (e.g., conservative vs. dynamic driving style) that could be selected by the drivers of that vehicles.

The present study did not reveal any differences on accepted time gaps regarding gender, which contradicts H4 and the findings by [7]. However, the differences might be explained by the predominantly young and female sample in the current study resulting from an online acquisition in contrast to a more balanced sample in [7].

## 4.1 Conclusion and Implications

To provide a safe, cooperative and comfortable driving style that is accepted by the driver and the surrounding traffic participants, established communication capabilities of manual drivers also need to be implemented in AVs. For instance, GA could serve as a parameter of implicit communication in left-turn scenarios. Different aspects of the encounter with other traffic participants and particular driver characteristics need to be considered in AVs. Besides the anticipated threat of the respective interaction partner, the speed of the approaching vehicle should be taken into account by AVs when selecting cooperative gaps as a form of implicit communication to execute different maneuvers. Due to the significantly conservative time gaps for older drivers, the application of different selectable driving style profiles (e.g., conservative vs. dynamic driving style) seems recommendable to meet the requirements of different personal preferences with regard to a comfortable driving style of AVs.

**Acknowledgements.** The research was funded by the Deutsche Forschungsgemeinschaft (DFG, German Research Foundation) – Project-ID 416228727 – SFB 1410.

# References

1. European Road Transport Research Advisory Council, Automated driving roadmap. http://www.ertrac.org/uploads/documentsearch/id48/ERTRAC_Automated_Driving_2017.pdf. Accessed 01 Feb 2021

2. Venkatesh, V., Thong, J.Y.L., Xu, X.: Consumer acceptance and use of information technology: extending the unified theory of acceptance and use of technology. MIS Q. **36**(1), 157–178 (2012)
3. Schieben, A., Wilbrink, M., Kettwich, C., Madigan, R., Louw, T., Merat, N.: Designing the interaction of automated vehicles with other traffic participants: design considerations based on human needs and expectations. Cogn. Technol. Work **21**(1), 69–85 (2018). https://doi.org/10.1007/s10111-018-0521-z
4. Elbanhawi, M., Simic, M., Jazar, R.: In the passenger seat: investigating ride comfort measures in autonomous cars. IEEE Intell. Transp. Syst. Mag. **7**(3), 4–17 (2015). https://doi.org/10.1109/MITS.2015.2405571
5. Färber, B.: Communication and communication problems between autonomous vehicles and human drivers. In: Maurer, M., Gerdes, J.C., Lenz, B., Winner, H. (eds.) Autonomous Driving 2016, pp. 125–144. Springer, Heidelberg (2016). https://doi.org/10.1007/978-3-662-488 47-8_7
6. Beggiato, M., Witzlack, C., Springer, S., Krems, J.F.: The right moment for braking as informal communication signal between automated vehicles and pedestrians in crossing situations. In: Stanton, N. (ed.) Advances in Intelligent Systems and Computing, vol. 597, pp. 1072–1081. Springer, Cham (2018). https://doi.org/10.1007/978-3-319-60441-1_101
7. Yan, X., Radwana, E., Guob, D.: Effects of major-road vehicle speed and driver age and gender on left-turn gap acceptance. Accid. Anal. Prev. **39**(4), 843–852 (2007)
8. Summala, H.: Towards understanding motivational and emotional factors in driver behaviour: comfort through satisficing. In: Cacciabue, P.C. (eds.) Modelling Driver Behaviour in Automotive Environments. Springer, London (2007). https://doi.org/10.1007/978-1-84628-618-6_11
9. Robbins, C.J., Allen, H.A., Chapman, P.: Comparing drivers' gap acceptance for cars and motorcycles at junctions using an adaptive staircase methodology. Transp. Res. Part F **58**, 944–954 (2018). https://doi.org/10.1016/j.trf.2018.07.023
10. Schleinitz, K., Petzoldt, T., Gehlert, T.: Drivers' gap acceptance and time to arrival judgements when confronted with approaching bicycles, e-bikes, and scooters. J. Transp. Saf. Secur. **12**(1), 3–16 (2019). https://doi.org/10.1080/19439962.2019.1591551
11. Hancock, P.A., Caird, J.K., Shekhar, S., Vercruyssen, M.: Factors influencing driver' left turn decisions. Proc. Hum. Factors Soc. Annu. Meet. **35**(15), 1139–1143 (1991). https://doi.org/10.1177/154193129103501525
12. Das, S., Manski, C.F., Manuszak, M.D.: Walk or wait? An empirical analysis of street crossing decisions. J. Appl. Economet. **20**, 529–548 (2005). https://doi.org/10.1002/jae.791
13. OCDE/OECD: Safety of vulnerable road users. DSTI/DOT/RTR/RS7(98)1/Final DSTI/DOT/RTR/RS7(98)1/Final. Organization for Economic Co-Operation and Development, Paris (1998)

# Smart and Inclusive Bicycling? Non-users' Experience of Bike-Sharing Schemes in Scandinavia

Michala Hvidt Breengaard[1], Malin Henriksson[2(✉)], and Anna Wallsten[2]

[1] University of Copenhagen, Copenhagen, Denmark
mbr@soc.ku.dk

[2] The Swedish National Road and Transport Research Institute (VTI), Linköping, Sweden
{malin.henriksson,anna.wallsten}@vti.se

**Abstract.** Being both affordable and sustainable, bike-sharing schemes have a promising potential of providing smart and sustainable mobility solutions for all. However, for bike-sharing to become part of a convenient, sustainable, and accessible mobility system, it must meet the needs of a wide range of users. Today, existing supply of bike-sharing schemes rarely take diversity into account: people who travel with kids, people who do not feel secure in biking or people who carry heavy luggage, do not have the opportunity to use the system. The lack of diversity in the contemporary bike-sharing supply presents a problem for visions of smart mobility for all. While a body of research points to differences in bicycling due to socio-economic factors and norms, there is little knowledge on how diverse mobility needs affect the attractiveness of using a bike-sharing scheme. This paper addresses non-users' perceptions of public bike-sharing schemes in Denmark and Sweden. The empirical material includes 14 in-depth interviews and two focus groups with non-users. Research questions include what everyday mobility needs the informants have, and if they can be meet by the local bike-sharing scheme, as well as how the bike-sharing scheme meets the diversity in restrictions, needs, and preferences of transport. The paper finds that non-use of bike-sharing schemes in Scandinavia can be explained through three overall narratives: 'I have my own bicycle', 'I travel with kids', and 'I don't feel safe.' It argues that obstacles of using bike-sharing schemes in part can be explained the 'one fits all' approach that dominates bike-sharing design today. By adding a perspective on diversity, the paper contributes to filling the research gap in new mobility solutions and diversity.

**Keywords:** Bike-sharing schemes · Biking · Gender · Diversity · Smart transport

## 1 Introduction

"If you've seen *rows of identical bikes* in your city, you may be observing a bike-sharing system. Such programs are now implemented in dozens of cities in the U.S. and more than a thousand cities worldwide"[1]

---

[1] Rick Leblanc, Sustainable business 2019: https://www.thebalancesmb.com/how-do-bicycle-sharing-systems-work-4176148. Italics added by authors.

© Springer Nature Switzerland AG 2021
H. Krömker (Ed.): HCII 2021, LNCS 12791, pp. 529–548, 2021.
https://doi.org/10.1007/978-3-030-78358-7_37

For more than 20 years, bike-sharing schemes have had a promising agenda in promoting green and accessible urban mobility. Biking has proven a good way to improve health by reducing the sedentary lifestyle of car usage as well as decreasing other health endangering issues such as air quality, transport safety and mental well-being (Woodcock et al. 2016; Wang and Zhau 2017). Bike-sharing is appointed in the EU Cycling Strategy as an essential component of any multi-modal transport system and an enabler of innovation at city level. This gives a broad framework for investments in the public bike-sharing sector, that is strategically positioned to be a major contributor to the success of the strategy. Bike-sharing schemes are part of ambitions of making Smart Cities, i.e., cities which meet the increasing urban challenges of population growth, increased population diversity, traffic congestion, and climate change. New mobility solutions are key in solving these problems making Smart Cities tied to the promise of a well-functioning smart mobility system (Albino et al. 2015). Yet, to present day, the advancement of bike-sharing schemes tends to focus on technological innovation in the development of tablets, APPs, GPS, electric bicycles, etc. In fact, the technological innovation in bike-sharing schemes is currently central for defining the system as 'Smart.' Although major technological advancements have been made throughout the bike-sharing generations, the bicycles themselves have remained largely the same. Today the main differences between bike-sharing schemes are whether they are electric or not. A key feature of present bike-sharing schemes is that the solution offers a fleet of identical bicycles. The resemblance of bicycles in the bike-sharing schemes is so consolidated today that it appears almost as a trademark of the system. This raises questions regarding how this uniform design can be adopted to users beyond the norm and meet equality demands (cf. Nixon and Schwanen 2019).

While bike-sharing schemes in many parts of the world are provided by private operators, the Scandinavian countries have made considerable public investments in offering the urban population a bike-sharing solution (Caggiani et al. 2021; Uteng et al. 2019). The potentials of bike-sharing schemes in making public transport easier, green, and more attractive, place bike share solutions as a politically interesting form of mobility. However, despite these promises, the Danish and Swedish bike-sharing schemes face problems of attracting users. Even though bike-sharing technologies are well-developed, the fact is, we are far from having a well-functioning market for bike-sharing solutions. The question is, what are we missing?

This paper addresses challenges of present bike-sharing schemes in meeting the needs of various potential users in the contexts of Copenhagen, Denmark, and Linköping, Sweden. It aims to advance the present understanding of bike-sharing schemes as a mobility solution that is designed to 'fit all.' Drawing upon interviews with people who do not use the bike-sharing schemes, the paper identifies three mobility narratives, shedding light on experiences and reflection of potential users, yet at present non-users, through which problems of the existing bike-sharing solutions are put forward and analyzed. The theoretical foundation in the paper is informed by studies on gender, diversity, and transport, highlighting the diversity of transport patterns and mobility needs of different social groups. The ambition is to contribute to knowledge about the social dimensions of biking and fill the gap that exists in both research, policy, and planning of bike-sharing schemes as well approaches to Smart Cities and Smart Mobility.

There are several reasons for studying equality aspects of bike-sharing in a Scandinavian context. First, the Nordic countries in general stands out when it comes to cycling, with large shares of cycling in the modal split in several cities. Copenhagen is even regarded as the best cycling city in the world, with a very specific cycling culture (Haustein et al. 2020; Freudendal-Pedersen 2015). The cycling cultures in Swedish cities might be different compared to the Copenhagen example, but there are explicit political goals on national and local levels to increase cycling (Balkmar 2020; Haustein et al. 2020). In Denmark and Sweden, a large proportion of daily life activities, such as transport between home and work, picking up children from childcare, as well as shopping errands are carried out by bicycle. Not least in the bigger cities. Secondly, gender equality has also been a part of the political agenda for several years in both Sweden and Denmark, and both countries score high on the gender equality index. Gender equality in decision making seem to push forward new norms in transport planning, which can support female-take up of cycling (Balkmar 2020; Haustein et al. 2020, Kronsell et al. 2020; Aldred et al. 2016). Moreover, research indicates that a high gender equality also means minimum differences in cycling between men and women (Prati 2018), which is the case in both Sweden and Denmark. In Copenhagen, as well as in Linköping, the proportion of men and women who cycle is largely the same. Generally, the cityscape of cyclists in Scandinavian cities is diverse, counting differences in social categories, such as gender, income, and age. There is a tradition of letting children down to the age of four bicycle themselves accompanied by a parent. The diversity in biking is also visible in an increasing innovation of private bicycles in the streets of especially Copenhagen, where new bike designs and especially various types of e-bikes and cargo bikes are gaining ground. In Linköping too, cargo bikes are a more common feature than previously, and the number of e-bikes are growing. There is a mix of privately-owned e-bikes, e-bikes used in care work and shared e-bikes.

How and if equality in cycling is regarded as an important issue in the Nordic context of policymaking and planning is however still not clear. Neither is it clear if policies on equality affect how bike-sharing schemes are set up and are operated. Last but most importantly, we lack knowledge on whether present bike-sharing schemes are actually supporting the everyday mobility needs of a variety of users as well as non-users. Tapping into these questions, the paper is structured so that first it describes the status of bike-sharing schemes in the contexts of Denmark and Sweden. Then it goes on to explain its theoretical and methodological departures, which are followed by three analytical sections on different aspects of non-usage. Finally, the paper concludes by a discussion on the yet unmet potentials of bike-sharing schemes.

## 2 Bike-Sharing in the Scandinavian Context

Denmark was one of the forerunners of bike-sharing schemes with Copenhagen as the first city to implement a formalized system in 1995 (Fishman and Allan 2019). In 2014, Copenhagen launched a new public bike-sharing scheme, called the white city bikes, which is the official system running in the Danish metropolitan area with an aim to offer a 'fourth leg' in public transport.[2] That is, whereas Copenhagen's former public bike-sharing scheme mainly addressed tourist, the new bike-sharing scheme targets commuters' last mile between trains or busses and workplace. The public bike-sharing scheme is docking based solution with an advanced technological system, being integrated in the Danish web and app-based travel planning ('Rejseplanen'), in which the shared bicycles appear as an option in trip planning. The Copenhagen system has 130 docking stations distributed in the areas of Copenhagen, Frederiksberg, and the suburban area of Rødovre. 46 of the docking stations are connected to trains and metro stations. Again, Denmark is a frontrunner by introducing a system consisting fully of electric bicycles.

Several of larger Danish cities, such as Aarhus and Odense, offer the population a bike-sharing scheme. Yet, today Copenhagen is the only city that runs a public bike-sharing scheme which coexists with several private providers. The goal for the public bike-sharing scheme in Copenhagen was that each bike should be used three times a day by local commuters. Instead, the bikes have only been used 0.8 times a day and mainly by tourists (Colville-Andersen 2015). The lack of success in getting commuters to use the Danish bike-sharing solution is particularly challenging due to the continuous growth in car transport. Car use in Denmark increased from 78.55% of total passenger transport in 2008 to 79.2% in 2018. In the second quarter of 2018, the Danish population used 475,503 $m^3$ of motor fuel, which is higher than the same time the year before (2017). A study of the attitude to climate change shows that the population is reluctant to live without the car.

In Sweden, bike-sharing has been part of metropolitan transport system for nearly two decades. In the three largest cities, different types of bike-sharing schemes have been launched with different success rates. A study by Nikitas et al. (2016) give insights on how users and non-users perceive bike-sharing in Gothenburg. Here, a self-service rental system was launched after a pilot study in 2010, consisting of 300 conventional "one size fits all" bicycles and 40 stations. Five years later the system had expanded and consisted of 1000 bicycles. The system had also been integrated with the public transport system, enabling travelers with a public transport pass to use a bike-sharing scheme. Despite modest usage rates, the system was well-received by the citizen who stated that it has a "pro-social value" even if it has not been useful for them personally. By pro-social is meant that a bike-sharing scheme symbolize pro-environmental values, and a city administration that support active mobility for the citizens. Reason for not using the system appears primarily to be due to bike-ownership and lack of a good bike-sharing infrastructure, but safety issues and the lack of a renting station in one's neighborhood is also mentioned (Nikitas et al. 2016). Since 2016, a bike-sharing system has been in place in Malmö. Today it has around 100 stations in the city, and a total of 1000 bicycles. It was procured by the city of Malmö and is financed through advertisement, which is also the case of the Stockholm city bicycles. According to Nikitas et al. (2016), bike-sharing in Stockholm did not manage to expand in the same rate as in Gothenburg due to political unwillingness and limited urban space.

In the last few years, bike-sharing schemes have been launched in medium-sized cities in Sweden and there is a general growing interest to develop business models that

---

[2] See more https://bycyklen.dk/om-bycyklen/

attract users while being economically viable (Caggiani et al. 2021; Henriksson and Wallsten 2020; Berg et al. 2019). A noteworthy addition to the Swedish bike-sharing schemes is LinBike in the medium-sized municipality Linköping. It was launched by the municipality in 2019 and offers 200 electrically assisted bicycles and 17 docking stations, mainly located close to larger working places and to public transport. While not prioritizing residential areas in general, two stations in low-income residential areas have been included in the BBS. It should be noted that since the stations are located at central nodes it also connects to the centrally located residential areas. The system applies advanced technology such as GPS and geo-fencing. After five months in service, 5 900 users had registered through the app and approx. 50 000 trips had been made. During a normal day, 130 bicycles are in circulation (LinBike 2020). LinBike has strong political support and is believed to push forward bike-sharing in favor of car trips and is regarded as a vital part of the city branding, where green entrepreneurship and smart technology are important aspects that are highly valued. In line with Nikitas et al. (2016) this is success factors, which make it interesting to see how non-users' reason about this system.

In a master thesis, Gelinder (2020) investigates justice consequences of bike-sharing schemes in the Öresund region, primarily in Copenhagen and Malmö. He shows that both public and private schemes are developed to satisfy what the operators believe to be the general demand of the public and prioritizes viable market solutions. In practice this means adjustments to fit the majority rather than recognizing diverse and unique needs. This, Gelinder argues, is in line with traditional transport planning which seeks to be cost-effective and to benefit "everyone". An alternative approach to planning is the transport justice approach, that focuses on needs rather than demand. Interestingly, Gelinder shows that there are traces of a transport justice approach in how operators talk about the schemes, in relation to the potential benefits of bike-sharing. In Malmö, there are also political goals that support an equity approach to cycling, and by which the city's bike-sharing system is legitimized. In practice, it is the 'economy of scale' that guide the function, design and business models for the studied Scandinavian bike-sharing schemes. Related, it is interesting to note that actors involved in cycling and bike-sharing in Denmark and Sweden indicate that they lack knowledge about the user groups that do not cycle today, and that these users are perceived as very far from adopting bike-sharing into their everyday mobility practices.[3] Thus, to further analyse the non-users' experiences is as important from a research perspective, as it is from a practice-oriented perspective.

## 3  Theoretical Approaches

Research on gender and diversity in transport is still a scattered field, which nevertheless demonstrates clear gender differences in the ways people travel as well as in their transport needs (Kronsell et al. 2016; Breengaard et al. 2007; Scholten and Joelsson 2019; Uteng et al. 2020). Studies show that women more often than men undertake 'care trips'

---

[3] We draw this conclusion from an initial round of telephone interviews we did as part of this study, to understand how equity and bike-sharing is approached by professional actors, as well as NGO:s that support cycling in the Oresund region.

where they accompany children – as well as other dependent family members – to activities, such as childcare institution, school, doctor appointments, etc. (Turner and Grieco 2000; de Madariaga and Zucchini 2019). Gendered key events such as the birth of a child affects travel behaviour in different life stages (Scheiner 2014). Thus, care tasks have an impact on women's mobility opportunities and needs at certain stages of their lives. In this article we are not so much concerned with how gender structures people's mobility needs and patterns, but how mobility needs structure and restrict the use of certain means of mobility, in this case bike-sharing schemes. The opportunities and restrictions to use certain forms of mobility are, of course, gendered – and that is an important point – but a point is also that the restrictions apply to everyone - across genders - who performs these tasks. Studies show that more and more men are taking part in tasks of picking up and bringing children as well as responsibility for other domestic mobility-related activities, such as shopping tasks.

Working from the perspective on mobility needs makes room for thinking more broadly about gender in terms of gender *and* diversity. Over the last decades, gender studies have tuned onto perspectives of diversity, pointing to differences across genders due to social categories of class, age, ethnicity, etc. These perspectives on intersecting and diversifying categories have also had an entry in transport research, e.g., in the European research project TInnGO.[4] The inclusion of diversity as a vital element in how people travel points out that there are large differences within each category of gender. A 22-year-old ethnic Danish woman without children will have different mobility patterns and needs than a 22-year-old minority woman with children. Likewise, an 83-year-old woman might have very different travel needs than both of the two 22-year-olds. As our approach to non-users of bike-sharing schemes is structured around mobility needs, the analysis will move across gender and include various diversity variables, mainly age, family type, and ethnicity. This approach is framed within theories of intersectionality, claiming that gender cannot be analysed alone, but is affected by other categories, such as age, ethnicity, income, or location (Crenshaw 1989). In transport studies, the intersectional perspective has opened up for seeing how transport patterns and mobility needs are dependent on, while in the same time influencing, different social groups' everyday mobility (Hansson 2010; Law 1999; Angeles 2017). Intersectionality thus pays attention to more explanations on how different people travel, including issues such as time, feelings of security and control as well as factors of convenience.

Although a lot of new perspectives have entered transport research in the last 10 years, perspectives of gender and diversity are still marginal in transport research, as well as transport planning, and design. Feminist research on the subject of transportation have pointed out that common infrastructure planning is based on a standard that is oriented around how to get fastest from A to B, especially by private car (Scholten and Joelsson 2019). This single perspective not only prioritize some types of mobility needs while making mobility more difficult for others, it also works from a gender and diversity-blind idea of 'one fits all' (Sing 2020). In our research on bike-sharing schemes, we have found that the idea of 'one fits all' is particularly dominant in the design of the bicycles, which is characterized by a bicycle fleet consisting of completely identical bicycles. Our purpose in this paper is to challenge and advance the idea in bike-sharing schemes of

---

[4] https://www.tinngo.eu/

'one fits all.' To do so, we focus on people who do not use the existing bike-sharing schemes.

## 4 Methodological Approaches

The paper builds on reflections, experiences, and accounts of non-users of bike-sharing systems, collected in two geographical settings where bike-sharing is incorporated in urban spaces. To talk to people as a mean of understanding their lives is a core approach in the social sciences (Hitchings 2012), and this include mobility research as well (Büscher et al. 2020). A challenge in designing the present study was how to ask people about something that they do not do, i.e., use bike-sharing schemes, and whether their answers would in fact provide any insights into their non-doings. Inspired by the mobility literature, we focused our questions about their daily mobility with a particular interest in travel needs. These questions spurred reflections on transport experiences, attitudes, and emotions, and allowed us to analyze how the complexities of everyday life structure people's mobility needs. When describing and reflecting upon everyday practices related to mobility, to reflect upon how other mobility practices, such as bike-sharing, might or might not work seems plausible. To enable reflections on how a bike-sharing system could be designed, we introduced some illustrations of different types of bicycles in the interviews. In most of the interviews, we asked the participants if a bike-sharing scheme would be attractive to them if it offered any of these bicycles. Our visualization of a bike-sharing solution characterized by different types of bicycles initiated various stories about imagined transport practices as well as openings for new mobility opportunities. Thus, we set out to contrast non-user narratives on mobility to what kind of mobility present bike-sharing systems prescribe and allow.

Analytically, we are combining theories of gender and diversity with the narrative approach. Narratives should here by understood as the stories people tell themselves and each other to make sense of their past experiences (Riessman 2008:8). The narrative approach argues that storytelling is part of social meaning making: "if stories are about our lives, the world, and its events, they also are part of society" (Gubrium and Holstein 2009:xv). To study narratives thereby means to analyze the meaning-making that takes place between individuals and society (Andrews et al. 2000:9). That means that we emphasize the mutual dependent structures which decide why people do not use the bike-sharing schemes. These are spelled out in overall storylines, which represents certain mobility needs, illustrating how these needs are not met by existing bike-sharing schemes. The storylines on mobility are summarized in three overarching narratives, which we regard as analytical categories. The analytical categories are constructed on the basis of narratives emerging in our interviews on why people do not use the present bike-sharing systems.

We will now describe the empirical material in more detail, focusing also on the geographical context in which the data was collected.

### 4.1 Copenhagen Data Collection

Copenhagen, the capital of Denmark, is a metropolitan area with a population of 632.340 in Copenhagen municipality and counting a population of 1.330.993 in the area of Greater

Copenhagen. The City of Copenhagen has continued investments in the city's bicycle infrastructure and a goal that 50% of trips between home and work will be made by bicycle in 2025. The figure is currently 49% (2019). The city of Copenhagen is experiencing a growing shortage of space, mainly a lack of bicycle parking and congestions on the bicycle lanes due to the large number of daily cycling commuters. The need to expand infrastructures must further be seen in the light of an overall growth in traffic as a result of population growth, as well as increasing car ownership.[5]

In Copenhagen, we conducted seven single interviews on the use of the city's bike-sharing scheme. Of the seven interviewees, three were men and four women. Three were in their twenties, two in the thirties, and the last two in the mid-forties and fifties. They all lived in the area of Greater Copenhagen, two were students, one was unemployed, and four worked in different white-collar jobs. Most of the interviewees had tried the white city bicycles a single time, a few had used it on a regular basis for some time but had now stopped using it due to various reasons. One of the interviewees had never used a bike-sharing bike. The interviews were conducted online by the use of a semi-structured interview guide. Questions covered topics on their view on bike-sharing solutions, if they ever used it and how the system eventually met their mobility needs and patterns. Each interview lasted about 30 min, they were all recorded and afterwards transcribed.

Further, we conducted two focus group interviews with in total seven ethnic minority women living in a suburban part of Copenhagen. They were interviewed in connection with a biking course established by the Red Cross for adults who wanted to learn to how to ride a bike. In contrast to our single interviews, these women were inexperienced cyclists and were not born in the Danish biking culture. They came from cultures where biking was not a widespread mode of transportation and where it sometimes was illegitimate for women to cycle. The interviewees were between 37 and 61 years old. They were from Morocco or Turkey, most had lived in Denmark for about 10–20 years, one had moved to Denmark three years ago. They worked as cleaners at the local schools, one just went on early retirement and one was attending courses at an adult education center. The two focus group interviews lasted about an hour each. They were recorded and transcribed.

## 4.2  Linköping Data Collection

Linköping is a middle-sized municipality in eastern Sweden. The population is 160 000. The municipality has a well-functioning public transport network and infrastructure for cycling. There are dedicated cycle paths, shared cycle/pedestrian paths and cycling in mixed traffic is reoccurring. Traffic volumes is generally low or medium and pedestrians and cyclists are common. Almost half of the population have five kilometers or closer to work, and the overall bicycle share is 27%, which is relatively high in comparison to other Swedish cities. This reflects the geographical conditions, infrastructure but also to the fact that the university attracts a lot of students. Still the municipality aims to increase the number of bicycle trips to a total of 40% of all trips. Based on this, bicycling can be described as an established practice in Linköping. On the other hand, a report (Linköping municipality 2018) about bicycling in a low-income area in Linköping, reveals that women with an immigrant background seldom bicycle. In the report, 60 women over

---

[5] https://www.kk.dk/sites/default/files/edoc/Attachments/23020624-32266360-1.pdf.

35 year in the area were interviewed. Among these, 15 persons stated that they could not cycle at all. All except one had an immigrant background. Many states that they want to bicycle more, but regard it as too burdensome. Many states that they have access to bicycles, but report that their bicycles are in bad shape, that it is time-consuming to cycle, and that thefts are common, as obstacles (Linköping municipality 2018). The report gives insights on how bicycling is perceived as different depending on your background and experiences and that bicycle culture varies between as well as within cities.

In the Linköping case, we interviewed persons who have access to the local bike-sharing scheme today, through their residential area or workplace, but they do not use the system. In all, we interviewed 11 non-users. We wanted to reach a broad group of non-users and therefore used a variety of recruiting strategies. We advertised for informants through a mobility management newsletter that target workplaces that promotes sustainable mobility for their employees. Through the newsletter we reached 7 persons (5 women, 2 men) that had access to the local bike-sharing scheme where a station was located close to their working place. These persons had an ethnic Swedish background, had white-collar jobs, and where in their forties (4 persons) and fifties (2 persons). One person, a woman, was 25. They all travel to work on an everyday basis, a majority of them cycled, and some travelled by car, but had little or no experience of bike-sharing. They were all familiar with the bike-sharing scheme. These non-user interviews were conducted online (due to the covid-19 pandemic) as one-to-one meetings and lasted approximately 30 min. They were all recorded and transcribed in full length. To reach more inexperienced cyclists in low-income areas, we also made an advert that was circulated in two of the areas that had access to the bike-sharing service, and on Facebook. The advertisement had a Swedish and an Arabic version. We reached one person, a 17-years old man with an immigrant background. As a result of the advert, we were invited to attend a bicycle course that targeted immigrant women that was set up by the Red Cross. We made observations on two occasions and did on-site interviews with three women between the ages 25 and 45. Two of them attended school for newly arrived on a daily basis. They had an origin in Lebanon and were both married with small children. The third woman, a care worker, had lived in Sweden for a longer time, and worked in her residential area. She was divorced and cared for her three children.

## 5  Narratives on Mobility and Cycling

In the next three sections, we analyze obstacles for using the existing bike-sharing schemes. The analysis departs in interviews with non-users in the context of Linköping (Sweden) and Copenhagen (Denmark). The analysis is structured around narratives on mobility, which we found in the interviews, and which relate to cycling abilities and mobility needs. In our analysis of the empirical material, we have identified three overall narratives: 'I have my own bicycle,' 'I travel with kids,' and 'I don't feel safe.'

**'I have my own bicycle'**

*"I always cycle around on my own bicycle."*

Throughout the interviews it became clear that the interviewees were in fact support-ive to the concept of bike-sharing, i.e. they were not against the idea, but just could not see the relevance to themselves. This reflects findings in previous research. 'I have my own bicycle' was a mobility narrative, repeated by various non-users, and perhaps the most straightforward and recognizable explanation for why people do not use a shared bicycle. Private bicycle ownership is very common in Scandinavia, and to use the bicycle in everyday life is regarded as 'normal.' As such it is also difficult to get people to talk about why they use their own bicycle and not a shared bicycle. However, a central story throughout the interviews was that using your private bicycle appears to be more con-venient than using a shared bicycle. A young man expresses the convenience of private cycling, saying:

> I have my own bicycle, and so... bike-sharing costs money, and it's also fair enough, but then you also have to go and find it. I do not know if there is so much to do about it. No, I just think that now I have my own bicycle and then it is easier just to use my bicycle (Man, 21, Copenhagen)

Several of the people we interviewed in the Danish setting, including this man, had tried the bike-sharing scheme once or also several times. This young man had used shared bicycles when he lived at home with his parents about 10 km from Copenhagen city center. When he had left his own bicycle at home, he rented a shared bicycle to move around in the city. However, he stopped using bike-sharing schemes when he moved into town. A very similar narrative came from another young man, who had used the public bike-sharing system when he did not live in Copenhagen, but occasionally came to visit friends. Now when he moved to Copenhagen, he prefer to use his own bicycle because it, as he says, is 'nicer':

> Definitely that I have a nice bicycle, it is good to ride, it is not so heavy, if you are too drunk to ride a bicycle, then you can also take it into an S-train, you know, it is not large. And then there is also something special about having your own bicycle, it's such a mental thing. Then you also take better care of it, like, so it's just like your own bicycle. (Man, 22, Copenhagen)

As such, a frequent argument among the cyclists is that they do not use shared bicycles since they simply prefer using their own bicycle rather than renting one. With little experience of bicycle sharing, questions around rules, unexpected costs and access arise. One respondent in Linköping who usually takes her own bicycle for work trips does not find Linbike to be attractive. She explains that it makes her feel uncomfortable if she during a meeting would occupy a bicycle that could be used by someone else, and adds that she generally associates renting with a sense of stress, where unexpected events risk leading to extra expenses. Another respondent worries about access:

> What will happen if I have planned to cycle, and then there is no bicycle available, and I will get late to work? (Woman, 25, Linköping)

As mentioned, this kind of reasoning is not surprising in the Scandinavian context, where the majority of people own one or more bicycles. Some of those who usually cycle, especially the ones who cycles year around and commute far distances, are reluctant of using a shared bicycle with reference to that they need to be certain of the bicycles' functionalities and that the bicycle is adjusted to their personal body sizes. For example, the ability to plan ahead is also mentioned in relation to artifacts used when bicycling. One respondent, a woman, who commute to work 8 km by bicycle, argues that preparations are essential for this to work and that she for example have bike-able clothing ready. She is hesitant to use the bike-sharing scheme in Linköping since they neither provide protective clothes nor helmets. This reasoning sheds light on that cycling is not regarded as spontaneous for everyone, since such trips demand access to equipment besides the actual bicycle:

> I believe that a helmet is something that many still want. I for one, will not get on a bicycle without a helmet. (Woman, 59, Linköping)

A consistent narrative in the category of 'I have my own bicycle' was that a bike-sharing scheme could in principle be attractive *if* it offered something that one's own bicycle could not. This 'in principle' had been started by us as interviewers and was not something that the interviewees themselves came up with. As previously noted, it is rather difficult to interview people about something they do not do or use. Using a shared bicycle was for many of them not even present in their consciousness. *"I do not really think about that they are there, so I do not think of that I could actually use them if my own bicycle was punctured,"* as a 26-year-old woman said. During the interviews we asked the interviewees to look at the document we had made with illustrations of different types of bicycles, asking if a bike-sharing scheme would be attractive to them if it offered any of these. Our visualization of a bike-sharing scheme characterized by different types of bicycles initiated various stories about imagined transport practices as well as openings for new mobility opportunities. Especially when it came to cargo bikes.

> If there were cargo bikes, not because I so often use a cargo bike, but for example I have been to a couple of bachelor parties, where we have had to use a cargo-bike. Or if I want to go for a picnic or if I want to take a ride with our nieces, so it is such an easy way to take them around town, instead of having to borrow a child seat from their parents and make sure they brought them in and connect it to our own bicycles, so if there were cargo bikes, I think I might as well consider using them. (Woman, 26, Copenhagen)

The cargo bike as part of a shared bicycle fleet would offer them something that their own bicycle does not. None of the persons who represent the narratives of 'I have my own bicycle' owned a cargo bike themselves, mainly because they did not have children and therefore did not need to transport anyone other than themselves on a daily basis. But sometimes, of course, they needed to move someone or something. For example, in the Linköping case, a man with a passion for cycling in general, mention that it is hard to transport soil to his allotment with his current bicycle. Another respondent argue that a cargo bike can be a valid alternative to the car:

*You could use one of those [a cargo bike] to the grocery store to be able to shop more. Really, you could load it with four or five grocery bags instead of taking the car. That could be an idea. (Man, 46, Linköping).*

The idea of more diversity of shared bicycles clearly increased the interest in using a bike-sharing solution among people with own bicycles. Cargo bikes, as well as other special bicycles, are expensive solutions that not everyone can afford or use to such an extent that it is worth investing in. While a shared solution of cargo bikes to some extent is offered today in Copenhagen, mainly in connection with municipal cultural centers or libraries, they typically only offer 2–3 cargo bikes that can be booked and rented or borrowed. Not everyone is aware of this service and as one of the interviewees told, the few cargo bikes available are not enough to meet the needs of an entire urban population.

*I have been interested in some of the cargo bikes which you can borrow at the libraries, but they just haven't been there when I wanted to borrow them. But now the need has been less because we have signed up for a car-sharing scheme and use it a lot for transporting things and since there is a car option for this in a somewhat similar way, you can say that it is not because the bike-sharing does not appeal to me, I can easily see the point in it. (Woman, 36, Copenhagen)*

The 'I have my own bicycle' narrative is strongly related to convenience, which is also reflected in previous research on bike-sharing (Berg et al. 2019). The problem with the public bike-sharing solution in Copenhagen is, according to more of the people we interviewed, that the bicycles have not been well maintained. When the technology of the bicycles failed, e.g., when the GPS did not work or the electric motor was broken, it became too inconvenient. Then the introduction of a new metro system became an alternative for everyday travel that outperformed the bike-sharing solution. Further, the public bike-sharing scheme has been heavily vandalized, which meant that most of the bicycles were taken away from the streets, leaving docking stations almost empty. The bike-sharing system in Linköping have also encountered problems with vandalism and several stations have been shut down during longer periods of time. There have also been extensive complaints on technological errors as wells as problems with the maintenance of the bicycles due to heavy use.

This section has shown that a reason of non-use of bike-sharing schemes is due to the fact that it is a solution that offers something that many already have access to – a bicycle. Furthermore, privately owned bicycles are considered better adapted to mobility needs and more easily accessible and therefore convenient. As one of the interviewees put it *"I think it would take quite a lot for me to stop having my own bicycle"* (Man, 22, no kids, Copenhagen). Yet, the attraction of a bike-sharing scheme might not so much be about getting rid of private bicycle ownership, but about acknowledging that differences in mobility needs requires differences in bicycle solutions. We will take a closer look at this in the next section.

**'I travel with kids'**

*"A bicycle where I could have a child would be a huge advantage for me."*

In the section above, we highlighted how bike-sharing schemes could become more attractive to people with their own bicycles if bike-sharing solution began to think more in terms of bicycle diversity. In this section, we look at how the 'one fits all' solution that characterizes bike-sharing schemes today actually works to exclude some people from using the system. This is especially true for the groups who travel with children as well as others who are dependent on them, such as older family members. This is a group who perform what we have referred to as care trips, i.e. where the purpose of the trip is to pick up and bring children to child care institutions or schools, to escort family members who cannot travel on their own to doctor's appointments, etc. Care tasks is further related to other household tasks, such as shopping errands and traveling with groceries. Of course, these responsibilities connected to the household define – as well as have impact on – people's mobility needs and opportunities. Narratives about mobility needs in relation to caring tasks were present in both the Danish and Swedish contexts. In particular, a Danish 39-year-old man who previously had used bike-sharing schemes in Copenhagen expressed several challenges in making everyday life work when using the existing bike-sharing scheme.

> *It is clearly a shortcoming of the white city bicycles that their basket is such a flap that needs to be pulled up and tightened. It is not very good, for example a shopping bag then things pop up. I have tried to shop on the way home, but then you have to tie your shopping bag to a bundle before it works, so it is not so good to carry things on, besides a bag that you have on your back, so a basket is completely what I miss. And also because I have a concrete need with my daughter, if I could go to a place close to my home and get a bicycle where I could have a child in, also a big child, then it would actually be a huge advantage for me, because then I could just keep the bicycle all the way to my work. (Man, 39, Copenhagen)*

He lives a little outside the center of Copenhagen and goes to work in the city center every day as well as he every other week, where he has his two children, brings them to school somewhere else in the inner city of Copenhagen. As his daughter feels unsafe in traffic and does not want to cycle herself, he first takes the children on public transport to school, after which he sometimes switched to a shared bicycle to continue to work. He prefers to just cycle and thus use the same form of mobility all the way, but since he does not own a cargo bike that the daughter can be transported in, this is not an option. As he says:

> *I could of course just invest in a cargo bike, but because I am a part time dad, there are many times where it does not make sense for me to ride around on a cargo bike by myself. (Man, 39, Copenhagen)*

For people who do not need to transport someone or something every day, a cargo bike is a big investment, as well as a bulky form of cycling, and therefore many will be reluctant to buy one. For this man, the possibility of sometimes renting a cargo bike as part of the shared bicycle solution would thus clearly be attractive. Furthermore, care trips often mean that different needs must be met on the same trip. Some go to school, some to work, some want to cycle themselves while others prefer not to cycle themselves.

*I live with my children every other week and their school is too far away to walk and they have different ages, so the oldest can cycle to school and the youngest can sometimes, but she is not so happy about it, so therefore I sometimes accompany her to school via public transport and then I have to go to work where I absolutely always prefer to cycle. (Man, 39, Copenhagen)*

Care trips are often characterized by multiple trip chains, meaning that there are multiple stops along the way. Here, bike-sharing schemes could be seen as a good opportunity as the solution is intended as a way to cover the last. But in a busy everyday life with kids, this might not be a particularly easy choice.

*On a normal day I cycle to work and back home. The days I am responsible for getting the kids to the preschool, I take them in the bicycle cart. It takes me five minutes to get there. Then I leave the cart at the preschool, and cycle to work. And then after work I will cycle to the pre-school again, and get the kids and the bicycle cart. I don't live like I did before now that I have moved out of the city and have small kids. I now have a bigger need for flexibility and can't go between fixed docking stations. But I think it [bicycle sharing] is great for those who move about those locations where the stations are. There would need to be a docking station in my garage at home, and one at work. It's not realistic. The same goes with the bicycle cart. The shared bicycles are not prepared for bicycle carts. (Woman, 40, Linköping)*

The above quote from an interview with a 40-years woman from Linköping clearly shows how care responsibilities as well as location of home and work impact on people's mobility needs. Although bike-sharing schemes are said to provide a flexible solution, it is not a form of flexibility that caters to families with children. The problem can be summed up quite simply as a question of 'how to bring children on a shared bicycle?' The flexibility of a bike-sharing scheme is also challenged by the fact that the shared bicycles in both Copenhagen and Linköping are connected to docking stations. For the solution to be attractive, there must be a docking station near the beginning and end of the journey. As the woman from Linköping tells, this is not the case for her. Research also shows that docking stations are unequally placed and mostly found in areas where predominantly men work (Uteng et al. 2019) or where affluent groups with an already wide choice of accessibility services live (Nixon and Schwanen 2019). More respondents mention that they could see themselves using shared bicycles but argue for that to be realistic they need to be located where they live, for example in the basements of multifamily dwellings. A respondent commuting by bus mentions that the docking stations in the Linköping scheme are not located to match her trip and that she would need to take an unnecessarily complicated detour to pass the docking stations. Another respondent mentions that her family situation with small kids requires a flexibility that is not possible to achieve if she is required to bicycle between predetermined locations.

To summarize, everyday care tasks often means that different travel needs must be met on the same journey. From this follows that some of the main features of bike-sharing scheme's do not suit people with care responsibilities. This is particular due to fixed stations far away from residential areas as well as the "one fits all" approach to design. Users with caring responsibilities often have established bicycle practices and own vehicles that suit their needs.

Our final narrative includes persons who to some extent feel insecure about practices of cycling, but are nevertheless potential cyclists as well as users of bike-sharing schemes.

### 'I don't feel safe'

*"It can be dangerous to ride the shared e-bikes in traffic without helmet."*

A third reoccurring narrative about everyday mobility included stories about feeling unsafe as a bicyclist. In our material this narrative was reoccurring in the interviews with ethnic minority women who had not cycled before, but currently are attending a bicycle course to learn how to cycle. For them, there are several obstacles to cycling, some very practical such as a lack of knowledge about how to ride a bicycle, but also related to cultural norms about cycling, which sometimes include a notion about cycling not suitable for women. For these women, to start cycling is interpreted as getting access to previously denied areas. To be able to bicycle can for example be a requirement for certain jobs that include work trips with bicycles, such as home care service. The women argue that bicycling allows freedom, to be less dependent on public transport or on their partners or children to drive them. That bicycling is regarded as normal and even required in the Scandinavian setting, is also acknowledged. A woman participating in one of the bicycle courses in Linköping jokingly says, *"In Sweden, everyone cycles, and everyone drinks water"*. To ride a bicycle is desired, but is also connected to feelings of lack of knowledge, unsafety and even shame.

To take on cycling as an adult is challenging. Sometimes dramatic changes in one's life can be a reason for daring to attend a bicycle course. One of the women in Linköping shares a story about how she always wanted to bicycle but has been too embarrassed to try. When she got severe cancer, she decided to divorce her husband and learn how to bicycle, *"many are embarrassed not being able to cycle, but I just don't care anymore"*, she says. She thinks it is difficult to get the balance right – an experience she shares with many of the ethnic minority women that learn to cycle as adults.

To feel unsafe as a cyclist is thus an obvious obstacle for bike-sharing. Another obstacle is knowledge about and access to the bike-sharing schemes. When we talked about shared bicycles as e-bikes in the focus group interviews we did in Denmark, the participants thought we were talking about e-scooters, which they know well. The bike-sharing scheme is apparently less known to them than the shared scooters, which were launched relatively recently in Copenhagen. In the Copenhagen context, the ethnic minority women in general lack knowledge of bike-sharing schemes. This has to do with that the solution is not available in the suburban area they live in. A 50-year-old ethnic minority woman says:

*How do you find a shared bicycle? For example, yesterday I went to Østerbro to visit The Little Mermaid. If I were to take a shared bicycle, how would I find one? (Woman, 50, Copenhagen)*

Even if many lack knowledge about the bike-sharing docking stations, the ethnic minority women from the Danish bicycle course were keen to use the shared bicycle system. However, the traffic situation in the central places form another obstacle for inexperienced bikers: *"At Nørreport and Vesterport it is really difficult. I don't dare to cycle there"*, a 61-year-old ethnic minority woman says. In the Linköping context, the women in the bicycle courses knew about the system since many lived in an area where a docking station was located. However, they had not tried the system themselves and were not interested in doing so. For them, the thought of riding a fast electrical bicycle was far away from how they imagine cycling, which some of the informants that owned a bicycle also mentioned. A 37-year-old woman says that she just wants a normal bicycle, with a basket, so that she can go to the forest and other nice outings with her children. Since she works in the same area as where she lives, she is not interested in cycling in her everyday life. To cycle is interpreted as being recreational, and can improve the quality of life.

As mentioned, the ethnic minority women in the Copenhagen context show interest in using a shared bicycle. Yet, they are reserved towards the electrical solution, which is associated with high speed and more accidents. They were, however, quite interested in possibilities for renting a tri-bike. Not so much because they have to transport others, as was the case in the two narratives above, but because their balance is still poor, and they are struggling to learn to ride the two-wheelers. The attractiveness of renting a tri-bike is also due to the fact that tri-bikes as well as the cargo-bikes are very expensive and are often stolen. A woman in Linköping reasons in similar terms when it comes to what kind of bicycle she prefers but is more positive towards the electric bicycle. Even if she still feels insecure cycling on a standard bicycle, she bought a cargo-bike, which she uses to ride her children to school and to the preschool. While the cargo-bike is more stable and thus easier to operate, it also a heavy vehicle, here electricity would be able to assist her.

Issues of safety is not only valid for unexperienced bicyclists. A young man that lives in a low-income area is familiar and positive towards the Linköping bike-sharing scheme. He describes how he has seen similar solutions in other cities and regards it as an urban quality. He indicates that e-bikes are attractive to him since they make cycling more comfortable. However, he has also noticed how fast the shared bicycles go and thinks of this as a safety issue. Since the bicycles share space with cars, and cars are allowed to dominate the traffic environment in Linköping, he would not use the shared bicycles without a helmet and as he says, *"I don't use a helmet, for different reasons"*. He also mentions that he does not like the bright orange color of the shared bicycles. A more neutral color would suit him better. This brings about a final aspect of the "secure"-narrative, which connects to the fact that using bike-sharing makes you visible in the urban space. To use a shared bicycle might include unwanted attention. How this young man reasons about the potential dangers of bike-sharing, related to being viable as a user, indicates that norms about being perceived as someone who values safety, prevents him from being a future shared bike-user. At the moment, he is prevented from using the shared bicycles due to an 18-years age limit, which is another barrier to usage.

## 6   Concluding Remarks: The Potentials of Bike-Sharing Schemes

The point of departure in this paper has been that new so called smart mobility solutions must be attractive for the whole population if they are to be incorporated into their daily lives. Today, the development of Smart Cities and the smart mobility solutions within these cities, is focused on technological solutions. Yet, smart transport solutions need to meet the needs of actual people living in these cities in order to succeed. That is, for bike-sharing to work, we need to understand the needs of potential users. The lack of perspectives on diversity is striking. Throughout research on bike-sharing as well in strategy papers and existing bike-sharing providers' platforms, the (potential) users of bike-sharing schemes are expressed in general terms with no specification about who these users are in terms of gender, age, family status as well as other variables that might affect their mobility needs. That makes bike-sharing a service that aims to attract more people, but without knowledge about what would attract them to use a shared bicycle. Tapping into this problem, the aim of this paper has been to gain an in-depth understanding of what prevents different people from using the present bike-sharing system.

Informed by theories on gender and diversity, that point to the need to adopt multiplicity when designing and implementing new 'smart' mobility solutions, we have identified three overall narratives on mobility, which describe different kind of non-user's experiences, norms, and habits around everyday mobility. The first narrative includes experiences of owning a bicycle, which is thought to offer more comfort than a shared bicycle. The second narrative draws attention to care trips, and how persons with caring responsibilities must attend to a variety of needs, which make it difficult for them to adopt a bike-sharing. These obstacles are mainly because of fixed docking stations and the uniform bicycle design, which offers little opportunity to bring children on the trip. Finally, the narratives told by primarily ethnic minority women show that for inexperienced cyclists, cycling might be a challenging form of mobility, which we often take for granted in Denmark and Sweden. Relating to these narratives, we ask if there is a potential to widen the approaches to bike-sharing?

One the one hand, our findings is in line with previous research, showing that having your own bicycle is the primary reason for not using bike-sharing systems (Nikitas et al. 2016), and that comfort is a driving force when choosing modes of transport, including cycling (Berg et al. 2019). While it seems from the narratives of "I have my own bicycle" as if bike-sharing schemes is less attractive than one's own bicycle when living in the city, this does not mean, however, that there are no openings for city dwellers to adopt bike-sharing. People's private bicycles are not always well-functioning – they might be punctured, or individuals do not have the opportunity to take their bicycles with them on the entire journey. Also, in areas where bicycle thefts and vandalism are common, to adopt a bike-sharing schemes can be a way of getting access to a well-functioning and high-quality bicycle, without worrying about thefts or extra maintenance. This, however, demands that the operators take extra notion about security and maintenance issues. Furthermore, the narratives within the narrative "I have my own bicycle" showed that private bicycle ownership is not the same as having all kinds of bikes and the possibility to rent a cargo-bike for certain activities, seemed attractive to this group.

The narrative of travelling with kids shed light on how care trips structures people's mobility and generate certain needs when it comes to cycling. Travelling with dependent family members, such as children, means that some of the present main features of bike-sharing schemes do not suit users with caring responsibilities. New technologies, such as GPS tracking and geofencing, might offer more flexibility when it comes to picking up and returning shared bicycles. Yet, these solutions might demand new approaches from the operators, which can conflict with previous setup when it comes to management and maintenance. More importantly, a more diverse bicycle fleet that enables care trips is vital if families are to adopt bike-sharing. A bike-haring fleet which includes cargo-bikes could meet the mobility needs when travelling with children. Furthermore, a bike-sharing scheme with cargo-bikes could help to mitigate the pressing urban challenges seen in Scandinavia of an increasing number of private cargo bikes on the bicycle paths, which create a shortage of space and thus increase the feeling of unsafety of persons travelling with children or prevent children from start cycling themselves (cf. Aldred et al. 2016).

In relation to the narrative of feeling unsafe, we have foremost highlighted ethnic minority women, even if we also raised how issues on safety and norms about cycling can be important for more experienced cyclists too. This narrative thus cut across several other categories. Older people and other persons with impaired mobility, as well as those with no experience of e-bikes, might need bicycles designed for their needs, such as tri-bikes and cargo bikes. As these bicycles are expensive, the inclusion of such bicycles in bike-sharing schemes might improve the mobility of people with impaired mobility radically and thus benefit equal opportunities in mobility. Also, offering a bike-sharing scheme of various bicycles would help problems of bicycle parking and congestion in urban areas.

While our analysis points to that a "one fits all" model is not an attractive solution for everyone and the need to think and design bike-sharing schemes with the notion of more diversity, most of the bike-sharing schemes today are still conforming towards this single model. As Gelinder (2020) points out, even if diversity might be regarded as an interesting topic for operators, the 'economy of scale' is deemed as more important. If bike-sharing operators continue a 'business as usual' logic, bike-sharing will most likely not be able to meet a variety of needs and have little potential to be inclusive. However, there are examples where operators have gone beyond the dominant bike-sharing designs (cf. Nixon and Schwanen 2019). The Kajteroz system in the polish city Chorzów offers bicycles for children as well as adults, and options with and without electricity. Also, there are examples of municipalities offering cargo-bikes and e-bikes as part of "bike libraries," free for citizens to try out. Third, that many citizens already have their own bicycle can be viewed as a potential for developing new models for bike-sharing. Cycle.Land is an example of a peer-to-peer sharing service where users get access to a variety of privately owned bicycles.[6] Whether innovations like these are able to complement or even challenge the present "one fits all" services, depend on the decisiveness of public actors.

---

[6] These examples are collected from a workshop on bike-sharing business models that can support diversity, arranged by the authors within the frame of TInnGO, a horizon 2020 program focusing on gender and smart mobility. More info about the workshop: https://transportgenderobservatory. eu/2021/01/25/scandinavian-hub-workshop-on-gender-and-diversity-in-biking/

This paper has argued that we need to include the various mobility needs of different people in the development and design of bike-sharing schemes in order to make it an attractive and 'smart' mobility solution for all. A social perspective on bike-sharing schemes could advance the systems, and support cities goals related to gender equality, diversity and inclusion. We suggest that for public actors to engage in new models for bike-sharing can be way to meet diversity and equality goals.

**Funding.** This research received funding from the European Union's Horizon 2020 research and innovation programme under grant agreement no. 824349.

# References

Albino, V., Berardi, U., Dangelico, R.M.: Smart cities: Definitions, dimensions, performance, and initiatives. J. Urban technol. **22**(1), 3–21 (2015)

Aldred, R., Woodcock, J., Goodman, A.: Does more cycling mean more diversity in cycling? Transp. Rev. **36**(1), 28–44 (2016)

Andrews, M., Sclater, S.D., Squire, C., Treacher, A., Denzin, N.K.: Lines of narrative: Psychosocial perspectives vol. 8, Psychology Press (2000)

Angeles, L.C.: Transporting difference at work. In: Cohen, M.G., Cohen, M.G. (eds.) Climate Change and Gender in Rich Countries: Work, Public Policy and Action, pp. 103–118. Routledge, Abingdon, New York, Routledge (2017)

Balkmar, D.: Cycling politics: imagining sustainable cycling futures in Sweden. Appl. Mobil. **5**(3), 324–340 (2020)

Berg, J., Henriksson, M., Ihlström, J.: Comfort first! Vehicle-sharing systems in urban residential areas: the importance for everyday mobility and reduction of car use among pilot users. Sustainability **11**(9), 2521 (2019)

Breengaard, M.H., Christensen, H.R., Oldrup, H.H., Poulsen, H., Malthesen, T.: TRANSGEN - gender mainstreaming European transport research and policies: building the knowledge base and mapping good practices. University of Copenhagen: Co-ordination for Gender Studies (2007)

Büscher, M., Freudendal-Pedersen, M., Kesselring, S., Kristensen, N.G.: Handbook of Research Methods and Applications for Mobilities. Edward Elgar Publishing (2020)

Caggiani, L., Camporeale, R., Hamidi, Z., Zhao, C.: Evaluating the efficiency of bike-sharing stations with data envelopment analysis. Sustainability **13**(2), 881 (2021)

Colville-Andersen, M.: Watching Copenhagen Bike Share Die. Copenhagenize.com (2015). http://www.copenhagenize.com/2015/02/watching-copenhagen-bike-share-die.html

Crenshaw, K.: Demarginalizing the Intersection of Race and Sex: A Black Feminist Critique of Antidiscrimination Doctrine, Feminist Theory, and Antiracist Politics, University of Chicago Legal Forum, pp. 139–67 (1989)

Fishman, E., Allan, V.: Bike share. Advances in Transport Policy and Planning **4**, 121–152 (2019)

Freudendal-Pedersen, M.: Whose commons are mobilities spaces? The case of Copenhagen cyclists. ACME Int. J. Crit. Geograph. **14**(2), 598 (2015). http://acmejournal.org/index.php/acme/article/view/1188

Gelinder, M.: Potential justice implications in system design of bicycle sharing systems. Radboud University Nijmegen (2020). https://theses.ubn.ru.nl/handle/123456789/10248?locale-attribute=en

Gubrium, J.F., Holstein, J.A.: Analyzing narrative reality. sage. (2009)

Hanson, S.: Gender and mobility: new approaches for informing sustainability. Gend. Place Cult. **17**, 5–23 (2010)

Henriksson, M., Wallsten, A.: Succeeding without success: Demonstrating a residential bicycle sharing system in Sweden. Transportation Research Interdisciplinary Perspectives **8**, 100271 (2020)

Hitchings, R.: People can talk about their practices. Area **44**(1), 61–67 (2012)

Haustein, S., Koglin, T., Nielsen, T.A.S., Svensson, Å.: A comparison of cycling cultures in Stockholm and Copenhagen. Int. J. Sustain. Transp. **14**(4), 280–293 (2020)

Kronsell, A., Dymén, C., Rosqvist, L.S., Hiselius, L.W.: Masculinities and femininities in sustainable transport policy: a focus on Swedish municipalities. NORMA **15**(2), 128–144 (2020)

Kronsell, A., Rosqvist, L.S., Hiselius, L.W.: Achieving climate objectives in transport policy by including women and challenging gender norms: the Swedish case. Int. J. Sustain. Transp. **10**(8), 703–711 (2016). https://doi.org/10.1080/15568318.2015.1129653

Law, R.: Beyond 'women and transport': towards new geographies of gender and daily mobility. Prog. Hum. Geogr. **23**(4), 567–588 (1999)

Linbike: Linbike – Linköpings elcykelpool. Uppföljning Sept. 2019–Jan. 2020. PPT by P. Semberg. [LinBike – Linköping bike-sharing service. Follow-up Sept. 2019–Jan. 2020 (2020)

Linköping Municipality: Cykellänken. Så blir den bättre för fler. Linköping municipality and Pedalista. [The Bicycle link. How it will be better for more people] (2018)

de Madariaga, I.S., Zucchini, E.: Measuring mobilities of care, a challenge for transport agendas. In: Integrating Gender into Transport Planning, pp. 145–173. Palgrave Macmillan, Cham (2019)

Nikitas, A., Wallgren, P., Rexfelt, O.: The paradox of public acceptance of bike sharing in Gothenburg. Proc. Inst. Civ. Eng. Eng. Sustain. **169**(3), 101–113 (2016). https://doi.org/10.1680/jensu.14.00070

Nixon, D.V., Schwanen, T.: Bike sharing beyond the norm. J. Transp. Geogr. **80**, 102492 (2019). https://doi.org/10.1016/j.jtrangeo.2019.102492

Prati, G.: Gender equality and women's participation in transport cycling. J. Transp. Geogr. **66**, 369–375 (2018)

Riessman, C.K.: Narrative methods for the human sciences. Sage. (2008)

Scheiner, J.: Gendered key events in the life course: effects on changes in travel mode choice over time. J. Transp. Geogr. **37**, 47–60 (2014)

Scholten, C.L., Joelsson, T. (eds.): Integrating Gender into Transport Planning: From one to Many Tracks. Springer, Cham (2019). https://doi.org/10.1007/978-3-030-05042-9

Singh, Y.J.: Is smart mobility also gender-smart? J. Gend. Stud. **29**, 832–846 (2020). https://doi.org/10.1080/09589236.2019.1650728

Turner, J., Grieco, M.: Gender and time poverty: the neglected social policy implications of gendered time. Transp. Travel Time Soc. **9**(1), 129–136 (2000)

Uteng, T.P., Espegren, H.M., Throndsen, T.S., Böcker, L.: The gendered dimension of multimodality: exploring the bike-sharing scheme of Oslo. In: Gendering Smart Mobilities, pp. 162–187. Routledge (2019)

Uteng, T.P., Christensen, H.R., Levin, L. (eds.): Gendering Smart Mobilities. Routledge (2020). https://doi.org/10.4324/9780429466601

Wang, M., Zhou, X.: Bike-sharing systems and congestion: Evidence from US cities. J. Transp. Geogr. **65**, 147–154 (2017)

# Electroencephalography Shows Effects of Age in Response to Oddball Auditory Signals: Implications for Semi-autonomous Vehicle Alerting Systems for Older Drivers

Melanie Turabian(✉) [iD], Kathleen Van Benthem[iD], and Chris M. Herdman[iD]

Carleton University, Ottawa, ON K1S 5B6, Canada
melanieturabian@cmail.carleton.ca, {kathyvanbenthem,
chrisherdman}@carleton.ca

**Abstract.** This research considers the efficacy of auditory alert systems in semi-autonomous vehicles (SAVs) from the perspective of the neurological processing of multi-modal information. While SAVs are growing in popularity, there is much to be discovered concerning driver safety. Understanding how the brain integrates multi-modal information is essential to determining the efficacy of auditory alerting systems in SAVs and whether or not they suffice as a method for conveying information to drivers. Investigating how younger and older groups process various types of auditory information while engaged in visuospatial tasks of different workload levels is a crucial step to take to optimize safety in SAVs. We report on how auditory processing of deviant and standard tones was impacted by age at decision-making areas of the brain using electroencephalography (EEG). EEG and behavioural data from 10 participants, five older (57–78) and five younger (18–26) were analyzed. Participants completed four rounds of a match-to-sample visuospatial task while paired tones (using an oddball protocol) were delivered through headphones. Regardless of age, deviant tones resulted in greater P200 components, which highlights the importance of auditory alert systems implementing novel alert sounds for emergencies, such as handover tasks. Results also showed that neural responses to salient auditory tones in the second position were attenuated in older adults in difficult conditions of a visuospatial task. Thus, the effects of age and visuospatial workload level on auditory processing are critical to consider, given that features related to SAVs, such as alerting systems, are still being developed.

**Keywords:** Semi-autonomous vehicles · Auditory alert systems ·
Electroencephalography · Human-computer interaction · Neuroscience · Aging ·
Workload

## 1 Introduction and Background

While completing a driving task, it is commonly known that engaging in unrelated activities, such as text messaging, is considered driver distraction and is profoundly dangerous. However, as autopilot features in semi-autonomous vehicles (SAVs) become more

© Springer Nature Switzerland AG 2021
H. Krömker (Ed.): HCII 2021, LNCS 12791, pp. 549–562, 2021.
https://doi.org/10.1007/978-3-030-78358-7_38

widespread, the required level of task awareness becomes less clear. Semi-autonomous vehicles currently available to the public demand that users remain engaged in autopilot driving tasks such that they can perceive and act on various forms of alerts if necessary [1]. Despite the need to keep the human-in-the-loop, drivers are less engaged in autopilot driving tasks than is recommended by manufacturers [2]. Furthermore, while in autopilot, drivers are more susceptible to fatigue and engaging in secondary tasks [2, 3]. Secondary, non-driving tasks include text messaging, talking to a passenger, interacting with the in-vehicle information system, or any other attentional deviations from the primary task of maintaining a safe driving interaction [4]. Considering the lack of clarity surrounding SAVs, and the tendency for engaging in secondary tasks, a better understanding of safety features and their efficacy is necessary.

Most SAVs have integrated auditory alerting systems to redirect driver attention to some aspect of the driving task, and are often activated during high-risk events, particularly those that require immediate driver action. However, little is known empirically regarding the effectiveness of these alerts and whether they have the necessary salience across a wide age-range of drivers. The present research considers the efficacy of auditory alerts from the perspective of the neurological processing of multi-modal information. Understanding how the brain integrates multi-modal information is essential to determining the efficacy of auditory alerting systems in semi-autonomous vehicles and whether or not they suffice as a method for conveying information to drivers. Investigating how younger and older groups process various types of auditory information while engaged in visuospatial tasks of different workload levels is a critical step in optimizing safety in semi-autonomous vehicles. Using electroencephalography (EEG), this research examined the neurological processing of standard and deviant auditory tones at decision-making areas of the brain. Additionally, the effects of older age on the processing of these auditory tones were also examined for key brain regions.

## 1.1 Semi-autonomous Vehicles

According to the Society of Automotive Engineers On-Road Automated Vehicle Standards Committee, there are currently six levels of automation, ranging from Level 0, no automation to Level 6, fully automated [5]. Levels 0–2 require the driver to be fully engaged in the task of driving, while Levels 3–5 are less clear about the degree of focus drivers require. The present research applies mainly to conditional and highly automated vehicles, or Levels 3 and 4 whereby drivers must be engaged enough to perceive and act on alerts [1]. While this research applies mainly to Levels 3 and 4, at this point, only Level 2 SAVs are available for public use, with more advanced SAVs remaining prototypes or concepts [6]. As such, most of the literature included in this research will cover Level 2 SAVs. Alerts in SAVs generally include visual and auditory information, and the literature indicates that the most common alerts experienced in a Level 2 SAV (Tesla Model S) include a notification to control the steering wheel (visual alert when low urgency, visual and auditory alert when urgency is high), warning of an impending collision (visual and auditory alert), or a notification that the user must immediately take over the driving task (visual and auditory alert) [6]. Seemingly, more urgent warnings include both visual and auditory alerts, which further solidifies the importance of understanding how auditory alerts are processed.

## 1.2 Handover Tasks and Disengagements

This research places a focus on the moments leading up to disengagements with autopilot, otherwise known as handovers [7]. Disengagement involves any instance whereby users are prompted through auditory and visual cues to take over the driving task. Some reasons for disengagement include software issues, inclement weather, or user judgement [8].

It has been found that as automation increases, there is also a steady incline in secondary task engagement [9], which is especially troubling given current disengagement rates. The State of California Department of Motor Vehicles [10] has an Autonomous Vehicle Testing program whereby SAV manufacturers testing vehicles in California must provide an annual disengagement report, which details the number of occurrences and reason for disengagement, as well as distance (miles) test-driven in autopilot. Dividing the number of miles driven in autopilot by the number of disengagements experienced yields the distance driven per disengagement [8]. Following the application of this calculation to the 2019 Disengagement Report, it became clear that disengagement rates vary based on the manufacturer, and should be cause for concern. While the values cannot be compared, as there was no control for test-driving in autopilot between manufacturers, the outcomes were still of interest. Mercedes Benz Research and Development North America reported that in 14238 miles test-driven, 2054 disengagements occurred, which roughly translates into one disengagement every 6.9 miles. Furthermore, Aurora Innovation Incorporated reported that of 13429 miles test-driven, 141 disengagements occurred, which approximates to one disengagement every 95 miles. These inconsistencies further verify that while semi-autonomous vehicles are improving, the improvements are not widespread and human assistance is still necessary.

## 1.3 Auditory Alert Systems

**Semi-autonomous Vehicle Auditory Alert Systems.** Auditory alerting systems in vehicles have long been considered superior to visual alerting systems for their passive and easy to receive nature [11]. While auditory alerts are common practice in all vehicles, SAV manufacturers are implementing various forms of alerts, from standard tones to multi-tone alerts, repetitive chimes, and even speech messages [12].

In an on-road driving task, it is guaranteed that there will be some level of interior noise in the form of engine, road, and aerodynamic noise, to name a few [13]. Moreover, there are likely to be other task-irrelevant sounds contributing to the interior noise, such as other passengers, vehicles, music, or auditory notifications from mobile phones. The literature suggests that interior noise within vehicles ranges from 100 and 4000 Hz [14], and considering this, the loudness of any auditory alert systems is an important factor in vehicle safety. In a virtual reality driving study, participants engaged in a highway driving scenario whereby they were presented with various auditory alert signals of different loudness [15]. It was found that alerts at a frequency of 1750 Hz were significantly more successful in reducing poor driving behaviours than alerts at both 500 Hz and 3000 Hz.

There has been some research conducted on the position of auditory signals concerning secondary-task noise. For example, a simulator study paired a task-irrelevant e-mail alert that participants must respond to vocally, with a task-relevant auditory alert that indicated that they must press on the brakes immediately [16]. Their findings displayed that pairing the e-mail alert with the brake alert improved response time, and the researchers postulated that these unexpected results were due to participants anticipating a brake event, in case it were to arise while distracted by the email notification.

**Aging and Auditory Alert Systems.** With age, there is an overall decline in auditory processing and its mechanisms [17]. Thus, the inhibitions that come with age must be considered when designing auditory alerting systems. Even with this decline in auditory processing, the literature presents that in-vehicle auditory alerts help compensate for all forms of age effects [18]. Furthermore, another study found that auditory alerts were more successful in preventing incidents in older adults compared to younger adults [19].

**Workload and Auditory Alert Systems.** Driving a vehicle, regardless of its level of autonomy, involves the perception of multi-modal stimuli and decision making based on the available information. There are many theories on attention and workload, however, one that resonates with this research is Norman and Bobrow's model of resource allocation, which suggests that humans are limited by their own resources, as well as the data that is available to them [20]. Thus, there is a certain point where improving performance on a task is no longer possible. While humans are certainly limited in workload and processing of information, as the information available, or data, improves, the level of performance attainable also improves. This is particularly promising when considering that one goal of SAVs is to require as little human interaction as possible. That said, it also highlights the importance of designing SAV safety features in a manner that takes into consideration age and workload.

### 1.4 Auditory Paired Tone Oddball Paradigm

The present study implements an experimental design called the auditory paired tone oddball paradigm. In this study, two tones are presented sequentially, with one deviant tone, in that it has a different sound quality and occurs at a different latency than standard tones. This experimental design is frequently used in EEG studies to evaluate auditory processing in relation to workload [21] and is growing as a methodology for aging studies [22].

**P200 Component.** The component of interest in this research includes the P200, which is the positive-going amplitude that occurs approximately 200 ms after the presentation of an auditory tone. The P200 component was chosen for this research because it represents later attentional processing [23], which is appropriate given that responses to SAV auditory alert systems rely on attentional processes. Furthermore, the P200 component is also responsive to individual differences, such as normal aging [24], which is a key aspect of auditory processing being evaluated in this study.

## 2    Method

### 2.1    Participants

This study consisted of 28 participants (16 females) divided into two groups, younger and older. The younger group was made up of undergraduate Psychology students at Carleton

University (n = 17, M = 19.24, range = 18–26), while the older group were obtained from the same institutions "learning in retirement" program (n = 11, M = 68.82, range = 57–78). Compensation included a 2% course credit for the younger group, and for the older group free refreshments valued at $20.00. The inclusion criteria for the older participant group included individuals above the age of 50, who had normal hearing and were English speakers. This study received ethics clearance from the University's Research Ethics Board.

## 2.2 Procedure

Upon arrival, participants were familiarized with the events of the study. After providing informed consent, they were given a pre-test questionnaire, which requested information about the participants' state, as well as any relevant health conditions or medical history that would impact the neural or behavioural data. Once set up with the EEG headset, participants sat in front of an LCD monitor where they were also equipped with over-ear headphones and a response pad. Throughout the experiment, they were to wear the headphones which delivered paired pure tones using an oddball protocol (1000 Hz and 1500 Hz [deviant]).

After watching a 5-min video of nature scenes to obtain a baseline, participants began a match-to-sample task (Fig. 1). Participants completed four rounds of the visuospatial task with two alternating levels of workload, low and high, encountering each workload level twice. The order of presentation was counterbalanced across participants. After completing the tasks, the headset was removed, and they were debriefed on the specifics of the study.

## 2.3 Study Design

A 2 (Workload: low vs. high) × 2 (Age: younger vs. older) factorial design was used. The order of the four conditions were counterbalanced to account for order effects.

## 2.4 EEG Pre-processing and Evoked Potentials

EEGs were recorded at 1000 Hz using a 128-channel dense array system and a GES 250 amplifier. To record and reduce the data to 250 Hz, the software Net Station 4.3.1 (Electrical Geodesics, Inc.) was used. Data were processed and artifacts were removed using EEGLAB v.14 [25]. The same software was used to extract values and create ERPs. Data were filtered offline with a 1 to 30 bandpass, and to further identify and remove non-brain artifacts (such as muscle movements, eye blinks, and cardiac induced activity), an independent component analysis was used. Triggers were inserted into the recording by the tone presentation software at the onset of the stimulus to mark the tones, S1 or S2. Epochs had a baseline of 100 ms and extended for 500 ms post-tone presentation. The Study function in EEGLAB computed grand averages of the epochs for older and younger groups at each electrode.

In this research, amplitude values were extracted from individual ERP's using EEGLAB's [25] Darbeliai plug-in. The time point of interest in this research includes

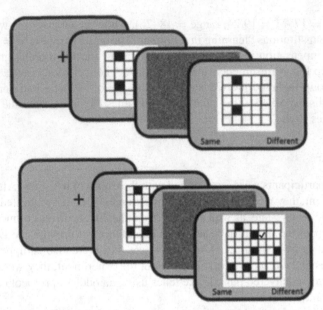

**Fig. 1.** Visuospatial match-to-sample task. The top image is the low-workload task, and the bottom image is the high-workload task.

200 ms after the onset of the tone, otherwise known as the P200. To obtain the amplitude of the P200 value, for each participant, the P100 and P200 values were extracted. The absolute difference was calculated to obtain the amplitude ($\mu$V) (see Fig. 2).

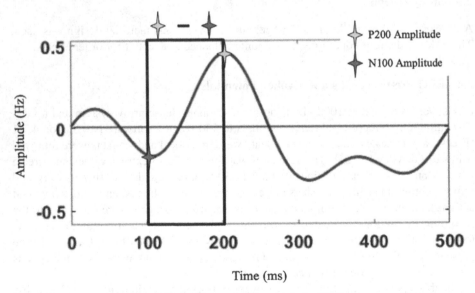

**Fig. 2.** Example of an ERP and the method in which the P200 value was obtained.

## 2.5 Brain Regions of Interest

The areas of interest were the decision-making areas of the brain. As seen in Fig. 3, the first area of the brain evaluated includes the dorsolateral prefrontal cortex, which falls under electrodes 15 and 16. These electrodes were averaged to obtain the P200 value. The primary auditory cortex was also of interest, which falls under electrode 103 [26].

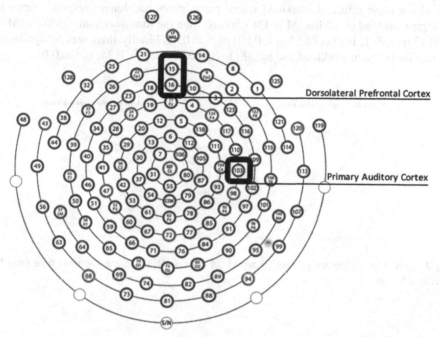

**Fig. 3.** Brain areas of interest and their corresponding electrodes on the EEG 128 Channel Dense Array Headset [26].

# 3   Results

## 3.1   Behavioural Measures

For the behavioural analyses, data were analyzed as a 2 (workload: low and high) × 2 (age: younger and older) mixed-factor ANOVA, with age as the within-subject factors. The p-value for each analysis was set at .05, and confidence intervals were set at 95%.

**Accuracy Rating.** There was a main effect of age on accuracy where the older group made significantly more errors in the match-to-sample task (M = 80.79%) than the younger group (M = 86.09%), $F(1, 108) = 15.26$, $p < 0.001$, $\eta_p^2 = 0.13$. There was also a main effect of workload where regardless of age, the high-workload condition resulted in significantly more errors in the match-to-sample task (M = 72.23%) than the low-workload condition (M = 95.79%), $F(1, 108) = 323.45$, $p < 0.001$, $\eta_p^2 = 0.75$.

Finally, there was also a significant interaction between workload and age; as shown in Fig. 4 [27], the older participant group were less accurate than the younger participant group only in the high-workload condition, $F(1, 108) = 7.29, p < 0.05, \eta_p^2 = 0.06$.

**Response Time.** As shown in Fig. 4 [27], there was a main effect of age on response time where the older group displayed longer response times (M = 1329.07 ms) than the younger group (1131.06 ms), $F(1, 108) = 28.03, p < 0.001, \eta_p^2 = 0.21$. There was also a main effect of workload where participants had longer response times in the high-workload condition (M = 1367.56 ms) than the low-workload condition (M = 1050.13 ms), $F(1, 108) = 75.54, p < 0.001, \eta_p^2 = 0.41$. Finally, there was no significant interaction between workload and age, $F(1, 108) = 0.91, p < 0.34, \eta_p^2 = 0.01$.

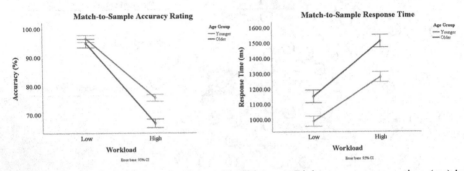

**Fig. 4.** Left. Mean accuracy percent by workload by age. Right. mean response time (ms) by workload by age [27].

## 3.2  EEG Results

For the EEG analyses, after filtering the data and determining the number of clean datasets, the participant group was reduced to 10 participants (7 female) split into younger (n = 5, M = 19.6, range = 18–21) and older groups (n = 7, M = 70.8, range = 65–78). This number of participants was reached because out of the original sample, only 5 older participants, and 10 younger participants had usable EEG data. To match the participant groups, 5 participants from the usable data of the younger group were chosen at random, to ensure that there was no bias.

The following sections present the extracted P200 values (in μV) from ERPs at different locations of the brain (dorsolateral prefrontal cortex, and primary auditory cortex) within different workloads (low and high), tone types (standard and deviant), and age groups (older and younger adults). Additionally, these results display the order of presentation (whether the tone was presented first or second). The EEG analyses implemented a 3-factor mixed-design ANOVA, with age as a between-subjects factor and tone type and workload as within-subjects factors. The p-value for each analysis was set at .05, and confidence intervals were set at 95%. Finally, extreme outliers identified in SPSS were winsorized, such that amplitude values were replaced by the next closest non-outlier [28].

**First Tone: Dorsolateral Prefrontal Cortex (Electrodes 15/16).** The first tone presentation, at electrodes 15 and 16 averaged, the 3-factor mixed-design ANOVA did not display any significant main effects of workload or age, nor were there any significant interactions. That said, tone type resulted in a significant main effect, $F(1, 8) = 9.75, p < .05, \eta_p^2 = .55$, such that regardless of age or workload level, the deviant tone resulted in larger P200 component amplitudes (Fig. 5).

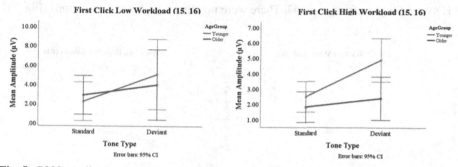

**Fig. 5.** P200 amplitude at electrodes 15 and 16 for the first tone, separated by age and workload.

**Second Tone: Dorsolateral Prefrontal Cortex (Electrodes 15/16).** For the second tone presentation at the dorsolateral prefrontal cortex, the 3-factor mixed-design ANOVA did not display any significant main effects of workload or age, however there was a main effect for tone type, $F(1,8) = 11.91, p < .05, \eta_p^2 = .59$, such that regardless of age or workload level, the deviant tone resulted in larger P200 component amplitudes. There was also a significant interaction of workload and age group, whereby the older participant group displayed larger amplitudes at the P200 in the low-workload task, but not the high-workload task, $F(1,8) = 5.01, p = .05, \eta_p^2 = .39$. There were no other significant interactions (Fig. 6).

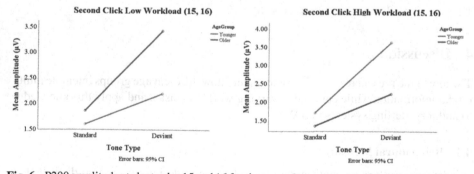

**Fig. 6.** P200 amplitude at electrodes 15 and 16 for the second tone, separated by age and workload.

**First Tone: Primary Auditory Cortex (Electrode 103).** For the first tone presentation at the primary auditory cortex, the 3-factor mixed-design ANOVA did not display any significant main effects of workload or age, however there was a main effect for tone type, $F(1,8) = 15.79, p < .05, \eta_p^2 = .66$, whereby the deviant tone resulted in larger P200 component amplitudes regardless of age or workload level. There was also a significant interaction between tone type and age group, whereby the older participant group displayed larger P200 component amplitudes than the younger group upon presentation of the standard tone, and smaller components in response to the deviant tone presentation $F(1, 8) = 6.99, p = .05, \eta_p^2 = .47$. There were no other significant interactions (Fig. 7).

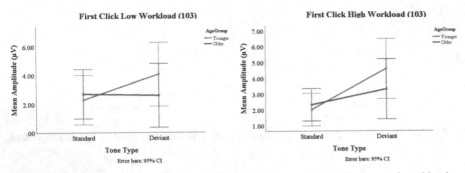

**Fig. 7.** P200 amplitude at electrode 103 for the first tone, separated by age and workload.

**Second Tone: Primary Auditory Cortex (Electrode 103).** For the second tone presentation at the primary auditory cortex, the 3-factor mixed-design ANOVA did not display any significant main effects of workload or age, however there was a main effect for tone type, $F(1, 8) = 10.57, p < .05, \eta_p^2 = .57$, whereby the deviant tone resulted in larger P200 component amplitudes regardless of age or workload level. There was also a significant 3-way interaction between workload, tone type and age group, whereby in the low-workload task, the deviant tone results in larger P200 component amplitudes in both age groups, however in the high-workload task, the younger group displays larger P200 component amplitudes than the older group, $F(1, 8) = 13.34, p = .63, \eta_p^2 = .47$. There were no 2-way interactions (Fig. 8).

## 4   Discussion

The aim of the present research is to understand how different age groups integrate multi-modal information while engaged in varying workload tasks, and apply this knowledge to auditory alerting systems in SAVs.

### 4.1   Behavioural Measures

The behavioural results confirm the planned research design, where two distinct (low and high) workload conditions were produced by the variations in the match-to-sample task.

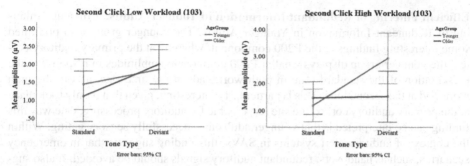

**Fig. 8.** P200 amplitude for electrode 103 for the second tone, separated by age and workload.

## 4.2  EEG Results

**Tone Type.** It was found that regardless of age, workload level, and the position of the tone, the presentation of the deviant tone consistently yielded larger P200 component amplitudes. The finding about tone type is particularly of interest given that in recent years, there has been a shift towards implementing pre-alerts in SAVs, particularly for handover situations. Van der Heiden and colleagues investigated the effect of introducing one, or multiple auditory pre-alerts to participants in a simulator, and found that this resulted in safer handover situations than those that did not include a pre-alert [29]. While the study supported that pre-alerts would help ensure safe and smooth handover situations, there was no investigation into the effect of repetitive stimuli that over time, become redundant. Therefore, in the future, it may be wise to conduct a similar study that implements deviant tones within the pre-alerts to maximize auditory processing and provide the safest experience possible.

**Compensation in Older Adults at the Second Tone.** Another finding of interest includes the attenuation of the P200 component amplitude in the older group at the second tone and high-workload condition of both the prefrontal cortex and the primary auditory cortex. One possible explanation for this result is the Compensation Related Utilization of Neural Circuits Hypothesis (CRUNCH), which suggests that older adults allocate more neural resources for lower workload tasks than would be seen in a young healthy brain [30]. Thus, as the task workload level increases, there are fewer resources to draw from, which might explain the attenuated P200 components in the high-workload tasks. In the context of handover scenarios in an SAV, the aforementioned result might indicate that while engaged in a high-level workload secondary task, auditory alerts may not be processed effectively. This suggests that age-related declines in sensory processing should be taken into consideration when developing policy surrounding SAVs.

Another possible interpretation is that the older adults are effectively sensory gating. Sensory gating refers to the neural ability to filter out redundant information [31], and in the context of this study, it seems as though the group is successful in inhibiting redundancies. That said, unexpectedly, the older neural responses also seem to be attenuating the novel tone. This might be explained by research that suggests that sensory gating is reflexive to individual differences [32].

**Efficient Filtering of Redundant Information in Younger Adults.** Efficient Filtering of Redundant Information in Younger Adults. The younger group also presented some interesting findings at the P200 component, whereby at the primary auditory cortex, the younger group displays smaller P200 component amplitudes in response to the presentation of the standard tone in both workloads at the first tone, and in the high workload at the second tone. This is particularly interesting, given that it only took place at the primary auditory cortex, the site responsible for auditory processing. One way this finding can be interpreted is that younger adults are successfully sensory gating. Within the context of auditory alert systems in SAVs, this finding suggests that in emergency scenarios, such as handovers, redundant auditory signals should be avoided. It also suggests that older adults have more activation to even redundant tones, which normally reflects some decline in inhibition within the central nervous system, leading to an overflow of information [33]. From the perspective of SAV alerting systems, however, the impaired inhibition might help in emergency scenarios when auditory alert systems do not take into consideration the intricacies of redundant alerts.

## 5 Conclusion

Analyses of P200 component amplitudes at decision making and auditory processing locations of the brain revealed interesting differences among auditory processing of standard and deviant tones. First, it was found that regardless of any other variables, the deviant tone consistently yielded greater activation at the P200 component. Furthermore, the older group displayed attenuated P200 responses to tones during the high-workload task, while the younger group displayed less activation at the P200 component for the second tone, in comparison to the standard tone, regardless of workload condition. These findings show how auditory processing varies across the lifespan, and suggest that the current design and research within auditory alerting systems in SAVs should be re-evaluated to consider the needs of different populations. While multi-tone alerting systems and varying tone types have proven to be successful in improving handover situations in driving simulators [29, 34], the issue of notifying drivers of emergency scenarios is multi-faceted, and current research pays little attention to age-differences. It is critical to evaluate all components of auditory alerting systems to enhance safety features within SAVs and ensure safe human-computer interactions.

## References

1. Knoefel, F., Wallace, B., Goubran, R., Sabra, I., Marshall, S.: Semi-autonomous vehicles as a cognitive assistive device for older adults. Geriatr. **4**, 63 (2019)
2. Morando, A., Gershon, P., Mehler, B., Reimer, B.: Driver-initiated Tesla autopilot disengagements in naturalistic driving. In: 12th International Conference on Automotive User Interfaces and Interactive Vehicular Applications (2020)
3. Jamson, A., Merat, N., Carsten, O., Lai, F.: Behavioural changes in drivers experiencing highly-automated vehicle control in varying traffic conditions. Transp. Res. Part C Emerg. Technologies. **30**, 116–125 (2013)

4. Jin, L., Guo, B., Jiang, Y., Wang, F., Xie, X., Gao, M.: Study on the impact degrees of several driving behaviors when driving while performing secondary tasks. IEEE Access **6**, 65772–65782 (2018)
5. 3016A: Taxonomy and Definitions for Terms Related to Driving Automation Systems for On-Road Motor Vehicles - SAE International. https://www.sae.org/standards/content/j3016_201609/
6. Banks, V., Eriksson, A., O'Donoghue, J., Stanton, N.: Is partially automated driving a bad idea? Observations from an on-road study. Appl. Ergon. **68**, 138–145 (2018)
7. McCall, R., et al.: A taxonomy of autonomous vehicle handover situations. Transp. Res. Part A Policy Pract. **124**, 507–522 (2019)
8. Lv, C., et al.: Analysis of autopilot disengagements occurring during autonomous vehicle testing. IEEE/CAA J. Automatica Sinica **5**, 58–68 (2018)
9. Carsten, O., Lai, F., Barnard, Y., Jamson, A., Merat, N.: Control task substitution in semiautomated driving. Hum. Factors J. Hum. Factors Ergon. Soc. **54**, 747–761 (2012)
10. Disengagement Reports - California DMV. https://www.dmv.ca.gov/portal/vehicle-industry-services/autonomous-vehicles/disengagement-reports/
11. Wogalter, M., Conzola, V., Smith-Jackson, T.: Research-based guidelines for warning design and evaluation. Appl. Ergon. **33**, 219–230 (2002)
12. Nees, M., Helbein, B., Porter, A.: Speech auditory alerts promote memory for alerted events in a video-simulated self-driving car ride. Hum. Factors J. Hum. Factors Ergon. Soc. **58**, 416–426 (2016)
13. Harrison, M.: Interior noise: assessment and control. Veh. Refin., 145–233 (2004)
14. Au, K.: Advances in Knitting Technology. Elsevier (2011)
15. Lin, C., et al.: Assessing effectiveness of various auditory warning signals in maintaining drivers' attention in virtual reality-based driving environments. Percept. Mot. Skills **108**, 825–835 (2009)
16. Wiese, E., Lee, J.: Auditory alerts for in-vehicle information systems: the effects of temporal conflict and sound parameters on driver attitudes and performance. Ergonomics **47**, 965–986 (2004)
17. Nagaraj, N., Kennett, S., Levisee, M., Atcherson, S.: Overview of central auditory processing deficits in older adults. Semin. Hear. **36**, 150–161 (2015)
18. Marshall, D., Chrysler, S., Smith, K.: Older drivers' acceptance of in-vehicle systems and the effect it has on safety. https://trid.trb.org/view/1331080
19. Baldwin, C., May, J., Parasuraman, R.: Auditory forward collision warnings reduce crashes associated with task-induced fatigue in young and older drivers. Int. J. Hum. Factors Ergon. **3**, 107 (2014)
20. Norman, D., Bobrow, D.: On data-limited and resource-limited processes. Cogn. Psychol. **7**, 44–64 (1975)
21. Ullsperger, P., Freude, G., Erdmann, U.: Auditory probe sensitivity to mental workload changes – an event-related potential study. Int. J. Psychophysiol. **40**, 201–209 (2001)
22. Pavarini, S., et al.: On the use of the P300 as a tool for cognitive processing assessment in healthy aging: A review. Dement. Neuropsychol. **12**(1), 1–11 (2018)
23. Boutros, N., Korzyukov, O., Jansen, B., Feingold, A., Bell, M.: Sensory gating deficits during the mid-latency phase of information processing in medicated schizophrenia patients. Psychiatry Res. **126**, 203–215 (2004)
24. Gmehlin, D., Kreisel, S., Bachmann, S., Weisbrod, M., Thomas, C.: Age effects on preattentive and early attentive auditory processing of redundant stimuli: is sensory gating affected by physiological aging? J. Gerontol. A Biol. Sci. Med. Sci. **66A**, 1043–1053 (2011)
25. Delorme, A., Makeig, S.: EEGLAB: an open source toolbox for analysis of single-trial EEG dynamics including independent component analysis. J. Neurosci. Meth. **134**(1), 9–21 (2004). https://doi.org/10.1016/j.jneumeth.2003.10.009

26. HydroCel Geodesic Sensor Net. https://www.egi.com/images/stories/manuals/Printed%20I
    FUs%20with%20New%20Notified%20Body/HC_GSN_channel_map_128_8403486-54_
    20181207.pdf
27. Turabian, M., Van Benthem, K., Herdman, C.M.: Impairments in early auditory detection
    coincide with substandard visual-spatial task performance in older age: an ERP study. In:
    Stephanidis, C., Antona, M., Ntoa, S. (eds.) HCII 2020. CCIS, vol. 1294, pp. 110–118.
    Springer, Cham (2020). https://doi.org/10.1007/978-3-030-60703-6_14
28. Hoffmann, U., Vesin, J., Ebrahimi, T., Diserens, K.: An efficient P300-based brain–computer
    interface for disabled subjects. J. Neurosci. Meth. **167**, 115–125 (2008)
29. van der Heiden, R., Iqbal, S., Janssen, C.: Priming drivers before handover in semi-
    autonomous cars. In: Proceedings of the 2017 CHI Conference on Human Factors in
    Computing Systems (2017)
30. Reuter-Lorenz, P., Cappell, K.: Neurocognitive aging and the compensation hypothesis. Curr.
    Dir. Psychol. Sci. **17**, 177–182 (2008)
31. Jones, L., Hills, P., Dick, K., Jones, S., Bright, P.: Cognitive mechanisms associated with
    auditory sensory gating. Brain Cogn. **102**, 33–45 (2016)
32. Zabelina, D., O'Leary, D., Pornpattananangkul, N., Nusslock, R., Beeman, M.: Creativity and
    sensory gating indexed by the P50: selective versus leaky sensory gating in divergent thinkers
    and creative achievers. Neuropsychologia **69**, 77–84 (2015)
33. Javitt, D., Freedman, R.: Sensory processing dysfunction in the personal experience and
    neuronal machinery of Schizophrenia. Am. J. Psychiatry **172**, 17–31 (2015)
34. Nees, M., Walker, B.: Auditory displays for in-vehicle technologies. Rev. Hum. Factors Ergon.
    **7**, 58–99 (2011)

# Audio-Based Interface of Guidance Systems for the Visually Impaired in the Paris Metro

Gérard Uzan[1] and Peter Wagstaff[2]([⊠])

[1] Laboratoire THIM, Université de Paris 8, Saint Denis, France
[2] Consulting Engineer, 22 bis Rue Laugier, 75017 Paris, France

**Abstract.** This article deals with the results of work carried out on the development of audio-based guidance system interfaces for the visually impaired at the THIM laboratory at the University of Paris 8. Research projects with the Paris Transport Authority (RATP) and many other partners have shown the importance of the structure and vocabulary of the instructions generated by the guidance system in guaranteeing the clarity, comprehension and effectiveness of the messages, particularly in conditions of high ambient noise levels. The choice of a vocabulary with optimum intelligibility and the use of structured messages that are as brief as possible ensures easier comprehension of the instructions. This allows the user more time to detect and integrate the audible and other clues to their situation from the local physical environment. This is a critical factor in increasing the safety of the user with visual deficiencies in the complex and unique environment of the Metro. A standard methodology and structure have been developed to integrate these principles in systems tested with success in the Paris Metro system.

**Keywords:** Visual deficiencies · Guidance systems · Audio interface · Intelligibility

## 1 Introduction

The increased awareness of the needs of persons with problems of mobility, coupled with new obligations to ensure access for all in urban environments has led to improvements in accessibility to public spaces and transport systems by removing many of the barriers to mobility and increasing the security of all users. Unfortunately, underground public transport systems in older cities such as Paris are one of the areas where much still remains to be done. Most of the basic infrastructure dates from concepts introduced in the 19th century and modifications to improve accessibility are difficult to achieve because of the lack of space and the means to redesign to modern standards. Despite many modifications to improve the situation, much of the basic infrastructure in the Paris Metro, which was built in the early 1900's remains essentially the same today. Open platforms, stairs, corridors and barriers were never designed for use by the elderly or travelers with disabilities.

An article in the newspaper the Guardian on the 21st September 2017 noted that only 3% of the 303 Paris metro stations are fully accessible to wheelchair users and

© Springer Nature Switzerland AG 2021
H. Krömker (Ed.): HCII 2021, LNCS 12791, pp. 563–576, 2021.
https://doi.org/10.1007/978-3-030-78358-7_39

these are the 9 stations of the newest fully automatic line 14, which was completed in 1998. In the London underground system 18% of the 270 stations were fully accessible, which may be contrasted with 88% for Tokyo. These differences are essentially linked to the technical difficulties of improving accessibility in the specific environment of each system. An additional problem in the Paris metro is the multiplicity of entrances and exits at many stations, the network density and close proximity of the stations, which are often served by several lines with diverse modes of transfer from one line to another. The existence of overhead sections of several lines and up to three different levels of lines underground, which are often in close proximity is an additional problem which increases the complexity of the passage from one line to another.

The introduction of a new law on accessibility in 2005 in France has led to many advances such as, the introduction of platform barriers on lines 1 and 13 in addition to those on line 14. The doors of access only open at the arrival of a train which has eliminated the risk of passengers falling onto the track. Figure 1 shows a traditional side platform station on line 6 where the line is above ground which may be compared with the station on the oldest line, line 1 which was recently equipped with barriers and converted to automatic operation. More attention has also been paid to the clarity of audio and visual public announcements which have now been included inside the trains to indicate the next station on many of the lines, as well as improving public announcements on the platform to indicate the train's destination. The development and modernization of the bus and tramway system in Paris has also been an important factor which means it is now totally accessible for wheelchair users.

**Fig. 1.** A side platform layout in a station on an overhead section of the traditional metro (line 6) (left) and a recently installed platform barrier on the oldest line of the metro, line 1 (right).

It has also resulted in an increased effort by public authorities to encourage projects of research in accessibility and particularly in the area of mobile aids and navigation systems for persons who are visually impaired. A review of navigation systems for persons with visual deficiencies is presented in reference [1]. At the THIM laboratory (Technologies Handicaps Interaction Multimodalities), in collaboration with various other partners, projects such as "Infomoville" and "RAMPE" [2, 3] were designed to aid persons with visual deficiencies to use bus and tramway networks. Bus and tramway navigation systems can use GPS positioning [4, 5] and dynamic information at bus and tram stops, but the problem of locating where you are in the metro is much more difficult and other methods have to be used such as inertial transducers as described in the ANR project DANAM in collaboration with the CEA and THIM [6]. The fusion of data from the transducers with the cartography of the network and the perception of the users permits a satisfactory validation of their true position. This is an example of tri-coherence. A subsequent project at THIM with the RATP called SIV (System Information Voyager) also calculated the position of the user with the aid of inertial transducers, but the emphasis was on improving the structure and quality of messages. Since the users of the system would also generally be using a cane or a guide dog, as well as listening to audible clues to check their position this may lead to doubts and hesitation if their announced position and their own impression of the local environment seem incompatible. These results also indicated that the structure and intelligibility of the audio messages used to guide the user could have an important influence on the errors in navigation. Subsequent projects such as the smartphone-based Sound Atlas have enabled these aspects to be refined, developed and formalised. A standard has been developed to improve the structure, pertinence, intelligibility and clarity of the messages of guidance systems for the metro which can also be useful for other types of situation and persons with other types of disability.

It should be clear that this research has been developed for the specific needs of transport systems in France using the French language, but can be transposed to other languages and environments. In this context it is interesting to compare our findings with the Open Standard Wayfindr [7] and the corresponding ITU-T recommendation F.921 [8]. After a brief summary of the functional requirements for guidance systems the importance of the choice of the vocabulary to enhance intelligibility and compre-hension is then discussed. The brevity and logical structure of the messages facilitates the detection and assimilation of audible clues to the users local environment.

## 2  Orientation Localization and Information

Design criteria for evaluating the requirements of persons with problems of mobility have already been proposed and evaluated for applications in public transport [9, 10]. The acronym SOLID, standing for "Safety, Orientation, Localization, Information and Displacement" has been used as a check list of the important parameters to be taken into account to ensure that all the needs of disabled travelers are provided for. It is unlikely that all of the stations of the older lines of the Paris metro will be totally accessible for persons in wheelchairs because of space limitations in the city environment, but some efforts have been made to respond to the needs of persons with other types of disability. Since our interest is particularly focused on the needs of persons with visual deficiencies who are at the greatest risk in the metro system, the discussion is essentially limited to their requirements. Over the years a number of advances to help travelers with visual deficiencies, as well as other disabilities, have been integrated in the Metro system. These include the increased use of audio and visual messages in trains and stations and tactile paving to indicate the edge of platforms and the presence of stairs, escalators, elevators or changes of direction. The difficulties of persons with visual deficiencies in the metro are principally Orientation, localization and Information after determining the fastest route to the destination and making their way to the nearest station. This requires navigating all the potential hazards safely from their starting point to the station, then searching for the correct platform, boarding and leaving the correct train and exiting their destination station, all with no visual support. Navigation systems with GPS have been available for many years to deal with needs of the visually impaired. The introduction of newer navigation systems using low energy Bluetooth beacons for periodic checks on position has replaced the older solutions of dead reckoning with inertial measurements and FM beacons or RFid tags to obtain more reliable verification of the user's position at points on their route. Much work still remains to integrate all these advances and produce a satisfactory solution for a complete metro system.

## 3   Guidance Systems

The most useful type of solution to the problems of travelers with visual deficiencies is the development of reliable individual guidance systems which could ensure almost complete autonomy for the user. In a previous publication on the criteria for the choice of grammar and expressions used in the audio messages for guidance systems [10], we have stressed the importance of making sure that the system does not induce overconfidence or undue reliance on the guidance system in situations of danger. In this, as in many other cases, the Principle of Precaution must apply, both for the user and for the designer of the human-machine-interface. In this previous publication we referred specifically to this problem in terms of the "humility" of the system which should be conveyed by the system's audio interface. This is ensured by the specific choice of the structure of the grammar and the expressions used to guide the user.

The expression "tri-coherence" has also been used to indicate that there is always the possibility that the guidance system can make an error of calculation and the user an error of comprehension. The users should always integrate the complementary information provided by their cane, the acoustic environment, air currents, tactile paving, or intuition, to validate the information and instructions given by the system. In cases of doubt the most sensible course of action is to retrace one's steps to the previous known position or verify one's position and orientation with other travelers. The traditional metro map is of no help in these circumstances, either for planning the route or for wayfinding during the trip and the same is true for the plans of the different exits.

This paper is concerned only with the user interface which includes these functions using audio messages with verbal commands to prompt the next stage or ask for complementary information. Our principal concern is the optimization of the message content, structure, timing and intelligibility, not the technical specifications of the navigation device.

## 4   Construction of the Messages

### 4.1   Basic Principles

We can distinguish between the language used and the practical implications of the construction of the words and phrases to define the different types of message which can be summarized below. In our work the target language is of course French, so the translation into English will not necessarily yield the optimum form of the words or messages which would be used if English was the target language, but they do illustrate the reasoning behind the final choice.

- assertive: the goal is to engage the person and conviction dominates; e.g. "He will come tomorrow".
- directive: the goal is to get the person to do something, the will, the desire; e.g. "Come with me".
- permissive: the goal is to engage the person in the performance of an action. Here the sincerity of intention predominates; e.g. "I would come".
- expressive: the goal is to express a psychological state where the content assigns a priority to the person; e.g. "Excuse me".
- declarative: the goal is to create a reality; e.g. "I declare war".

For guidance and descriptions, as described in our previous paper on criteria for the choice of language and grammar of guidance systems [11], we chose to base the statements on an assertive mode, and to limit the messages to the target objects or milestones to be achieved using a minimum of additional descriptors to avoid surplus information. As an illustration of the principle used here, this philosophy has already been adopted for the public announcements inside the trains to indicate the next station in the Paris metro system. Instead of using superfluous phrases such as, "the next station is" or "the train is arriving at", the message simply announces the name of the station with a rising pitch and then repeats it with a falling pitch to enable the listener to confirm the name of the station if it was not clear the first time. In the case of messages for a mobile guidance system it is also desirable to choose words with two syllables if possible, since they are easier to recognize and fairly short. As an example of this, in English we could replace the word "stairs" in a message by "stairway" which would be easier to recognize in a noisy environment because the second syllable reinforces and confirms the recognition of the word.

In French, as in English, the dictionary is relatively rich in be derived from other languages which have been adopted in common speech. This often enables us to choose an alternative word for guidance messages if the original French word could lead to confusion or cause difficulties of intelligibility. As a simple example of this we can cite the word "escalier" for stairs or stairway and "escalier mécanique" for escalator. Since we need to distinguish clearly between the two words in the message without adding the second word, our choice has been to use the word escalator in French. Facilitating the recognition of the primary word or target and shortening the rest of the message to the minimum number of superfluous words or qualifiers avoids potential risks which could be introduced by using words that are less intelligible or giving a more complete description. These are:

- Increasing the possibility of misunderstanding the message.
- Monopolizing the attention of the users for a longer time with the risk of saturating their memory.
- Increasing the possibility of missing audible clues from the local environment.

- Introducing a discrepancy between the user's mental image of their situation, that transmitted by the system and the true environment (tri-coherence criterion).

The choice of the words used to construct the guidance messages is made by prioritizing the criteria of intelligibility and brevity whenever possible and favouring easily discriminated phonemes in the generally noisy environment of a public transport system. Not all phonemes have the same capacity to be easily identified in heterogeneous noisy environments, which implies that the selection of the words used in the messages should be made, if possible, by preferring those which can be recognized most easily in situations with high ambient noise. The choice of the vowels used in the words coupled with the most intelligible consonants are a key factor in the comprehension of each syllable in the keywords of the message. The Table 1 presents a selection of words in French used to illustrate the phonetic equivalents of the 14 vowel sounds and 17 consonant sounds plus the 3 semi-consonant sounds found in the French language.

**Table 1.** Words with the phonetic equivalents of vowel sounds and consonants in French

| Vowels | Consonants | Semi-consonants |
|---|---|---|
| [i] il, vie | [p] pou, père | [j] paille, pied |
| [e] blé, jouer | [t] vitre, tarama | [w] oui, nouer |
| [ɛ] lait, merci | [k] carat, kanak | [ɥ] huile, lui |
| [a] plat, patte | [b] bonbon, robe | |
| [ɑ] bas, pâte | [d] dadais, dans | |
| [ɔ] mort, donner | [g] gare, draguer | |
| [o] mot, eau | [f] photo, faon | |
| [u] genou, roue | [s] ceci, salami | |
| [y] rue, truc | [ʃ] choir, chêne | |
| [ø] deux, peu | [v] voyage, vous | |
| [œ] peur, meuble | [z] maison, zozo | |
| [ ] le, premier | [ ] je, gageure | |
| [ɛ̃] matin, brin | [l] alors, tralala | |
| [ɑ̃] sans, vent | [R] raison, rare | |
| | [m] mamie, mais | |
| | [ɲ] gagner, agneau | |
| | [ŋ] camping, ping-pong | |

The formants or broadband maxima of the frequency components of the acoustic signal produced by each of these sounds are critical factors in determining the intelligibility of the vowels, consonants and syllables used to form each word. The information that humans require to distinguish between different sounds can be represented quantitatively by specifying these peaks in the amplitude or frequency spectrum. The formant with the lowest frequency is defined as F1, the second as F2 and the third as F3, but normally the first two formants are sufficient to recognize a vowel. High pitched vowel sounds such as [e], [i] and [y] have their first formant or frequency in the range of 2000 Hz and above and are less easily detected or understood in a noisy environment, whereas low pitched vowels such as [a] and [o] have a first frequency maximum in the range of 1000 Hz where the human ear is most sensitive and are easier to recognize.

In reference [12] it is noted that the intelligibility of the phonetic vowel [a] is better than that of the vowels [i] and [y]. Another study by Meyer et al. [13] on the intelligibility of speech between two persons outside as the distance is increased between them showed that the phonetic vowel [a] was recognized most accurately with 97% of correct responses at 23 m. For the consonants, the [s], [ʃ] and [z] are recognized most easily as the distance is increased and for the nasal consonants the [m] is easier to identify than [ɲ] and [ŋ]. In noisy environments therefore we prefer to choose critical words with these lower pitched vowels and more intelligible consonants to obtain maximum reliability. As mentioned previously it is also preferable to use words of at least two syllables, for the most significant words where possible, since the second syllable reinforces and confirms the comprehension of the first. Using the intelligibility criteria for vowels and consonants the use of the expression "metro 6" rather than "ligne 6" in French is doubly justified since the [m] tr and [o] is more intelligible than the [l], [i] and [ŋ]. The ability to comprehend the messages in a noisy environment is of critical importance, since we are dealing with persons with visual deficiencies, they will also depend on using their hearing to provide complementary information on their immediate environment. The option of using noise cancelling headphones to listen to the messages is therefore out of the question. Bone conducting earphones can be used, as suggested by Wayfindr (www. Wayfindr.net), but often there is a preference to minimise additional visible apparatus and use the smartphone or navigation device directly in the breast pocket or with an unobtrusive speaker and microphone clipped to their clothing.

## 4.2  Direction

The choice of systems for indicating direction was studied during the project SIV (Systems for Information of Voyagers) in collaboration with the RATP. The use of the clock-face analogy in the guidance messages to indicate the direction to travel on a route rather than the expressions left, right, forward etc. The table below compares the French expressions and their equivalents for these egocentric instructions. The guidance system was simulated using a series of 20 directional messages to permit 9 persons who were blind to take a route from the entrance of the metro to a platform at the station Nation. The expressions front and rear are used to indicate where the user is positioned on the platform. It should be noted that in France the metro runs on the right and the main trainlines run on the left, so when arriving on a metro platform the train will arrive from the left on a standard side-platform station (Table 2).

The results of these trials showed no significant differences between the two formats in terms of the number of times the users asked for the instruction to be repeated or made an error and the number of times they asked for a description of their position. Five of the messages were understood perfectly, ten with an error of 3.5% and the remainder with

**Table 2.** Clock face and classic direction instructions in French and English

| Mode mots | Classical mode | Mode cadran | Clockface mode |
|-----------|----------------|-------------|----------------|
| Tout à gauche | Sharp left | A 8h | 8 o'clock |
| A gauche | Left | A 9h | 9 o'clock |
| En face à gauche | Ahead left | A 10h ou 11h | 10 or 11 o'clock |
| En face | Ahead | A midi | 12 o'clock |
| En face à droite | Ahead right | A 1h et 2h | 1 or 2 o'clock |
| A droite | Right | A 3h | 3 o'clock |
| Tout à droite | Sharp right | A 4h | 4 o'clock |
| En tête | Front of train | | |
| En queue | Rear of train | | |

an error of 8%. However the differences may simply be due to the local environmental conditions where they were employed. In terms of the time taken to complete the route however the difference was more significant. Table 3 shows the time taken for nine of the test subjects to complete the route using the two different systems.

**Table 3.** Time taken for route using the classical directional word mode or clockface directions

| Subjects | $t_1$ (min) | $t_2$ (min) |
|----------|-------------|-------------|
| s1 | 38 | 22 |
| s2 | 26 | 12 |
| s3 | 21 | 14 |
| s4 | 26 | 12 |
| s5 | 26 | 11 |
| s6 | 28 | 9 |
| s7 | 40 | 23 |
| s8 | 28 | 9 |
| s9 | 26 | 11 |
| Average | 28.77 | 13.66 |
| Standard deviation | 5.807 | 4.94 |

Clearly the clockface method results in a more rapid completion of the route for all the participants with an average time saving of 50%. This was initially thought to be due partly to the brevity of the statement formed only by two syllables (literally "three hours" in French), as much as the use of a reference system which is easily understood by the user. However, the recommendation of this method in other references using English which has an extra syllable, shows that the clockface method is intrinsically valid for those who can adapt to it. The option of using the classical expressions "left", "right" etc., remains necessary for guide dog users or persons who have difficulties creating the mental image of an analogue clock face. For people using a guide dog, the option of using the classical expressions has the advantage of not requiring conversion or translation by the user, since it is exploiting the dog's own keywords, so there should always be the option to take into account this aspect of "canine-compatibility".

## 4.3  The Structure and Content of the Messages

The experiments we have performed in the past to test different approaches to optimise the construction and grammar of the messages for guidance systems have enabled us to adopt a standard structure which ensures that the traveller has the correct and relevant information in a familiar and easily recognised format. Each vocalized message is composed of 7 segments which are organised as follows:

1. A supporting descriptor
2. A danger signal
3. Reference information
4. The target object or milestone
5. Properties of the target object
6. Orientation
7. Distance

The different segments may be explicit, implicit or eliminated if superfluous. The following points illustrate typical situations.

### Segment 1: The Support Descriptor
This introduces the message but can be optional. Up to 3 descriptors at the most can be included, each one succeeding the other sequentially in a message. In French it is always before the target object (segment 4), but other languages may require a different format.

### Segment 2: Danger Signal (Alert or Presence of Danger)
This defines a danger zone and is only used if required. The danger signal selected after multiple trials has a total duration of 750 ms. It is broken down as follows: a steep-slope envelope signal consisting of three frequencies (250, 600 and 1200 Hz) over a duration of 250 ms then a silence of 125 ms, then the same signal for 375 ms. The signal (an Mp3 file) is available from the THIM team. In metro stations, the only critical moment of danger is the arrival on a platform. In an open environment, it can also be used to announce a difficult crossing of a road (a crossing in two stages for example).

### Segment 3: Reference Information
This always refers to the previous target object or a milestone on the route and allows the users to situate and orientate themselves at the start of the next route sequence. The segment is optional and placed before the next target object.

Some target objects are more significant as references than others (e.g., stairs, lifts, escalators, barriers). Examples of this in English such as "after exit from lift" or "after exiting ticket barrier" are in fact "dos à l'elevateur" (with back to the lift) and "dos au contrôle" (with back to the ticket barrier) in French.

## Segment 4: The Target Object or milestone

This segment is mandatory and defines an object that will be referred to as one of a sequence of target objects used in the description of the journey, such as platforms, stairways, escalators, etc.

## Segment 5: Target Object Properties

The properties of the target object cannot exist alone and must be accompanied by the target object. It can include up to 3 properties associated with the target object.

Some object properties will be optional or will be implicit and muted. For example, a staircase or escalator goes either up or down, but as soon as the word "stairway" or "escalator" is pronounced it is assumed to be downwards first and the user should be alerted to the possibility of falling and ready to react quickly to the potential danger. To shorten the message, the word "down" can be implicit, but the word "up" is always pronounced when applicable to enable the user to react accordingly.

## Segment 6: Orientation

This element is mandatory but can be implicit and silent in certain cases (e.g., for "12 o'clock" or "in front"). It is always placed after the target object and before the distance (element 7). It is given only once per message and the user can choose between the clock face or canine compatible versions.

## Segment 7: Distance

This segment is optional if the target is close and is always the last segment of the message. It is used once per message or is implicit if very close. Since it is difficult to estimate distances accurately, a set of approximate distances has proved to be the most reliable and useful indicator during previous tests and evaluations. When the distance is very short, it can be replaced by an expression of time such as "Immediately" or can be implicit. When the distance is short (about 10 m), it can be expressed as "about 10 m". Beyond, it is expressed by "about n meters". This leaves us with the following expressions for distance.

- Between 0 and 6 m, the distance can be mute or replaced by "immediately.
- Between 6 and 15 m, the system uses the expression "about 10 m";
- Between 15 and 25 m, the system uses the expression "about twenty meters";
- Between 25 and 35 m, the system uses the expression "about thirty meters";
- Between 35 and 65 m, the system uses the expression "about fifty meters";
- Beyond 65 m, the system uses the expression "a hundred meters".

When entering the target object data, each property has a flag determining its position relative to the target object (before or after) e.g., stairs "up" or "long" corridor.

**Silences:** Between each segment, a pause or "silence" will be respected. The silences are inter-segments and inter-elements. They are essential to avoid saturating the user's memory and to maintain awareness of clues from environmental noises.

Inside the support segment, the elements are separated by "slnc 1" type silences of 100 ms. Between the target object and its properties, and between the properties, we are also in a "slnc 1" type of silence. Between the segment of the last property and that of the orientation, there is a silence "slnc 2" equal to 3 times "slnc 1" (i.e., 300 ms). Between the support segment and the next segment, a silence of type "slnc 3" equal to 6 times "slnc 1" (i.e., 600 ms) giving the following periods of silence by default.

- between 1 and 2: 600 ms
- between 2 and 3: 100 ms
- between 3 and 4: 100 ms
- between 4 and 5: 100 ms
- between 5 and 6: 300 ms
- between 6 and 7: 300 ms

## 4.4 Display and Description

Each message corresponds to a standard vocabulary and structure described above making it possible to automate the construction of messages. As mentioned before, whenever a word is implicit, it will be muted to reduce the burden of memorisation. In the application, the full message appears on the screen, so items that are silent are also shown, but not pronounced. In addition, it is also essential to provide supporting descriptors to provide additional information on the local environment in certain situations. The supporting descriptors are important when the travellers start their route from the platform (for example when leaving a train) or when they are confronted with a station where several lines meet.

We can distinguish three situations: on a platform, in the main hall and elsewhere. The supporting descriptors are obtained using the verbal command "describe", to describe the local environment or warn of any risk of error. As an example, for the platform, most stations are with side platforms with 1 or several openings. The descriptor can be "side platform n openings" and the correct exit can be indicated by "i'th opening from front of train". A central platform usually only has an entry/exit at one end (front/rear of train) or both, which are given the descriptors "central platform 1 opening front of train" or "central platform 2 openings front and rear of train".

## 5   Constructing Messages and Recording the Route Data

As mentioned previously, the choice of a logical structure and a specific vocabulary enabling us to describe each stage on the route means that entering the data to describe a route from a specific start point and endpoint can be achieved easily by using an operator to walk the same route. The appropriate descriptors at each milestone or "target object" can be entered with the help of prompting from appropriate pop-up menus. This process requires no competence in programming nor any specific knowledge or data on the environment and can be carried out by a member of staff of the transport authority fairly rapidly after a short familiarisation with the interface. The application was developed by Urbilog using the framework Ionic 2 and is a hybrid application based on Angular using a plugin Cordova to implement the Text to Speech (TTS) requirement for the messages. The data to describe the route and the grammar was formatted in JSON and stored in a base MongoDB.

The application is an archive apk installed and functioning in Android, so it can be installed on any Android smartphone. This has the advantage of permitting users to employ their own smartphone or notepad as their guidance system and does not require them to get familiar with a new device.

They can of course also continue to use their telephone to stay in contact with friends and relatives and also respond to business calls during their trip.

The details of the process of programming, organising the data and storing the data model of a route are outside the scope of this paper, but the application, has been designed so that it can also be extended to other languages relatively easily using the principles already described above.

## 6 Evaluation of the Guidance System

The application was tested by the RATP in 2018 using their own protocols for evaluation of the system with different types of user. These tests were carried out independently with no participation of the authors in order to ensure a lack of bias.

The metro station Pyramids on lines number 14 and 7 was chosen as the site of the tests, since, although the line 7 was constructed in 1916, it is now wheelchair accessible because it was modernised to link directly with the fully accessible and automatized line 14 constructed in 1998. The line 14 is equipped with platform barriers in the two side platform arrangement. Three groups of users were evaluated during these tests, one group of persons who were blind, but with no other disabilities, one group of wheelchair users who were all capable of seeing and reading their smartphone and a control group of users with no significant disabilities. Three representative routes were chosen to reproduce the typical tasks required when using the metro: (a) Entering the station going to the correct platform and boarding a train: (b) Arriving on a train and going to another line to board another train: (c) Leaving a train and going to the appropriate exit point. The routes for the wheelchair users were of course different from those chosen for the users who were blind, since the latter could use the stairs, escalators and elevators whereas the wheelchair users had to locate where the elevators were for their transfers between different levels.

The results of these trials were conclusive, with both groups who were using the application reporting an increased ease of mobility with the help of the guidance system and a decrease in the number of failures or times they required assistance to reach the end target. Of the 20 or so blind users of the system the number of fails was less than an average of 3 for a third of the group and two thirds of the group had no problems whatsoever. Additionally, the speed of their progress was almost comparable to that of the control group. In the case of the wheelchair users their speed was of course lower, but there were no failures at all in this group. The advantage of not having to search for the location of the elevators and having the most direct route programmed for them directly on their smartphone also enabled them to travel to their destination much quicker than they would normally.

The smartphone application "Sound Atlas" which was developed by Urbilog, used an interface to collect data for a specific route. This is entered directly into the device by a trained member of staff who walks the same route and enters the appropriate data into the device at each milestone or "target object" on the route. The operator is reminded by the application to enter the specific properties of each target object (is the escalator up or down?) and enter changes of direction, distance to the next milestone or target object, the presence of stairs, platforms, openings, barriers or any other features of importance for each segment of the route using the popup menus. The application automatically associates the responses with the translator which directly stores the correct form of the message to be given to the user of the device. The end result is a smartphone-based guidance system which automatically integrates the optimum choice of grammar and expressions to guide the user as well as the principle of maximum security by taking into account the natural limitations and reactions of the user.

## 7  Conclusion

This paper has dealt with the methodology and philosophy behind the creation of audio interfaces for mobile guidance systems which can be integrated on a smartphone to assist blind persons when they are using the Paris metro system. The importance of defining a specific choice of grammar and vocabulary associated with a logical, economical and intelligible structure for the system messages has been discussed in this context to ensure maximum safety within the parameters of accuracy of the system. Although persons with visual deficiencies normally acquire improved audible acuity and comprehension of normal and synthesized speech, the choice of words with maximum intelligibility in noisy environments improves comprehension and reduces the possibility of errors. The brevity and logical structure of the messages also ensures that there is less possibility that audible and other environmental clues confirming the local situation of the user will be missed while the user is concentrated on trying to understand the messages.

The system was implemented on an Android smartphone and evaluated in the Paris metro system using protocols developed and applied by the RATP. Results were extremely convincing for travellers with visual deficiencies, who could navigate the route with very few errors and the same system also proved useful for sighted travellers in wheelchairs.

Future developments could include testing with foreign tourists since versions can also be made available for English and other languages. Work at the THIM laboratory is also being carried out on methods of automatically calculating the route and generating the instructions from data on the network which could greatly increase the utility of the system. The message structure used by the system to guide the user has proved to be an efficient and secure method of helping visually impaired travellers on the Paris metro and it is hoped to develop future systems based on the same principles described here.

**Acknowledgements.** The authors gratefully acknowledge the support and collaboration of the RATP, the SNCF, the TCL, the ANR and all partners from other institutions who have collaborated on the projects for the development and validation of the ideas presented in this paper.

# References

1. Lakde, C.K., Prasad, P.S.: Review paper on navigation systems for visually impaired people. Int. J. Adv. Res. Comput. Commun. Eng. 4(1), 166–168 (2015). https://doi.org/10.17148/IJA RCCE.2015.4134
2. Baudoin, G., Venard, O., Uzan, G., et al.: The RAMPE project: interactive, auditive information system for the mobility of blind people in public transport. In: Proceedings of the 5th International Conference on ITS Telecommunications (ITST), Brest, France, 27–29 June 2005 (2005)
3. Giudice, N.A., Legge, G.E.: Blind navigation and the role of technology. In: The Engineering Handbook of Smart Technology for Aging, Disability (2008). https://doi.org/10.1002/978047 0379424.ch25
4. Pretorius, S., Baudoin, G., Venard. O.: Real time information for visual and auditory impaired passengers using public transport - technical aspects of the Infomoville project. In: Handicap 2010, IFRATH, Paris (2010)
5. Coldefy, J.: The Mobiville project: large scale experimentation of real-time public transport information service providing satellite based pedestrian guidance on smartphones. In: Proceedings of the 16th ITS World Congress, Stockholm (2009)
6. Uzan, G., Lamy-Perbal, S., Carrel, A., Saragaglia, S., Isabelli, G., Malafosse, B.: Localisation et orientation des aveugles: dispositif de guidage dans le métro. In: IFRATH, Handicap 2010, Paris (2010). ISBN 978-2-953-689-0-7
7. Wayfindr. Open Standard for Audio-based Wayfinding, Version: Recommendation 2, 31 January 2018. www.wayfindr.net
8. Recommendation F.921: Audio based Network navigation system for persons with visual impairment. ITU-T, March 2017
9. Uzan, G., Wagstaff, P.R.: A model and methods to solve problems of accessibility and information for the visually impaired. In: STHESCA, Krakow (2011)
10. Uzan, G., Hanse, P-C., Seck, M., Wagstaff, P. R.: Solid: a model of the principles, processes and information required to ensure mobility for all in public transport systems. In: Proceedings of the 19th Triennial Congress of the IEA, Melbourne, 9–14 August 2015 (2015)
11. Uzan, G., Bodard, J., Clair, G., Wagstaff, P.R.: New criteria for the design and assessment of systems for the guidance of persons with difficulties of orientation in transport hubs and public spaces. In: Proceedings of the 19th Triennial Congress of the IEA, Melbourne, August (2015)
12. Benoit, C., Mohamadi, T., Kandel, S.: Effects of phonetic context on audio-visual intelligibility of French. J. Speech Lang. Hear. Res. 37(5), 119 (1994)
13. Meyer, J., Dentel, L., Meunier, F.: Intelligibilité de la parole à plusieurs distances dans un bruit naturel. In: 10ème Congrès Français d'Acoustique, HAL, Lyon, 12–16 April 2010 (2010)

# Author Index

Printed in the United States
by Baker & Taylor Publisher Services

Printed in the United States
by Baker & Taylor Publisher Services